Oxidative Stress in Dermatology

CLINICAL DERMATOLOGY

Series Editor

Alan R. Shalita, M.D.

Professor and Chairman
Department of Dermatology
State University of New York
Health Science Center at Brooklyn
Brooklyn, New York

1. Cutaneous Investigation in Health and Disease: Noninvasive Methods and Instrumentation, *edited by Jean-Luc Lévêque*
2. Irritant Contact Dermatitis, *edited by Edward M. Jackson and Ronald Goldner*
3. Fundamentals of Dermatology: A Study Guide, *Franklin S. Glickman and Alan R. Shalita*
4. Aging Skin: Properties and Functional Changes, *edited by Jean-Luc Lévêque and Pierre G. Agache*
5. Retinoids: Progress in Research and Clinical Applications, *edited by Maria A. Livrea and Lester Packer*
6. Clinical Photomedicine, *edited by Henry W. Lim and Nicholas A. Soter*
7. Cutaneous Antifungal Agents: Selected Compounds in Clinical Practice and Development, *edited by John W. Rippon and Robert A. Fromtling*
8. Oxidative Stress in Dermatology, *edited by Jürgen Fuchs and Lester Packer*

ADDITIONAL VOLUMES IN PREPARATION

Diseases of Connective Tissue in the Skin, *edited by Charles M. Lapière and Thomas Krieg*

Oxidative Stress in Dermatology

edited by

Jürgen Fuchs
Johann Wolfgang Goethe University
Frankfurt, Germany

Lester Packer
University of California
Berkeley, California

Marcel Dekker, Inc. New York • Basel • Hong Kong

Library of Congress Cataloging-in-Publication Data

Oxidative stress in dermatology / edited by Jürgen Fuchs, Lester Packer.
p. cm. -- (Clinical dermatology; 8)
Includes bibliographical references and index.
ISBN 0-8247-9049-9 (alk. paper)
1. Skin--Pathophysiology. 2. Free radicals (Chemistry)--Physiological effect. 3. Active oxygen in the body. 4. Dermatotoxicology. I. Fuchs, Jürgen. II. Packer, Lester. III. Series.
[DNLM: 1. Skin Diseases--physiopathology. 2. Oxygen--metabolism. 3. Oxygen--adverse effects. 4. Free Radicals. W1 CL69L v. 8 1993 / WR 140 0975 1993]
RL96.095 1993
616.5'07--dc20
DNLM/DLC
for Library of Congress 93-1195
CIP

The publisher offers discounts on this book when ordered in bulk quantities. For more information, write to Special Sales/Professional Marketing at the address below.

This book is printed on acid-free paper.

MARCEL DEKKER, INC.
270 Madison Avenue, New York, New York 10016

Current printing (last digit):
10 9 8 7 6 5 4 3 2 1

PRINTED IN THE UNITED STATES OF AMERICA

To Sabine and Anne

Series Introduction

During the past decade there has been a vast explosion in new information relating to the art and science of dermatology as well as fundamental cutaneous biology. Furthermore, this information is no longer of interest only to the small but growing specialty of dermatology. Scientists from a wide variety of disciplines have come to recognize both the importance of skin in fundamental biological processes and the broad implications of understanding the pathogenesis of skin disease. As a result there is now a multidisciplinary and world-wide interest in the progress of dermatology.

With these factors in mind, we have undertaken a new series of books specifically oriented to dermatology. The series will be purposely broad in focus and will range from pure basic science to practical, applied clinical dermatology. Thus, while there will be something for everyone, all editions in the series should ultimately prove to be valuable additions to the dermatologist's library.

The current volume, edited by Drs. Jürgen Fuchs and Lester Packer, is an excellent example of the basic science focus of this series, yet it maintains clinical relevance. Significant advances in our understanding of the formation of free radicals and their role in the pathophysiology of cutaneous disease should make this volume not only timely but of interest to a broad range of physicians and other scientists.

Alan R. Shalita
SUNY Health Science Center
Brooklyn, New York

Preface

Clinical dermatology is a fascinating area of cutaneous biology. Most skin diseases are obvious to the patient and physician, although the underlying pathology remains obscure in many cases and therapy is frequently empirical. The high prevalence of skin disease and its frequently unresolved pathogenesis make dermatology an urgent priority for intensive research efforts. For historical reasons, the puzzling world of skin disease is complicated by a mostly descriptive terminology and a multitude of synonyms. Classification of skin diseases is mainly based on the prevalent macromorphology, rather than on a rationale influenced by pathophysiological considerations. However, owing to significant advances in the life sciences, the descriptive morphological viewpoint of dermatology has changed remarkably. Molecular biology and immunology have had a great impact on the understanding of skin pathophysiology and have opened up new avenues in dermatological research. Also, free radical biology and medicine is a newly emerging and rapidly developing research area that is now exerting great influence on scientific developments in dermatology.

We have sought here to develop a state-of-the-art volume on cutaneous free radical biology with particular relevance to clinical dermatology. Leading researchers and clinicians have contributed chapters on the relevance of oxidants and antioxidants in cutaneous biology. The concept of oxidative damage and its repair is bound to become a significant determinant in dermatopathophysiology and is a novel and intriguing subject that will improve our understanding of skin, aging, and disease.

Jürgen Fuchs
Lester Packer

Contents

Contributors

Rajesh Agarwal, Ph.D. *Department of Dermatology, Case Western Reserve University, Cleveland, Ohio*

Junichi Akiyama, Ph.D. *Department of Dermatology, Shinshu University School of Medicine, Asahi, Matsumoto, Japan*

Michael F. Angel, M.D. *Division of Plastic Surgery, The Johns Hopkins School of Medicine, Baltimore, Maryland*

Homer S. Black, Ph.D. *Department of Dermatology, Baylor College of Medicine, and Veterans Affairs Medical Center, Houston, Texas,*

Graeme F. Bryce, Ph.D. *Preclinical Dermatology Research Group, Hoffmann-La Roche, Inc., Nutley, New Jersey*

Michael Buslau, M.D. *Department of Dermatology, Johann Wolfgang Goethe University, Frankfurt, Germany*

Pedro Buc Calderon, Ph.D. *Unité de Biochimie Toxicologique et Cancerologique, Ecole de Pharmacie, Université Catholique de Louvain, Brussels, Belgium*

Jürgen Fuchs, M.D., Ph.D. *Zentrum der Dermatologie und Venerologie, Johann Wolfgang Goethe University, Frankfurt, Germany*

Marie-Anne Gougerot-Pocidalo, M.D. *Department of Immunology and Hematology, Hôpital Bichat, Paris, France*

Herbert F. Haberman, M.D., FRCP(C) *Clinical Sciences Division, University of Toronto, Toronto, Ontario, Canada*

Barry Halliwell, D.Sc. *Division of Pulmonary-Critical Care Medicine, University of California at Davis Medical Center, Sacramento, California*

Jeff Harvell, M.D. *Department of Dermatology, University of California, San Francisco, San Francisco, California*

M. E. Hatcher, M.S. *San Francisco State University, San Francisco, California*

Gerd Hoffmann, M.D. *Institute for Sports Medicine, Johann Wolfgang Goethe University, Frankfurt, Germany*

Klaus Hohlmaier, M.D., Ph.D. *Department of Dermatology, Schwaerzberg Clinic for Allergies, Asthma, and Psoriasis, Bad Rappenau, Germany*

Kowichi Jimbow, M.D., Ph.D., FRCP(C) *Division of Dermatology and Cutaneous Sciences, Faculty of Medicine, University of Alberta, Edmonton, Alberta, Canada*

Irene E. Kochevar, Ph.D. *Department of Dermatology, Wellman Laboratories of Photomedicine, Massachusetts General Hospital, and Harvard Medical School, Boston, Massachusetts*

Daniel Maes, Ph.D. *Department of Biological Research, Estee Lauder Research Laboratories, Melville, New York*

Howard I. Maibach, M.D. *Department of Dermatology, University of California, San Francisco, San Francisco, California*

Paul N. Manson, M.D. *Division of Plastic Surgery, The Johns Hopkins School of Medicine, Baltimore, Maryland*

Kenneth Marenus, Ph.D. *Department of Biological Research, Estee Lauder Research Laboratories, Melville, New York*

I. Aravind Menon, Ph.D. *Departments of Medicine and Opthalmology, University of Toronto, Toronto, Ontario, Canada*

Rainer Milbradt, M.D. *Zentrum der Dermatologie und Venerologie, Johann Wolfgang Goethe University, Frankfurt, Germany*

Yoshiki Miyachi, M.D., Ph.D.* *Tenri Hospital, Nara, Japan*

Paul Motchnik, Ph.D. *Department of Molecular and Cell Biology, University of California, Berkeley, California*

Hasan Mukhtar, Ph.D. *Department of Dermatology, Case Western Reserve University, Cleveland, Ohio*

F. R. Ochsendorf, M.D. *Zentrum der Dermatologie und Venerologie, Klinikum der Johann Wolfgang Goethe University, Frankfurt, Germany*

Ryohei Ogura, M.D., Ph.D. *Department of Medical Biochemistry, Kurume University School of Medicine, Kurume, Japan*

Lester Packer, Ph.D. *Department of Molecular and Cell Biology, University of California, Berkeley, California*

Jennifer G. Peak, Ph.D. *Biological and Medical Research Division, Argonne National Laboratory, Argonne, Illinois*

Meyrick J. Peak, Ph.D. *Biological and Medical Research Division, Argonne National Laboratory, Argonne, Illinois*

Edward Pelle, M.S. *Department of Biological Research, Estee Lauder Research Laboratories, Melville, New York*

W. Z. Plachy, Ph.D. *San Francisco State University, San Francisco, California*

**Present affiliation*: Department of Dermatology, Gunma University School of Medicine, Maebashi, Gunma, Japan

Narendranath S. Ranadive, Ph.D. *Department of Pathology, University of Toronto, Toronto, Ontario, Canada*

Krzysztof Reszka, Ph.D. *Laboratory of Molecular Biophysics, National Institute of Environmental Health Sciences, Research Triangle Park, North Carolina*

Jean-Pierre Revillard, M.D. *Laboratory of Immunology, University of Lyon, Lyon, France*

M. B. Roberfroid, Ph.D. *Unité de Biochimie Toxicologique et Cancerologique, Ecole de Pharmacie, Université Catholique de Louvain, Brussels, Belgium*

Nanna Schürer, M.D. *Hautklinik der Heinrich Heine Universität Dusseldorf, Dusseldorf, Germany*

Yasuko Shindo, M.D., Ph.D. *Department of Dermatology, Shinshu University School of Medicine, Asahi, Matsumoto, Japan*

Masayasu Sugiyama, Ph.D. *Department of Medical Biochemistry, Kurume University School of Medicine, Kurume, Japan*

Yoshio Takase, M.D., Ph.D. *Department of Dermatology, Shinshu University School of Medicine, Asahi, Matsumoto, Japan*

John R. Trevithick, Ph.D. *Department of Biochemistry, Faculty of Medicine, University of Western Ontario, London, Ontario, Canada*

Donald A. Vessey, Ph.D. *Department of Veterans Affairs Medical Center, and Departments of Medicine and Pharmacology, University of California, San Francisco, California*

Eric H. Witt, Ph.D. *Department of Molecular and Cell Biology, University of California, Berkeley, California*

Oxidative Stress in Dermatology

I

OXIDANTS, ANTIOXIDANTS, AND OXIDATIVE INJURY IN DERMATOPATHOLOGY

Part 1 presents basic information on the generation of free radicals and reactive oxygen species in skin, the detailed composition of the skin antioxidant system, and the mechanisms of peroxidative skin damage. In particular, the role of reactive oxidants in ultraviolet-mediated skin pathology and cutaneous carcinogenesis is highlighted.

1

Reactive Oxygen Species in Pathology with Special Reference to the Skin

Barry Halliwell
University of California at Davis Medical Center,
Sacramento, California

I. INTRODUCTION

The biomedical literature continues to abound with claims that "free radicals" are involved in different human diseases. They have been implicated in over 100 disorders, ranging from rheumatoid arthritis and hemorrhagic shock to gastrointestinal ischemia and AIDS (reviewed in Refs. 1–4). Some skin disorders in which free radicals have been suggested to play a role are listed in Table 1. This wide range of disorders implies that free radicals are not something esoteric, but that their increased formation accompanies tissue injury in most, if not all, human diseases (1–5). Sometimes they make a significant contribution to the disease pathology; at other times they may not (5). Telling the difference requires specific assays of free radical activity that are applicable to humans: the lack of such assays has greatly impeded progress in our understanding of the role played by free radicals in normal physiology and in human disease. Thus, as summarized in Table 2, demonstrating that free radicals are important in skin diseases (or indeed any other diseases) involves much more than a mere demonstration of their formation.

The skin is particularly prone to insult: it can be exposed (to a degree greater than any other body tissue) to excess heat or cold and UV radiation, which not only has direct damaging effects but also can trigger free radical chain reactions that may deplete skin antioxidants (6). Toxic chemicals in

Table 1 Some of the Skin Disorders That May Involve Reactive Oxygen Species

Autoimmune symptoms (e.g., "butterfly rash" of lupus)
Skin cancer
Frostbite
Skin transplantation–reoxygenation injury
Skin wrinkling due to cigarette smoke
Photosensitizing drugs/products (including plant products such as hypericin)
Contact dermatitis
Skin inflammation generally (including acne)
Porphyria
Psoriasis
Thermal injury
UV-induced skin damage
Exposure to ionizing radiation

water, dust, plant materials, applied creams, perfumes, deodorants, etc., as well as atmospheric pollutants, constantly touch the skin surface.

Many of these problems are discussed in detail in subsequent chapters. The purpose of this introductory chapter is to set the scene by reminding readers of some basic definitions and principles.

II. WHAT ARE FREE RADICALS?

Electrons in atoms move within regions of space known as *orbitals*. Each orbital holds a maximum of two electrons. A free radical is simply defined

Table 2 Criteria for Implicating Any Agent as a Significant Mechanism of Tissue Injury in Human Disease

1. The agent should always be present at the site of injury.
2. Direct application of the agent to the tissue at concentrations within the range found in vivo should reproduce most or all of the damage observed.
3. Removing the agent or inhibiting its formation should diminish the injury to an extent approximately proportional to the degree of removal of the agent or inhibition of its formation.

1 and 3 obviously rely on accurate assay methodology. *Free radicals are difficult to measure*: it is rarely clear what percentage of a radical formed is actually trapped when a trapping method is used and "end products" of free radical damage are rapidly removed by excretion from the body and/or metabolism. *Molecular targets of free radical damage are numerous* and their relative importance can change markedly as a result of changes in radical concentration or cell type. Thus it is hard to know what targets of injury should be measured under 2. Free radical damage can, for example, occur without measurable lipid peroxidation (1,40).
Source: Ref. 40.

as *any species capable of independent existence that contains one or more unpaired electrons*, an unpaired electron being one that is alone in an orbital. Examples of free radicals are superoxide ($O_2^{\cdot -}$) and hydroxyl ($OH^{\cdot}$), both oxygen-centered radicals, thiyl ($RS^{\cdot}$, a sulfur-centered radical), trichloromethyl ($CCl_3^{\cdot}$, a carbon-centered radical), and nitric oxide ($NO^{\cdot}$), in which the unpaired electron is delocalized between both atoms. The dot designates the presence of one or more unpaired electrons. Most biological molecules are nonradicals, containing only paired electrons. When nonradicals and radicals react together, free-radical chain reactions often result. For example, when highly reactive hydroxyl or trichloromethylperoxyl radicals react with lipids, lipid-derived radicals are produced and the free radical chain reaction of *lipid peroxidation* is initiated. Many drugs and toxins are metabolized in vivo to produce free radicals (1).

III. FREE RADICALS AND REACTIVE OXYGEN SPECIES IN VIVO: BOTH USEFUL AND DANGEROUS

It is now well established that free radicals (especially $O_2^{\cdot -}$) and other reactive oxygen species (such as H_2O_2) are continuously produced in vivo (1–6). The term *reactive oxygen species* (ROS) is a collective one that includes not only oxygen-centered radicals such as $O_2^{\cdot -}$ and $OH^{\cdot}$, but also some nonradical derivatives of oxygen, such as hydrogen peroxide (H_2O_2), singlet oxygen $^1\Delta g$, and hypochlorous acid (HOCl). Hydroxyl radical is also produced by splitting of water caused by "background" levels of ionizing radiation (7). In consequence, organisms have evolved not only antioxidant defense systems to protect against reactive oxygen species, but also repair systems that prevent the accumulation of oxidatively damaged molecules (reviewed in Refs. 1, 7–9).

Hydroxyl radical may be produced in living organisms by at least two mechanisms in addition to the effects of ionizing radiation (7): the reaction of transition metal ions with H_2O_2 (reviewed in Ref. 10) and the decomposition of peroxynitrite formed by reaction of $O_2^{\cdot -}$ with $NO^{\cdot}$ (11). Hydroxyl radical is a fearsomely reactive species that can attack all biological molecules, usually setting off free radicals chain reactions (1,10). A wide range of transition metal ions, including iron and copper, can react with H_2O_2 to form $OH^{\cdot}$, but the reactions are complex and other oxidizing species are formed as well (10). An important physiological antioxidant defense in body tissues and extracellular fluids is the sequestration of transition metal ions into "safe" forms unable to catalyze free radical reactions (1,12,13). Sweat, an excretory fluid, has been shown to contain significant levels of iron and copper ions "catalytic" for free radical reactions (Table 3) (14,15). This not only provides metal ions to facilitate the growth of skin bacteria, but it may worsen the consequences of oxidative damage to the skin. For example, neutro-

Table 3 Content of "Catalytic" Iron and Copper in Sweat from Different Body Sites

	μmol/L (mean ± SD) of	
Site studied	Fe	Cu
Abdomen	8.8 ± 7.0 (14)[a]	14.0 ± 9.9 (8)
Arm	5.0 ± 2.5 (10)	8.2 ± 7.5 (4)
Chest	8.9 ± 10.8 (15)	11.5 ± 11.6 (8)
Back	3.6 ± 3.2 (19)	8.8 ± 6.8 (14)

[a]Numbers of samples analyzed are given in parentheses.

Subjects were healthy male adults (age range 18–45 yr, mean age 21 yr). Sweat was collected during exercise and centrifuged to remove skin cells and bacteria. Catalytic iron was measured by the bleomycin assay, copper by the phenantholine assay.

Source: Ref. 14.

phils present at sites of inflammation (e.g., as in acne) make $0_2^{\cdot -}$ and H_2O_2, which could become converted to $OH^{\cdot}$ in the presence of iron or copper ions. Bissett et al. (16) showed that iron ion chelators could delay the onset of skin damage induced by UVB radiation in albino hairless mice. Formation of singlet O_2 by photosensitization reactions mediated by drugs, toxins, or constituents of cosmetics is a particular nuisance in the skin (see Table 1), as is discussed in detail in subsequent chapters.

Some of the $O_2^{\cdot -}$ production that occurs in vivo appears to be a "chemical accident" due to autoxidation reactions and the "leakage" of electrons from electron transport chains on to oxygen (9,17). Yet much $O_2^{\cdot -}$ is also made deliberately: this radical plays an essential role in the killing of bacteria by activated phagocytes (18). Superoxide is also made, to a lesser extent, by other cell types, such as fibroblasts and lymphocytes (19–21). Removal of excess $O_2^{\cdot -}$ by superoxide dismutase (SOD) enzymes is an important physiological antioxidant defense mechanism in aerobic organisms (6,17,22), although too much SOD (in relation to the activities of H_2O_2-removing enzymes such as catalase and glutathione peroxidase) may sometimes be deleterious (e.g., Refs. 23, 24).

SOD enzymes convert $O_2^{\cdot -}$ into O_2 and H_2O_2: the latter may also have some metabolic functions (e.g., Refs. 1, 12, 25, 26). H_2O_2 might conceivably act as a metabolic signal under certain circumstances, possibly by oxidizing specific protein thiol groups and triggering intracellular events (1). H_2O_2 crosses cell membranes easily (1), and it can attack, apparently directly, a few cellular targets. For example, high levels of H_2O_2 can inactivate the glycolytic enzyme glyceraldehyde-3-phosphate dehydrogenase in mammalian cells (27,28).

On the whole, however, $O_2^{\cdot -}$ and H_2O_2 have limited chemical reactivity. For example, increased generation of $O_2^{\cdot -}$ and H_2O_2 within cells often leads to DNA damage, but neither of these species reacts directly with DNA (reviewed in Ref. 29). Interest has, therefore, focused on the ability of $O_2^{\cdot -}$ and H_2O_2 to generate more dangerous species, such as $OH^{\cdot}$, in vivo. Most (1,10,29), but not all (11), of these proposed reactions involve transition metal ions, whose availability in sweat may tend to exacerbate oxidative damage involving the skin (14–16). Because some ROS play useful metabolic roles, antioxidant defenses are not 100% efficient. This means that it is easy to injure organisms by depleting antioxidants and/or by raising ROS formation. *Oxidative stress* is said to occur when ROS formation and defenses against ROS are imbalanced in favor of the former (8).

IV. WHAT ROLE DO REACTIVE OXYGEN SPECIES PLAY IN HUMAN DISEASE?

Some human diseases may be *caused* by oxidative stress. Thus ionizing radiation generates $OH^{\cdot}$ by splitting water molecules and many of the biological consequences of excess radiation exposure may be due to free radical damage to proteins, DNA, and lipids (reviewed in Ref. 7). The symptoms produced by chronic dietary deficiencies of selenium (e.g., Keshan disease) or of tocopherols (neurological disorders seen in patients with defects in intestinal fat absorption) might also be mediated by oxidative stress. Selenium is an essential component of H_2O_2-destroying glutathione peroxidase enzymes, although this is unlikely to be its sole biological function (30,31), and tocopherols are important chain-breaking antioxidants that protect membrane lipids and circulating lipoproteins against peroxidation (32).

However, in most human diseases, oxidative stress is a *consequence* and not a cause of the primary disease process. Tissue damage by infection, trauma, toxins, abnormally low or high temperatures, and other causes usually leads to formation of increased amounts of putative "injury mediators," such as prostaglandins, leukotrienes, interleukins, and TNFs. All of these have at various times been suggested to play important roles in different human diseases. Currently, for example, there is much interest in the roles played by TNFα and interleukins in adult respiratory distress syndrome and septic shock. Although clearly stated some 8 years ago (5), it has not yet been appreciated in all quarters that ROS can be placed in the same category, i.e., *tissue damage will usually lead to increased ROS formation*. Tissue injury can, for example, cause recruitment and activation of phagocytes (producing $O_2^{\cdot -}$ H_2O_2, HOCl, and $NO^{\cdot}$), release of "catalytic" iron ions and heme proteins (which can also accelerate oxidative damage in the presence of H_2O_2 [33]) from injured cells, disruption of electron transport chains so that more electrons leak to O_2 to form $O_2^{\cdot -}$, and activation of the enzyme xanthine oxidase. Although xanthine oxidase may not be important in human myocardial reoxygenation injury (34), it might play a role in joint inflammation (35) and in Dupuytren's contracture (36), and so it should not be ignored in considerations of human disease (41).

If, in most human diseases, ROS are produced in increased amounts as a consequence of tissue injury, do they then make a significant contribution to the disease pathology, or is their formation of little or no consequence? The answer probably differs in different diseases (5). Indeed, the same question could be asked about any putative mediator of tissue injury (see Table 2). In order to show that ROS are important in a particular disease, it is necessary to demonstrate that:

ROS are formed at the injury site.
The time course of their formation is such that they could play a role in disease pathology.
Removal of ROS or prevention of their formation has beneficial effects.
Direct application of ROS at concentrations found in vivo reproduces most or all of the injury.

These criteria must be borne in mind when considering proposed roles of ROS in skin disease. However, the required criteria are not easy to fulfill. For example, suppose that giving α-tocopherol to patients with a disease produces no discernible therapeutic benefit. This could mean that ROS are not important tissue-injury mechanisms in that disease. It could also mean that:

Not enough antioxidant reached the correct site of action.
The oxidative injury involves damage not to lipids, but to protein and/or DNA. Oxidative stress frequently causes damage to DNA and to specific proteins, and such damage may often be of more biological consequence than is damage to lipids (1,5,27,29,37,38).

In the same way, suppose that giving an antioxidant to patients *did* improve their disease. This could mean that 1) ROS do make a significant contribution to disease pathology or 2) the antioxidant has other beneficial pharmacological effects unrelated to its action on ROS. For example, many chain-breaking antioxidant inhibitors of lipid peroxidation also inhibit lipoxygenase enzymes, important in synthesis of leukotrienes (39). What is needed is good *methodology* to *measure* free radical damage and ROS in human disease, and such measurements must accompany clinical tests of the effects of antioxidants to show that they actually did or did not remove the ROS. It is important to determine the precise target of attack by ROS in each case (see Table 2). Fortunately, assays for measuring ROS are improving rapidly, as recently reviewed (1,40). For example, molecular "fingerprints" of oxidative damage to the purine and pyrimidine bases of DNA exist: they can, for example, be used to study whether or not the DNA damage that occurs when cells and tissues are subjected to oxidative stress is due to direct attack of ROS upon the DNA (29).

V. CONCLUSION

Oxidative stress occurs in most, if not all, human diseases, including skin disorders. Our understanding of the role played by ROS in disease pathology is still weak, largely due to the lack of accurate methodology applicable to humans. Antioxidant treatment still has promise as a therapeutic approach, particularly in diseases in which other treatments are ineffective (42). Its evaluation must, however, be combined with accurate methods of measuring oxidative damage, which are now beginning to become available (40).

REFERENCES

1. Halliwell B, Gutteridge JMC. Free radicals in biology and medicine. 2nd ed. Oxford: Clarendon Press, 1989.
2. Halliwell B, Cross CE. Reactive oxygen species, antioxidants and acquired immunodeficiency syndrome. Arch Int Med 1991; 157:29–31.
3. Parks DA. Oxygen radicals; mediators of gastrointestinal pathophysiology. Gut 1989; 30:293–8.
4. Southorn PA. Free radicals in medicine. II: involvement in human disease. Mayo Clin Proc 1988; 63:390–408.
5. Halliwell B, Gutteridge JMC. Lipid peroxidation, oxygen radicals, cell damage and antioxidant therapy. Lancet 1984; 1:1396–8.
6. Kagan V, Witt E, Goldman R, Scita G, Packer L. Ultraviolet-light-induced generation of vitamin E radicals and their recycling. A possible photosensitizing effect of vitamin E in skin. Free Radical Res Commun 1992; 16:51–64.
7. Von Sonntag C. The chemical basis of radiation biology. London: Taylor and Francis, 1987.
8. Sies H. Oxidative stress: oxidants and antioxidants. New York: Academic Press, 1991.
9. Fridovich I. Superoxide dismutase: an adaptation to a paramagnetic gas. J Biol Chem 1989; 264:7761–4.
10. Halliwell B, Gutteridge JMC. Role of free radicals and catalytic metal ions in human disease: an overview. Meth Enzymol 1990; 186:1–85.
11. Beckman JS, Beckman TW, Chen J, et al. Apparent hydroxyl radical production by peroxynitrite: implications for endothelial injury from nitric oxide and superoxide. Proc Natl Acad Sci 1990; 87:1620–4.
12. Halliwell B, Gutteridge JMC. Oxygen free radicals and iron in relation to biology and medicine. Some problems and concepts. Arch Biochem Biophys 1986; 246: 501–14.
13. Halliwell B, Gutteridge JMC. The antioxidants of human extracellular fluids. Arch Biochem Biophys 1990; 280:1–8.

14. Aruoma OI, Reilly T, MacLaren D, Halliwell B. Iron, copper and zinc concentrations in human sweat and plasma; the effect of exercise. Clin Chim Acta 1988; 177:81–8.
15. Gutteridge JMC, Rowley DA, Halliwell B, Cooper DF, Heeley DM. Copper and iron complexes catalytic for oxygen radical reactions in sweat from human athletes. Clin Chim Acta 1985; 145:267–73.
16. Bissett DL, Chatterjee R, Hannon, DP. Chronic ultraviolet radiation-induced increase in skin iron and the photo-protective effect of topically applied iron chelators. Photochem Photobiol 1991; 54:215–23.
17. Imlay JA, Fridovich I. Assays of metabolic superoxide production in *Escherichia coli*. J Biol Chem 1991; 266:6957–65.
18. Babior BM, Woodman RC. Chronic granulomatous disease. Semin Hematol 1990; 27:247–59.
19. Murrell GAC, Francis MJO, Bromley L. Modulation of fibroblast proliferation by oxygen free radicals. Biochem J 1990; 265:659–65.
20. Maly FE. The B-lymphocyte: a newly-recognized source of reactive oxygen species with immunoregulatory potential. Free Radical Res Commun 1990; 8:143–8.
21. Meier B, Radeke H, Selle S, et al. Human fibroblasts release reactive oxygen species in response to treatment with synovial fluids from patients suffering from arthritis. Free Radical Res Commun 1990; 8:149–60.
22. Touati D. The molecular genetics of superoxide dismutase in *E. coli*. An approach to understanding the biological role and regulation of SODS in relation to other elements of the defence system against oxygen toxicity. Free Radical Res Commun 1989; 8:1–8.
23. Groner Y, Elroy-Stein O, Avraham KB, et al. Down syndrome clinical symptoms are manifested in transfected cells and transgenic mice overexpressing the human Cu/Zn-superoxide dismutase gene. J Physiol (Paris). 1990; 84:53–77.
24. Scott MD, Eaton JW, Kuypers FA, et al. Enhancement of erythrocyte superoxide dismutase activity: effects on cellular oxidant defense. Blood 1989; 74: 2542–9.
25. Dupuy C, Virion A, Ohayon R, et al. Mechanism of hydrogen peroxide formation catalyzed by NADPH oxidase in thyroid plasma membrane. J Biol Chem 1991; 26:3739–43.
26. Shapiro BM. The control of oxidative stress at fertilization. Science 1991; 252: 533–6.
27. Hyslop PA, Hinshaw DB, Halsey WA, Jr, et al. Mechanisms of oxidant-mediated cell injury. The glycolytic and mitochondrial pathways of ADP phosphorylation are major intracellular targets inactivated by hydrogen peroxide. J Biol Chem 1988; 263:1665–75.
28. Baker MS, Bolis S, Lowther DA. Oxidation of articular cartilage glyceraldehyde-3-phosphate dehydrogenase (G3PDH) occurs in vivo during carrageenin-induced arthritis. Agents Actions 1991; 32:299–305.
29. Halliwell B, Aruoma OI. DNA damage by oxygen-derived species. Its mechanism and measurement in mammalian systems. FEBS Lett 1991; 281:9–19.
30. Levander OA. A global view of human selenium nutrition. Ann Rev Nutr 1987; 7:227–50.

31. Burk RF. Molecular biology of selenium, with implications for its metabolism. FASEB J 1991; 5:2274–9.
32. Muller DPR, Goss-Sampson MA. Neurochemical, neurophysiological and neuropathological studies in vitamin E deficiency. Crit Rev Neurobiol 1990; 5:239–65.
33. Harel S, Kanner J. The generation of ferryl or hydroxyl radicals during interaction of haemoproteins with hydrogen peroxide. Free Radical Res Commun 1988; 5:11–19.
34. Eddy LJ, Stewart JR, Jones HP, Engerson TD, McCord JM, Downey JM. Free radical-producing enzyme, xanthine oxidase, is undetectable in human hearts. Am J Physiol 1987; 253:H709–11.
35. Allen RE, Outhwaite J, Morris CJ, Blake DR. Xanthine oxido-reductase is present in human synovium. Ann Rheum Dis 1987; 46:843–5.
36. Murrell GA, Francis MJ, Bromley L. Free radicals and Dupuytrens contracture. Br Med J 1987; 295:1371–5.
37. Cochrane CG. Mechanisms of oxidant injury of cells. Mol Aspects Med 1991; 12:137–47.
38. Halliwell B. Oxidants and human disease: some new concepts. FASEB J 1987; 1:358–364.
39. Laughton MJ, Evans PJ, Moroney MA, et al. Inhibition of mammalian 5-lipoxygenase and cyclo-oxygenase by flavonoids and phenolic dietary additives. Biochem Pharmacol 1991; 42:1673–1681.
40. Halliwell B, Gutteridge JMC, Cross CE. Oxidants, antioxidants and human disease: Where are we now? J Lab Clin Med 1992; 119:598–620.
41. Yokoyama Y, Beckman JS, Beckman TK et al. Circulating xanthine oxidase: potential mediator of ischemic injury. Am J Physiol 1990; 258:G564–70.
42. Halliwell B. Drug antioxidant effects: a basis for drug selection? Drugs 1991; 42:569–605.

2

Solubility-Diffusion of Dioxygen in the Skin

M. E. Hatcher and W. Z. Plachy
San Francisco State University, San Francisco, California

I. INTRODUCTION

The skin is the organ that is most exposed to the dangers of the environment. The skin's exceptionally low permeability to most molecules enables it to regulate the transport of molecules between the body and the environment. Understanding of the skin's permeability barrier function began as early as 1853 when Homalle (1) observed that the formation of a blister indicated that the epidermis was less permeable to water than the dermis, suggesting that the barrier resided in the epidermis.

The conclusive results that led to wide-scale acceptance that the barrier was localized in the outermost layers of the epidermis, the stratum corneum (SC), were shown by Scheuplein and coworkers. Scheuplein and Ross (2) found diffusion through the skin to be passive and similar to diffusion through a composite media. They enzymatically removed the epidermis from the SC and observed the SC to be 3–5 times less permeable than the epidermis (3). Scheuplein and Blank found that permeation through the skin obeyed Fick's law and could be modeled as diffusion through three layers in series (3).

Scheuplein et al. (4) measured the diffusion coefficients through the SC for various molecules. The diffusivities ranged from 10^{-9} to 10^{-13} cm^2/s versus 10^{-5} to 10^{-6} cm^2/s for the same molecules in bulk water. Scheuplein also measured the diffusivities of oxygen and carbon dioxide gas and found them to be on the order of 10^{-7} cm^2/s, which is in the same range as the diffusion of these gases in polymers (3).

The "brick and mortar" model of the SC proposed by Michaels et al. (5) accounts for the rapid diffusion with low activation energy of lipophilic molecules traversing the SC. This model also proposes a method for diffusion of polar solvents, but it is inconsistent with the high affinity of the SC for small polar molecules. The bifunctional solubility of the SC is thought to arise from the coexistence of aqueous and lipid regions. The smaller diffusion coefficients and larger activation energies of the polar molecules seem to suggest that these molecules are effectively bound in the aqueous regions (6).

The skin is the organ most exposed to the atmosphere, but its barrier function prevents it from imbibing oxygen as efficiently as the lung. However, it is of interest to study oxygen uptake by the skin as an alternate source of oxygen for the viable epidermis. Wound healing creates additional demands for oxygen (7). The optimum molecular oxygen pressure for keratinocyte growth is 130 mmHg (8), and after a wound, epidermal regeneration requires an increased supply of oxygen. This added requirement could be partially supplied by transcutaneous oxygen, TC-O_2 (7). Hyperbaric oxygen therapy, HBO, has been used for years to increase the partial pressure of oxygen, Po_2, in the body (9–11). In HBO therapy, patients breathe high-pressure pure oxygen and a systemic increase in Po_2 results via the cardiopulmonary system. This indeed increases the Po_2 at the wound site but also increases the risk of internal damage caused by oxygen radicals. This is of particular concern for sensitive organs such as the lumen of the lung and the cornea of the eye (10). By applying the HBO therapy directly at the wound site, the partial pressure of O_2 at the wound site is increased without endangering other organs. The drawback to transcutaneous HBO is the inefficient oxygen uptake by the skin. Oxygen is a lipophilic molecule (12) and should diffuse through the SC lipids easily (3). However, according to the brick and mortar model, the keratinocytes act as obstacles, which force the oxygen to travel a tortuous path (5). This increase in the path length required to reach the viable layers of the epidermis results in a lower apparent diffusion coefficient since this coefficient is proportional to the square of the distance traveled per unit time (13). This diffusion rate can be increased by increasing the fluidity of the lipids in the skin. This increase in fluidity can be achieved by the application of chemicals that further disorder the lipid regions in the SC (2,14–18).

II. ELECTRON PARAMAGNETIC RESONANCE STUDIES

Hatcher (19) studied the permeability barrier of the stratum corneum as it pertained to oxygen diffusion. Electron paramagnetic resonance, EPR, was used in this study to determine the oxygen diffusion coefficient. EPR gives information about the motional dynamics of a spin system in a particular environment (20), thus providing information about that environment. The effects of permeation-enhancing drugs in diphosphatidylcholine bilayers (16) were studied using EPR in an attempt to understand the effects of these drugs in the skin. Tanaka et al. (21) used the EPR spin probe method to study the changes in membrane fluidity in hexadecane-induced hyperproliferative epidermis.

The EPR line width of nitroxide spin labels is sensitive to the spin exchange interaction between paramagnetic molecular oxygen and the spin label (22–27). Studying this time-independent exchange broadening provides a way to study the diffusion of oxygen in solvents where the solubility of oxygen is known (12). The product of the diffusion coefficient and the solubility of oxygen using spin exchange was measured by Povich in 1975 (28), and Windrem and Plachy used this method to determine the oxygen diffusion-solubility in lipid bilayers (29). Backer et al. (30) used spin exchange to follow the uptake of oxygen by respiring mitochondria.

The solubility of oxygen in aqueous dimyristoyl phosphatidylcholine dispersions was studied by Smotkin et al. (31), who found that molecular oxygen was three to four times more soluble in the liquid crystalline state (>24°C) than in the gel state (<24°C). The molecular oxygen solubility was not found to be strongly temperature dependent except for the solubility increase noted at the phase transition. Windrem and Plachy (29) obtained the dioxygen Bunsen coefficient, α_{O_2}, times the dioxygen diffusion coefficient, D_{O_2}, in dialkyl lecithin bilayers. By using a family of doxyl fatty acid spin probes, they were able to show that the diffusion-solubility product was a function of position within the bilayer. Subczynski and Hyde (32) obtained similar results using EPR T1 measurements on the same family of spin probes.

The time-dependent spin exchange line broadening can be observed at a constant magnetic field as a decrease in line height in the second derivative spectrum. Packer and coworkers (33) used time evolution of second derivative EPR lines to monitor the concentration of molecular oxygen in photosynthetic systems.

The time evolution of the second derivative spectrum was used in our study to determine the molecular oxygen diffusion coefficient in the stratum corneum and to study the effects of various permeation-enhancing drugs on this diffusion coefficient. In this mode, the O_2 diffusion kinetics are independent of the O_2 solubility.

III. MATERIALS AND METHODS

The diffusion of oxygen into the stratum corneum and stratum corneum model lipid systems was studied using EPR. The oxygen diffusion coefficient was determined by measuring the time dependence of the line broadening caused by spin exchange between oxygen and the nitroxide spin labels that had been applied to the SC samples.

In our study we used the biomimetic lipid spin probe, d_8 ^{15}N 17-doxyl stearic acid (d_8 ^{15}N 17-DSA) (34). The spin probe was ^2H and ^{15}N substituted to de-

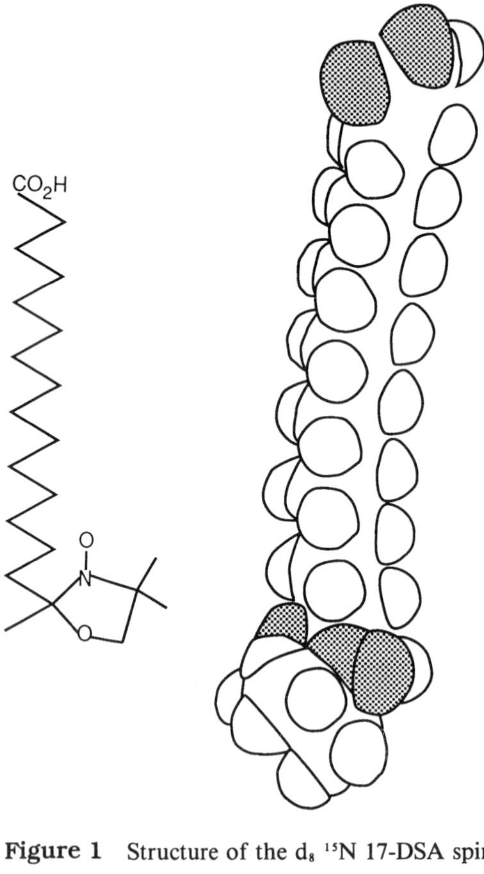

Figure 1 Structure of the d_8 ^{15}N 17-DSA spin-probe used in this study.

crease the effect of hyperfine interaction and thus increase the spectral resolution (35). The structure of the spin probe is shown in Figure 1.

The oxygen diffusion was measured via a gas switch experiment. The sample was bathed in an inert diamagnetic nitrogen atmosphere until the start of the experiment. To start the experiment, the gas was rapidly switched to paramagnetic oxygen, taking less than 0.3 second. The stratum corneum experiments were performed on samples hydrated to 75% relative humidity by bathing the sample with nitrogen gas that had been bubbled through saturated sodium chloride.

The decrease in the second derivative line height due to heterospin exchange was observed. Figure 2 shows the line height decrease with time due

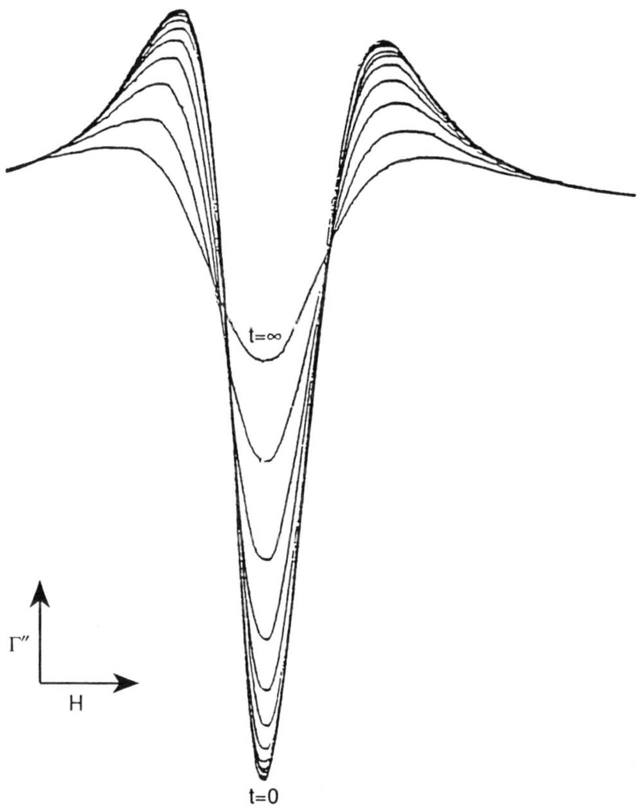

Figure 2 The decrease with time of the second derivative line height of d_8 ^{15}N 17-DSA in silicon oil in the presence of oxygen ($t = 0$ is the start of the experiment when the sample is bathed in nitrogen and $t = \infty$ is the time when the sample is completely bathed in oxygen).

to increasing amounts of oxygen in a sample of d_8 ^{15}N 17-DSA in film of silicon oil. This example of the probe in oil is used to demonstrate the oxygen effect because the gas diffusion rate into the stratum corneum is too rapid to allow a magnetic field sweep. The line height decrease as a function of time in the stratum corneum and model lipid samples was observed by

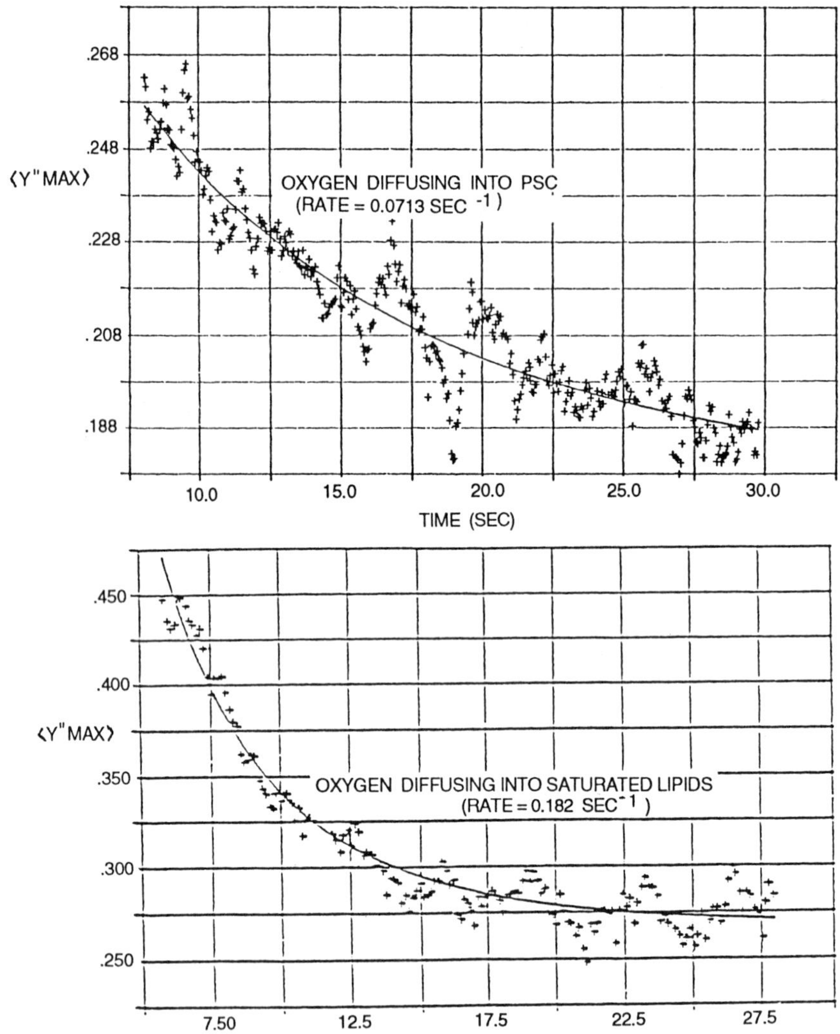

Figure 3 The kinetic curve of the average of the second derivative line height as a function of time.

setting the spectrometer at a constant field directly on the second derivative peak. The rate of molecular oxygen diffusion was then determined using the Asystant Software (36), which had been programmed with the equation for this line height decrease due to spin exchange. Figure 3 shows examples of a rate curve obtained from Asystant for a saturated lipid sample and a human SC sample. The rate, R (s^{-1}), obtained from these plots of the average line height as a function of time, was used to determine the oxygen diffusion coefficient (cm^2/s) according to:

$$D_{O_2} = \frac{4L^2R}{\pi^2}$$

where L is the thickness of the sample (cm) (13). This thickness was determined by measuring the mass of the sample, dividing it by the density and dividing the result by the area.

The d_8 ^{15}N 17-DSA spin probe and permeation enhancers were applied to the samples (20% by weight) using an aerosol spray. Experiments on samples treated with the relatively volatile enhancer dimethyl sulfoxide (DMSO) were run immediately after application and repeated every hour for 4 hours. The experiments on samples treated with the large molecular weight enhancers oleic acid, decylmethyl sulfoxide (DeMSO), and 1-dodecylazacycloheptan-2-one (Azone) were incubated overnight in a nitrogen atmosphere at 37°C and run every day for 4 days.

IV. MODEL LIPID EXPERIMENTS

Experiments were first carried out on synthetic SC lipid model systems as described by Lampe et al. (37). A marked difference was observed between the probe spectra of the saturated and unsaturated lipid mixtures in nitrogen (Fig. 4). The saturated lipid mixture possessed a much larger line width than the unsaturated mixture (ΔH_{pp} = 3.5 vs. 0.8 G). This larger line width is characteristic of a less fluid environment for the spin probe. The EPR spectrum of a spin-labeled system is sensitive to the nature and rate of the motions of the probe. The narrow spectral line width of the unsaturated sample as compared to the saturated sample is consistent with previous studies demonstrating that unsaturation results in less efficient packing and, thus, increased fluidity (defined as the inverse of the viscosity) (37,38). The spectra from the saturated sample reveal EPR signals that are unsymmetrical and broad. The broad EPR lines seen in the saturated lipids are likely the result of relatively tight packing of the lipid acyl chains (37–39).

The decreased line widths observed for the unsaturated lipids in N_2 (see Fig. 4) result from the larger fluidity of this model lipid system. This increase

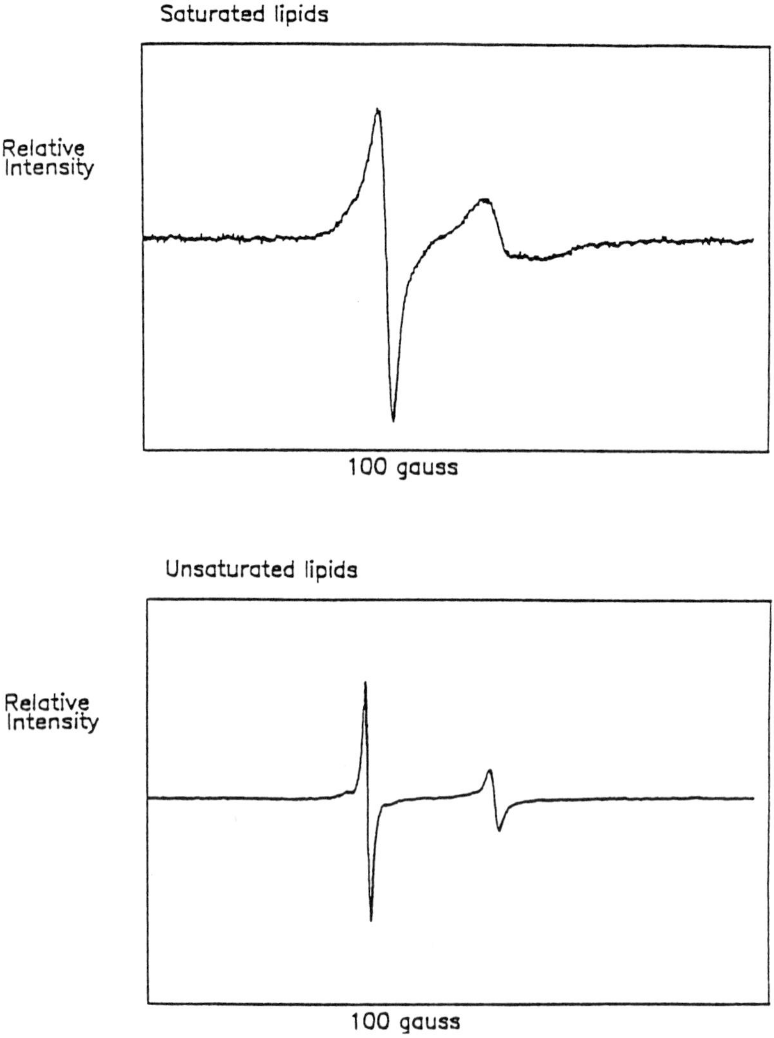

Figure 4 The EPR spectra of model lipid systems labeled with d_8 ^{15}N 17-DSA under nitrogen.

in fluidity resulted in an increased oxygen diffusion coefficient, D_{O_2}. The D_{O_2} for the unsaturated lipids at 37°C was measured to be seven times larger than that for the saturated lipids ($D_{O_2} = 0.2 \times 10^{-5}$ vs. 0.03×10^{-5} cm²/s). This infers that increased lipid fluidity results in increased oxygen diffusion. This is consistent with studies that show increased SC permeability with increased lipid fluidity (16–18,37–40).

Permeation-enhancing chemicals were added to the saturated lipids in an attempt to increase the oxygen diffusion coefficient. Measurements of the line widths showed no change with the application of the lipophilic enhancers, oleic acid, and azone, and these enhancers produced no change in the D_{O_2}. A small decrease in the probe line width in N_2 (ΔH_{pp} = 3.0 vs. 3.5 G) was observed in the presence of the hydrophilic enhancers DeMSO and DMSO. We found that the presence of these enhancers increased the D_{O_2} by a factor of two. This increase of D_{O_2} was similar to the increase of D_{O_2} in the unsaturated lipids observed in the presence of these enhancers (see Table 2). It can therefore be inferred that these enhancers produce the same relative disruption to the packing of both saturated and unsaturated lipid mixtures.

V. STRATUM CORNEUM EXPERIMENTS

Stratum corneum experiments were performed on human and porcine stratum corneum. These experiments were done on top-to-bottom fused layers of SC to increase the sample size and, thus, increase the signal-to-noise ratio. This increased thickness also made the gas switch time much less than the time of the experiment. The D_{O_2} remained constant over varying numbers of layers, indicating that the SC barrier function is constant across the cell layers (Table 1).

The EPR spectrum of the d_8 ^{15}N 17-DSA spin probe in the heterogeneous stratum corneum possesses two distinct probe sites (Fig. 5). The probe in one of these sites reveals rapid and isotropic motion ($\tau_c \leqslant 1$ ns). The other probe site has an anisotropic spectrum ($\tau_c \geqslant 20$ ns), indicating that the spin probe is immobilized.

The oxygen diffusin coefficient, D_{O_2}, for human stratum corneum was found to be 3×10^{-7} cm²/s at 37°C. This value is approximately four times that observed for porcine stratum corneum (7×10^{-8} cm²/s). However, the line width in nitrogen of the fluid line in porcine SC was the same as that for human SC (0.9 G). This would seem to indicate that the decrease in D_{O_2} is

Table 1 Oxygen Diffusion Coefficient Over Layers of Porcine SC

Number of layers	D_{O_2} (cm²/sec) ± 20%
2	6×10^{-8}
4	7×10^{-8}
6	7×10^{-8}
8	9×10^{-8}

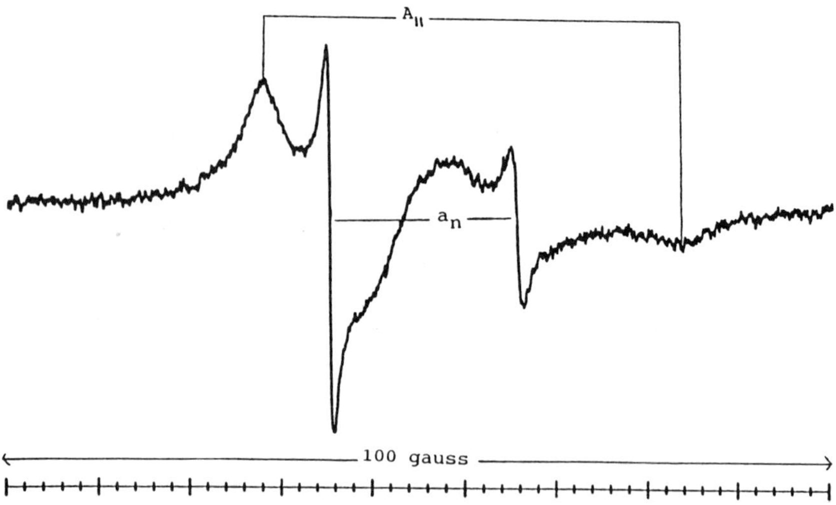

Figure 5 The spectrum of porcine stratum corneum labeled with d_8 ^{15}N 17-DSA (a_n is the isotropic hyperfine splitting and A_{11} is the anisotropic hyperfine splitting).

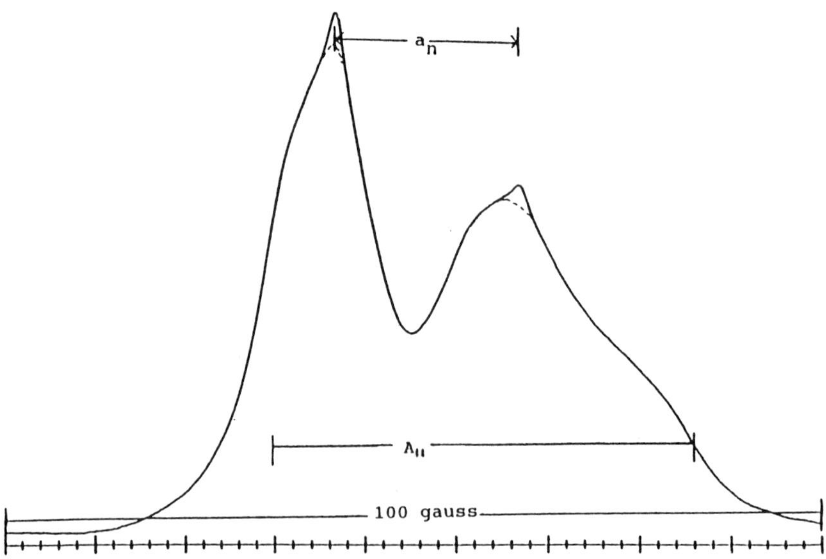

Figure 6 The integrated spectrum of porcine stratum corneum labeled with d_8 ^{15}N 17-DSA. The immobilized signal is below the dashed line. Notice how the immobilized signal dominates the integrated spectrum. (A_{11} and a_n defined in Figure 5.)

Table 2 Effect of Permeation Enhancers on D_{O_2}

Biomedia	$D_{O_2} \times 10^5$ (cm²/s)	P.E.	$D_{O_2} \times 10^5$ (cm²/s)[a]
Saturated lipids	0.03	Azone	0.04
		DeMSO	0.07
		Oleic acid	0.03
		DMSO	0.05
Unsaturated lipids	0.2	Azone	0.2
		DeMSO	0.44
		DMSO	0.46
Human SC	0.03	DeMSO	0.06
		DMSO	0.09
		Oleic acid	0.03

[a] ±20%.

due to the nature of the immobilized region of the porcine SC. Figure 6 shows that the EPR spectrum of porcine SC does, in fact, possess a greater fraction of immobilized spin probe. Less dramatic differences were also observed between samples of the same species. The differences in the spectra may be, therefore, due to regional variations within a species, variations between species, or methods of sample preparation.

The oxygen diffusion coefficient, D_{O_2}, for human stratum corneum was increased twofold in the presence of DeMSO and threefold in the presence of DMSO. There was no change in the D_{O_2} with the application of oleic acid. The oxygen diffusion coefficients for human SC, saturated lipids and unsaturated lipids, both with and without permeation enhancers, are given in Table 2.

VI. CONCLUSIONS

The stratum corneum is a heterogeneous biomedia, which makes studying its physical properties both interesting and challenging. By studying the dioxygen diffusion across the stratum corneum, we were not only able to determine this important diffusion coefficient, but we were able to learn about the role of the lipid and protein domains in the SC.

The EPR spectra of the stratum corneum demonstrated that at least some of the SC lipids are fluid. The line width in nitrogen of the fluid signal in human SC was found to be quite narrow (0.9 G). This is very close to the value obtained for the unsaturated lipid model system (0.8 G). This indicates that the lipids of the SC contain some degree of unsaturation (15,41).

Dioxygen is a hydrophobic molecule (12). Dioxygen diffusion in the SC, therefore, is well described using the "brick and mortar" model proposed by Michaels et al. (5). In this model the proteins serve as barriers forcing oxygen to travel a tortuous diffusion pathway. Calhoun et al. (46) have established that dioxygen has reduced access through protein. This tortuosity effect is seen in the difference between the oxygen diffusion coefficients in unsaturated lipids and human stratum corneum.

As was stated earlier, the line width of the fluid domain in the stratum corneum is comparable to the unsaturated lipid mixture. This would make the unsaturated mixture, rather than the saturated mixture, the obvious choice for the model for the SC lipids. The oxygen diffusion coefficient for the unsaturated lipids was found to be seven times as large as that for the human SC (2×10^{-7} vs. 3×10^{-8} cm^2/s). Recalling that the diffusion coefficient is dependent upon the square of the distance traveled per unit time, the tortuosity produced by the keratin increases the distance traveled by the oxygen $\approx \sqrt{7}$. The increased amount of immobilized probe observed in the porcine SC spectrum suggests an increase in the tortuosity. The D_{O_2} in porcine SC was found to be 0.007×10^{-5} cm^2/s as compared to the 0.03×10^{-5} cm^2/s found for the human SC.

The oxygen diffusion coefficient can be increased by increasing the fluidity of the stratum corneum lipids using permeation enhancers. The larger molecular weight enhancers, oleic acid and azone, did not produce observable changes in the D_{O_2}. The smaller molecular weight permeation enhancers, DMSO and DeMSO, showed marked increases in the D_{O_2}. These enhancers have previously been found to increase solute flux across the stratum corneum by fluidizing the lipids (16,17).

Transport of molecules in the lipid regions of membranes is important because the lipid bilayers provide cellular organelles with protection in the form of a diffusion barrier (42). This is also true for the stratum corneum. Models for the diffusion of small molecules through lipids have been proposed (43,44). Trauble (43) proposed that these small molecules enter the free volume formed by kinks in the chains and then diffuse along with the kinks. Pace and Chan (44) proposed that the molecules hop from kink to kink rather than traveling with the kink. Subczynski et al. (45) studied the transport of molecular oxygen in phosphotidylcholine-cholesterol membranes using EPR. They determined that these kinks were pockets produced by gauche-*trans* isomerism of alkyl chains and by packing defects. They concluded that the oxygen diffused both by jumping from pocket to pocket and by migrating with the pocket itself and that an increased number of kinks could increase the transport of oxygen.

It is clear that in order to increase the diffusion of oxygen across the stratum corneum the lipids must be disrupted. We have shown that this disruption can be achieved through the application of dimethyl sulfoxide, DMSO, and decylmethyl sulfoxide, DeMSO. This increase in transcutaneous oxygen diffusion may be useful in satisfying the oxygen demand for epidermal regeneration.

The SC dioxygen diffusion results from this laboratory to date are encouraging but should be viewed as preliminary. More work is in progress.

ACKNOWLEDGMENTS

We wish to thank Drs. H. Maibach, R. Potts, and M. Francouer for providing us with tissue samples, and S. J. Rehfeld for helpful discussions. This work was supported by NIH grant #1R15GM44281-01 and AFOSR/UES grant #F49620-88-C-0053/SB5881-0378.

REFERENCES

1. Homalle A. Experiences physiologiques sur l'absorption par la tegument externe chez l'homme dans la bain. Union Med 1853; 7:462–3.
2. Scheuplein RJ, Ross L. Effects of surfactants and solvents on the permeability of epidermis. J Soc Cosmet Chem 1970; 21:853–73.
3. Scheuplein RJ, Blank IH. Permeability of the skin. Physiol Rev 1971; 51:702–47.
4. Scheuplein RJ, Blank IH, MacFarlane DJ. Mechanism of percutaneous absorption. III. J Invest Dermatol 1967; 49:582–9.
5. Michaels AS, Chandrasekaran SK, Shaw JE. Drug permeation through human skin. J Am Inst Chem Eng 1975; 21:985–96.
6. Scheuplein RJ. Mechanism of percutaneous absorption. II. J Invest Dermatol 1976; 67:31–8.
7. Winter GD. Oxygen transport to tissue. Adv Exp Med Biol 1977; 94:673–8.
8. Horikoshi T, Balin AK, Carter DM. Effect of oxygen on the growth of human epidermal keratinocytes. J Invest Dermatol 1986; 86:424–7.
9. Fisher BH. Topical hyperbaric oxygen treatment of pressure sores and skin ulcers. Lancet 1969; 2:405–9.
10. Gruber RP, Heitkamp DH, Billy LJ, Amato JJ. Skin permeability to oxygen and hyperbaric oxygen. Arch Surg 1970; 101:69–70.
11. Gruber RP, Heitkamp DH, Amato JJ. Systemic absorption of topically applied hyperbaric oxygen. Texas Rep Biol Med 1970; 28:547–52.
12. Wilhelm E, Buttino R. Thermodynamic functions of the solubilities of gases in liquids at 25°C. Chem Rev 1973; 73:1–9.
13. Crank J. Mathematics of diffusion. London: Oxford, 1975:44–69.
14. Kligman AJ. Topical pharmacology and toxicology of dimethyl sulphoxide. JAMA 1965; 193:796–804.
15. Rehfeld SJ, Plachy WZ, Hou SYE, Elias PM. Localization of lipid microdomains and thermal phenomena in murine stratum corneum and isolated membrane complexes. J Invest Dermatol 1990; 95:217–23.

16. Gay CL, Murphy TM, Hadgraft J, Kellaway IW, Evans JC, Rowlands CC. An electron spin resonance study of skin penetration enhancers. Int J Pharm 1989; 49:39–45.
17. Hey MJ, Taylor DJ, Derbyshire W. Water sorption by human callus. Biochim Biophys Acta 1978; 540:518–33.
18. Patel R, Vasavada RC. Effect of pH on enhancement of in vitro percutaneous transport of isoproterenol HCl by azone. Dev Ind Pharm 1990; 16:1365–74.
19. Hatcher ME. Dioxygen diffusion in the stratum corneum. MS thesis, San Francisco State University, 1991.
20. Hubbell WL, McConnell HM. Orientation and motion of amphiphilic spin labels in membranes. Proc Natl Sci USA 1969; 64:20–7.
21. Tanaka T, Ogura R, Hidaka T, Sugiyama M. Changes of electron spin resonance membrane fluidity in hexadecane-induced hyperproliferative epidermis. J Invest Dermatol 1989; 93:682–5.
22. Davis JH. The influence of membrane proteins on lipid dynamics. Chem Phys Lipids 1986; 40:223–58.
23. Pake E, Tuttle TR. Anomalous loss of resolution of paramagnetic resonance hyperfine structure in liquids. Phys Rev Lett 1959; 3:403–25.
24. Edelstein N, Kwok A, Maki AH. Effect of hydrostatic pressure and temperature on spin exchange between free radicals in solution. J Chem Phys 1964; 41: 3473–8.
25. Plachy W, Kivelson D. Spin exchange in solutions of di-tert-butyl-nitroxide. J Chem Phys 1967; 47:3312–8.
26. Hyde JS, Subczynski WK. Spin label oximetry. Biol Magn Reson 1989; 8:399–425.
27. Swartz HM, Glockner JM. Advanced EPR in biology and biochemistry. New York: Elsevier, 1989:753–82.
28. Povich MJ. Measurement of dissolved oxygen concentrations and diffusion coefficients by electron spin resonance. Anal Chem 1975; 47:346–7.
29. Windrem DA, Plachy WZ. The diffusion-solubility of oxygen in lipid bilayers. Biochim Biophys Acta 1980; 600:655–65.
30. Backer JM, Budker V, Eremenko SI, Molin YN. Detection of the kinetics of biochemical reactions with oxygen using exchange broadening in the ESR spectra of nitroxide radicals. Biochim Biophys Acta 1977; 460:152–6.
31. Smotkin ES, Moy FT, Plachy WZ. Dioxygen solubility in aqueous phosphatidylcholine dispersions. Biochim Biophys Acta 1991; 1061:33–8.
32. Subczynski WK, Hyde JS, Kusumi A. Oxygen permeability of phosphatidylcholine-cholesterol membranes. Proc Natl Sci USA 1989; 86:4474–8.
33. Belkin S, Melhorn RJ, Packer L. Determination of dissolved oxygen in photosynthetic systems by nitroxide spin-probe broadening. Arch Biochem Biophys 1987; 252:487–95.
34. J. Windle. USDA. Albany, CA.
35. Beth AH, Perkins RC, Venkataramu SD, Pearson DE, Park and CR, Park JH. Advantages of deuterium modification of nitroxide spin labels for ibological epr studies. Chem Phys Lett 1970; 69:24–8.

36. Asyst Software Technologies, Inc. Rochester, NY.
37. Lampe MA, Burlingame AL, Whitney J, Williams ML, Brown Be, Roitman E, Elias PM. Human stratum corneum lipids: characterization and regional variations. J Lipid Res 1983; 24:120–30.
38. Golden GM, Guzek DB, Kennedy AH, McKie JE, Potts RO. Stratum corneum lipid phase transitions and water barrier properties. J Pharm Sci 1985; 26:2382–8.
39. Eikenbary EF, Tate MW, Turner DC, Shymsunder E, Gruner SM. Nonbilayer phases of membrane lipids. Chem and Phys Lipids 1991; 57:147–64.
40. Garson JC, Doucet J, Leveque JL, Tsoucaris G. Oriented structure in human stratum corneum revealed by x-ray diffraction. J Invest Dermatol 1991; 96:43–9.
41. Small D. The physical chemistry of lipids. New York: Plenum Press, 1986:276–81, 395–473.
42. Albert B, Bray D, Lewis J, Raff M, Roberts K, Watson JD. Molecular biology of the cell. 2d ed. New York: Garland, 1988:300–4.
43. Trauble H. The movement of molecules across lipid membranes: a molecular theory. J Membr Biol 1971; 4:193–208.
44. Pace RJ, Chan SI. Molecular motions in lipid bilayers III, lateral and transverse diffusion in bilayers. J Chem Phys 1982; 76:4241–7.
45. Subczynski WK, Hyde JS, Kusumi A. Effect of alkyl chain unsaturation and cholesterol intercalation on oxygen transport in membranes. Biochemistry 1991; 30:8578–90.
46. Calhoun DB, Vanderkooi JM, Woodrow III GV, Englander SW. Penetration of dioxygen into proteins studied by quenching of phosphorescence and fluorescence. Biochemistry 1983; 22:1526–32.

3

Evidence for UV Light as an Oxidative Stressor in Skin

Eric H. Witt, Paul Motchnik, and Lester Packer
University of California, Berkeley, California

I. INTRODUCTION

Solar ultraviolet (UV) light has been implicated as a factor in melanoma and nonmelanoma skin cancer (1–5), photoaging (6), photosensitization (7), and other light-related skin pathologies. One unifying hypothesis for these various cutaneous effects of UV light is that 1) UV light induces the formation of free radicals in skin, 2) skin antioxidant defenses nullify free radicals, but if the dose of UV light is great enough antioxidant defenses will be overwhelmed, and 3) the resulting free radical load will cause damage to cellular components such as proteins, lipids, and DNA, resulting in altered cell function and eventual pathology. Predictions of this hypothesis are: 1) free radicals should be detectable in irradiated skin; 2) the concentrations and/or redox status of cutaneous antioxidants should be altered by UV irradiation; 3) the concentration of oxidatively damaged molecules should increase in UV-irradiated skin, and the increase should correlate with changes in antioxidant status; and 4) manipulation of skin antioxidants should affect the increase of damage markers and disease—depletion should increase damage and supplementation should inhibit damage.

Prediction 1 of this hypothesis has been tested to some degree, and it has been found that free radicals are produced from UV exposure of whole skin

(8,9) and skin homogenates (10) (see Chapter 4). However, it is difficult to conclusively test this portion of the hypothesis in vivo, due to technical obstacles involved in detection of the extremely short-lived radical species that are expected to be produced by UV radiation (e.g., hydroxyl radicals). In this chapter we summarize work from our laboratory in which we have tested predictions 2, 3, and 4. We have used as a model the hairless mouse, using both excised skin and an in vivo system, and irradiating with either UVA, UVB, or a solar simulator which emits UVA and UVB in a pattern very similar to natural sunlight. We have measured UV-induced changes in enzymic and nonenzymic antioxidants, lipid peroxidation, and the effects of dietary supplementation with the major chain-breaking lipophilic antioxidant, α-tocopherol, on UV-induced skin damage.

Taken together, our results and those of others provide convincing evidence that UV light causes free radical formation in skin, which, even in a single dose, overwhelms skin antioxidant defenses and causes significant damage, which can be prevented by supplementation with antioxidants.

II. STUDIES ON EXCISED SKIN

We first studied excised skin, irradiated with either UVA or UVB.

A. UVA Irradiation

Skin was removed from female hairless mice (strain HRS/J, from Jackson Laboratory) after euthanasia by CO_2 inhalation followed by decapitation. Subcutis and fascia were removed, and the skin was divided in half longitudinally. One half was irradiated and the other half was kept at room temperature. The light source was an Oriel 1000 W Xe lamp, filtered to allow only UVA and visible light. Total dose was 6 J/cm^2. After irradiation, skin sheets were cut up and immediately homogenized in ice-cold buffer. Either whole homogenates or supernatants were used for antioxidant assays (see Ref. 11 for complete experimental details).

Enzymic antioxidants assayed were catalase (12), glutathione reductase (13,14), glutathine peroxidase (15), and superoxide dismutase (16). The nonenzymic antioxidants assayed were α-tocopherol, ubiquinol and ubiquinone (17), glutathione (18), and ascorbate (19). Of the enzymes, only catalase and glutathione reductase decreased in activity with irradiation (27%, $p < 0.01$, and 15%, $p < 0.05$, respectively); glutathione peroxidase showed a slight decrease, which was not statistically significant, and superoxide dismutase remained unchanged (Fig. 1). Of the nonenzymic antioxidants, the water-soluble antioxidants glutathione (reduced and oxidized forms) and

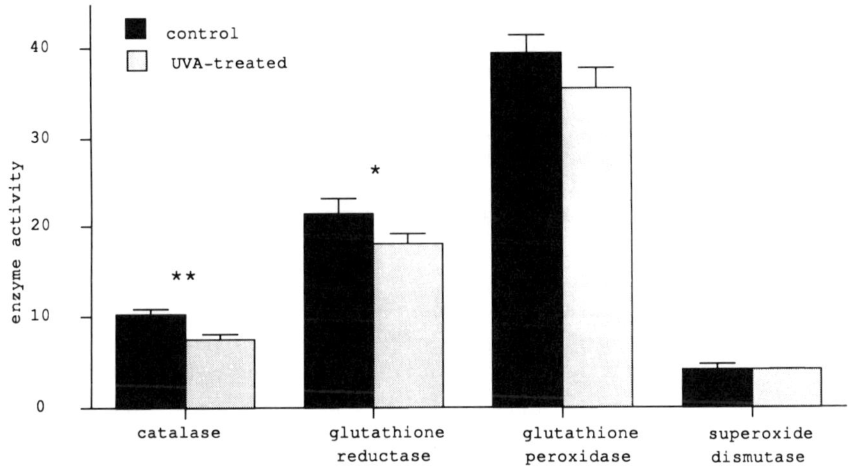

Figure 1 Enzyme activities in homogenates of control and UVA-treated skin. Catalase, μmol oxygen/mg protein/min; GSH reductase, nmol GSH/mg protein/min; GSH peroxidase, nmol NADPH/mg protein/min; SOD, U/mg protein. Mean values ± SEM are shown. Statistically lower than unirradiated controls: $^{**}p < 0.01$; $^{*}p < 0.05$.

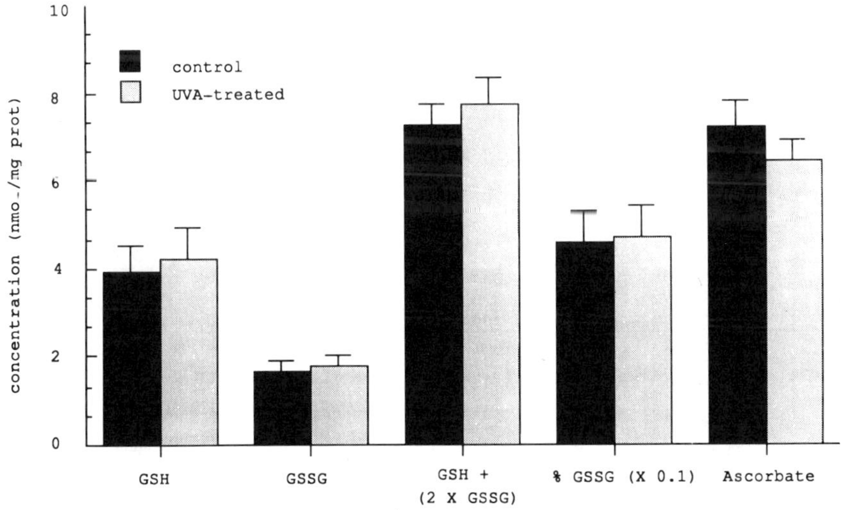

Figure 2 Water-soluble antioxidants of control and UVA-treated skin. GSH = reduced glutathione; GSSG = oxidized glutathione. Mean values ± SEM are shown.

ascorbate were unaffected by irradiation (Fig. 2). The lipophilic antioxidants α-tocopherol, ubiquinol, and uniquinone exhibited a tendency toward a decrease in all cases with irradiation, but this was not statistically significant (Fig. 3).

B. UVB Irradiation

Skin was prepared and irradiated as in Experiment A, except the filters were chosen to allow only UVB through, and the dose was 0.3 J/cm². Antioxidants were assayed as in A, and in addition lipid peroxidation was measured by assaying for malondialdehyde (20; see Ref. 21 for experimental details). Enzymic antioxidants exhibited the same pattern as seen with UVA irradiation, with catalase and superoxide dismutase decreasing significantly (32% and 18%, respectively) ($p < 0.05$), glutathione peroxidase decreasing slightly and nonsignificantly, and superoxide dismutase remaining unchanged (Fig. 4). However, there were greater changes in nonenzymic antioxidants with UVB irraidation than with UVA irradiation. Reduced glutathione (GSH) declined 37% ($p < 0.05$), and oxidized glutathione (GSSG) increased slightly and nonsignificantly; the net effect was a decrease in total glutathione (Fig. 5). Tocopherol concentration declined 20% while ubiquinol concentration exhibited a dramatic decrease of 62%, and ubiquinone decreased 43% (p

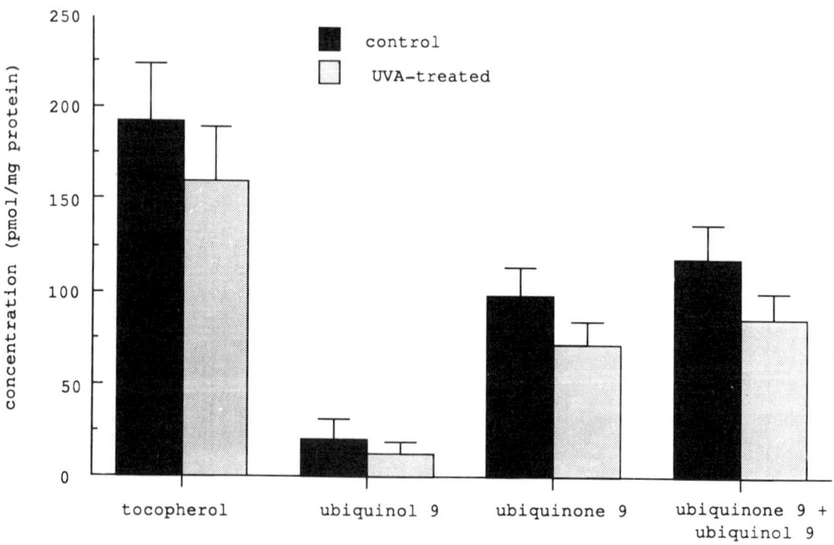

Figure 3 Lipid-soluble antioxidants in control and UVA-treated skin. Mean values ± SEM are shown.

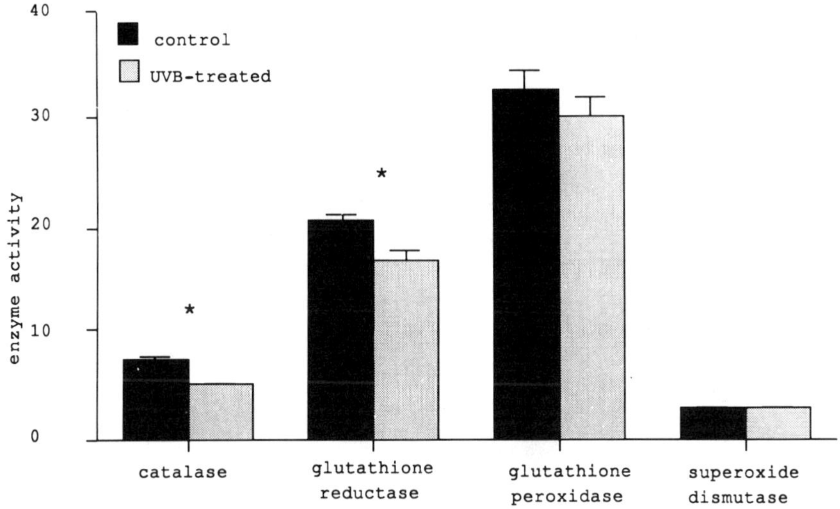

Figure 4 Enzyme activities in homogenates of control and UVB-treated skin. Catalase, μmol oxygen/mg protein/min; GSH reductase, nmol GSH/mg protein/min; GSH peroxidase, nmol NADPH/mg protein/min; SOD, U/mg protein. Mean values ± SEM are shown. Statistically lower than unirradiated controls: *p < 0.05.

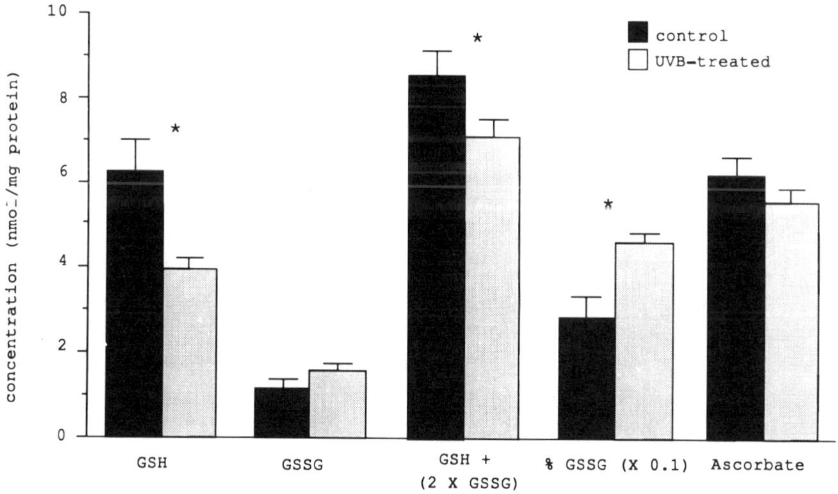

Figure 5 Water-soluble antioxidants of control and UVB-treated skin. GSH = reduced glutathione; GSSG = oxidized glutathione. Mean values ± SEM are shown. Statistically lower than unirradiated controls: *$p < 0.05$.

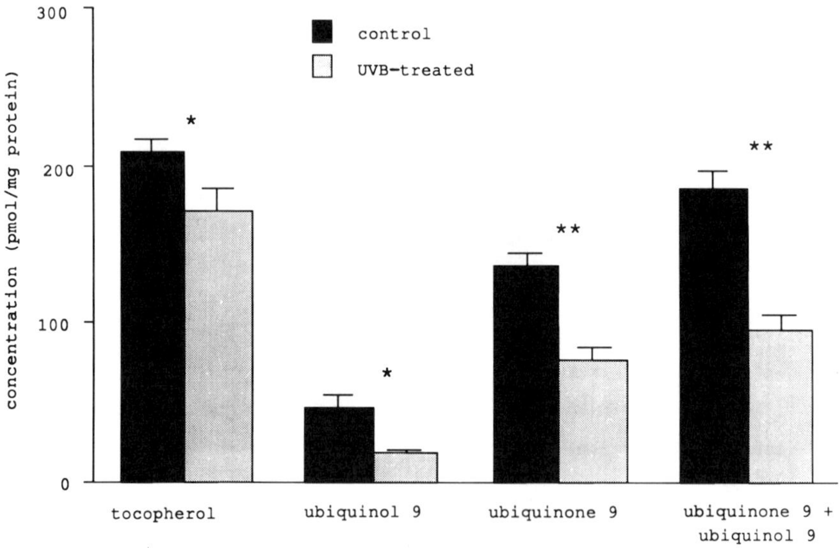

Figure 6 Lipid-soluble antioxidants in control and UVB-treated skin. Mean values ± SEM are shown. Statistically lower than unirradiated controls: $^{**}p < 0.01$; $^{*}p < 0.05$.

< 0.05 for all) (Fig. 6). Malondialdehyde concentrations were the same in irradiated and nonirradiated skin (data not shown).

C. Conclusions from Studies on Excised Skin

The effects of UVA and UVB on enzymic antioxidants were quite similar, with catalase and glutathione reductase most affected. The exact origin of the damage to these enzymes is unknown, although it is known that catalase is directly destroyed by visible light (22). The decrease in these enzymes indicates that even a single dose of UVA or UVB can leave the skin antioxidant enzyme systems somewhat compromised; in the case of UVA, this might partially account for the carcinogenic potential of UVA (23). However, damage to skin enzymic antioxidants is not in itself support for the free radical hypothesis of UV-induced skin damage, since the damage is not necessarily mediated by free radicals.

The effects of UVA and UVB on nonenzymic antioxidants were strikingly different—UVA had no significant effect on either water-soluble or lipophilic antioxidants, while UVB caused a marked decrease in all except ascorbate. Both ubiquinol (24) and α-tocopherol (25) have absorbence maxima at about 295 nm, in the UVB region, and it is possible that part of the destruction of these lipophilic antioxidants may be a direct result of irradiation. This cannot, however, explain the destruction of glutathione, which does not absorb in this region. Thus, the destruction of these three antioxidants with a single dose of UVB light provides strong indirect support for the idea of UV-induced free radical production.

The lack of lipid peroxidation in the UVB study could have a number of causes. It is known that lipid peroxidation does not begin in in vitro membrane systems until α-tocopherol is depleted (26) and α-tocopherol levels remained high after irradiation, at least with this dose of UV light. Several studies have shown a time lag of one to several hours after irradiation to occur before a measurable increase in cutaneous lipid peroxidation can be detected (27–29); since skin was processed immediately in this study, lipid peroxidation may not have been detectable. Finally, the TBARS assay is notoriously fraught with artifact and has a relatively high background (30), and "noise" levels may simply have been too high to detect peroxidation. Results from later experiments (see below) indicate that the final possibility is most likely.

Overall, these early experiments indicated that, at the very least, skin antioxidant systems are compromised by a single dose of UV light, and that free radical production probably occurs with UVB irradiation, depleting cutaneous antioxidants (prediction 2 of the free radical hypothesis). They were inconclusive concerning free radical-induced damage (prediction 3) and did not investigate antioxidant supplementation (prediction 4). Later, in vivo studies extended these experiments.

III. IN VIVO STUDIES

To cleary implicate free radical processes in solar light-induced skin pathologies, it is necessary to 1) study effects in vivo, 2) use UV light that is similar to sunlight, 3) examine doses of UV light that are physiological for the animal employed as a model, and 4) assay for cellular damage that is caused by free radicals and has some link to pathologies.

Accordingly, in the experiments described below, we used anesthetized hairless mice irradiated with light from a solar simulator (Solar Light Co., Philadelphia, PA, model 14S) whose output closely matches the UVA and UVB spectrum of natural sunlight (Fig. 7). In experiment A, a series of doses was used. Dose was measured by three meters with overlapping sensitivities (UVX-25,-31,-36, UVX Radiometers, UVP, Inc., San Gabriel, CA); because of the overlap the total from the three meters is probably an overestimate. According to these meters the doses used in our dose-response experiment ranged from 0.25 to 25 J/cm^2. Using these meters it was found that the highest

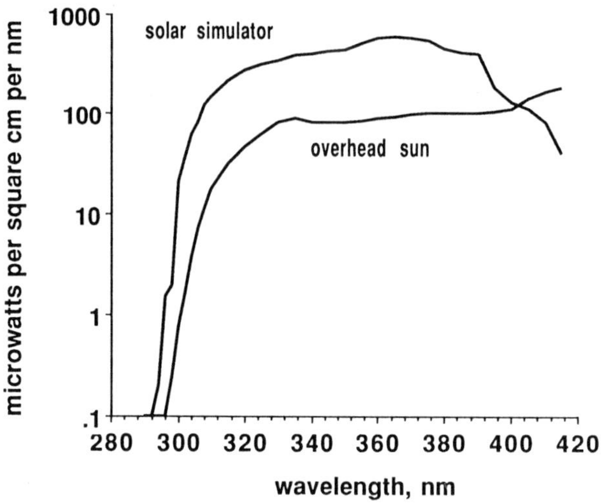

Figure 7 Comparison of the output of solar simulator and natural sunlight. The intensity of the solar simulator output can be adjusted by changing the distance from the simulator to the sample; for the distance in this figure, the output of the solar simulator is more intense than natural sunlight in order to facilitate comparison of the two.

dose was the equivalent of exposure to 4–5 hours of natural autumn sunlight at our latitude (38°N). This dose was approximately 10 MED (minimum erythemal dose) for these mice. Thus, in the dose-response experiment, mice received doses of (UVA + UVB) ranging from the equivalent of a few minutes exposure to sunlight (about 0.1 MED) to several hours of exposure (about 10 MED), which is a range of doses normally encountered in a natural setting.

In experiment B, a single dose (the highest from the dose-response experiment) was used, and lipid hydroperoxides and lipophilic antioxidants were measured simultaneously on single skin samples. Lipid peroxyl radicals have been linked to carcinogenesis in skin induced by chemical treatment (31) and by UV light (32).

A. Dose-Response for Lipophilic Antioxidants

Mice (female, SKH-1, from Simonsen) were anesthetized by intraperitoneal injection of sodium pentobarbital (50 mg/kg body weight), and a plate with a 2.3-cm diameter hole was fixed on the right side of the mouse and irradiated. Light from the solar simulator passed through the hole and hit the skin. After irradiation, mice were killed by cervical dislocation, and irradiated skin on the right side and control skin from the contralateral area on the left side were removed immediately, scraped free of adhering subcutis and fascia, homogenized, and assayed for α-tocopherol, ubiquinol, and ubiquinone, as described above.

At doses as low as 2.5 J/cm² (about 1 MED) a 10–15% loss of α-tocopherol and ubiquinol 9 was observed; at higher doses ubiquinol 9 became almost completely depleted while α-tocopherol losses reached 50% (Fig. 8). These data suggest that, in vivo, skin antioxidant defenses are compromised by even low doses of UV light. At high doses the destruction of some cutaneous antioxidants (e.g., ubiquinol) is complete. This represents a severe drain of the skin's capacity to prevent oxidative damage. Since one response to a large

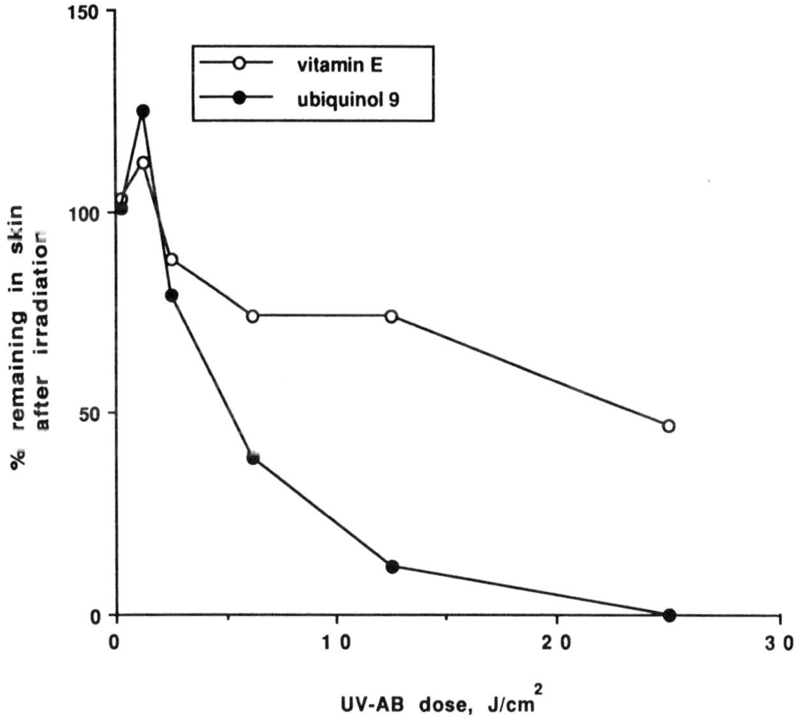

Figure 8 Response of cutaneous lipophilic antioxidants to increasing doses of simulated sunlight (combined UVA and UVB). The lowest dose (0.25 J/cm²) is the equivalent of about 0.1 MED; the highest dose (25 J/cm²) is the equivalent of about 10 MED.

dose of UV light is inflammation, with its attendant oxidant load, such a loss of antioxidants may leave the skin open to further oxidative damage resulting, not from the UV exposure itself, but from the subsequent physiological events. Thus, free radicals do not necessarily need to be generated by the UV exposure itself for them to play a role in skin pathology; nor does destruction of skin antioxidants necessarily need to be due to UV light-induced free radicals—direct destruction of antioxidants by UV light, coupled with later inflammation, may also cause cutaneous pathology. This may partially explain why susceptibility to some forms of skin cancer appears to be greater in people who have experienced severe sunburns at certain critical periods in childhood (33). It can be speculated that the initiating event may be not only sun exposure per se, but also free radicals producing during subsequent inflammation which overwhelm the already compromised antioxidant systems.

It is interesting that ubiquinol was depleted much more rapidly than α-tocopherol; ubiquinol was also more depleted in the previous UVB irradiation of excised skin (see Fig. 6). It has been suggested that ubiquinol acts to regenerate α-tocopherol from the tocopheroxyl radical (34), thus regenerating tocopherol at the expense of ubiquinol; it has also been suggested that ubiquinol acts as an antioxidant directly and is the first line of defense among lipid-soluble antioxidants (35). These data are consistent with either of these explanations and suggest that supplementation with ubiquinone might be more effective than supplementation with α-tocopherol in preventing UV-induced antioxidant loss.

B. Lipid Hydroperoxides and Effects of E Supplementation

In these experiments we used a new method to directly detect the formation of lipid hydroperoxides in skin immediately after irradiation, rather than indirect methods that were previously used, such as detection of malondialdehyde. We also dietarily supplemented hairless mice with vitamin E at various doses and found that it was incorporated in the skin and reduced UV-induced lipid hydroperoxide formation 60–70% compared to nonsupplemented animals. This is in contrast to a recently published report (28).

1. Experiment 1

Six female hairless mice (strain SKH-1, Simonsen Laboratories, Gilroy, CA) were irradiated on one flank with UV light from the solar simulator. Total dose of UV-AB light was 25 J/cm^2 (the highest dose used in the dose-response experiments, about 10 MED). Immediately after irradiation, skin was removed from the irradiated flank and from the contralateral, nonirradiated flank. The skin was extracted into hexane, and the dried hexane extract was redissolved in methanol/ethanol. This sample was used for simultaneous quantitation of α-tocopherol, ubiquinol 9, ubiquinone 9, and lipid hydroperoxides.

Lipid hydroperoxides were separated by HPLC and detected by chemiluminescence (36): The reaction solution for the postcolumn assay contained microperoxidase (25 mg/L) and isoluminol (177.2 mg/L) and was pumped at a flow rate of 1.5 ml/min. A fluorometer with the excitation source turned off was used as a photon detector. This assay measures the hydroperoxy groups themselves and not indirect indices of lipid peroxidation such as conjugated dienes or breakdown products of lipid hydroperoxides. The detection limit is 1 pmol. Chemiluminescence also detects ubiquinols. To confirm that any chemiluminescence observed in this assay was due to a hydroperoxide, not a hydroquinone, some samples were reduced with triphenyl phosphine and rerun. Since the chemiluminescence response of hydroperoxides, but not of hydroquinones, is eliminated by triphenyl phosphine, the disappearance of chemiluminescence peaks in the treated samples indicated that chemiluminescence observed in this assay was due to hydroperoxides and not hydroquinones.

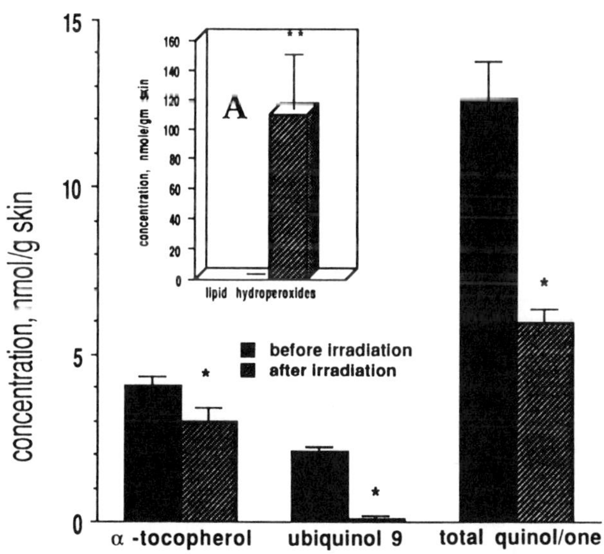

Figure 9 Concentrations of cutaneous lipophilic antioxidants after a single large dose (25 J/cm²) of simulated sunlight. Total quinol/one = ubiquinol + ubiquinone. $n = 6$. A) Concentration of lipid hydroperoxides in the same skin samples, irradiated and nonirradiated (none detectable in nonirradiated samples). Statistically different than unirradiated controls: $^{**}p < 0.01$; $^{*}p < 0.05$.

Lipophilic antioxidants were separated by HPLC and detected electrochemically (α-tocopherol and ubiquinols 9 and 10) and by UV absorbence (ubiquinones 9 and 10), as described previously (17).

The concentrations of α-tocopherol, ubiquinol 9, and total (ubiquinols + ubiquinones) were less in skin from the irradiated flank compared to the nonirradiated flank ($p < 0.01$), as seen previously (Fig. 9). The content of lipid hydroperoxides increased from undetectable levels to 110.6 nmol/gm skin ($p < 0.01$) (Fig. 9A). α-Tocopherol is considered the major chain-breaking antioxidant in membranes (26,37), and ubiquinol has also been suggested as a membrane antioxidant (34,35). It has been suggested that lipid peroxidation is a causal factor in cutaneous carcinogenesis (31,32). These results are consistent with prediction 3 of the free radical hypothesis of UV light–induced damage—that as antioxidants are depleted, molecular cutaneous damage leading to pathogenesis will occur.

Representative chromatograms from chemiluminescent detection of lipid hydroperoxides are shown for one mouse in Figure 10. In control (nonirra-

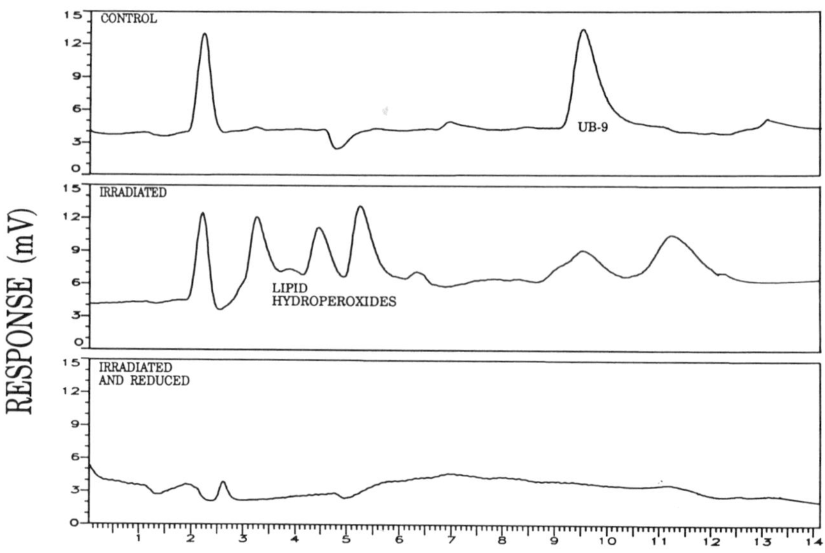

Figure 10 Lipid hydroperoxides in skin of hairless mice analyzed by HPLC with chemiluminescent detection. Representative chromatograms are shown for one mouse. Top: nonirradiated. Middle: Irradiated. Bottom: Irradiated, treated with triphenyl phosphine.

diated) skin (top chromatogram), two peaks are visible; the first is an unknown (possibly hydrogen peroxide), and the second is ubiquinol 9 (UB-9). In irradiated skin (middle chromatogram), several lipid hydroperoxide peaks appear. The bottom chromatogram is of the irradiated sample after treatment with triphenyl phosphine. As the chemiluminescence response of hydroperoxides, but not of hydroquinones, is eliminated by reducing agents such as triphenyl phosphine, such reduction was used to confirm that any chemiluminescence observed in this assay is a hydroperoxide, not a hydroquinone. Note the presence of several distinct peaks in the illuminated sample, which opens the possibility of identifying specific irradiation-induced damage products using this technique.

In general, previous studies have either reported no increase in lipid peroxidation products in skin after UV irradiation (21) or an increase only several hours after irradiation (27–29). Furthermore, in previous studies the background level of lipid peroxidation products has been high, even in control, nonirradiated skin. In contrast, in the present study lipid hydroperoxides were undetectable before irradiation, but appeared at very high concentrations (over 25 times the concentration of α-tocopherol, for example) in skin immediately after irradiation. These discrepancies may be due to differences in technique. It is well known that the malondialdehyde method of detecting lipid peroxidation suffers from many possibilities for artifact (30), thus nonirradiated samples can exhibit apparent high levels of lipid peroxidation products even if there are none, reducing the sensitivity of the assay. The HPLC-chemiluminescence assay used in the present study, however, is specific for lipid hydroperoxides and has less potential for artifact. Also, lipid hydroperoxides may have appeared immediately after irradiation in this study because the irradiation time was about one hour; consequently, if lipid peroxidation started early in the irradiation, it would have had time to continue and amplify throughout the course of the irradiation.

The very high level of hydroperoxides seen in this study after a single exposure to a physiological dose of UV irradiation demonstrates that even a single dose may cause a large amount of cellular damage. Furthermore, since these measurements were done on whole skin, and epidermis is the layer which is first exposed to environmental insult and is the layer in which the most serious form of skin damage, carcinogenesis, occurs, it is probable that the concentration of products of oxidation damage is even higher in epidermis. This idea is supported by the fact that vitamin E, measured in whole skin samples, only decreased by about 26% after irradiation, but it is known that, in vitro, membrane α-tocopherol must be completely depleted before lipid peroxidation begins (26), implying that some membranes in the skin (presumably in the outermost, epidermal layers, which received the full force of the irradiation) were completely depleted in vitamin E and had levels of lipid hydroperoxides even higher than those seen here, which were averaged over the entire skin. However, since the outermost epidermal cells are fully differentiated and destined to be sloughed off, the damage may not be as great as it would seem at first glance. Further studies are necessary to determine the extent of photodamage to epidermis compared to dermis and to various layers of the epidermis.

2. *Experiment 2*

Female hairless mice were fed a diets containing either 30 IU/kg diet α-tocopherol ($n = 8$) or 10,000 IU/kg diet α-tocopherol ($n = 12$). Periodically, mice from each group were irradiated with UV light, and their skin content of lipophilic antioxidants and lipid hydroperoxides was analyzed as described for Experiment 1.

The cutaneous concentration of α-tocopherol increased in animals receiving supplemental dietary α-tocopherol, compared to the group receiving normal levels, reaching a level about eightfold higher in the supplemented animals after 4 weeks (Fig. 11). The concentration of lipid hydroperoxides in

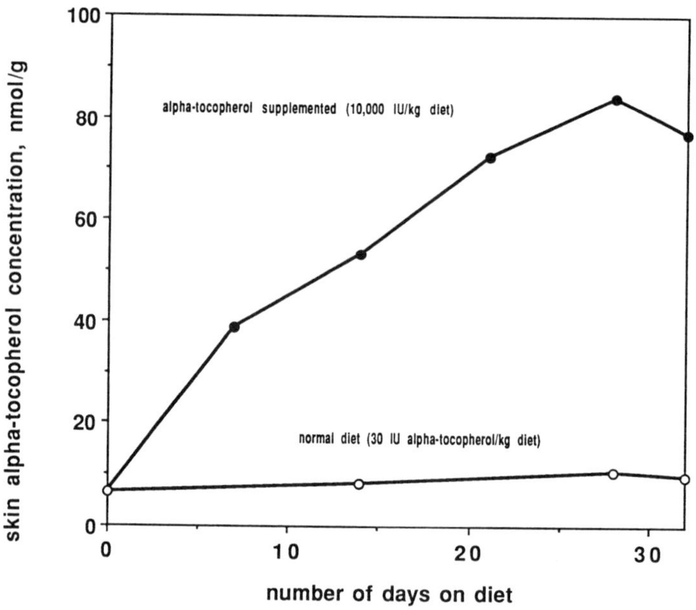

Figure 11 Skin α-tocopherol concentrations in mice fed normal (30 IU α-tocopherol/kg diet) or supplemented (10,000 IU/kg diet) diets. Each point represents one or two mice.

irradiated skin of α-tocopherol–supplemented mice was about one-third that seen in unsupplemented mice ($p < 0.05$) (Fig. 12), suggesting that this antioxidant protects against UV-induced skin damage.

To further probe the connections between lipophilic antioxidants, UV light, and lipid peroxidation, we examined the relationship between the UV-induced decrease in antioxidants and lipid hydroperoxide formation, pooling data from Experiment 1 and from unsupplemented mice in Experiment 2. This was possible because the HPLC-chemiluminescence method of measurement of lipid hydroperoxides allows measurement of hydroperoxides in the same sample in which lipophilic antioxidants are measured, allowing correlation of degree of depletion of the latter with the appearance of the former. Surprisingly, there was no correlation between the decrease in α-tocopherol concentration and lipid hydroperoxide formation, but there was a correlation between the decrease in (ubiquinol + ubiquinone) concentration and lipid hydroperoxide formation (Fig. 13). One explanation for this is that the pool of reduced quinone may act as an antioxidant "buffer" for α-tocopherol, recycling it from its radical form (34,35). If this is so, changes in ubiquinol/one concentration may serve as a more sensitive indicator of oxi-

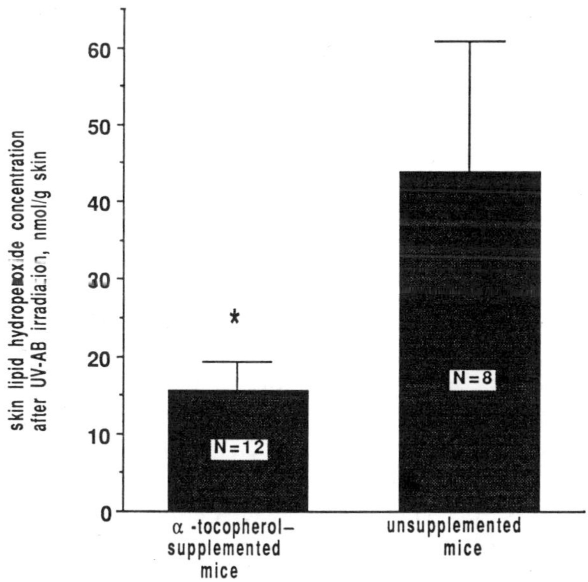

Figure 12 Lipid hydroperoxide concentration in irradiated skin from control and α-tocopherol–supplemented mice. *Supplemented concentration of lipid hydroperoxides was lower, $p < 0.05$.

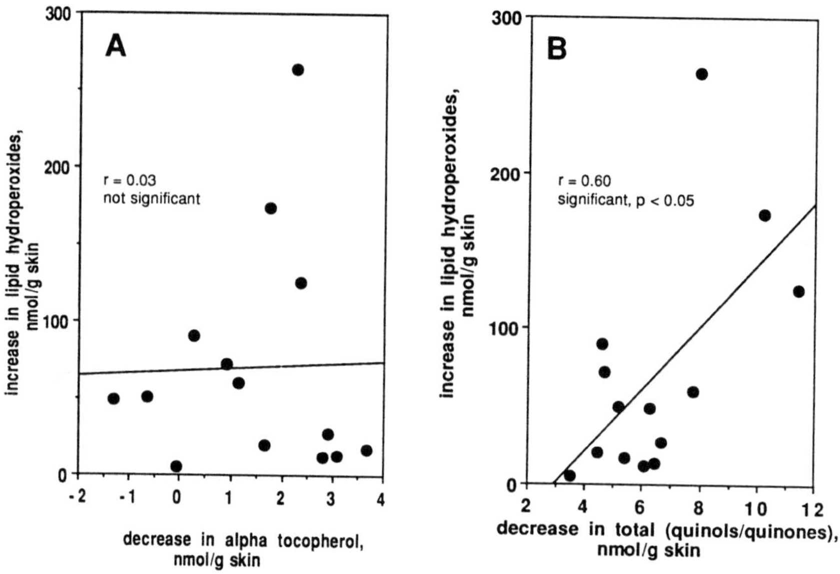

Figure 13 Relationships between lipophilic antioxidants and lipid hydroperoxides in irradiated skin. A) There was no relationship between decrease in α-tocopherol and appearance of lipid hydroperoxides. B) There was a significant ($p < 0.05$) relationship between decrease in total Q (quinols + quinones) and appearance of lipid hydroperoxides.

dative stress than changes in α-tocopherol concentration, and supplemental ubiquinol/one may be even more effective than α-tocopherol in preventing UV-induced damage.

IV. CONCLUSIONS

These studies, and those detailed elsewhere in this volume, provide compelling evidence for the free radical hypothesis of UV light-induced cutaneous pathology—antioxidant levels decrease on irradiation, as would be expected with a free radical load, lipid peroxidation increases tremendously, and supplementation with α-tocopherol prevents this damage. Future studies will clarify the relationships between cutaneous antioxidants during UV irradiation in the dermis and epidermis, providing probes of mechanism and suggesting possible avenues of prevention or treatment.

ACKNOWLEDGMENTS

Supported by 1R55CA53812-01A1 of the National Cancer Institute, National Institutes of Health, Department of Health and Human Services, and by 93030, Johns Hopkins University Center for Alternatives to Animal Testing.

REFERENCES

1. Rundel RD, Nachtwey DS. Skin cancer and ultraviolet radiation. Photochem Photobiol 1978; 28:345–56.
2. Epstein JH. Photocarcinogenesis, skin cancer, and aging. J Am Acad Dermatol 1983; 9:487–502.
3. Sober AJ. Solar exposure in the etiology of cutaneous melanoma. Photo-Dermatol 1987; 4:23–31.
4. Swerdlow AJ, English JS, MacKie RM, O'Doherty CJ, Hunter JA, Clark J, Hole DJ. Fluorescent lights, ultraviolet lamps, and risk of cutaneous melanoma. Br Med J 1988; 297:647–50.
5. Roza L, Baan RA, van der Leun JC, Kligman L, Young AR. UVA hazards in skin associated with the use of tanning equipment. J Photochem Photobiol B:Biol 1989; 3:281–7.
6. Gilchrest BA, Szabo G, Flynn E, Goldwyn RM. Chronologic and actinically induced aging in human facial skin. J Invest Dermatol 1983; 80:81s–85s.
7. Moleiere P. Drug-induced photosensitivity: phototoxic and photoallergic reactions—a few molecular aspects. Biochemie 1986; 68:849–55.
8. Norrins AL. Free radical formation in the skin following exposure to ultraviolet light. J Invest Dermatol 1962; 39:445–8.
9. Pathak MA, Stratton K. Free radicals in human skin before and after exposure to light. Arch Biochem Biophys 1968; 123:468–76.
10. Nishi J, Ogura R, Sugiyama M, Hidaka T, Kohno M. Involvement of active oxygen in lipid peroxide radical reaction of epidermal homogenate following ultraviolet light exposure. J Invest Dermatol 1991; 97:115–19.
11. Fuchs J, Hufleit ME, Rothfuss LM, Wilson DS, Carcoamo G, Packer L. Acute effects of near ultraviolet and visible light on the cutaneous antioxidant defense system. Photochem Photobiol 1989; 50:739–44.
12. Del Rio LA, Ortega MG, Lopez AL, Gorge JL. A more sensitive modification of the catalase assay with the Clark oxygen electrode. Meth Enzymol 1977; 105: 409–15.
13. Karni L, Moss SJ, Tel-Or E. Glutathione reductase activity in heterocysts and vegetative cells of cyanobacterium *Nostoc muscorum*. Arch Microbiol 1984; 140:215–17.
14. Akerboom TPM, Sies H. Assay of glutathione, glutathione disulfide and glutathione mixed disulfides in biological samples. Meth Enzymol 1981; 77:373–82.
15. Flohé, Günzler WA. Assays of glutathione peroxidase. Meth Enzymol 1984; 105:114–21.

16. Flohé L, Otting F. Superoxide dismutase assays. Meth Enzymol 1984; 105: 93–105.
17. Lang JK, Gohil K, Packer L. Simultaneous determination of tocopherols, ubiquinols, and ubiquinones in blood, plasma, tissue homogenates, and subcellular fractions. Anal Biochem 1986; 157:106–16.
18. Reed DJ, Babson JR, Beatty PW, Brodie AE, Ellis EE. High performance liquid chromatography of nanomole levels of glatathione, glutathione disulfide and related thiols and disulfides. Anal Biochem 1980; 106:55–62.
19. Roe JH. Chemical determination of ascorbic, dehydroascorbic and diketogulonic acids. In: Glick D, ed. Methods of biochemical analysis. vol 1. New York: Interscience, 1954:115–39.
20. Asakawa T, Matsushita S. Thiobarbituric acid test for detecting lipid peroxides. Lipids 1979; 14:401–6.
21. Fuchs J, Huflejt ME, Rothfuss LM, Wilson DS, Carcamo G, Packer L. Impairment of enzymic and nonenzymic antioxidants in skin by UVB irradiation. J Invest Dermatol 1989; 93:769–73.
22. Cheng LYL, Packer L. Photodamage to hepatocytes by visible light. FEBS Lett 1979; 97:124–8.
23. van Welden H, de Gruiji FR, van der Leun JC. Tumors induced by UVA in mice. Photochem Photobiol 1983; 37:S79.
24. Dawson RMC, ed. Data for biochemical research. Oxford: Clarendon Press, 1962.
25. Baxter JG, Robeson CD, Taylor JD, Lehman RW. Natural α-, β-, and γ-tocopherols and certain esters of physiological interest. J Am Chem Soc 1943; 65: 918–24.
26. Burton GW, Ingold KU. Vitamin E: application of the principles of physical organic chemistry to the exploration of its structure and function. Acc Chem Res 1986; 19:194–201.
27. Punnonen K. Autio P, Kiistala U, Ahotupa M. In-vivo effects of solar-simulated ultraviolet irradiation on antioxidant enzymes and lipid peroxidation in human epidermis. Br J Dermatol 1991; 125:18–20.
28. Record IR, Dreosti IE, Konstantinopoulos M, Buckley RA. The influence of topical and systemic vitamin E on ultraviolet light-induced skin damage in hairless mice. Nutr Cancer 1991; 219–25.
29. Khettab N, Amory MC, Briand G, Bousquet B, Combre A, Forlot P, Barey M. Photoprotective effect of vitamins A and E on polyamine and oxygenated free radical metabolism in hairless mouse epidermis. Biochemie 1988; 70:1709–13.
30. Halliwell B, Gutteridge JMC. Free radicals in biology and medicine. 2d ed. Oxford: Clarendon Press, 1989.
31. Marnett LJ. Peroxyl free radicals: potential mediators of tumor initiation and promotion. Carcinogenesis 1987; 8:1365–73.
32. Black HS, Lenger A, Gerguis J, Thornby JI. Relation of antioxidants and level of dietary lipid to epidermal lipid peroxidation and ultraviolet carcinogenesis. Cancer Res 1985; 45:6524–9.
33. Koh HK, Kligler BE, Lew RA. Sunlight and cutaneous malignant melanoma: evidence for and against causation. Photochem Photobiol 1990; 51:765–79.

34. Kagan VE, Serbinova EA, Packer L. Antioxidant effects of ubiquinones in microsomes and mitochondria are mediated by tocopherol recycling. Biochem Biophys Res Comm 1990; 169:851–7.
35. Frei B, Kim MC, Ames BN. Ubiquinol-10 is an effective lipid-soluble antioxidant at physiological concentrations. Proc Natl Acad Sci USA 1990; 87:4879–83.
36. Yamamoto Y, Brodsky MH, Baker JC, Ames BN. Detection and characterization of lipid hydroperoxides at picomole levels by high-performance liquid chromatography. Anal Bioch 1987; 160:7–13.
37. Tappel AL. Vitamin E as the biological lipid antioxidant. Vitam Horm 1962; 20:493–510.

4

Reactive Oxidants in Skin: UV-Induced Lipid Peroxidation

Ryohei Ogura and Masayasu Sugiyama
Kurume University School of Medicine, Kurume, Japan

I. INTRODUCTION

The skin is the part of the body that receives the most exposure to exogenous environmental factors, such as sunlight, heat, cold, chemical products, carcinogens, and irradiation. There is increasing recognition of the importance of the biochemical activities of reactive oxidants following exposure of skin to such exogenous factors. Sunlight, especially ultraviolet (UV) light, is known to produce severe skin damage, including lipid peroxidation, sunburn, inflammation, pigmentation, and skin tumors (1). Lipid peroxide in the skin has long been suspected of being the source of the cellular damage induced by the oxidative stress of UV light. Lipids in the cellular plasma membrane and skin surface contain a large amount of polyunsaturated fatty acids, which are easily oxidized to lipid peroxide by UV light. Lipid peroxide is known to be potentially deleterious to cellular function, having cytotoxic effects, stimulatory or inhibitory effects on enzymes, cell membrane damages, and carcinogenic effects (2,3).

Recent observations indicate that lipid peroxidation is mediated by free radicals. Lipids are generally assumed to be the primary target for free radical attack. The formation of lipid peroxide is mainly the result of peroxidation of membrane lipids by active oxygen. The process of lipid peroxidation is thought to be as follows: the UV irradiation leads to an initiation process and produces a lipid radical (alkyl radical, $L^{\cdot}$) from a polyunsaturated fatty acid (LH). The lipid radical reacts rapidly with oxygen to form a lipid peroxy radical ($LOO^{\cdot}$). The lipid peroxy radical can then react with another polyunsaturated fatty acid (LH) to generate lipid peroxide (LOOH). On the other hand, polyunsaturated fatty acid forms another lipid radical (propagation). Lipid molecules can be peroxidized in this manner.

$$LH \longrightarrow L^{\cdot}$$
$$L^{\cdot} + O_2 \longrightarrow LOO^{\cdot}$$
$$LOO^{\cdot} + LH \longrightarrow LOOH + L$$

Although lipid peroxidation is widely known to occur among polyunsaturated fatty acids in the membrane, peroxidation is also observed in cholesterol (4,5). We will discuss here the endogenous formation of lipid peroxide from polyunsaturated fatty acids in skin exposed to ultraviolet light. Radical reactions were examined by means of the electron spin resonance (ESR) technique.

II. LIPID PEROXIDE IN LIVING SKIN

After a single exposure of skin to UV light, lipid peroxidation occurs in the epidermis. Lipid peroxide content in the epidermis was determined colorimetrically by the thiobarbituric acid method and was expressed in terms of nmol of malondialdehyde per protein (6). In one of our experiments (7,8), hairless mice were exposed to a single dose of UV light (5.6×10^6 erg/cm^2). The lipid peroxide content was significantly elevated 3–6 hours after exposure (Table 1).

Table 1 Lipid Peroxide Content of Hairless Mice Epidermis After a Single Exposure to Ultraviolet Light

	Non-exposure	Immediately after UV	3 h	6 h	16 h	24 h
Control	50.4 ± 8.2	54.2 ± 7.7	69.8 ± 18.4	75.2 ± 13.7	40.9 ± 9.3	49.8 ± 5.0
CoQ liposome	50.4 ± 6.7	41.9 ± 2.5	53.7 ± 5.2	60.8 ± 5.6	38.4 ± 2.4	39.5 ± 7.0

UV: 10 min, 5.6×10^6 erg/cm^2 per min. Lipid peroxide: TBA value, 10^{-3} nmol/per g protein. CoQ liposome group received a topical application of 1% CoQ_{10} liposome lotion (0.5 ml) 6 and 3 hours prior to UV exposure.

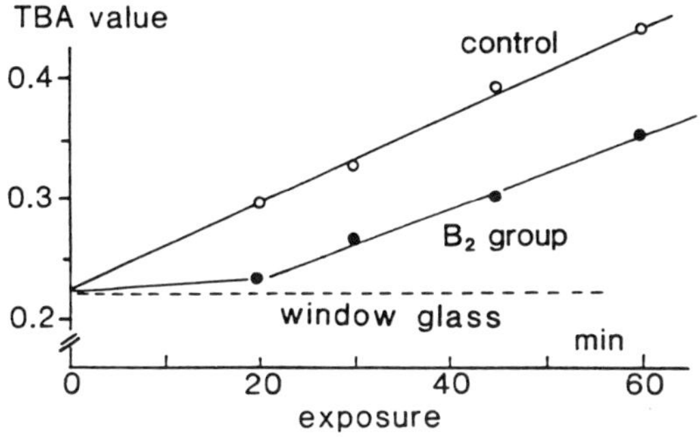

Figure 1 Effect of UV light exposure (5.6×10^3 J/m²/min) on lipid peroxide content in rat epidermis. TBA value in nonexposed epidermis: 0.22 nmol/mg protein. B_2 group rats received 20 mg of riboflavin butyrate per kg body weight intraperitoneally for one week.

In another series of experiments (8–10), we found that the lipid peroxide content of guinea pig epidermis exposed to UV light was elevated 48 hours after exposure. The exposure of living skin to UV light initiates lipid peroxidation in the epidermis, however, lipid peroxide content in the living skin did not increase immediately after UV exposure. The elevation of lipid peroxide appeared after a certain interval of time. This delay may be because of the action of the antioxidants distributed in the epidermis. Black and Mathews-Roth (5) observed an inverse relationship between UV-induced cholesterol epoxide formation and skin antioxidant content following exposure of the mouse skin to UV light. The pattern of lipid peroxide formation in epidermal homogenate differs from that of living skin. Epidermal homogenate placed in a Petri dish was exposed to UV light. The lipid peroxide content elevated immediately and increased in proportion to the time of exposure (Fig. 1).

III. ULTRAVIOLET LIGHT WAVELENGTH

To identify the UV wavelength responsible for peroxidation of membrane lipids, monochromatic wavelengths were applied to a rat liver mitochondrial suspension at intervals of 20 nm between 240 and 320 nm with 11.2×10^6 erg/cm² energy (11). A high-pressure xenon arc grating monochromater (Japan Spectroscope Co. Ltd., CRM-FA) was used. The wavelengths between 260 and 300 nm were found to be most efficient for prodcucing lipid peroxide (Fig. 2). As seen in Figure 2, UV light reduced the respiratory control ratio (RCR) in the exposed mitochondria. It is interesting to note that

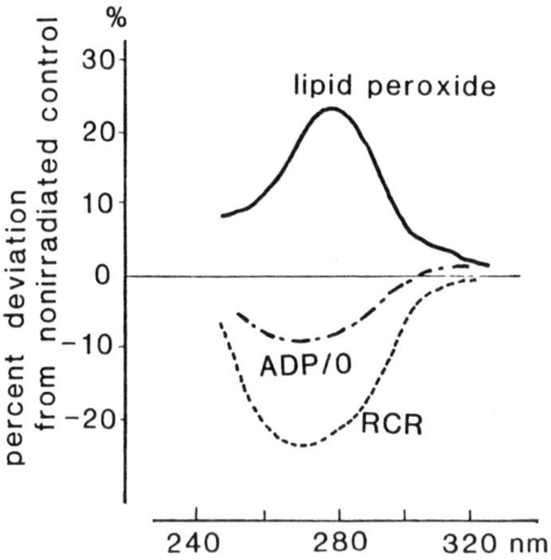

Figure 2 Effect of exposure of monochromatic wavelength ultraviolet light on the mitochondrial suspension. UV: 11.2×10^6 erg/cm^2.

an inverse relationship between lipid peroxide formation and the mitochondrial respiratory responses (RCR, ADP/O) was observed. Utilizing mitochondria obtained from rats that had been administered 8-methoxypsoralen phototoxic compound intraperitoneally for 2 weeks, the wavelengths between 320 and 340 nm, in addition to the original range of 260–300 nm, were found to be effective in producing lipid peroxide (12). These wavelengths (320–340 nm) correlated with the known erythema action spectrum in skin treated with 8-methoxypsoralen.

IV. ANTIOXIDANTS AND LIPID PEROXIDE

It is believed that antioxidants must function as photoprotectants, because antioxidation is known to play a role in protection against the formation of lipid peroxide. CoQ was first used as an antioxidant for UV oxidation. Skin of hairless mice was pretreated with a topical application of 1% CoQ_{10} liposome lotion (0.5 ml) 6 and 3 hours prior to UV exposure (7). Following a single UV exposure, the lipid peroxide content in the epidermis remained in the normal range until 24 hours postexposure (Table 1). Topical application of this antioxidant was shown to be effective in preventing the formation of UV-induced lipid peroxide. In a separate experiment, we intraperitoneally administered 20 mg of riboflavin butyrate per kg body weight to albino rats. The epidermis obtained from the animals after one week of administration

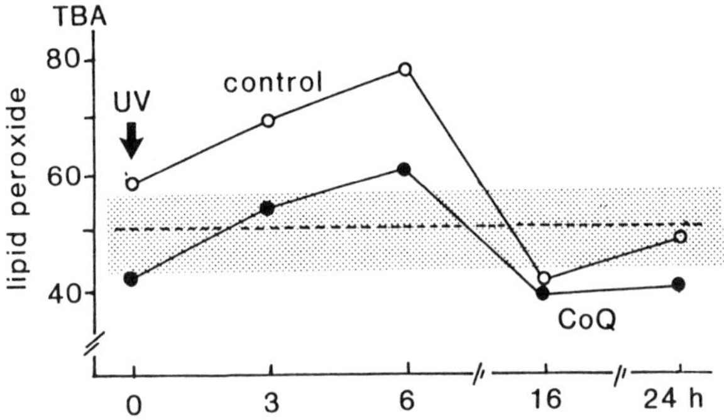

Figure 3 Effect of topical application of CoQ_{10} liposome lotion on the lipid peroxide content of epidermis of hairless mice receiving a single exposure of ultraviolet light. UV: 10 min, 5.6×10^6 erg/cm²/min. 1% CoQ_{10} liposome lotion (0.5 ml) was applied topically to the skin 6 and 3 hours prior to UV exposure.

showed that formation of the UV-induced lipid peroxide has been suppressed (see Fig. 1). Black et al. (5,13) demonstrated that antioxidant dietary supplements provide a protection against UV-induced cholesterol epoxide formation in mice skin. In addition, antioxidants markedly reduced the severity of UV-induced skin tumors. Antioxidants were demonstrated to impede the formation of UV-induced photooxidation products.

	Number of animals bearing tumors	Total number of actinic lesions
Regular diet	5/21	33
Supplemental diet	0/13	4

In another experiment (14), we exposed an isolated rat liver mitochondrial suspension to UV light. The irradiation induced an increase of lipid peroxide content in the mitochondria and resulted in mitochondrial respiratory dysfunctions. However, when mitochondria obtained from rats fed a diet supplement with 3 mg% riboflavin for 8 weeks (Table 2), a marked degree of protection against both lipid peroxidation and respiratory dysfunction was observed (10,15). A diet supplemented with 5 mg% glutathione yielded similar results (16). Thus, antioxidants were shown to protect against or minimize formation of the lipid peroxide induced by UV exposure. Lipid peroxidation was proven to be a free radical–mediated response.

Table 2 Effect of Riboflavin or Glutathione Administration on Lipid Peroxide Content and Respiratory Function of a Rat Liver Mitochondrial Suspension Exposed to Ultraviolet Light

	Lipid peroxide (nmol/mg protein)	Respiratory responses	
		RCR	ADP/O
Control	0.59 ± 0.03	4.20 ± 0.10	1.93 ± 0.08
UV exposure			
regular chow	1.26 ± 0.05	2.08 ± 0.03	1.80 ± 0.04
B_2 group	0.75 ± 0.05	3.38 ± 0.18	1.93 ± 0.34
GSH group	0.90 ± 0.03	2.40 ± 0.27	1.96 ± 0.04

The control group of rats received regular laboratory chow, and the B_2 group received the regular chow supplemented with 3 mg% riboflavin butyrate for 8 weeks. The glutathione group received the same chow with 5 mg% glutathione. The isolated liver mitochondrial suspensions was exposed to UV light at 5.6×10^6 erg/cm² per 10 min.

Using a topical application of an active oxygen scavenger, the involvement of active oxygen species in the process of lipid peroxidation was demonstrated. Cu/Zn superoxide dismutase (reduction of $O_2^{\cdot-}$ to H_2O_2, SOD) liposome cream was applied topically to guinea pig skin before UV exposure (8,9). The clipped-back skin of the guinea pigs was treated with 0.3 g of 0.05% SOD liposome cream 24 and 18 hours before UV exposure. Lipid peroxide content in the epidermis from the control animals increased 48 hours following a single UV exposure, from 24.3 to 36.9 nmol/g wet weight epidermis. However, lipid peroxide content in the treated epidermis remained in the normal range (Table 3). Next, instead of the topical application prior to UV

Table 3 Effect of Topical Application of Superoxide Dismutase Liposome Cream on the Lipid Pereoxide Content of Guinea Pig Epidermis Exposed to Ultraviolet Light

	None	Topical application	
		SOD-liposome cream	Liposome cream alone
Control	24.3 ± 4.2		
18 hours	25.4 ± 7.9	27.4 ± 5.9	29.5 ± 4.7
48 hours	36.9 ± 2.3	22.9 ± 1.7 (23.7 ± 4.0)	33.1 ± 5.7

A single dose of UV exposure: 4.8×10^7 erg/cm². SOD liposome cream (0.3 g) was given 2 times at 24 and 18 hours before UV exposure. The value in parentheses indicates a posttreatment following UV exposure. These applications were made immediately after exposure and again 6 hours later.

Table 4 Effect of Superoxide Dismutase on Respiratory Responses of Mitochondria Affected by UV-Exposed Methyl Linoleate

	RCR	ADP/O	State 4	Surface potential of inner membrane
Control (nonexposed methyl linoleate)	5.0 ± 1.2	1.7 ± 0.1	10.9 ± 1.9	0.34 ± 0.03
UV-methyl linoleate	2.6 ± 0.5	1.6 ± 0.1	41.0 ± 4.9	0.09 ± 0.02
UV-methyl linoleate with SOD	3.4 ± 1.4	1.6 ± 0.1	10.9 ± 2.5	0.17 ± 0.02

UV-exposed methyl linoleate (150 μg) emulsion was incubated for 20 minutes at 26°C with mitochondrial suspension (5 mg protein). Surface potential of inner mitochondrial membrane was determined by absorbance change (OD at 540–465 nm) of ethidium bromide.

exposure, SOD liposome cream was applied immediately after UV exposure and then again 6 hours later. The lipid peroxide content 48 hours after exposure was found to be in the normal range (see Table 3). In addition to the use prior to exposure, use immediately after exposure was also effective in preventing UV-induced lipid peroxidation. Although SOD enzyme is distributed throughout the epidermis (17), topical application was required to protect against UV-induced photoxidation. From these SOD experiments, it is suggested that superoxide radicals are involved in the UV-induced lipid peroxidation process.

Lipid peroxides are known to have deleterious effects on the cellular membrane. We demonstrated the direct effect of UV-formed lipid peroxide on membrane function. A mitochondrial suspension was incubated for 20 minutes at 26°C with methyl linoleate or UV-photooxidized methyl linoleate (18). Disorders of respiratory function were observed following incubation with UV-exposed methyl linoleate. However, respiratory disorders were less severe when incubated together with SOD liposome. It was suggested that SOD may play a role in preventing cellular damages induced by UV exposure (Table 4).

V. RADICAL REACTIONS IN LIPID PEROXIDATION

To elucidate the principal processes of UV-induced lipid peroxidation through radical reactions, the presence of radical reactions in the skin was proved by the ESR method (19). A quartz ESR glass tube (5-mm diameter) was filled with epidermal homogenate and inserted in the ESR cavity. The ESR tube was irradiated with UV light through a quartz lens installed in the front of the ESR cavity using a UV lamp (ES-UV-05H UV lamp, JEOL, 260–600 nm, 19 × 10^3 J/m^2/min). The ESR spectra were recorded on an x-band spectrometer (JES-FE3X, JEOL). The ESR cavity temperature was controlled by a variable temperature controller (JES-VT-3A, JEOL) with a liquid nitrogen flow.

No ESR signal was detected from epidermal homogenate without UV exposure at room temperature or even at the low temperature of −150°C. However, following UV exposure, an ESR signal appeared at 2.0064 g under −150°C conditions (Fig. 4). The ESR signal increased in size with increasing time of exposure, maintaining the original g value of 2.0064. The ESR signnal was assigned as that of free radicals derived from the lipid radical (L˙). The U V irradiation was found to increase the cellular concentration of free radicals in the skin.

Methyl linoleate was then used instead of epidermal homogenate, since linoleate is one of the major polyunsaturated fatty acids in membrane phospholipids. An ESR signal appeared at 2.0050 g, following UV-exposure at −150°, and the ESR spectra obtained from UV-exposed samples were similar in shape to those obtained from epidermal homogenate (20).

To detect the involvement of oxygen in the process of lipid peroxidation, the temperature of the ESR cavity was elevated gradually from −150°C to room temperature following a single exposure of UV light using the JEOL variable temperature contoller. The epidermal homogenate in the ESR tube was irradiated with a single exposure of UV light, and ESR spectra were recorded at −150, −100, −40, and −20°C (Fig. 5A). The ESR spectra varied in shape and g value. The g value ranged from 2.0064 to 2.0095 at −100°C, and from 2.0095 to 2.014 at −40°C. The shape of the ESR spectra

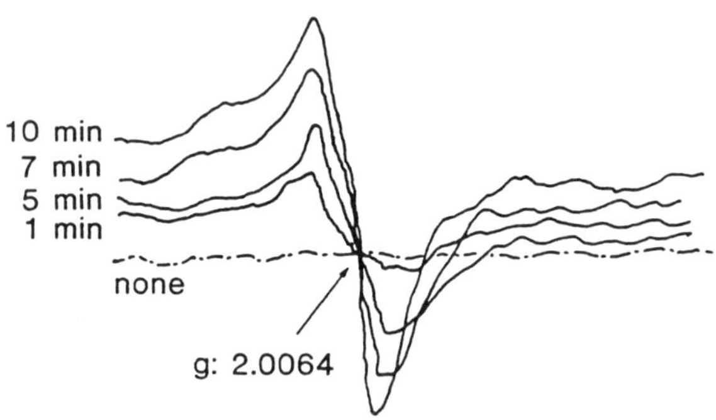

Figure 4 ESR spectra of epidermal homogenate exposed to UV light with different times of exposure under −150°C condition. UV: 19 × 10^3 J/m^2/min.

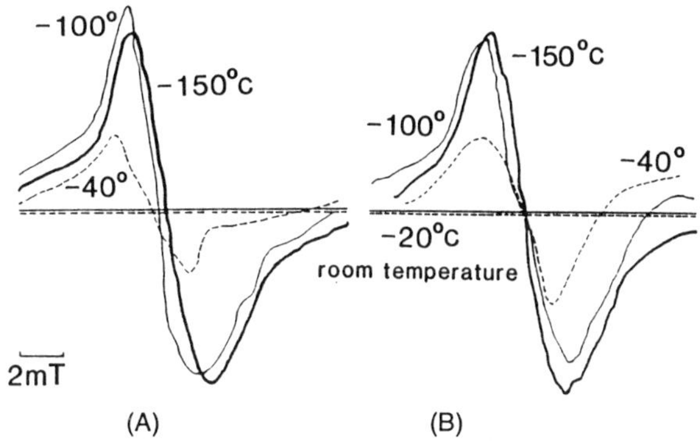

Figure 5 Effects of temperature elevation (from −150°C to −20°C) on ESR spectra of the UV-exposed epidermal homogenate. Temperature of ESR cavity was changed gradually with a variable temperature controller from −150°C to room temperature. The thick line indicates the original spectrum, which appeared at −150°C following a single exposure of UV (10 min). (A) Control (aerobic condition), (B) Degassed condition.

became sharper and smaller, and finally disappeared at −20°C because of molecular motion.

In addition, to determine the involvement of oxygen in the radical reaction indicated in the ESR spectrum, an anaerobic environment in an ESR tube filled with homogenate was prepared using a vacuum pump (10^{-3} mmHg). The degassed sample was exposed to UV light (Fig. 5B). The ESR signal (L$^{\cdot}$) appeared immediately after a single exposure of UV light at −150°C, as observed in the control. However, the change of ESR spectra following the elevation of temperature was quite different from that of the control. The ESR signal did not change in shape, and the g value maintained its original value, as the cavity temperature was elevated. These results prove that the modification of the ESR signal following elevation of temperature is associated with oxygen.

$$L^{\cdot} + O_2 \longrightarrow LOO^{\cdot}$$

A similar ESR spectrum was observed following the elevation of cavity temperature in a separate experiment using methyl linoleate (20).

VI. ACTIVE OXYGEN SPECIES

The electron spin resonance technique with spin trapping was used to determine what kinds of active oxygen species are involved in the formation of the lipid radical (L$^{\cdot}$). The spin trap agents were 5,5′dimethyl-1-pyrroline-N-oxide (DMPO) or α-phenyl-N-tert-butylnitrone (PBN).

Epidermal homogenate was mixed with DMPO, and the mixed solution was taken into a quartz ESR flat cell (160 μl, ES-LC12, JEOL). The ESR flat cell was irradiated under an ultraviolet lamp (Toshiba photochemical UV lamp H400P, 260–585 nm, 5.6×10^3 J/m^2/min). The ESR flat cell was then inserted into the ESR cavity and the ESR spectra were recorded.

No ESR signal appeared without UV expousre (Fig. 6A). The irradiation of homogenate with UV light led to an enlargement of the ESR spectrum which increased with the time of exposure (Fig. 6B). As shown in Figure 7, a sextet line signal was observed with hyperfine splitting (21) (aN = 1.55 mT, aH = 2.27 mT), which was most suggestive of the trapped alkyl radical, accompanied by a spectrum of DMPO-hydroxyl radical adduct (DMPO-OH, aN = aH = 1.55 mT). The sextet signal was also detected in the UV-exposed Folch solution extract obtained from homogenate (Fig. 6C). In addition, the ESR spectrum of the PBN spin adduct observed in the UV-exposed epidermis revealed the similar six-line signal, aN = 1.6 mT, aH = 0.36 mT

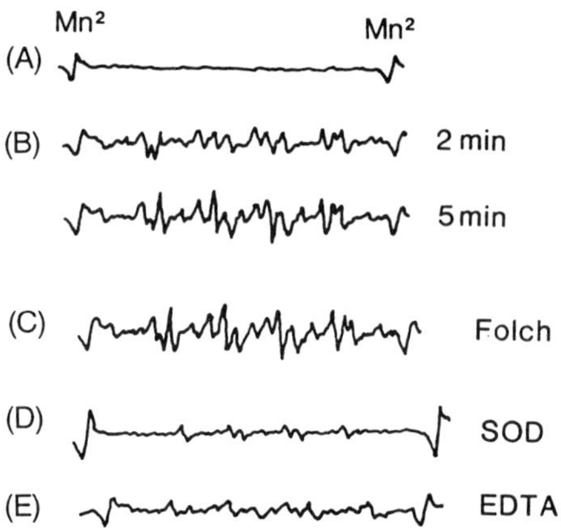

Figure 6 ESR spectra of the DMPO spin adduct obtained from epidermal homogenate exposed to UV light. (A) DMPO alone. (B) UV exposure: 2 min, 5 min. (C) UV exposure of Folch solution extract (chloroform-alcohol mixture 2:1). (D) UV exposure in the presence of SOD. (E) UV exposure in the presence of EDTA. ESR signals (Mn^{2+}) at both sides of spectrum are derived from manganese. The manganese reference (MnO) enclosed in a sealed glass tube equipment (ES-DM1 JEOL Ltd) was installed in the ESR cavity. The Mn^{2+} signal appears at an uniform height in the ESR signal (g: 2.034, 1.981; at the interval of 8.69 ± 0.01 mT).

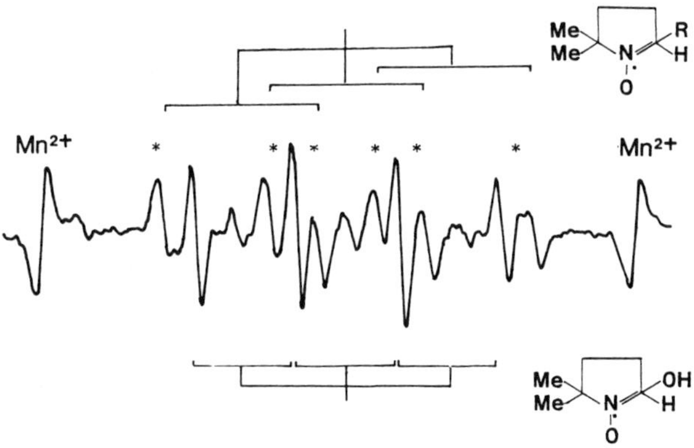

Figure 7 Hyperfine splitting analysis of the DMPO spin adduct spectrum that appeared from epidermal homogenate following UV exposure. DMPO-R: aN = 1.55 mT, aH = 2.27 mT. DMPO-OH: aN = aH = 1.55 mT.

(Fig. 8). The signal intensity increased proportionately with increasing time of exposure. This evidence indicates that the trapped adduct is associated with lipid radicals (19).

In considering the involvement of active oxygen in lipid peroxidation, we first examined the possible role of the superoxide radicals. The UV experi-

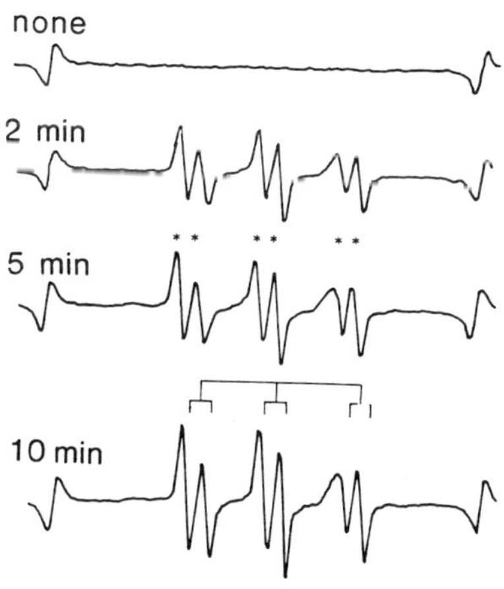

Figure 8 ESR spectrum of PBN spin adduct obtained from epidermal homogenate exposed to UV light with different times of exposure. UV: 19×10^3 J/m^2/min. Spin trapping agent: PBN. PBN-R: aN = 1.6 mT, aH = 0.36 mT.

ment was carried out in the presence of SOD (22). The addition of SOD to the homogenate resulted in the reduction of the ESR spectrum (see Fig. 6D). Heat-denatured SOD, however, had no perceptible influence on the ESR spectrum. Superoxide radicals were found, therefore, to be involved in the formation of lipid radical. However, the spectrum of the DMPO-superoxide radical (DMPO-OOH) adduct could only barely be detected in the ESR spectrum, as seen in Figure 7.

Next, we carried out an experiment to determine the participation of metal in lipid peroxidation. In Figure 6E, the ESR spectrum of DMPO-L$^{\cdot}$ is greatly reduced in the presence of the iron-chelating agent, EDTA, as compared with that in the nonchelating sample. The metal, as well as superoxide anion, is involved in the formation of lipid radicals (22).

As a result of the ESR spin trapping experiments, it is clearly understood that superoxide radicals and hydroxyl radicals are related to the formation of lipid radicals in the epidermis exposed to UV light. This brings up the question of which kind of active oxygen species is actually involved in abstracting the hydrogen from an unsaturated fatty acid. A simplified system was designed without UV exposure in which the epidermal homogenate was mixed with a superoxide radical generation system (xanthine oxidase system) or hydroxyl radical generation system (Fenton reaction system). We found that the spectrum of the DMPO-lipid radical adduct appeared only when the Fenton reaction system was used (Fig. 9B-2). A further experiment using methyl linoleate instead of epidermal homogenate showed a similar result (22). It was confirmed that the lipid radicals were formed by the hydroxyl radicals, not the superoxide radicals.

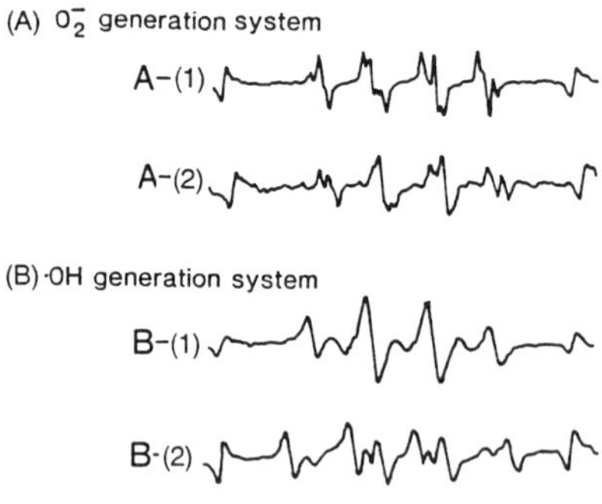

Figure 9 ESR spectrum of DMPO spin adduct obtained from the mixture of epidermal homogenate and active oxygen generation system. (A) xanthine oxidase system ($O_2^{\cdot -}$): (1) enzyme system alone, (2) epidermal homogenate with enzyme system. (B) Fenton reaction system: (1) reaction system alone, (2) epidermal homogenate with reaction system.

The next problem was to examine the possible role played by the superoxide radicals in the UV-induced lipid peroxidation. The combination use of the xanthine oxidase and Fenton reaction systems was employed. Figure 10 shows that the ESR spectrum of the DMPO-OH adduct enlarges severalfold during combination use (Fig. 10B) as compared with that in the Fenton reaction system alone (Fig. 10A). However, the spectrum of the DMPO-OOH adduct can barely be detected during combination use. In addition, as shown in Figure 10C, the presence of SOD in the combination use reduced the ESR signal intensity to the same level as observed in the Fenton reaction alone. This suggests that most of the superoxide radicals formed in the epidermis must be used for recycling the Fenton reaction. However, the source of the superoxide radicals formed in the epidermis remains unclear at the present time.

Figure 11 shows the proposed mechanism for the UV-induced lipid peroxidation in the epidermis. Superoxide radicals produced in the epidermis are scavenged to hydrogen peroxide by SOD enzyme distributed in the epidermis (17).

$$2O_2^{\cdot -} + 2H^+ \longrightarrow H_2O_2 + O_2$$

Hydroxyl radicals are generated from hydrogen peroxide with ferrous iron by the Fenton reaction. On the other hand, the ferric iron formed by the Fenton reaction is reduced to ferrous iron by superoxide radicals for recycling the Fenton reaction. The hydroxyl radicals formed in the epidermis are the primary active oxygen species leading to the formation of lipid radicals.

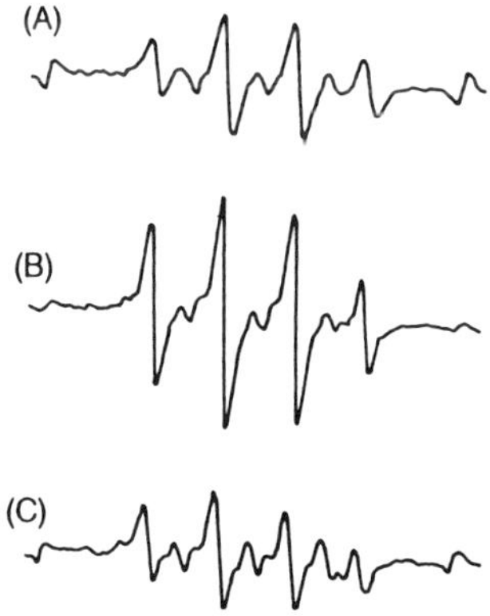

Figure 10 ESR spectrum of DMPO spin adduct obtained from the combination use of xanthine oxidase and Fenton reaction. (A) Fenton reaction alone. (B) Fenton reaction and xanthine oxidase system. (C) SOD is added to the combination system (B).

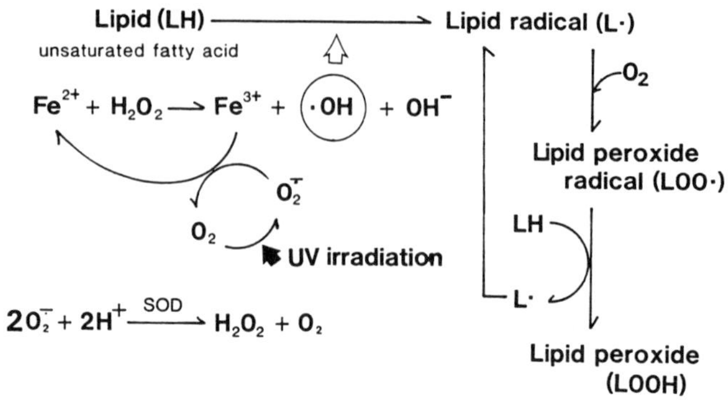

Figure 11 The proposed mechanism for generation of lipid peroxide in the epidermis exposed to UV light.

VII. MEMBRANE FLUIDITY

Recently the properties of plasma membranes in guinea pig epidermal cells have been determined by means of the ESR labeling technique (23,24). The keratinization process in the epidermis was found to be accompanied by a decrease of membrane fluidity and by a decline of Na^+, K^+-ATPase activity, as shown in Table 5 (25–27). The influence of UV light on membrane fluidity was examined in cultured B-16 melanoma cells (28). The lipid peroxide content of melanoma cells exposed to UV light increased in direct proportion to the radiant energy. However, following the incubation of UV-irradiated B-16 melanoma cells with tissue culture medium, the lipid peroxide increased rapidly diminished to less than the control level, followed

Table 5 Fluidity Parameters Obtained with 5-Doxylstearic Acid ESR Spin Labeling, and Na^+-K^+-ATPase Activities on Guinea Pig Epidermal Cells of the Three Regions Separated by Percoll Gradient

	Order parameter S value with 5-DSA	Na^+-K^+-ATPase
Upper epidermal cells	0.638 ± 0.002	1.07 ± 0.18
Middle epidermal cells	0.625 ± 0.009	1.26 ± 0.15
Lower epidermal cells	0.611 ± 0.013	1.94 ± 0.45

Order parameter S value was calculated from the ESR spectrum, according to the formula of Gaffney. The range of S value is from 0 to 1. As the membrane fluidity decrease, S value approaches 1. ATPase activity was expressed as nmoles inorganic phosphate liberated into the acid-soluble supernatant per mg protein·min.

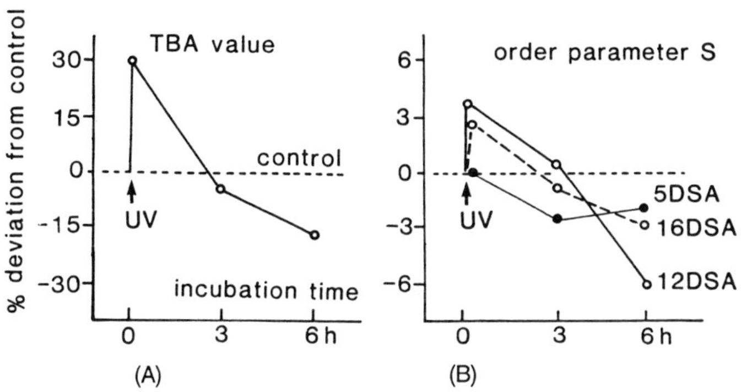

Figure 12 Effect of postincubation on the lipid peroxide content and membrane fluidity of B-16 melanoma cells, following single exposure of UV light. (A) Lipid peroxide, after exposure of UV light for 10 min (1.6×10^3 J/m^2), B-16 melanoma cells were incubated with culture medium. (B) Membrane fluidity. Three kinds of doxyl stearic acid (DSA) spin probes were used: 5-DSA, 12-DSA, and 16-DSA.

by a return to normal range (Fig. 12A). On the other hand, UV exposure resulted in an immediate decrease of membrane fluidity (increase of order parameter S value) in the 12- and 16-doxyl-stearic acid regions of the plasma membrane. After that, the fluidity in all regions rapidly increased to more than the control level 6 hours after exposure (Fig. 12B). It is of interest to note that addition of exogenous *t*-butyl hydroperoxide to intact tissue culture cells induced a fall in membrane fluidity in which the alteration pattern was similar to that found in the UV-exposed tissue culture cells (28). These experiments indicate that the UV-induced lipid peroxidation is closely related to plasma membrane fluidity. These changes of membrane fluidity involve the reduction of membrane-bound enzyme activity, such as that of Na^+-K^+ ATPase (26). It was demonstrated that UV-induced lipid peroxide induces physical changes in the membrane.

VIII. DNA LESIONS

The mechanism of sunlight-induced damage to the skin is one of the most urgent problems to be clarified in the field of skin tumors (5,29). UV light produces various kinds of DNA lesions besides pyrimidine dimers, i.e., single-strand breaks, DNA-protein cross-links, DNA double-strand breaks, and thymine glycol (30,31). The etiological role of UV radiation in skin cancer induction is well accepted. It is important to note that number of UV-mediated cutaneous responses can be diminished by systemic administration of antioxidants (5). However, it is still not clear whether or not there is a causal relationship between active oxygen species or lipid peroxide and the genetic and mutagenic development of skin tumors.

IX. CONCLUSION

From the results described here, it is clear that ultraviolet light is an important source of reactive oxidents, which give oxidative stress to the skin. The exposure of skin to UV light induces lipid peroxidation through radical reactions. It appears that lipid peroxide attached to the skin causes various kinds of cell damage, and leads to skin disease and skin tumors. Hydroxyl radicals formed in the skin act as the most biologically active oxygen species. Antioxidants, which are powerful inhibitors of lipid peroxidation, can reduce the deleterious effects of UV-induced lipid peroxide.

REFERENCES

1. Kligman AM, Takase Y, eds. Cutaneous aging. Tokyo: University of Tokyo Press, 1988.
2. McBrien DCH, Slater TF, eds. Free radicals, lipid peroxidation and cancer. New York: Academic Press, 1982.
3. Yagi K, ed. Lipid peroxides in biology and medicine. New York: Academic Press, 1982.
4. Smith LL, Teng JI, Lin YY, Seitz PK, McGehee MF. Lipid peroxidations of cholesterol. In: Yagi K, ed. Lipid peroxides in biology and medicine. New York: Academic Press, 1982:89–105.
5. Black HS, Mathews-Roth MM. Protective role of butyrated hydroxytoluene and certain carotenoides in photocarcinogenesis. Photochem Photobiol 1991; 53:707–16.
6. Ogura R, Sakanashi T, Nagata O, Sugiyama M, Kajiyama K, Nakagawa T, Shin G, Hidaka T. Assay for lipid peroxide content in mitochondria by the thiobarbituric acid reaction. Kurume Med J 1987; 34:53–8.
7. Ogura M, Iwahashi H, Sakanashi T, Suematsu T, Ogura R. Effect of CoQ_{10} liposome lotion on lipid peroxidation in the epidermis exposed to ultraviolet light. J Kurume Med Assoc (in Japanese) 1986; 49:64–9.
8. Ogura R, Sugiyama M, Sakanashi T, Hidaka T. Role of oxygen in lipid peroxide of skin exposed to ultraviolet light. In: Hayaishi O, Imamura S, Miyachi Y, eds. The biological role of reactive oxygen species in skin. Tokyo: University of Tokyo Press, 1987:55–63.
9. Sugiyama M, Mishima Y, Hino Y, Murakata S, Ogura R. Effect of superoxide dismutase to lipid peroxidation in the epidermis exposed to ultraviolet light. J Kurume Med Assoc (in Japanese) 1984; 47:233–7.
10. Ogura R, Sugiyama M. Cellular membrane damage and lipid peroxide following exposure of the skin to ultraviolet light. In: Kligman AM, Takase Y, eds. Cutaneous aging. Tokyo: University of Tokyo Press, 1988:459–71.

11. Ogura R, Luo RP, Sayanagi H, Nagata O. Effect of monochromatic ultraviolet light on lipoperoxide formation and the function of isolated rat liver mitochondria. Kurume Med J 1977; 24:237–9.
12. Yamauchi K, Luo RP, Ogura R. Effect of ultraviolet light exposure on lipid peroxide formation of isolated mitochondria from rat liver administrated 8-methoxypsoralen. Kurume Med J 1978; 25:169–73.
13. Black HS, Chan JT. Suppression of ultraviolet light-induced tumor formation by dietary antioxidants. J Invest Dermatol 1975; 65:412–14.
14. Ogura R, Sayanagi H, Noda K, Hara H, Shimada T, Murakami M. Biochemical and morphological changes in isolated rat liver mitochondria exposed to ultraviolet light. Kurume Med J 1978; 25:65–78.
15. Sayanagi H, Hara H, Luo RP, Ogura R. Effect of riboflavin free-form and riboflavin tetrabutyrate-form on mitochondria damage exposed to ultraviolet light. J Kurume Med Assoc (in Japanese) 1977; 51:513–18.
16. Sayanagi H, Hara H, Yamauchi K, Ogura R. Effect of glutathione on the lipid peroxide and respiratory chain of the isolated mitochondria exposed to ultraviolet light. Nutr Food (in Japanese) 1978; 31:341–6.
17. Ogura R, Ueta H, Sugiyama M, Hidaka T. Distribution of superoxide dismutase activity in the epidermis—measurement with ESR spin trapping. J Invest Dermatol 1990; 94:227–9.
18. Ogura R, Murakata M, Sakata T, Chiba R. Effect of superoxide dismutase on the surface potential disorders of mitochondria treated with ultraviolet light exposed methyl linoleate. Kurume Med J 1981; 28:1–8.
19. Nishi J, Ogura R, Sugiyama M, Hidaka T, Kohno M. Involvement of active oxygen in lipid peroxide radical reaction of epidermal homogenate following ultraviolet light exposure. J Invest Dermatol 1991; 97:115–19.
20. Sugiyama M, Kajiayama K, Hidaka T, Kumano S, Ogura R. Lipid peroxidation and radical formation in methyl linoleate following ultraviolet light exposure. J Dermatol 1984; 5:455–9.
21. Buettner GR. Spin trapping, ESR parameter of spin adducts. Free Rad Biol Med 1987; 3:259–303.
22. Ogura R, Sugiyama M, Nishi J, Haramaki N. Mechanism of lipid radical formation following exposure of epidermal homogenate to ultraviolet light. J Invest Dermatol 1991; 97:1044–1047.
23. Marsh D. Electron spin resonance: spin labels. In: Grell E, ed. Membrane spectroscopy. Berlin: Springer-Verlag, 1980:51–42.
24. Ogura R, Sugiyama M, Sakanashi T, Ninomiya T. ESR spin-labeling method of determining membrane fluidity in biological materials—tissue culture cells, cardiac mitochondria, erythrocyte and epidermal cells. Kurume Med J 1988; 35:1–8.
25. Tanaka T, Sakanashi T, Kaneko N, Ogura R. Spin labeling study on membrane fluidity of epidermal cell (cow snout epidermis). J Invest Dermatol 1986; 87: 745–7.
26. Tanaka T, Hidaka T, Ogura R, Sugiyama M. Changes of membrane fluidity and Na^+,K^+-ATPase activity during cellular differentiation in guinea pig epidermis. Arch Dermatol 1988; 280:29–32.

27. Tanaka T, Ogura R, Hidaka T, Sugiyama M. Changes of electron spin resonance membrane fluidity in hexadecane-induced hyperproliferative epidermis. J Invest Dermatol 1989; 93:682–6.
28. Ogura R, Sugiyama M, Sakanashi T, Suematsu T, Hidaka T, Morikawa F, Kon Y. Membrane response of B-16 melanoma cells to single exposure of ultraviolet light. Arch Dermatol 1989; 280:481–6.
29. Tyrrell RM. Mutagenic action of monochromatic UV radiation in the solar range on human cell. Mutat Res 1984; 129:103–10.
30. Peak MJ, Peak JG, Jones CA. Different mechanisms for the induction of DNA-protein crosslinks in human cells by far- and near-ultraviolet radiations (290 and 450 nm). Photochem Photobiol 1985; 42:141–6.
31. Matsumoto K, Sugiyama M, Ogura R. Non-dimer DNA damage in Chinese hamster V-79 cells exposed to ultraviolet-B light. Photochem Photobiol 1991; 54:389–92.

5

Vitamin E Prevention of Ultraviolet-Induced Skin Damage

John R. Trevithick
University of Western Ontario, London, Ontario, Canada

The focus of this chapter will be narrow, and the introduction somewhat cursory, since several other chapters in this book deal with such topics as the antioxidant system of the skin, its depletion and repletion, antioxidants and sunscreens, and so on.

I. SUNBURN AND ULTRAVIOLET IRRADIATION

The exposure of skin to ultraviolet radiation (UV) has from ancient times caused sunburn. It is also a problem for some domestic animals. For instance, pigs allowed to run loose on the range in the spring after being confined to barns for the winter often develop severe sunburn (1).

The immediate symptoms of sunburn are a reddening of the skin (erythema). This is observed at 3–4 hours after 15 minutes to 1 hour of exposure to full sunlight. The erythema continues to develop and persists for a period of up to 5–6 days, depending on the intensity of the exposure (1,2). Accompanying the erythema is increased skin sensitivity to touch. In severe sunburn this may even cause pain and discomfort during movements that flex the swollen and edematous skin. The edema leads to a puffy appearance, or in skh-1 hairless mice, to visible edema (3–5). After a period of 6–10 days the old skin epidermis peels off in a generalized desquamation or flakes of skin in the phenomenon commonly called "peeling."

The UV exposure leads to production of free radicals in the skin (6–8) and produces leukocyte chemoattractants (9). Although a number of leukocyte chemoattractants have been suggested to be active in recruiting leukocytes to the sunburned skin, histamine and PGE_2 appear to be most strongly established as chemoattractants (9).

Later changes to the skin occur by 24 hours as the process of inflammation develops, leading to the observation that the sunburned skin is an inflamed skin (10). The inflammation in turn results in further oxidative stress to the skin, leading to an autocatalytic multistep process of degeneration. This can lead to accelerated aging of the skin (11) and increased risk of skin cancer (12). It has been suggested that a contributing factor in the development of skin cancer might be the loss of immune surveillance, which occurs as a result of the destruction of the skin Langerhans cells (13,14).

Packer's group (15) has reported that a decrease in ubiquinone and tocopherol in skin occurs after UVB irradiation. This appears to be a response to the increased oxidative stress initiated by the UV irradiation. This stress results in the production of oxygen free radicals and the components of the oxidative cascade such as superoxide, peroxide, and hydroxyl radicals. Reaction of endogenous antioxidants in the skin with these radicals and their products such as lipoperoxides is the most probable cause of the decreased concentration of antioxidants in the skin after UV irradiation (15).

Although tocopherol may be regenerated from the tocopheroxy radical by ascorbate (16), depletion of the antioxidant defenses of the skin may place the skin cells at risk for subsequent stages of the inflammatory process when increased numbers of leukocytes are observed associated with the later stages of the erythema such as those observed at 24 hours. These may be responsible for the production of other damaging oxidizing species such as hypochlorite, which is produced by the action of the white cell myeloperoxidase. If this is the case, it suggests that following the initial UV-induced stress on the skin, which is lesser in magnitude, the inflammatory release of oxidizing substances during the prolonged period of inflammation results in an order of magnitude more damage than the initial UV-induced damage, which acts as a trigger for the inflammation.

This can be tested by developing strategies that will prevent the loss of the antioxidant defenses of keratinocytes and, in fact, all skin cells or enable the refurbishing of these defenses by replacing the lost antioxidant levels.

II. ABSORPTION OF TOCOPHEROL BY THE SKIN

Kamimura's group (17,18) showed that topically applied tocopherol acetate was taken up by the skin. There are two main routes of absorption from the skin surface by which vitamin E is transported to the dermal layer. One makes use of a pathway through the horny layer, the epidermis, and the borderline between the dermis and epidermis to the dermis. The second uses the hair follicles to penetrate to the dermis, by way of the pilo-sebaceous canal, the inside of the hair follicles, the inner and outer root sheaths, and the connective tissue sheaths. Thus the main routes make use of direct percutaneous uptake, as well as the hair follicles of the skin to reach the dermis. In these studies Kamimura and Matsuzawa used tocopherol acetate but did not establish whether the form detected by radioactivity in the skin was the tocopherol acetate or whether free tocopherol was released during the process of uptake by the skin.

Tocopherol functions as an antioxidant when the phenolic hydroxyl of its chromanol ring is free (unesterified) (19). The free hydroxyl can function as a scavenger of free radicals or singlet oxygen, usually being itself oxidized to the semiquinone or quinone. This group can be protected from oxidation by esterification with the carboxyl group of organic acids, forming esters such as the acetate or succinate derivatives, which in theory reduces the antioxidant activity of the tocopherol acetate or succinate esters to zero. Some workers (11) have shown that tocopherol acetate can act as a scavenger of superoxide free radicals in vitro, which does not agree with the theoretical inability of the phenolic hydroxyl to react with free radicals.

Although tocopherol is subject to air oxidation if stored, its acetate ester, tocopherol acetate, by contrast, does not oxidize when stored. Tocopherol acetate is hydrolyzed by enzymes in the gut prior to being absorbed at the intestinal wall, as Burton and Ingold showed (19). Dietary supplements of tocopherol are usually dispensed in gelatin capsules containing 200–400 IU of vitamin E acetate each. These dissolve in the stomach, allowing the tocopherol acetate to be hydrolyzed by the intestinal esterase enzymes to free tocopherol, which can be absorbed by the intestine.

In contrast to this, tocopherol acetate, a liquid oil when applied to the skin, does not encounter hydrolytic enzymes on the skin as it would in the gut. Because it is a lipid or fat, it can diffuse across skin cell membranes, enter cells, and be associated with other hydrophobic membranous structures present in the cells, for instance, mitochondria, and membranes of the endoplasmic reticulum and nuclear membranes.

Kamimura and Matsuzawa (17) in 1968 did not show that conversion of the acetate ester into the free tocopherol occurred during its absorption by the skin, although they suggested that it might occur. Since their paper, another investigation of prevention of drug toxicity by tocopherol and its esters in tissue culture cells has also suggested that hydrolysis of the esters of tocopherol could occur intracellularly (20), although it has not been proven by the experiments performed. This study suggests that the uptake of vitamin E by the skin may be influenced by the esterified chemical form used for administration to cells. In this study Pascoe and Reed (20) found that another tocopherol ester, the succinate form, was better able to protect cells in tissue culture than the free tocopherol when tested against oxidative stress induced by adriamycin, even though the cellular concentrations of tocopherol were identical after both treatments. They proposed that this preferential protection by an ester might be due to intracellular hydrolysis of the succinate ester to free tocopherol: this would occur so that the free tocopherol was released at the intracellular site at which the damaging oxidative species released by adriamycin was produced. In contrast to this, the free tocopherol, with a greater lipid solubility, was suggested to be retained at the cell membrane, where it would not be able to reach its desired site of action. This difference would result in increased biological effectiveness for the ester form compared to the free tocopherol.

Mitton and Redae of our laboratory have showed in preliminary experiments that 24 hours following topical application of tocopherol acetate to the skin of skh-1 mice the free tocopherol per gram of skin is markedly increased. This result is consistent with the suggestion made by Pascoe and Reed (20) that intracellular hydrolysis of tocopherol esters occurs after uptake of the ester by the cell.

Dr G. LeCak (21) followed the penetration of free tocopherol (as a 5% solution in ethanol) through rat skin. He confirmed that the free vitamin also penetrates over a period from 1 to 16 hours through the exterior and horny layer, increasing in the inner layers of skin from 19 to 123 $\mu g/cm^2$ after 16 hours.

III. ANTIINFLAMMATORY ACTION OF TOCOPHEROL

Kamimura (18) showed a decrease of inflammation in two model systems: (1) in inflammation induced intramuscularly by injection of a solution of dextran by injected tocopherol and (2) in two chemically induced examples of dermatitis, inflammation or irritation was reduced by tocopherol acetate. In none of his examples was the inflammation caused by UV light. Although these procedures caused erythema, the processes that caused the erythema are not mechanistically the same as those induced by UVB light.

IV. ANTIOXIDANTS AND FREE RADICAL SCAVENGERS REDUCE UV-INDUCED DAMAGE

Several workers have shown that antioxidants scavenging superoxide can protect against UV-induced damage of various kinds, ranging from erythema to chronic skin damage in the hairless mouse skin and in the human skin. Khettab et al. (22) showed that significant reduction of the "free radical" levels as estimated by the TBA reaction was obtained by application of solutions of either butylated hydroxytoluene (BHT, 0.02%, 56%) or vitamin E (0.25%, 60%) or vitamin A (0.02%, 30%) or β-carotene (0.02%, 63%) to the mouse back in repeated doses at 1 hour before and 1 and 2 hours after irradiation.

V. EFFECTS OF ANTIOXIDANTS AGAINST UV-INDUCED ERYTHEMA: COMPARISON OF FREE TOCOPHEROL AND TOCOPHEROL ACETATE

Several research groups have tested the protective effect of a number of superoxide-scavenging antioxidants against UV-induced erythema, sunburn cell formation, chronic damage to the skin, and tumorigenesis in the skh-1 mouse skin. In most experiments vitamin E was applied topically to the skin prior to UV irradiation (21–29). If free tocopherol was used, it had a protective effect (21–29), while tocopherol acetate was reported to be ineffective or only marginally effective (25,28).

The exact time course of the development of the erythema is still somewhat controversial, which may be since the effects depend on the UV dose (5). One group (28,29) indicates that the peak of erythema occurs at 30 hours, while Potapenko's group (26–28) evaluated changes at 20–24 hours, and Gilchrest et al. (14) showed an effect at 24 hours, but did not sample again until 72 hours when a decline was noted. In our studies of skh-1 mice, the change in skin redness, measured using an erythema meter (5), and the skin thickness (a measure of edema or skin thickening), measured by magnetic resonance imaging, increased linearly when subjected to linear regression analysis for a period of 2–3 days after the irradiation. Later, as the inflammation subsided and desquamation occurred, the erythema and skin thickness returned towards normal, but the skin thickness had not reached normal by 8 days postirradiation except in the case of vitamin E acetate treatment.

As long ago as 1978 De Rios et al. (23), and Potapenko's group (26) showed that butylated hydroxytoluene (BHT) reduced the UV-induced erythemic action of UV light. Potapenko's group, by measuring the change in the mechanoelectrical properties of the skin using its ability to transmit radiofrequency waves through the skin as a measure of the erythema, extended the experimental protocol to include a test of DL-α-tocopherol (26). They found that tocopherol and BHT, but not tocopherol acetate (at a concentration of 10^{-10} mol/cm^2) showed a strong inhibition of the erythemal response to UV light when applied either before or 2 minutes after irradiation. In further experiments, they confirmed that for UVB irradiation (280–320 nm) tocopherol inhibits the dark reactions occurring after the irradiation. By comparison, for PUVA treatment (dermatological therapy using Psoralens such as 8-methoxypsoralen [8-MOP] combined with UVA irradiation), tocopherol inhibited directly the reaction of 8-MOP with the skin (28). Thus for the two systems different mechanisms operate, but in both systems tocopherol prevents the oxidative changes leading to skin damage. Potapenko's group states that the concentration of tocopherol used on the skin was too small to have any sunscreen effect, mediated by absorbing the incident UV light at the skin surface (26).

VI. IN VITRO EXPERIMENTS SUPPORTING THE ROLE OF FREE RADICALS IN SUNBURN CELL PRODUCTION

The first in vitro experiments in this area that provided support for free radical scavenger protection against UV were performed by Miyachi et al. in 1983 (8). Although tocopherol was not used, these experiments were germinative since they indicated that scavengers of free radicals such as superoxide dismutase (SOD), catalase, xanthine, and mannitol could suppress sunburn cell formation when pieces of skin were irradiated in the presence of the agents and then incubated in vitro with these agents. Since the scavenging agents were present during the irradiation period as well as postirradiation during the incubation, the question of whether the scavenging was effective during the postincubation period was not specifically addressed. In fact, these authors ascribed the strong protective effect observed for xanthine to its strong absorption of UVB light during the irradiation. They did observe that sunburn cell production was more strongly inhibited in the group that was irradiated and incubated in the presence of the scavengers, as compared to the group that had scavengers present only during the irradiation.

VII. ANTIOXIDANT EFFECTS ON ERYTHEMA

Ohzawa et al. (24) showed that application of DL-α-tocopherol or BHT either 60 minutes before UV light of 280–365 nm or 2 minutes after was effective in increasing the minimum erythema dose (MED) of UV by a factor of 40% to approximately 70%. There was no significant difference between vitamin E and BHT, and the effect was diminished if the application was made postirradiation at later times: 5 hours (12% or 4%, respectively, which was not significant) or 10 hours (11% or 0%, again not significant). Even when applied at 60 minutes prior or 2 minutes after, tocopherol acetate was not effective in increasing the MED.

Several other papers originating with the group led by Wallat have shown a protective effect of D-α-tocopherol against skin damage and erythema induced in several models, including the human and the hairless mouse (28, 29). In their experiments three preparations were compared: 67% pure D-α-tocopherol, a mixture of crude tocopherols including α-, β-, and γ-tocopherols, and D-α-tocopherol acetate (28). When applied in a 1% solution in ethanol before the UVB irradiation, the free tocopherol showed a 65% inhibition of the increase in skin thickness at 30 hours after the initial irradiation, while the effect was progressively less for the tocopherol mixture (40% inhibition) and much weaker and barely significant for the tocopherol acetate.

VIII. TOCOPHEROL EFFECTS ON SUN PROTECTION FACTOR (SPF)

No application after the irradiation was tested in these experiments. In the human model (29), the effect of topical application before irradiation was evaluated using the sun protection factor (SPF, given by the ratio of the MED for the treated/MED of the untreated skin). Again using the preparations described above, the sun protection factors were as follows: D-α-tocopherol in alcohol, 5% (1.34), 20% (2.03), D-tocopherol mixture in alcohol, 20% (1.96), tocopherol acetate, 20% (1.29).

When 1% of the light blocker Eusolex 6300 (EX) was tested in combination with 5% D-α-tocopherol in emulsions, the sun protection factor (SPF) of the mixture (2.3) was elevated by comparison to either the 5% tocopherol alone (2.03) and 1% EX (1.98). These data indicated that α-tocopherol can increase the sun protection factor (SPF) of a sunscreen when added to it. In further experiments (29) increased SPF factors were found for creams with 5% D-α-tocopherol (3.6 or 3.2) or 5% mixed tocopherols (3.7).

IX. VITAMIN E EFFECTS ON SKIN EDEMA

Several models were used to test the effect on UV-induced edema of sunburn when the test substance was applied *after* the irradiation. While a 1% solution of D-α-tocopherol gave a 25–30% reduction in skin edema in the hairless mouse, tocopherol acetate gave no visible effect (30). Although human experiments did not yield as clear a result for application after the irradiation, these authors (29) state that the same trends were found for the effect of topical vitamin E, and only free tocopherol was found to have any perceptible or visible effect.

Wallat and associates make the statement (29, Section 4.4), "The protective effect of vitamin E against UV-light has been demonstrated on UV-edema in hairless mice." Reference is made (29) to a previous paper (28) above to substantiate this claim. Although these authors state in this section (29, Section 4.4) that topical application of vitamin E either before or after UV irradiation produced significant inhibitory effects with respect to edema development, they do not make any claim for tocopherol acetate being effective if applied after UV irradiation. They show (29) that tocopherol acetate applied *before* UV-irradiation can inhibit edema developing at 30 hours after irradiation.

In order to determine whether it was known that tocopherol acetate is capable of inhibiting the edema developing after UV irradiation, it is necessary to refer to their first paper (28). In this paper, these authors state that, if applied 2 minutes after the irradiation, vitamin E as well as BHT (butylated hydroxytoluene) showed clearly measurable decreases in edema, while tocopherol acetate showed hardly any effect on the MED (minimum erythema dose) determination using a model of UV-irradiated rabbits. The other example discussed, the hairless mouse, showed inhibition of edema only when applied *prior* to UV-irradiation. Application *prior* to UVB irradiation resulted in a 19% inhibition of edema measured after 30 hours. This was much less than the 65% inhibition found for free tocopherol or the 40% inhibition found for a mixture of α-, β-, γ-, and δ-tocopherols.

In humans, the inhibition of edema was also tested for *prior* application of tocopherol acetate (28) and showed weaker effects than either D-α-tocopherol or the mixture of isomers mentioned above.

Application *after* the UVB irradiation in the hairless mouse model is discussed (28). These authors did not find any effect when tocopherol acetate was applied after UVB irradiation. They state D-α-tocopherol acetate resulted in no perceptible action in this test. In contrast to this observation, the inhibitory effects of 1, 2, and 5% D-α-tocopherol solutions *after* the UVB irradiation were "clearly perceptible."

X. EFFECTS OF CONCENTRATED TOCOPHEROL ACETATE APPLIED TOPICALLY

Thus it appears that neither Potapenko's group (25–27) nor the Wallat group (28) was able to observe the effect that we have observed of topical tocopherol acetate preventing sunburn damage to the skin when applied after the UV irradiation. When we applied pure vitamin E acetate to the skin after UV or sham irradiation, however, we found large increases in free tocopherol in the skin when analyzed 24 hours after the topical applicaiton. Parallel to this, we demonstrated (see below) a significant effect on several of the indicators of sunburn-associated skin damage (30).

In our experiments, exposure of the skin of the back of skh-1 hairless mice to UVB (310-nm peak) irradiation at doses of 0.115 or 0.23 J/cm^2 results after 24–48 hours in an erythema that can be quantified using an erythema meter, providing a useful model of sunburn. Application of pure D-α-tocopherol acetate, a thick oil, to the skin immediately following the exposure to UVB significantly reduces the increase in erythema index by 40–55%. At the lower dose (0.115 J/cm^2), skin thickness (associated with edematous swelling of the sunburned skin) was measured by a novel noninvasive technique not previously reported for this purpose: magnetic resonance imaging (MRI). In two experiments the UVB-induced increase in skin thickness was significantly reduced at 24 hours by 29 and 54%, and at 48 hours by 26 and 61%. After 8 days the untreated irradiated mouse skin still showed a significant increase in thickness (24%) compared to the untreated unirradiated control, while the treated irradiated control was not significantly thicker than the unexposed control. Skin sensitivity was tested using a modification of the technique of esthesiometry, by observing rapid avoidance responses of the mouse to a pressure of 0.96 g/cm^2 exerted by applying to the skin the tip of a nylon esthesiometer fiber extended to 60 mm in length. The untreated irradiated mice were more sensitive ($p < 0.07$, Wilcoxon test) than the treated irradiated mice and also significantly different from the untreated unirradiated control mice ($p < 0.04$, Wilcoxon test), but the treated irradiated mice were not significantly differently sensitive when compared to the unirradiated controls ($p < 0.32$). Taken together these data indicate that the erythema, edema, and skin sensitivity commonly associated with UVB-induced sunburn are significantly reduced by topical application of tocopherol acetate even after the exposure has occurred. This observation suggests that treatment of sunburn may be possible, even after the irradiation has stopped. by a derivative of D-α-tocopherol that is stable to autooxidation but may be converted to free tocopherol once absorbed by the skin.

XI. LONGER-TERM EFFECTS

Prior application of tocopherol and other superoxide-scavenging antioxidants such as ascorbate, propyl gallate, and Trolox (the tocol nucleus) was effective in preventing chronic skin damage (11). The chronic skin damage was evaluated by skin wrinkling and skin sagging, presence of tumors, and histology of formalin-fixed skin. The greatest reductions were found for 5% tocopherol (84%), ascorbate (80%), and 5% Trolox (81%). Skin penetration of these agents after 24 hours except for the water-soluble ascorbate (65 $\mu g/cm^2$ for human, 15 $\mu g/cm^2$ for hairless mouse) was roughly similar to their effects: α-tocopherol (227 $\mu g/cm^2$ for human, 153 μ/cm^2 for mouse), tocopherol acetate (840 $\mu g/cm^2$ for mouse), or tocopherol succinate (637

$\mu g/cm^2$ for mouse). The possible conversion of tocopherol esters to free tocopherol during penetration was not estimated.

Assessment of superoxide scavenging by inhibition of cytochrome *c* reduction using a xanthine/xanthine oxidase mixture revealed parallel scavenging of superoxide approximately followed the protective effects on chronic skin damage (11) when the two were compared. The best inhibitor of superoxide production at 0.1 mM was ascorbate (94%) followed by tocopherol acetate (77%) and tocopherol succinate (46%), while propyl gallate showed an effect of similar magnitude (56%). Unfortunately, tocopherol does not seem to have been tested. Just as for the topical applicaiton, BHT, glutathione, and pantethine were not effective superoxide scavengers, while urea, histidine, and β-carotene were ineffective in preventing skin damage but were not tested for superoxide scavenging. The activity of the tocopherol acetate and succinate as scavengers of superoxide, which Bissett's group reports (11), is somewhat surprising: Burton and Ingold (19) and theoretical considerations would argue against these esters being active against free radicals, since the chromanol hydroxyl group of the vitamin E should not be available for reaction in the esters.

XII. TUMOR RISK REDUCTION

Black (31) and Black and Chan (32) early showed diminished tumor incidence, and the same group showed erythema reduction (23) in hairless mice by a dietary mixture of antioxidants including 1.2% ascorbic acid 0.5% BHT, 0.2% dl-α-tocopherol, and 1% reduced glutathione. Testing of these individual components against erythema induction by UV (23) showed that the BHT alone was effective in producing the erythema reduction. These authors state that topical application of the mixture of the components individually showed a lesser effect than that of the dietary administration, but the effects were not compared at levels of administration, which resulted in equal levels of the test components in the skin. The actual skin levels of the components were not determined by these authors.

Gensler and Magdaleno (12) used topical tocopherol to reduce the incidence of skin tumors in the skh-1 mouse model. They studied the effect of topical application of free tocopherol (25 mg in 200 μl acetone, three times per week, 30 minutes after the daily period of irradiation). Treatments with UV were given 5 days per week for 27 weeks using mainly UVB (Westinghouse SF40 sunlamps) at a dose rate of 0.6 $mJ/cm^2/s$ for a total dose of 146 J/cm^2. Using this protocol, the incidence of skin tumors was approximately halved from 81 to 42% at 33 weeks from the first UV exposure. Gensler and Magdaleno (12) investigated the basis for this result by comparing the levels of immunosuppression by passive transfer of splenocytes from normal and UV-irradiated mice, with and without topical vitamin E treatment. When splenocytes from UV-irradiated, VE-treated mice were used, recipients rejected the UV-induced tumor challenge; those receiving splenocytes from UV-irradiated mice did not reject this UV tumor challenge. The splenocyte Ia expression, downregulated by UV irradiation, was maintained by VE treatment; VE also increased the proportion of lyt-2^+ and $13T4^+$ splenocytes.

XIII. DIETARY VITAMIN E COMPARED TO TOPICAL VITAMIN E

In addition to Black (31) and Black and Chan (32), Record et al. (33) compared dietary and topical vitamin E treatment. They treated hairless mice 3 weeks before a single exposure equivalent to 1 MED. The skin damage, assessed by thymidine incorporation into DNA and by lipid peroxidation, was reduced by topical application of 1% tocopherol either 1 or 24 hours before radiation. Less protection was shown by dietary tocopherol, which did not reduce lipid peroxidation but did maintain thymidine incorporation at normal levels.

XIV. POSSIBLE MECHANISMS OF ACTION

As other chapters have suggested, a number of mechanisms of action for vitamin E in reducing the harmful effects of sunburn are possible. Beginning with the most immediate effects, it seems possible that by prior application of tocopherol, a sunscreening effect caused by the direct absorption of UV light must be considered, since tocopherol absorbs UV light maximally at approximately 279 nm.

Following the exposure to UV light, it appears that an increase in concentration of "free radical" species occurs (detected as thiobarbituric reactive substances TBARS), which Khettab et al. (22) showed was prevented by the topical application, in an emulsion containing 50 ppm begaptene, of 0.25% vitamin E or 0.02% BHT, or 0.2% β-carotene or 0.02% vitamin A. This level effect peaked at 3 hours, after which a steady decline to normal levels of TBARS was observed. A later effect, the elevation of polyamines such as putrescine, spermine, and spermidine, showed a peak at a much later time, 18 hours after irradiation. These levels were significantly reduced by BHT, vitamin E, and difluromethylornithine (an inhibitor of polyamine production by its inhibition of the enzyme ornithine decarboxylase). Putrescine was significantly decreased by β-carotene. Also acting as a scavenger of free radicals, VE may act by inhibiting the synthesis of prostaglandin E_2 (PGE_2) (34,35). The possible elevation of leukotriene B_4 (LTB_4) after UV irradiation was not confirmed (34,35). These signal molecules such as PGE_2 may act as initial chemoattractants for leukocytes, which (36) serve as part of a compartmentalized immunological circuit dedicated to immune surveillance and protection of the skin. The UV-exposed skin causes a decrease in splenic antigen-presenting cells (APC) (36). Splenic Ia cells in the skin are highly mobile and, after UV exposure, enter the skin in larger numbers, then eventually become located in draining peripheral lymph nodes. The role of various epidermal cytokines (36), such as epidermal cell-derived thymocyte-activating factor (ETAF, IL-1), a potent modulator of both inflammatory and antigen-specific immune response, may be important in the inflammation occurring after sunburn. In the UV-exposed skin, epidermal hyperplasia leads to an increase in the number of keratinocytes (36); this in turn may cause a greater amount of ETAF to be produced per square unit of area of the skin than in nonirradiated skin sites, perhaps leading to an autocatalytic phenomenon, in which more leukocytes are attracted, so that this inflammation damages the tissue, releasing more ETAF, and so on. In tumorigenesis, these changes presumably lead to an environment in the skin in which chronic antigen stimulation leads to antigen-specific immune suppression rather that the desired rejection or elimination response. This suppression permits the growth of the tumor tissue (12).

The role of tocopherol in reducing risk of tumor incidence or growth (12) would be to reduce the probability of the tumor antigen-specific immune suppression, by reducing the degree of immune stimulation. This apparently can occur by reducing the initial production of leukocyte chemoattractants. Consistent with this, we have observed that topical VE acetate treatment leads, after 24 hours, to an approximately 10-fold decrease in the numbers of mobile leukocytes adhering to venule walls in the UV-irradiated mouse ear skin. These leukocytes appear to pass through the capillary walls and to be the precursors of the leukocytes that appear in large numbers in the inflammation of the skin in UV-induced erythema. This result suggests that VE treatment has reduced the concentration of leucocyte chemoattractants in the skin, likely as a result of the large increase in free tocopherol found in skin after the topical application of tocopherol acetate. Another mechanism that has been suggested is the VE inhibition of the activation of phospholipase A_2 (29). Another possible mechanism is the stabilization of membranes by VE, which in turn may reduce the release of histamine and hydrolytic enzymes from mast cells and lysosomes (29).

REFERENCES

1. Penny RHC, Muirhead MR. In: Lenman A, Straw B, Glock RD, Mengelny WL, Penny RHC, Schoel E, ed. Diseases of swine. Ames: Iowa University Press, 1986: 82–4.

2. Kligman LH, Kligman AM. The nature of photoaging: its prevention and repair. Photodermatology 1986; 3:215–27.
3. Lever WF. Histopathology of the skin. Philadelphia: Lipincott, 1990.
4. Young AR. The sunburn cell. Photodermatology 1987; 4:127–34.
5. Diffey BL, Oliver RJ, Farr PM. A portable instrument for quantifying erythema induced by ultraviolet radiation. Br J Dermatol 1984; 111:663–72.
6. Norins AL. Free radical formation in the skin following exposure to ultraviolet light. J Invest Dermatol 1962; 39:445–8.
7. Pathak MA, Stratton K. Free radicals in human skin before and after exposure to light. Arch Biochem Biophys 1968; 123:468–76.
8. Miyachi Y, Horio T, Imamura S. Sunburn cell formation is prevented by scavenging oxygen intermediates. Clin Exper Dermatol 1983; 8:305–10.
9. Luger TA. UVL and epidermal cell cytokine production. Photodermatology 1986; 3:123–4.
10. Roberts LK, Samlowski WE, Daynes RA. Immunological consequences of ultraviolet radiation exposure. Photodermatology 1986; 3:284–98.
11. Bissett DL, Chatterjee R, Hannon DP. Photoprotective effect of superoxide-scavenging antioxidants against ultraviolet radiation-induced chronic skin damage in the hairless mouse. Photodermatol Photoimmunol Photomed 1990; 7: 56–62.
12. Gensler HL, Magdaleno M. Topical vitamin E inhibition of immunosuppression and tumorigenesis induced by ultraviolet radiation. Nutr Cancer 1991; 15:97–106.
13. Humm SA, Cole S. Changes with time in Langerhans cell number, ATPase reactivity and morphology in murine epidermis after exposure to UVB. Photodermatology 1986; 3:174–8.
14. Gilchrest BA, Soter NA, Stoff JS, Mihm MC. The human sunburn reaction: histological and biochemical studies. J Am Acad Dermatol 1981; 5:411–22.
15. Fuchs J, Huflejt ME, Rothfuss LM, Wilson DS, Carlans G, Packer L. Acute effects of near ultraviolet and visible light on the cutaneous antioxidant defense system. Photochem Photobiol 1989; 50:739–44.
16. Niki E, Tsuchiya R, Tanimura R, Kamiya Y. Regeneration of vitamin E from α chromanoxy radical by glutathione and vitamin C. Chem Lett 1982; 789–92.
17. Kamimura M, Matsuzawa T. Percutaneous absorption of α-tocopherol acetate. J Vitaminol 1968; 14:150–9.
18. Kamimura M. Antiinflammatory activity of vitamin E. J Vitaminol 1972; 18: 201–9.
19. Burton GW, Ingold KU. Vitamin E as in vitro and in vivo antioxidant. Ann NY Acad Sci 1989; 570:1–22.
20. Pascoe GA, Reed DJ. Vitamin E protection against chemical-induced injury. II: Evidence for a threshold effect of cellular α-tocopherol in prevention of adriamycin toxicity. Arch Biochem Biophys 1987; 256:159–66.
21. Djerassi D, Machlin LE, Nocka C. Vitamin E: biochemical function and role in cosmetics. D & CI 1986:29–34, 78.
22. Khettab N, Amory MC, Briand G, Bousquet B, Combre A, Forlot P, Barey M. Photoprotective effect of vitamins A and E on polyamine and oxygenated free radical metabolism in hairless mouse epidermis. Biochimie 1988; 70:1709–13.

23. De Rios G, Chan JT, Black HS, Rudolph AH, Knox JM. Systemic protection by antioxidants against UVL-induced erythema. J Invest Dermatol 1978; 70: 123–5.
24. Ohzawa S, Arai M, Takeda Y. Vitamin E in topical preparations. Nippon Koshin Kakakkaishi (BJJAEV) 1984; 8:18–27.
25. Roshchupkin DI, Piszov MY, Potapenko AY. Inhibition of ultraviolet light-induced erythema by antioxidants. Arch Dermatol 1979; 266:91–4.
26. Potapenko AJ, Piszov MJ, Abijev GA, Pliquett F. α-Tocopherol, ein Inhibitor von durch UV-Strahlung induzierten Veraenderungen mechanoelektrisher Haut-eigenschaften. Dermatol Monatsschr 1983; 169:300–4.
27. Potapenko AY, Abijev GA, Pistov MY, Roshchupkin DI, Pliquett F, Ermolayev AV, Sarycheva IK, Evstigneeva RP. PUVA-induced erythema and changes in mechanoelectrical properties of skin. Inhibition by tocopherols. Arch Dermatol Res 1984; 276:12–16.
28. Moeller H, Potokar M, Wallat S. Vitamin E als Kosmetischer Wirkstoff. Parfuemerie und Kosmetik 1987; 68:688–94.
29. Moeller H, Ansmann A, Wallat S. The effects of Vitamin E on the skin in topical applications. Fett-Wissenschaft Technologie 1989; 8:295–315.
30. Trevithick JR, Xiong H, Lee S, Shum DT, Sanford SE, Karlik SJ, Norley C, Dilworth G. Topical tocopherol acetate reduces post-UVB, sunburn-associated erythema, edema, and skin sensitivity in hairless mice. Arch Biochem Biophys 1992; 296:575–582.
31. Black HS. Effects of dietary antioxidants on actinictumor prodiction. Res Commun Chem Pathol Pharmacol 1974; 7:783–6.
32. Black HS, Chan JT. Suppression of ultraviolet light-induced tumor formation by dietary antioxidants. J Invest Dermatol 1975; 65:412–14.
33. Record IR, Dreosti IE, Konstantinopoulos M, Buckley RA. The influence of topical and systemic vitamin E on ultraviolet light-induced skin damage in hairless mice. Nutr Cancer 1991; 162:219–25.
34. Ruzicka T, Przybilla B. Eicosanoids in skin UV inflammation—lack of leukotriene B_4 elevation in UVB-induced dermatitis. Photodermatology 1986; 3:306–7.
35. Kobza Black A, Barr RM, Wong E, et al. Lipoxygenase products of arachidonic acid in human inflamed skin. Br J Clin Pharmacol 1985; 20:185–90.
36. Roberts LK, Samlowski WE, Daynes RA. Immunological consequences of ultraviolet readiation exposure. Photodermatology 1986; 3:284–98.

6

The Cutaneous Antioxidant System

Donald A. Vessey
Department of Veterans Affairs Medical Center and University of California, San Francisco, California

I. INTRODUCTION

Our biology has evolved in an oxidative environment and is grounded in the utilization of the potentially dangerous oxygen molecule. As a result, all cells constantly generate a variety of reactive oxygen species which are capable of oxidatively destroying a number of cellular components. In addition, the number of oxidative risks from our external environment is proliferating, and this is particularly significant with regard to the skin because of its direct exposure to the external environment.

Oxidative damage can occur by direct chemical oxidation, or it can involve a more insidious chain reaction involving molecules that are oxidatively reactive due to the presence of a single unpaired electron, i.e., free radicals. The primary nonradical reactive oxygen species are peroxides and singlet oxygen, but other strong oxidants are also found in our environment. Free radicals frequently arise during one-electron transfers involving heavy metal ions associated with coordination complexes and proteins. Another primary free radical-generating source is UV light, which participates in a slow reaction in which hydrogen peroxide is split with the generation of the extremely reactive hydroxyl radical ($^{\cdot}OH$). Light is also able to excite endogenous and exogenous photosensitizers, which can then go on to produce free radical species. Further, the environment contains free radicals arising from combustion and also chemicals that can interact with biological systems to produce free radicals. Thus, the threat of free radical-induced oxidative damage is every present.

Biological systems are protected from the threat of oxidative assault by a diversity of mechanisms designed to suppress pernicious oxidative pathways. These antioxidant processes work by direct scavenging of the initiating prooxidant species, by sequestering of heavy metals thereby preventing initiation or propagation of free radical reactions, by scavenging secondary free radical species which terminates ongoing chain reactions, or by the return of oxidized groups to their reduced state. In addition to these approaches, cells also have mechanisms for the repair of damaged macromolecules and for enhancing cell renewal. These repair mechanisms will not be considered in this chapter.

A healthy cell is one in which the antioxidant systems effectively keep the level of prooxidants below a critical threshold. When the rate of regeneration of antioxidants is exceeded by their rate of consumption, a state of oxidative stress is induced in which cell function is compromised in a number of ways. A state of oxidative stress is characterized by both reversible and irreversible cell damage and can, with time and sufficient severity, lead to cell death.

In considering cutaneous antioxidants, separate attention needs to be given to the dermis and epidermis. Where possible, this is done in this chapter. However, the epidermis is the first line of defense and of greatest importance. Therefore, a great deal of work has been done with isolated epidermis. Also useful have been studies with cultured keratinocytes since approximately 97% of the epidermis is composed of keratinocytes at various stages of differentiation. Whole skin is also used occasionally but constitutes a very heterogeneous system.

This chapter does not deal with the relative oxidative potential of different oxidative agents nor with the relative effectiveness of different antioxidants. Such an undertaking would require a chapter in itself as the responses are frequently dependent on subcellular localization and specific prooxidant/antioxidant interactions. Further, there is too little information available on the interaction of different oxidants with skin. At this early point it is only possible to document which potential antioxidants are present in skin, in some cases to indicate their tissue localization, and finally to indicate situations in which their effectiveness has been shown to be enhanced or compromised. Hopefully a comparable version of this chapter will be written in the future when more information is available.

In this chapter I have divided the various antioxidant systems into categories. I will first discuss enzyme-catalyzed detoxification reactions. These systems have the greatest capacity for suppressing specific reactive oxygen species. Then I will discuss small molecule antioxidants (both water soluble and lipid soluble). I only consider those molecules of sufficient stability to survive long enough to be regenerated enzymatically either directly or via an intermediate redox agent. These antioxidants are generally nonspecific and are particularly valuable because of their ability to react with a wide variety of reactive species. Finally, I would like to acknowledge that the bulk of the background material for this chapter was gleaned from the general reference *Free Radicals in Biology and Medicine* by Halliwell and Gutteridge (1).

II. DETOXIFICATION ENZYMES

A. Superoxide Dismutases

Partial reduction of molecular dioxygen by a one-electron transfer gives rise to the superoxide anion ($^{\cdot}O_2^-$, referred to here just as superoxide). Superoxide arises biologically by the transfer of a single electron to molecular oxygen facilitated by photosensitive compounds or by components of the cellular electron transport chains. Superoxide production is a useful function of activated phagocytic cells, but in other cells the electrons leaking from the respiratory chain generate unintended sources of superoxide, and contribute to the constitutive oxidative stress accompanying aerobic metabolism. Superoxide per se does not appear to be extremely toxic; rather its toxicity is largely related to its ability to reduce transition metals, such as ferric ion, which then participate in the generation of the extremely reactive hydroxyl radical.

The primary pathway for the elimination of superoxide is via enzymes called superoxide dismutases (SOD, E.C. 1.15.1.1). These enzymes catalyze the reaction:

$$2H^+ + {}^{\cdot}O_2^- + {}^{\cdot}O_2^- \longrightarrow H_2O_2 + O_2 \qquad (1)$$

Since H_2O_2 can also have toxic consequences (see below), the action of these enzymes must be coupled to the action of the two principal hydrogen peroxide-metabolizing enzymes, catalase and glutathione peroxidase. It has been shown that when, through genetic engineering, the superoxide dismutase activity is made to exceed the activity of the latter two, then superoxide becomes more toxic (2).

Two immunologically distinct forms of SOD are found in animal tissues. The predominant forms of SOD are Cu^{+2}/Zn^{+2}-containing dimeric enzymes of molecular weight 31,000–33,000, which are located predominantly in the cytoplasm. There are also tetrameric Mn^{+2}-containing forms of molecular weight 80,000–90,000, which are found in mitochondria. Iron-containing forms of SOD also exist but have not yet been found in animals. Enzyme activity measured in cellular homogenates is primarily an expression of the activity of both forms. Although assay of the soluble cell fraction measures primarily the Cu,Zn forms, the Mn forms are released from damaged mitochondria and are present particularly when cells are opened by sonication, which is often done in dermatology research because of the inadequacy of homogenization. The relative amount of each form has been determined by immunoassay, selective purification, or the less convincing utilization of cyanide as a specific inhibitor of the Cu,Zn forms.

1. Constitutive Activity in Skin

SOD activity has been found in whole skin of every animal examined. The level of activity in skin is generally 5–10 times lower than in other tissues. The major portion of the activity resides with Cu,Zn-SOD but the Mn-SOD also is present (usually ca. 25% of total) in whole skin and epidermis from humans and most animals (3–5). The known exception is epidermis from newborn or adult albino CD-1 mice, both of which lack Mn-SOD (3). The iron-containing form has been looked for in skin but has not been found (3). Cu,Zn-SOD activity in human skin remains constant from birth until 30 years of age; it increases during the 30- to 39-year age bracket, and then returns to the same level from age 40–60 (5). Mn-SOD appeared to drop in the 20- to 29-year and 50- to 59-year periods (5). In contrast, for rat skin a steady decline with age has been reported (6).

SOD activity is not significantly different in dermis and epidermis (7). Cultured fibroblasts display higher activity than keratinocytes (8). Melanocytes have extremely low SOD activity (8). Studies of the subepidermal distribution of SOD activity in cow snout revealed that the activity is greatest in the lower layer of cells and decreases as the cells move to the surface (9). The SOD activity in human epidermis did not appear to differ significantly between blacks and caucasians (3) and is not significantly different in cultured keratinocytes from blacks as compared with caucasians (8). The SOD activity of adult human epidermis is much higher than in mouse or in newborn guinea pig epidermis (3).

2. Factors Alter SOD Activity

The SOD activity of skin has been found to be responsive to exogenous factors. SOD values tended to be higher from exposed areas of the skin as compared to unexposed areas of the skin in most age groups (5). When skin is exposed to oxidative stress, such as in inflammatory diseases characterized by an increase in oxygen radical production by neutrophls and in burn lesions or open wounds, there is often a compensatory increase in SOD activity (10). Such amplification has been seen in patients with mild symptomology with the following diseases: solar dermatitis, housewives' eczema, pernio, contact dermatitis, and atopic dermatitis. However, patients with severe cases of these diseases displayed normal SOD activities (10).

In hyperproliferative epidermal dermatoses, SOD activities are almost always decreased. Reduced SOD activity is a general feature of a wide range of noncutaneous cancers (11) and for skin cancers as well. Ohkuma et al. (12) found reduced SOD activity in squamous cell carcinomas and basal cell epitheliomas. Borello et al. (13) found normal SOD levels in basal cell epitheliomas, but significantly decreased SOD activity in intermediate level and squamous cell epitheliomas. In hyperproliferative epidermis of psoriasis and seborrheic keratoses, the specific activity of SOD is also reported to be decreased (12). Borello et al. (13) pointed out that a large percentage of this decrease in specific activity may be due to the accumulation of keratin and that after correction for this keratin accumulation the adjusted decreases in specific activity are small. In tape-stripping–induced hyperproliferative epidermis (12) and in phorbol ester–induced keratinocyte hyperproliferation (14,15), SOD activity is clearly decreased. The decrease in SOD activity accompanying phorbol ester treatment is associated with a decrease in the amount of mRNA coding for SOD (15). This indicates that the phorbol ester effect is due at least in part to genetic downregulation. Another consideration arises from the following facts: 1) SOD is slowly inactivated by H_2O_2 (16) and 2) H_2O_2 accumulates in phorbol ester–treated skin (17). Thus, part of the inactivation may be the result of a direct effect of H_2O_2 on SOD. Finally, part of the apparent decrease in specific activity may be artifactual (much as discussed above) due to the tremendous increases in protein concentration induced by phorbol ester treatment (18).

Epidermal hyperplasia is also induced by UVB irradiation (19). Using mouse skin, Miyachi et al. reported a UVB-induced decrease in SOD activity, which reached its lowest value 24 hours after irradiation (20). Using pig epidermis, Hashimoto et al. (21) found an initial 48-hour hypoproliferative period during which there was no decrease in SOD activity and then an ensuing 48-hour period of hyperproliferation in which SOD activity decreased markedly. It should be noted that there is no immediate direct effect of UV on the SOD activity of skin (15,21,22). Punnonen et al. (23) studied the effect of UVB on human keratinocytes in culture. They found that SOD activity was decreased by 33% after 0.5 hour of UVB radiation. This is a much faster response than noted in the studies on whole skin. These authors also studied the effect of UVA on SOD activity in keratinocytes (24) and found no effect except at extremely high intensity (15 J/cm^2), and even then there was only a 12% decrease in activity. However, the inclusion of phorbol ester in the medium greatly potentiated the effect of UVA on SOD activity. Xeroderma pigmentosum patients, who have increased sensitivity to sunlight, were found to have a constitutive level of SOD in the epidermis which is 50–75% lower than in normals (25).

B. Glutathione Peroxidases

1. Selenium-Dependent Glutathione Peroxidase

Glutathione peroxidase (EC 1.11.1.9) metabolizes both hydrogen peroxide and organic hydroperoxides. Hydrogen peroxide is formed biologically by a number of enzyme-catalyzed reactions, including those catalyzed by superoxide dismutases (see above) and by a variety of oxidases. As mentioned above, superoxide is generated in activated phagocytic cells but because of the poor membrane permeability of the superoxide anion, its dismutation product, hydrogen peroxide, is more likely to diffuse out into surrounding cells (1). Hydrogen peroxide is also formed by the autoxidation of certain compounds, many of which are formed by cytochrome P450 during hydroxylation and dealkylation. Organic hydroperoxides are common pharmaceutical agents and are also found as contaminants in pharmaceutical preparations. They are frequent environmental contaminants, and they are also formed in vivo by peroxidation of biological molecules. The toxicity of peroxide is dependent both upon the availability of oxidizable heavy metal (e.g., Fe^{+2}) with which it can react to form the highly reactive hydroxyl radical as well as a means of rereducing the heavy metal so it can be recycled in the reaction (e.g., by superoxide).

Glutathione peroxidase is a tetrameric selenoenzyme of molecular weight 84,000. The selenium atom is incorporated into each of the subunits during protein synthesis as selenocysteine. The reduction of peroxides by glutathione peroxidase utilizes reducing equivalents suppled by glutathione. The reaction is:

$$\mathrm{ROOH + 2\,GSH \longrightarrow ROH + GSSG} \tag{2}$$

$$\mathrm{H_2O_2 + 2\,GSH \longrightarrow H_2O + GSSG} \tag{3}$$

The true K_m value for GSH cannot be determined since the enzyme cannot be saturated with GSH; the limiting extrapolated K_m for GSH is approximately 130 μM with H_2O_2 and 17–30 μM with organic hydroperoxides (26). The K_m for organic hydroperoxides is generally in the range of 10–50 μM (depending upon the concentration of GSH) and for hydrogen peroxide approximately 10-fold lower (26). The catalytic reaction involves an active site selenium atom working in conjunction with an active site sulfhydryl group. The GSSG formed is reoxidized by the enzyme glutathione reductase (see below). Glutathione peroxidase is found in most tissues and is present in the cytoplasmic compartment and to a lesser extent in the mitochondrial matrix space. The enzyme is generally thought to be responsible for the metabolism of the bulk of the extra peroxisomal hydrogen peroxide in the cell. Catalase (see below) is mostly responsible for intraperoxisomal hydrogen peroxide metabolism.

a. Constitutive Activity in Skin Glutathione peroxidase activity has been found in both whole skin and in keratinocytes of every animal investigated. The specific activities reported for skin are generally lower than in liver but are comparable to heart, brain, and lung. The specific activity is more than twice as high in epidermis as compared with dermis (27,28). We have found that cultured human keratinocytes have higher levels of both glutathione peroxidase and glutathione reductase in nonprolierative differentiated keratinocytes as compared with proliferating undifferentiated basal cells (D. A. Vessey and K-H. Lee, unpublished). Glutathione peroxidase activity is higher in fibroblasts and keratinocytes than in melanocytes and is not significantly different in melanocytes from blacks and caucasians (8).

b. Factors That Affect Glutathione Peroxidase Activity The level of glutathione peroxidase in skin cells has been found to vary in response to oxidative stress induced by tumor promoters. Treatment with 12-O-tetradecanoylphorbol-13-acetate or mezerein results in a biphasic effect on enzyme activity (29,30). There is a brief initial increase in activity, which is followed by a decline in activity to levels significantly below that of untreated controls. This decline can be eliminated by a variety of agents (30), of which the most effective single agent was selenocystamine. Sodium selenite was almost as effective. These effects may be related to the fact that the enzyme contains selenium, although since selenium cannot be incorporated into the enzyme posttranslationally, the basis for this protection is not clear. GSH, and to a lesser extent cysteine, are also effective in limiting inactivation. The effects of these two groups of agents were additive, and the best protection was offered by a combination of GSH and selenocystamine. Similarly, treatment of hairless mice with the photosensitizer 8-methoxypsoralen in combination with UVA radiation results in the same biphasic effect on glutathione peroxidase activity (27). This biphasic effect on activity is shared equally between dermis and epidermis (27).

The tumor-promoting activity of phorbol esters can be blocked by garlic and onion oils. Interestingly, these oils also produce a significant increase in glutathione peroxidase activity in isolated epidermal cells, and this increase is long lasting (31). These oils are also able to block the inhibitory effect of phorbol esters on enzyme activity.

Because glutathione peroxidase is selenium dependent, it is expected that activity would also be reduced in selenium-deficient animals. Although this has been demonstrated in a number of tissues, it has not yet been documented for skin. Glutathione peroxidase activity has been investigated in a variety of epidermal tumors and found not to be significantly different from that of normal tissue (13).

The efficient functioning of the glutathione peroxidase reaction is dependent upon the constant supply of reduced glutathione (GSH). The concentration of GSH in cultured keratinocytes is approximately 1–2 mM, but during an oxidative stress this level can fall dramatically, thereby considerably reducing the rate of metabolism of peroxides (31a). Anything that causes a depletion of the cellular glutathione pool would be expected to curtail the effectiveness of the glutathione peroxidase. Agents known to deplete the glutathione pool in keratinocytes are discussed in Section III.B. The oxidized glutathione (GSSG) formed in the peroxidase reaction needs to be reduced back to GSH in order to ensure a constant supply of reducing equivalent. This is accomplished by the action of the enzyme glutathione reductase. Glutathione reductase levels are nearly as high in skin as they are in liver, and the enzyme is found in both the dermis and the epidermis (27,28). The reducing equivalents for the glutathione reductase reaction are in turn provided by reduced nicotinamide adenine dinucleotide phosphate (NADPH). Thus, adequate recycling of GSSG is ultimately dependent upon the cell's ability to provide NADPH. NADPH is generated directly in the cytoplasmic compartment by the pentose phosphate shunt. Skin has been shown to have a highly active pentose phosphate shunt (32). The rate-controlling enzyme for this pathway, and one of the sources of NADP reduction, is glucose-6-phosphate dehydrogenase (G6PDH), which catalyzes the first step of the pathway. Another source of NADPH is from mitochondria via the isocitrate shuttle system in which NADPH is generated in the cytoplasmic compartment by the action of the enzyme isocitrate dehydrogenase (ICDH). A third source is via the mitochondrial malate shuttle with cytoplasmic NADPH being generated by the action of malic enzyme. In liver, ICDH is considered to be the most important source of cytoplasmic-reducing equivalents, although there are situations in which the pentose phosphate shunt might be more important (33). We have found both ICDH and G6PDH activity in cultured human keratinocytes (31a). However, in contrast to liver, the G6PDH activity is higher than ICDH activity in keratinocytes.

Glutathione reductase has been shown to be inactivated by N,N-bis(2-chloroethyl)-N-nitrosourea (BCNU), and in vivo treatment with BCNU inactivates both cytoplasmic and mitochondrial glutathione reductase in liver (33). We have found that treatment of cultured human keratinocytes with 100 μM BCNU for 45 minutes leads to complete inactivation of glutathione reductase (34). Glutathione reductase activity is also reduced by treatment of skin with a combination of UVA and psoralen (28). UVA and UVB alone were without effect on the activity of the enzyme in this study, but a modest decrease in reductase activity was noted with UVB treatment in the study of Fuchs et al. (22).

2. Other Glutathione Peroxidases

A second selenium-containing glutathione peroxidase, a monomeric enzyme of molecular weight 22,000 (35), has been identified in a number of tissues (36). It appears to be specific for phospholipid hydroperoxides and has been termed phospholipid hydroperoxide glutathione peroxidase. Activity is low and is very hard to detect in most tissues. The enzyme has not yet been found in skin.

Glutathione peroxidase activity is also associated with certain isoforms of the enzyme glutathione S-transferase. This activity is, however, not manifested by the pi form of the enzyme and only the pi form is found in keratinocytes (37). Alpha forms of the enzyme, which do have organic peroxidase activity, have been reported for whole skin (38) but must have originated from the dermis, perhaps from fibroblasts.

C. Catalase

Catalase (EC 1.11.1.6) catalyzes the metabolism of hydrogen peroxide. The reaction is:

$$2\,H_2O_2 \longrightarrow 2\,H_2O + O_2 \qquad (4)$$

In the presence of an adequate level of hydrogen peroxide, the enzyme can also catalyze the reaction:

$$RCH_2OH + H_2O_2 \longrightarrow RCHO + 2\,H_2O \qquad (5)$$

Most aerobic cells contain catalase. Cellular catalase activity appears to be located predominantly, if not strictly, inside of peroxisomes. Peroxisomes are the site of a number of oxidases which generate hydrogen peroxide as an end product. Catalase is therefore important for intraperoxisomal hydrogen peroxide metabolism. Because of its location and extremely high K_m for hydrogen peroxide, catalase would not be expected to contribute to the metabolism of extraperoxisomally-generated hydrogen peroxide unless cytoplasmic concentrations were high, as would result if the glutathione peroxidase system were overwhelmed.

1. Constitutive Activity

Catalase is found in the skin of all animals. The specific activity values reported for skin are considerably lower than the values for liver but are comparable to brain and heart. In cultured human cells, catalase activity is twice as high in fibroblasts than in keratinocytes, which in turn have twice the activity of melanocytes (8). Interestingly, there was no difference in catalase activity between melanocytes from blacks and caucasians (8). Muglia et al. (39) have found that the catalase activity in melanocytes is so low as to be inconsequential in the defense of melanocytes against exogenous H_2O_2. Reiners et al. have developed a technique for obtaining subpopulations of keratinocytes of various stages of differentiation from trypsinized epidermis (40). They found catalase activity to decrease in the following order: granular cells > spinous cells > basal cells. This is, however, inconsistent with studies of cultured human keratinocytes in which enzyme activity was higher in preconfluent cells grown in low calcium than in postconfluent differentiated cells grown in high calcium (41).

That catalase activity can be important in oxidatively stressed skin has been shown by Amstad et al. (2), who constructed transfectants of mouse epidermal cell lines with increased complements of either SOD or catalase or both enzymes. When these cells were challenged with high levels of externally generated superoxide, the cells with the high SOD activity had poor viability. However, this poor viability was not seen in the cells which also had increased catalase activity. Glutathione peroxidase levels were constant across the cell lines.

2. *Factors Which Alter Catalase Activity*

Catalase activity has been found to be lower in cultured human tumor cells than in normal cultured human keratinocytes (42). Further, the catalase deficiency appears to be related to the intensity of tumor phenotype expression (42). Also, catalase activity is markedly reduced in skin biopsies from xeroderma pigmentosum (XP) patients and in cultured XP cell strains (43).

Because of its peroxisomal localization, it is expected that the cellular content of catalase would be increased by agents that induce peroxisomal proliferation. The HMG-CoA reductase inhibitors lovastatin and fluindostatin were found to induce peroxisome proliferation in cultured human keratinocytes (41). The increased content of peroxisomes was accompanied in each case by significant increase in catalase activity. Lovastatin was able to increase the catalase activity in both preconfluent and postconfluent cell cultures.

Catalase activity is severely decreased in photooxidative stress (22–24). Fuchs et al. (22) found an immediate reduction in catalase activity in mouse skin following a large single fluence of UVB radiation. Punnonen et al. found that irradiation of keratinocytes with either UVA (24) or UVB (23) radiation also produced an immediate decrease in catalase activity. Activity had returned to normal in the UVA study by 24 hours postirradiation. As expected, higher intensities of UVA were required to give the same inhibition as seen in the UVB studies. Catalase activity in keratinocytes can be irreversibly inhibited in a time-dependent process by the mechanism-based inhibitor aminotriazole (31a).

Phorbol ester treatment also results in a decrease in the specific activity of catalase (14,40). Birnboim et al. (18) contend that these decreases are largely artifactual, being the result of a selective increase in the rate of accumulation of a noncatalase protein. However, Crawford et al. (15) have shown that there is an actual decrease in the amount of catalase mRNA during phorbol ester treatment and argue that this is part of the same genetic downregulation encompassing SOD.

D. NAD(P)H:Quinone Reductase

Cellular electron transfer flavoproteins are known to catalyze the one-electron reduction of quinones. The resulting semiquinone radicals are potentially dangerous because they are able to pass this one electron to O_2 to form superoxide. Yet, some of these reductions are necessary, e.g., the mitochondrial reduction of ubiquinone (CoQ) as part of the normal electron transfer chain. However, for xenobiotic quinones, the resulting semiquinone radicals are sufficiently persistent to move about in the cell, and in addition to their ability to generate superoxide, they can catalyze the oxidation of sulfhydryl groups on proteins. Maintaining the cellular pool of xenobiotic quinones in their two-electron reduced hydroquinone form decreases the risk of these toxic reactions. The enzyme that catalyzes this two-electron reduction of xenobiotic quinones is called NAD(P)H:quinone reductase (NQR) or historically DT-diaphorase.

1. Constitutive Activity in Skin

NQR activity in epidermis is surprisingly high, of the same magnitude as in liver (44). This high level implies an important role for this enzyme. In human skin and in keratinocytes growing in culture, this activity is also very high. For cultured keratinocytes, activity is much higher in proliferating cells than in differentiated cells (45).

2. Factors Affecting NQR Activity

Studies in rat skin revealed that NQR activity was inducible by a number of agents (46). Thus, 3-methylcholanthrene, Aroclor 1254, and coal tar preparations (46) increased NQR activity by up to 385% in rat and mouse epidermis. Extension of this work (45) in murine skin and cultured human keratinocytes revealed that NQR is a member of the aromatic hydrocarbon-responsive Ah-gene battery and is thus induced by compounds such as 3-methyl cholanthrene and benzanthracene. Cutaneous NQR is also induced by butylated hydroxyanisole, which does not bind to the Ah receptor (45). This indicates that NQR can be induced by two different mechanisms.

Alternatively, NQR activity in skin can be inhibited by a number of agents. In both mouse skin and human keratinocytes, NQR activity was extensively inhibited by treatment with dicoumarol, indomethacin, or rutin (45). Also, the antipsoriatic drug anthralin leads to short-term inhibition of NQR activity, but over the long term it is metabolized to danthron, which is an inducer of NQR activity (44).

E. Thioprotein Reductases

Epidermis contains another free radical scavenging system(s) in addition to SOD. This system has been found to selectively reduce a number of free radical species, such as nitroxide and superoxide radicals (47,48). This system is inhibited by sulfhydryl reagents and can be heat denaturated. It is thus presumed to be a thioprotein. One thiol-sensitive free radical–reducing enzyme known to be present in the cytoplasm of keratinocytes and on the cell surface is thioredoxin reductase (48). Aside from its important role in DNA synthesis, this enzyme has been shown to reduce certain spin-labeled quaternary ammonium compounds (spin-labeled quat). Reduction of spin-labeled quat at the cell surface is inhibited by known inhibitors of thioredoxin reductase, i.e., sulfhydryl reagents and the alternate substrate thioredoxin. Thus, thioredoxin reductase has been posited as the primary free radical scavenger at the skin surface (48). Fuchs et al. (47) have developed a cationic nitroxide free radical (Prox X1), which is not reduced by thioredoxin reductase, and find that it is still reduced in skin biopsies and further that this reduction is inhibited by sulfhydryl reagents and heat denaturation. They attribute this activity to an alternate thiol-dependent mechanism unrelated to thioredoxin reductase.

Intracellular nitroxide reduction could also involve the enzyme thioredoxin reductase and/or the disulfides of thioredoxin. However, nearly complete inhibition of nitroxide reduction in skin homogenates can be obtained by treatment with ascorbate oxidase, which suggests that ascorbate is of greater importance (47).

The thioredoxin reductase/thioredoxin system has been implicated in a number of other antioxidant reactions (48). These other reactions are the reduction of methionine sulfoxide residues in proteins to methionine, the reduction of sulfite to sulfide, the reduction of insulin, the reduction of vitamin K epoxide via vitamin K epoxide reductase, the regulation of melanin biosynthesis through the inhibition of tyrosinase by reduced thioredoxin, the reactivation of hydrogen peroxide–inactivated 3-phosphoglyceraldehyde dehydrogenase by reduced thioredoxin, and activation of methionine sulfoxide peptide reductase.

Immunohistochemical staining for thioredoxin reductase in skin locates the enzyme only in the stratum germinativum and melanocytes, Langerhans cells, hair follicles, and sweat glands (49). Study of cultured human keratinocytes revealed that cultures of adult stratified cells have twice the activity of neonatal cells (50). The activity of the enzyme is inhibited by calcium, and this inhibition can be seen both at the cell surface and intracellularly in cultured keratinocytes grown in a high concentration of calcium. Punch biopsies of psoriatic skin revealed a significantly higher level of thioredoxin reductase (51).

The nitroxide-scavenging capacity of keratinocytes is inhibited by both UVA and UVB light, and this has been proposed to be the result of the inhibition of thioredoxin reductase (52). However, Fuchs et al. (53) contend that thioredoxin reductase is not reduced by UVA and UVB and argue that this is in fact evidence for an alternate system.

III. WATER-SOLUBLE ANTIOXIDANTS

A. Ascorbate

Ascorbic acid has a strong reducing potential, which enables it to take part in both one-electron and two-electron processes. Ascorbate has been shown to scavenge singlet oxygen and react rapidly with superoxide, hydroxyl radical, and hydrogen peroxide. Thus, in most instances ascorbate is considered a primary antioxidant. Ascorbate has also been shown to reduce the vitamin E free radical back to vitamin E and thus to assist vitamin E (see below) in scavenging membrane-bound free radicals.

The reaction of ascorbate with free radicals gives rise to the semidehydroascorbate radical:

$$\text{Ascorbate} + \text{R}^{\cdot} \longrightarrow \text{Semidehydroascorbate}^{\cdot} + \text{RH} \qquad (6)$$

Semidehydroascorbate is not highly reactive and survives long enough to be enzymatically reduced back to ascorbate. Both NADH- and GSH-utilizing ascorbate reductases have been found in mammalian tissues. Semidehydroascorbate can also react with itself nonenzymatically in a disproportion reaction:

$$2\,\text{Semidehydroascorbate}^{\cdot} \longrightarrow \text{Dehydroascorbate} + \text{ascorbate} \qquad (7)$$

Dehydroascorbate can be enzymatically reduced back to ascorbate or, due to its instability, it can decompose.

Ascorbate has also been found to function as a prooxidant in a variety of systems. However, the prooxidant activity of ascorbate acid is almost always dependent upon the existence of free metal ions (62), and in most cells the level of free metal ion is kept relatively low (54). Thus, intracellular ascorbic acid can be expected to function predominantly as an antioxidant.

The level of ascorbate plus dehydroascorbate in whole skin is substantial, being approximately 6 nmol/mg protein (22). A critical role for ascorbate as a skin antioxidant is indicated by the finding of nearly complete elimination of nitroxide reduction in skin homogenates by addition of ascorbate oxidase (47). Further, Fuchs et al. consider that ascorbate may play an important role in the reduction of free radicals at the surface of keratinocytes (47). Treatment of whole mouse skin with UVB radiation led to a small decrease in the level of ascorbate as measured by a nonspecific assay (22). Darr et al. topically treated pig skin with ascorbate to achieve levels that could be directly measured and found that UVB treatment led to a 70% decrease from the supplemented level (55). They also found that the ascorbate provided some protection against UVB and excellent protection against UVA.

B. Glutathione

As discussed above, glutathione plays a major role in suppression of oxidative stress by serving as a cofactor for the enzyme glutathione peroxidase. However, reduced glutathione (GSH) also functions as an independent small molecule antioxidant. GSH is a tripeptide (γ-Glu-Cys-Gly), and it is the -SH group on the cysteine that serves as an antioxidant. Glutathione is the predominant soluble intracellular thiol. GSH, and many sulfhydryl-containing compounds in general, are capable of the following reactions. The first is the reaction of GSH with free radicals:

$$R^{\cdot} + GSH \longrightarrow {}^{\cdot}GS^{-} + RH \qquad (8)$$

Among the free radicals GSH has been found to react with is superoxide. A major question concerning this reaction is the fate of the glutathionyl radical formed in the reaction. It is ultimately reduced back to GSH, but the reaction mechanism is in doubt. The simplest pathway is by the direct dimerization of two glutathionyl radicals to form GSSG. Since GSSG does appear to be the ultimate end product, it appears more likely that an oxygen-dependent pathway is involved (56).

A second important reaction of GSH is the reduction of protein sulfhydryl groups that have become oxidized:

$$RSSR' + GSH \longrightarrow RSSG + R'SH \qquad (9)$$

$$RSSG + GSH \longrightarrow GSSG + RSH \qquad (10)$$

Oxidation of protein thiols has been correlated with increased toxicity of several agents (57,58). In both of these processes, the GSSG formed is reduced back to GSH by the action of the enzyme GSH reductase (see page 88). In the case of a severe oxidative stress, GSSG can accumulate and reverse the process, i.e., it blocks protein sulfhydryl groups (reverse of reaction

[10]). This has been shown to lead to inhibition of enzymes such as phosphofructokinase (59).

Finally, GSH is able to inactivate singlet oxygen (1). Other thiols, for example cysteine, are equally effective. Methionine is also a very effective scavenger for singlet oxygen.

The concentration of glutathione in skin is as high as that in most other tissues, but lower than in liver. In the epidermis the concentration of glutathione can be estimated to be approximately 1 mM (27). The concentration of GSH is twice as high in epidermis as compared with dermis. On the other hand, the concentration of GSSG is fourfold higher in dermis than in the epidermis (27). This indicates that the epidermis is more proficient at maintaining GSH in the reduced state. We have noted that in cultured human keratinocytes the concentration of glutathione is higher in postconfluent cultures containing differentiated cells than in preconfluent cultures with undifferentiated cells (D. A. Vessey and K-H. Lee, unpublished). We take this as an indication that basal cells have a lower concentration of glutathione than cells in the intermediate layers of the epidermis.

A variety of treatments that generate an oxidative stress have been found to reduce the level of total glutathione and/or dramatically decrease the ratio of GSH/GSSG. Treatment of mouse skin with UVA or UVB produces a decrease in GSH (22,27) which is enhanced by psoralen (27). Treatment of mouse epidermal cells with the tumor promoters mezerein or 12-O-tetradecanoylphorbol-13-acetate (TPA) leads to a marked increase in GSSG from approximately 5% of the total glutathione to 20% (29,60). A number of other phorbol esters were less effective at increasing GSSG and less effective as tumor promoters (29). This increase in GSSG could be largely prevented by a variety of thiols such as cysteine and glutathione esters (60,61). Since these agents all led to an increase in the intracellular GSH concentration, it is not clear whether they block GSSG accumulation due to an increase in the cellular concentration of GSH or via a nonspecific thiol effect. These studies also revealed that the cellular concentration of GSH in unstressed cells can be increased more than twofold by a variety of thiol delivery agents, e.g., glu + gly + cyscys > glu + gly + cys > oxothiazolidine carboxylate > glutathione esters = N-acetylcysteine (60,61). Buthionine sulfoxime (BSO), a known inhibitor of glutathione biosynthesis, has also been found to be an effective inhibitor in cultured human keratinocytes. Treatment for 20 hours with 4 mM BSO produced a 90% depletion of total glutathione in less than 24 hours (34).

C. Uric Acid

Uric acid exists as an end product of the degradation of purines. However, it has been found to effectively scavenge singlet oxygen, ozone, and free radicals. In fact, uric acid is considered by some investigators to be the primary free radical scavenger in plasma (62). Uric acid also has the capacity to tightly bind iron and copper ions and block their free radical–initiating activity. This has led to the consideration of uric acid as a major antioxidant. The level of uric acid in plasma is quite high, and one would expect skin to reflect this. Indeed, uric acid is readily detectable in skin, and its concentration in guinea pig epidermis is approximately 15 nmol/mg protein (63). In psoriasis, serum uric acid levels are frequently elevated (64), and this is presumably indicative of higher skin levels.

IV. LIPID-SOLUBLE ANTIOXIDANTS

A. Tocopherols

Tocopherols (vitamin E) are the primary membrane-bound antioxidants. Tocopherols are lipid soluble and are found in the lipid phase of biological membranes, where they are oriented with the redox active chromanol moiety located near the bilayer surface (65). As such, tocopherols are primarily responsible for suppressing lipid free radicals and are thus important for preventing lipid peroxidation. The predominant tocopherol in mammalian tissues and also the most effective free radical scavenger is α-tocopherol (generally a mixture of both the d- and l-optical isomers). Tocopherols possess an -OH group attached to the aromatic ring, which will readily donate a hydrogen atom to a radical:

$$\text{Toc-OH} + \text{R}^{\cdot} \longrightarrow \text{Toc-O}^{\cdot} + \text{RH} \tag{11}$$

Because tocopherol reacts with free radical products following their formation by some external process, tocopherol is termed a chain-breaking antioxidant. The tocopheryl radical is stable, relatively unreactive, and it survives long enough to be reduced back to tocopherol. This is primarily accomplished (1,62) in a chemical reaction with ascorbic acid:

$$\text{Toc-O}^{\cdot} + \text{ascorbate} \longrightarrow \text{Toc-OH} + \text{semidehydroascorbate radical} \tag{12}$$

In ascorbate deficiency, tissue tocopherol levels decline. Maguire et al. (65a) have proposed that reduction of the α-tocopheroxyl radical occurs via the transmembrane electron transport systems with ubiquinone as reductant.

α-Tocopherol is the major lipid-soluble antioxidant in mouse skin (22). In human skin it is present at this same level but CoQ levels (see below) are equally high (66). Tocopherol levels are reported to be 3- to 10-fold higher in human basal cell epitheliomas than in normal epidermis (66).

Tocopherol levels in the skin of hairless mice are decreased immediately after UVB irradiation (22). In a study of long-term UV exposure in mouse skin, the total lipid-soluble antioxidant capacity (estimated as the amount of product formed from reaction with 1,1-diphenyl-2-picrylhydrazyl) is reported to decrease with chronic (up to 24 weeks) UV exposure (67). Study of human skin from exposed and unexposed areas revealed two- to threefold higher tocopherol and CoQ levels in exposed skin (67).

B. Coenzyme Q

Coenzyme Q (ubiquinone + ubiquinol) is a lipid-soluble redox molecule whose main function is as a redox component of membrane electron transport systems such as the mitochondrial respiratory chain. Ubiquinone can can be two-electron reduced at the quinoid nucleus to form ubiquinol. Ubiquinol, in contrast to ubiquinone, has also been shown to be as effective as α-tocopherol in preventing lipid peroxidation (68). Ubiquinol has been found to work synergistically with α-tocopherol in preventing lipid peroxidation. It is not clear if this is due to the different locations of the two in the membrane or to recycling of the tocopherol radical by ubiquinol (see above).

Skin contains significant levels of ubiquinone and ubiquinol, with over 70% of the CoQ in the ubiquinone form (22). Mouse and rat skin have been reported to contain ubiquinone-9 (9 isoprenoid units), with the levels being almost as high as for α-tocopherol (22,69). Human epidermis contains approximately equivalent concentrations of coenzyme Q (66). However, it is in the form of ubiquinone-10 (10 isoprenoid units) with much higher levels in skin from exposed areas as compared with unexposed areas (66).

When mouse skin was treated with UVB radiation, there was an immediate and dramatic decrease (ca. 50%) in both ubiquinone and ubiquinol; the ratio of ubiquinone to ubiquinol did not change appreciably (22). Topically applied coenzyme Q has been found to be taken up by skin (69) and may be an effective agent in the prevention of UV damage to skin.

C. β-Carotene

β-Carotene is a carotenoid pigment and a bioprecursor of vitamin A. It is also a known quencher of singlet oxygen and has been proposed as a radical trapping agent and a general antioxidant (70). Interestingly, the antioxidant properties of β-carotene are significant only at relatively low oxygen tension (<150 torr). Thus, it would not be expected to be very effective in the outer layers of the epidermis but might be effective near the basal layer. There is as yet no evidence for a role for β-carotene as a skin antioxidant and indeed it was found to be only minimally effective in preventing sunlight-induced erythema in normal human skin (71).

V. MELANINS

Melanin exists in keratinocytes in two forms. In the basal and suprabasal layers melanin exists in a granular form inside of melanosomes. In the prickle cell layer melanosomes degranulate and melanin is deposited in a colloidal form. There are two types of melanin, pheomelanin and eumelanin. Pheomelanin is a red or yellow pigment found in the skin of red-headed individuals. Eumelanin is the brown or black photoprotective agent in skin. Eumelanin is able to protect skin from UV radiation by absorbing the radiation and transforming it into heat energy (72,73). Further, eumelanin is able to scavenge a variety of free radicals including superoxide (which is commonly produced in photoreactions) (72,74,75). Eumelanin is very effective in photoprotection of skin because its interactions with light and radicals are reversible and nondestructive. Pheomelanin, on the other hand, is readily photodegraded with the production of superoxide (76) and thus is not a photoprotective agent.

REFERENCES

1. Halliwell B, Gutteridge JMC. Free radicals in biology and medicine. 2d ed. Oxford: Clarendon Press, 1989.
2. Amstad P, Peskin A, Shah G, Mirault M, Moret R, Zbinden I, Cerutti P. The balance between Cu,Zn-superoxide dismutase and catalase affects the sensitivity of mouse epidermal cells to oxidative stress. Biochemistry 1991; 30:9305–13.
3. Carraro C, Pathak MA. Characterization of superoxide dismutase from mammalian skin epidermis. J Invest Dermatol 1988; 90:31–6.
4. Ohkuma N, Matsuo S, Iizuka H, Ohkawara A. Superoxide dismutase in epidermis. J Dermatol (Tokyo) 1987; 14:218–23.
5. Ueda H, Sugiura K. Lipid peroxide and superoxide dismutase in human skin. In: Takase Y, Kligman AM, ed. Cutaneous aging. Tokyo: University of Tokyo Press, 1988:449–57.
6. Im MJ, Hooper JE. Age-dependent decreases in superoxide dismutase activity in rat skin. J Invest Dermatol 1984; 82:437A.
7. Kim YP, Lee SC. Superoxide dismutase activities in the human skin. In: Hayaishi O, Imamura S, Miyachi Y, ed. The biological role of reactive oxygen species in skin. Tokyo: University of Tokyo Press and New York: Elsevier, 1987: 225–30.
8. Yohn JJ, Norris DA, Yrastorza DG, Buno IJ, Leff JA, Hake SS, Repine JE. Disparate antioxidant enzyme activities in cultured human cutaneous fibroblasts, keratinocytes, and melanocytes. J Invest Dermatol 1991; 97:405–9.
9. Ogura R, Ueta H, Sugiyama M, Hidaka T. Distribution of superoxide dismutase activity in the epidermis: measurement with electron spin resonance spin trapping. J Invest Dermatol 1990; 94:227–9.

10. Niwa Y. Oxygen radical-related inflammatory diseases. In: Hayaishi O, Imamura S, Miyachi Y, ed. The biological role of reactive oxygen species in skin. Tokyo: University of Tokyo Press and New York: Elsevier, 1987:151–61.
11. Oberley LW, Buettner GR. The role of superoxide dismutase in cancer: a review. Cancer Res 1979; 39:1141–9.
12. Ohkuma N, Kajita S, Iizuka H. Superoxide dismutase in epidermis: its relation to keratinocyte proliferation. J Dermatol (Tokyo) 1987; 14:562–8.
13. Borello S, Seccia A, Galeotti T, Bartoli GM, Farallo, Serri, F. Protective enzymes in human epidermal carcinomas and psoriasis. Arch Dermatol Res 1984; 276:338–40.
14. Solanki V, Rana RS, Slaga TJ. Diminution of mouse epidermal superoxide dismutase and catalase activities by tumor promoters. Carcinogenesis 1981; 2: 1141–6.
15. Crawford DR, Amstad PA, Yin Foo DD, Cerutti PA. Constitutive and phorbol-myristate-acetate regulated antioxidant defense of mouse epidermal JB6 cells. Mol Carcinog 1989; 2:136–43.
16. Pigeolet E, Corbisier P, Houbion A, Lambert D, Michiels C, Raes M, Zachary M, Remacle J. Glutathione peroxidase, superoxide dismutase, and catalase inactivation by peroxides and oxygen derived free radicals. Mech Ageing Devel 1990; 51:283–97.
17. Perchellet EM, Abney NL, Perchellet J-P. Stimulation of hydroperoxide generation in mouse skins treated with tumor-promoting or carcinogenic agents in vivo and in vitro. Cancer Lett 1988; 42:169–77.
18. Birnboim HC, Morrison DP, Joyce TL. Measurement of enzyme activities in mouse epidermis following phorbol ester treatment: a potential artifact. Carcinogenesis 1986; 7:495–7.
19. Epstein JH, Fukuyama K, Fye K. Effects of ultraviolet radiation on the mitotic cycle and DNA, RNA and protein synthesis in mammalian epidermis in vivo. Photochem Photobiol 1970; 12:57–65.
20. Miyachi Y, Imamura S, Niwa Y. Decreased skin superoxide dismutase activity by a single exposure of ultraviolet radiation is reduced by liposomal superoxide dismutase pretreatment. J Invest Dermatol 1987; 89:111–12.
21. Hashimoto Y, Ohkuma N, Iizuka H. Reduced superoxide dismutase activity in UVB-induced hyperproliferative pig epidermis. Arch Dermatol Res 1991; 283: 317–20.
22. Fuchs J, Huflejt ME, Rothfuss LM, Wilson DS, Carcamo G, Packer L. Impairment of enzymic and nonenzymic antioxidants in skin by UVB irradiation. J Invest Dermatol 1989; 93:769–73.
23. Punnonen K, Puntala A, Jansen CT, Ahotupa M. UVB irradiation induces lipid peroxidation and reduces antioxidant enzyme activities in human keratinocytes in vitro. Acta Derm Venereol (Stockh) 1991; 71:239–73.
24. Punnonen K, Jansen CT, Puntala A, Ahotopa M. Effects of in vitro UVA irradiation and PUVA treatment on membrane fatty acids and activities of antioxidant enzymes in human keratinocytes. J Invest Dermatol 1991; 96:255–9.

25. Schallreuter KU, Pittelkow MR, Wood JM. Defects in antioxidant defense and calcium transport in the epidermis of xeroderma pigmentosum patients. Arch Dermatol Res 1991; 283:449–55.
26. Wendel A. Glutathione Peroxidase. In: Jakoby WB, ed. Enzymatic basis of detoxification. Vol I. New York: Academic Press, 1980:333–53.
27. Wheeler LA, Aswad A, Connor MJ, Lowe N. Depletion of cutaneous glutathione and the induction of inflammation by 8-methoxypsoralen plus UVA radiation. J Invest Dermatol 1986; 87:658–62.
28. Connor MJ, Wheeler LA. Depletion of cutaneous glutathione by ultraviolet radiation. Photochem Photobiol 1987; 46:239–45.
29. Perchellet J-P, Perchellet EM, Orten DK. Schneider BA. Decreased ratio of reduced/oxidized glutathione in mouse epidermal cells treated with tumor promoters. Carcinogenesis 1986; 7:503–6.
30. Perchellet J-P, Abney NL, Thomas RM, Guislain YL, Perchellet EM. Effects of combined treatments with selenium, glutathione, and vitamin E on glutathione peroxidase activity, ornithine decarboxylase induction, and complete and multistage carcinogenesis in mouse skin. Cancer Res 1987; 47:477–85.
31. Perchellet J-P, Perchellet EM, Abney NL, Zirnstein JA, Belman S. Effects of garlic and onion oils on glutathione peroxidase activity, the ratio of reduced/oxidized glutathione and ornithine decarboxylase induction in isolated mouse epidermal cells treated with tumor promoters. Cancer Biochem Biophys 1986; 8:299–312.
31a. Vessey DA, Lee K-H. Inactivation of enzymes of the glutathione antioxidant system by treatment of cultured human keratinocytes with peroxides. J Invest Dermatol 1993; In press.
32. Nguyen DT, Keast D. Energy metabolism and the skin. Int J Biochem 1991; 23:1175–83.
33. Reed DJ. Regulation of reductive processes by glutathione. Biochem Pharmacol 1986; 35:7–13.
34. Vessey DA, Lee K-H, Blacker KL. Characterization of the oxidative stress initiated in cultured human keratinocytes by treatment with peroxides. J Invest Dermatol 1992; In press.
35. Duan Y-J, Komura S, Fiszer-Szafarz B, Szafarz D, Yagi K. Purification and characterization of a novel monomeric glutathione peroxidase from rat liver. J Biol Chem 1988; 263:19003–8.
36. Zhang L, Maiorino M, Roveri A, Ursini F. Phospholipid hydroperoxide glutathione peroxidase: specific activity in tissues of rats of different age and comparison with other glutathione peroxidases. Biochim Biophys Acta 1989; 1006:140–3.
37. Blacker KL, Olson E, Vessey DA, Boyer TD. Characterization of glutathione S-transferase in cultured human keratinocytes. J Invest Dermatol 1991; 97:442–6.
38. Del Boccio G, DiIlio C, Alin P, Jornvall H, Mannervik B. Identification of a novel glutathione transferase in human skin homologous with class alpha glutathione transferase 2-2 in the rat. Biochem J 1987; 244:21–5.
39. Muglia JJ, Tonnesen MG, Osborn RL, Kissinger RM, Norris DA. Ineffective antioxidant defenses in human melanocytes. J Invest Dermatol 1986; 86:496A.

40. Reiners Jr JJ, Hale MA, Cantu AR. Distribution of catalase and its modulation by 12-O-tetradecanoylphorbol-13-acetate in murine epidermis and subpopulations of keratinocytes differing in their stages of differentiation. Carcinogenesis 1988; 9:1259–63.
41. Williams ML, Menon GK, Hanley KP. HMG-CoA reductase inhibitors perturb fatty acid metabolism and induce peroxisomes in keratinocytes. J Lipid Res 1992; 33:193–208.
42. Rabilloud T, Asselineau D, Miquel C, Calvayrac R, Darmon M, Vuillaume M. Deficiency in catalase activity correlates with the appearance of tumor phenotype in human keratinocytes. Int J Cancer 1990; 45:952–6.
43. Vuillaume M, Calvayrac R, Best-Belpomme M, Tarroux P, Hubert M, Decroix Y, Sarasin A. Deficiency in the catalase activity of xeroderma pigmentosum cell and simian virus 40-transformed human cell extracts. Cancer Res 1986; 46:538–44.
44. Bonnekoh B, Jugert F, Merk H, Mahrle G. NAD(P)H:quinonereductase (NQR) in human keratinocytes: induction by B-naphtoflavone (B-NF), tert-butylhydroxyanisole (BHA), danthron (D) and inhibition by anthralin (A). J Invest Dermatol 1989; 92:406A.
45. Merk H, Jugert F, Bonnekoh B Mahrle G. Induction and inhibition of NAD(P)H: quinone reductase in murine and human skin. Skin Pharmacol 1991; 4:183–90.
46. Khan WA, Das M, Stick S, Javed S, Bickers DR, Mukhtar H. Induction of epidermal NAD(P)H:quinone reductase by chemical carcinogens: a possible mechanism for the detoxification. Biochem Biophys Res Commun 1987; 146:126–33.
47. Fuchs J, Mehlhorn RJ, Packer L. Free radical reduction mechanisms in mouse epidermis skin homogenates. J Invest Dermatol 1989; 93:633–40.
48. Schallreuter KU, Wood JM. Free radical reduction in the human epidermis. Free Radical Biol Med 1989; 6:519–32.
49. Rozell B, Hansson HA. Immunohistochemical localization of thioredoxin and thioredoxin reductase in adult rats. Eur J Cell Biol 1985; 38:79–86.
50. Schallreuter KU, Pittelkow MR, Wood JM. Free radical reduction by thioredoxin reductase at the surface of normal and vitiliginous human keratinocytes. J Invest Dermatol 1986; 87:728–32.
51. Schallreuter KU, Pittelkow MR. Anthralin inhibits elevated levels of thioredoxin reductase in psoriasis. Arch Dermatol 1987; 123:1494–8.
52. Sundaram C, Koster W, Schallreuter KU. The effect of UV radiation and sun blockers on free radical defence in human and guinea pig epidermis. Arch Dermatol Res 1990; 282:526–31.
53. Fuchs J, Packer L. Photooxidative stress in the skin. In Sies H, ed. Oxidative stress: oxidants and antioxidants. New York: Academic Press, 1991:559–83.
54. Halliwell B, Gutteridge JMC. Oxygen free radicals and iron in relation to biology and medicine: some problems and concepts. Arch Biochem Biophys 1986; 246: 501–14.
55. Darr D, Combs S, Dunston S, Manning T, Pinnell S. Topical vitamin C treatment inhibits ultraviolet radiation induced damage to porcine. J Invest Dermatol 1991; 96:590A.

56. Ross D, Albano E, Moldeus P. The generation and fate of glutathionyl radicals in biological systems. In: Poli G, Cheeseman KH, Dianzani MU, Slater TF, ed. Free radicals in liver injury. Oxford: IRL Press Ltd, 1987:17–20.
57. DiMonte D, Bellomo G, Thor H, Nicotera P, Orrenius S. Menadione-induced cytotoxicity is associated with protein thiol oxidation and alteration in intracellular Ca^{++} homeostasis. Arch Biochem Biophys 1984; 235:343–50.
58. Pascoe GA, Olafsdottir K, Reed DJ. Vitamin E protection against chemical-induced cell injury. 1. Maintenance of cellular protein thiols as a cytoprotective mechanism. Arch Biochem Biophys 1987; 256:150–8.
59. Paetkau V, Lardy HA. Phosphofructokinase. Correlation of physical and enzymatic properties. J Biol Chem 1967; 242:2035–42.
60. Perchellet J-P, Owen MD, Posey TD, Orten DK, Schneider BA. Inhibitory effects of glutathione level-raising agents and D-α-tocopherol on ornithine decarboxylase induction and mouse skin tumor promotion by 12-O-tetradecanoylphorbol-13-acetate. Carcinogenesis 1985; 6:567–73.
61. Perchellet EM, Maatta EA, Abney NL, Perchellet J-P. Effects of diverse intracellular thiol delivery agents on glutathione peroxidase activity, the ratio of reduced/oxidized glutathione, and ornithine decarboxylase induction in isolated mouse epidermal cells treated with 12-O-tetradecanoylphorbol-13-acetate. J Cell Physiol 1987; 131:64–73.
62. Stocker R, Frei B. Endogenous antioxidant defences in human blood plasma. In Sies H, ed. Oxidative stress: oxidants and antioxidants. New York: Academic Press, 1991:225–43.
63. Ogura R, Kumano S. Free nucleosides and bases in the hyperkeratotic epidermis. Current Probl Dermatol 1983; 11:145–58.
64. Eisen AZ, Seegmiller JE. Uric acid metabolism in psoriasis. J Clin Invest 1961; 40:1486–94.
65. Perly B, Smith IC, Hughes L, Burton GW, Ingold KU. Estimation of the location of natural alpha tocopherol in lipid bilayers by 13C-NMR spectroscopy. Biochim Biophys Acta 1985; 819:131–5.
65a. Maguire JJ, Wilson DS, Packer L. Mitochondrial electron transport-linked tocopheroxyl radical reduction. J Biol Chem 1989; 264:21462–5.
66. De Simone C, Rusciani L, Venier A, Larussa FM, Littarru GP, Oradei A, Lippa S, Serri F. Vitamin E and coenzyme Q_{10} content in human epidermis and basal cell epithelioma. J Invest Dermatol 1987; 89:317A.
67. Maeda K, Naganuma M, Fukuda M. Effects of chronic exposure to ultraviolet-A including 2% ultraviolet-B on free radical reduction systems in hairless mice. Photochem Photobiol 1991; 54:737–40.
68. Mellors A, Tappel AL. The inhibition of mitochondrial peroxidation by ubiquinone and ubiquinol. J Biol Chem 1966; 241:4353–6.
69. Giovannini L, Bertelli AAE, Scalori V, Dell'Osso L, Alessandri MG, Mian M. Skin penetration of CoQ_{10} in the rat. Int J Tiss Reac 1988; X:103–5.
70. Burton GW, Ingold KU. β-Carotene: an unusual type of lipid antioxidant. Science 1984; 224:569–73.

71. Mathews-Roth MM, Pathak MA, Parrish J, Fitzpatrick TB, Kass EH, Toda K, Clemens W. A clinical trial of the effects of oral beta-carotene on the responses of human skin to solar radiation. J Invest Dermatol 1972; 59:349–53.
72. Tomita Y, Tagami H. The scavenging and filter effect of melanin against superoxide anion produced by photoirradiation. In: Hayaishi O, Imamura S, Miyachi Y, ed. The biological role of reactive oxygen species in skin. Tokyo: University of Tokyo Press and New York: Elsevier, 1987:95–9.
73. Pathak MA, Carraro C. Reactive oxygen species in cutaneous photosensitivity reactions in porphyrias and PUVA photochemotherapy and in melanin pigmentation. In: Hayaishi O, Imamura S, Miyachi Y, ed. The biological role of reactive oxygen species in skin. Tokyo: University of Tokyo Press and New York: Elsevier, 1987:75–94.
74. Goodchild NT, Kwock L. Melanin: a possible cellular superoxide scavenger. In: Rodgers MAJ, Posers EL, ed. Oxygen radicals in chemistry and biology. New York: Academic Press, 1981:645–8.
75. Pathak MA. Sunlight and melanin pigmentation. In: Smith KC, ed. Photochemical and photobiological reviews. New York: Plenum Press, 1976:211–39.
76. Chedekel MR, Smith SK, Post PW, Pokora A, Vessell DL. Photodestruction of pheomelanin: role of oxygen. Proc Natl Acad Sci USA 1978; 75:5395–9.

7

The Effects of UV Radiation on Skin Connective Tissue

Graeme F. Bryce
Hoffmann-La Roche Inc., Nutley, New Jersey

I. INTRODUCTION

It has been recognized for over a century that excessive sun exposure has a profoundly harmful effect on skin. More recently, socioeconomic and cultural forces have refocused attention on this phenomenon. Migration to southern latitudes of populations with pigmentation more suitable to the North and adherence of many of the younger generation to the cult of sun worship have created a new level of awareness of the hazards of chronic sun exposure (1) and its close association with the incidence of epithelium-based tumors, carcinoma, and malignant melanoma (2–4). Another dimension has been added to the problem in the form of artificial sources of ultraviolet radiation for either therapeutic use, e.g., PUVA for psoriasis, or for recreational use, e.g., tanning beds.

A confusion exists in terminology that is used to describe the effects of chronic sun exposure. Some degree of sun exposure is inevitable, and the differences between exposed and unexposed skin are readily apparent. Sun-protected skin retains a healthy appearance until an advanced age, whereas exposed skin shows a wide variety of features such as dryness, laxity, discoloration, telangiectasia, sagging, and deep wrinkling (5). All of these properties can be attributed to sun exposure without invoking an aging component; within reasonable limits there are few age-specific alterations. It is becoming appreciated that excessive sun exposure does not simply accelerate the natural process of aging, but rather provokes a set of dermal alterations conveniently termed photoaging or, if aging is to be explicitly eliminated, photodamage. Newer terms that may be encountered are "dermatoheliosis" (6), which describes the end stage of actinic damage, and "heliodermatitis" (7), coined to denote the chronic inflammatory condition in photodamaged skin, which is usually not visible clinically but which is a midstage condition and a forerunner of photodamage.

Although many regions of the nonionizing radiation spectrum making up the "actinic storm" may contribute to dermal damage (8), the present review will be limited to ultraviolet radiation, specifically, UVA (320–400 nm) and UVB (290–320 nm); solar UVC (<290 nm) does not reach the earth's surface, and its effects will not be considered.

II. HUMAN STUDIES

A. Skin Penetration

Although UVA and UVB differ relatively little in the energy of their quanta, the response of biological systems, such as erythema, differ considerably. UVA is 800–1000 times less erythemogenic than UVB but is present in noontime solar radiation in approximately 200-fold greater amounts. More significantly, UVA penetrates deeper into the dermis than UVB. Epidermal specimens have been prepared from suction blister roofs (9), separation from whole skin by NaBr treatment (10), or other chemical methods (11). Transmission of UV radiation through these preparations depends upon many factors such as degree of hydration, stratum corneum thickness, and whether or not the skin is tanned. Actual values of % transmission of UVB through whole epidermis were quoted to be 3–44% (11) and 13–43% (9); values for tanned epidermis were 8–12% (9). Thus, a significant amount of incident UVB penetrates into the dermis. The transmission of UVA is substantially greater than for UVB, the exact degree depending upon wavelength. A "standard" skin specimen (10), with a stratum corneum of 20 μm and a viable layer of 50 μm, will transmit 19% at 365 nm, 9.5% at 313 nm, 2.4% at 297 nm, and 0.27% at 290 nm.

B. Histology

At the histological level, the most conspicuous feature of photodamaged skin is the increase in elastic fibers. This elastic hyperplasia begins early in facial skin and progresses with continual exposure to a condition of elastosis in which there are tangled masses of abnormal elastin fibers (12,13). Concomitantly, there is a loss of mature collagen and greatly increased production of ground substance (glycosaminoglycans [GAGs]) and damage to the dermal microvasculature (14). In its final stage massive loss of elastic fibers takes place, which, together with collagen loss, leads to dissolution of the matrix.

Early biochemical studies (15) reported an increase in soluble collagen but a loss of insoluble and total collagen. Elastin increased in actinically damaged skin and had an amino acid composition similar to normal human dermal elastin, a finding that helped establish that the elastotic material was principally composed of elastin and not, as had been suggested (13), collagen that had been altered in some way so as to acquire the staining properties of elastin.

Numerous investigators have described the changes that occur with aging in exposed and unexposed skin at the level of electron microscopy (13,16–19). Increased amounts of hollow, 100–120 Å microfibrils were present in the papillary dermis of unexposed aged skin together with fibroblasts containing dilated cisternae and numerous coated vesicles. Both of these markers are found in fibroblasts engaged in elastogenesis. Exposed skin contained even greater amounts of these microfibrils. Mature elastin consists of microfibrils dispersed within an amorphous component. In models of elastogenesis the microfibrillar component is synthesized first. It has been suggested (16) that the increased amount of elastin microfibrils with age represents either an attempt by the skin to replace degraded elastin with new material or, possibly, is an indication of the inability of old skin to produce the amorphous component necessary for the assembly of mature elastin. In severely damaged skin, a notable feature was the Grenz zone lying just below the dermal-epidermal junction (7,16). In this region elastic fibers were absent and electron microscopy revealed the presence of bundles of collagen fibers arranged parallel to the skin surface. Positive staining for reticulin indicated new collagen synthesis in this band, which has been termed by some a microscar (16). As will be seen later, actinic damage will spontaneously repair itself to some extent when the insult of UV irradiation is removed, and the result is the appearance of a Grenz zone.

Reduplication of basement membrane components, particularly the lamina densa and anchoring fibrils, is observed, suggestive of attempts on the part of the skin to repair itself.

C. Biochemical Studies

It is now clearly established that the elastotic material is based on elastin. Using antibodies to a variety of matrix compounds, Chen et al. (20) examined severely sun-damaged skin from the posterior neck. Material that was visualized with Verhoeff's–van Gieson's stain gave intense staining in a fibrillar pattern with antibodies to elastin, microfibrillar protein, and fibronectin.

Immunostaining for the interstitial collagens types I, VI, and the aminopropeptide of type III revealed a fine network of collagen fibers within the elastotic area. The basement membrane proteins, laminin and nidogen, as well as collagen type IV were localized around vessels, appendages, and along the dermal/epidermal junction in a pattern similar to that seen in normal skin. Another study (21) confirmed the immunostaining with antibodies to elastin and further showed that there was no reaction with antisera to protease inhibitors such as α-1 antitrypsin, α-2 antichymotrypsin, or α-2 macroglobulin, which have been found in close association with immature elastic fibers in other types of elastosis, indicating that the elastotic fibers are mature.

Cross-linking has been implicated as a mechanism for the decreased solubility of collagen with age. Histidino-hydroxylysinonorleucine (HHL) is a newly described, stable cross-link that occurs in skin collagen, but not in other connective tissue (22). It is derived from the condensation of a lysine aldehyde, hydroxylysine, and histidine residue of type I collagen and increases with age in humans. In a study of nine individuals over the age of 60, the HHL contents of skin from exposed (wrist) and unexposed (buttock) areas were compared (23). Exposed skin contained only about 20% of the HHL in unexposed skin (0.13 ± 0.07 vs. 0.69 ± 0.17 mol/mol collagen, $p < 0.001$). However, acute irradiation of purified cross-linked peptide in an aqueous buffer with either 250 J/cm^2 UVA or 1 J/cm^2 UVB did not lead to any degradation; under these same conditions there was significant degradation of another nonreducible cross-link, pyridinoline. Possible explanations for these findings are: 1) the collagen molecules in the fibers must be in the correct stereochemical orientation for cross-linking to occur and these conditions are not met as a result of actinic damage or 2) a direct destruction of the cross-links may be mediated by reactive oxygen species generated by UV radiation.

D. UVA

The above-mentioned properties of actinically damaged skin have been ascribed to the UVB portion of the solar spectrum. The effects of UVA are of increasing concern in view of the prevalent use of UVB sunscreens and the accompanying false sense of security against photodamage. UVA is a higher proportion of the solar flux in the early morning and late afternoon because of less atmospheric absorption and has well characterized acute effects, such as erythema, melanogenesis, and is active chronically in photocarcinogenesis, all of which are beyond the scope of this review.

In a limited study of two volunteers, Kumakiri et al. (24) used a filtered xenon arc source (WG345 filter, 50% of its peak at 345 nm) emitting less than 0.01% in the range 315–320 nm. Over the course of about a year the subjects received totals of 1900 and 2600 J/cm^2. Significant findings were mast cell degranulation and the formation of Zebra bodies, which are possibly derived from collagen; there was also extensive damage to the vasculature with extravasation of blood cells. PUVA therapy gave progressive changes in elastin (25), first with damage to the amorphous elastin, then followed by breakdown of the microfibrils.

The effects of chronic UVA irradiation in humans are not well characterized. Inflammatory cell infiltrates have been observed with relatively modest doses (2.5 MEDs) of UVA together with mediators of inflammation (26). It is likely that proteases are also increased, which may contribute to degradation of matrix components. However, animal studies (see below) have provided some interesting results regarding the effects of UV irradiation on connective tissue components.

E. Surface Features

Skin surface features are the most readily apparent indicators of changes to the underlying structure. Wrinkling accompanies actinic damage and progresses with accumulated sun exposure. Wrinkles develop from folds and from underlying muscle contraction (27) and, once established, occur at the same position each time the skin flexes. Skin surface patterns were recorded by allowing a drop of cyanoacrylate to set on the surface; after setting the material was removed and, with it, a sheet of the superficial horny layer (28). Under stereomicroscopy, a latticework of squares and triangles was evident. The pattern was found, with slight variations, in the skin of the upper inner arm and buttocks, and there were only minor differences between the samples taken from young and old subjects. Exposed areas such as the cheek or back of the hand showed the greatest differences between young and old samples, which reflects the degree of accumulated sun-induced dermal damage.

At the light microscopy level there are no histological correlates of wrinkles (27,29); there are no discernible differences between the wrinkle and the immediately surrounding area. A feature found in all deep wrinkles is the loss of the fine candelabralike patterns of elastic fibers, which terminate close to the basal lamina; in sun-damaged skin this is replaced by tangled masses of amorphous material. The conclusion is that the damage to the elastic fiber network prevents the skin from snapping back to its original state and greatly compromises the elastic properties.

III. ANIMAL STUDIES

A. UVB

1. *Histology*

Despite the acknowledgment in some quarters of the deleterious effects of chronic sun exposure on skin, it was only recently that animal models were used to produce actinic damage with artificial UVB sources. Sams et al. (30) irradiated shaved mice with Westinghouse FS20 sun lamps (peak irradiance at 313 nm), which resulted in histological changes similar to those characteristic of human sun-damaged skin, namely, increased elastin, often in focal tangled masses, and prominently increased mucopolysaccharides. Biochemical analysis showed a significant decrease in insoluble collagen after UV irradiation, but no differences in elastin content were detected. Essentially similar results were obtained in rats (31). During the weeks of irradiation of the mice, significant inflammation was observed, which subsided only when the exposure was terminated. Elastic fibers decreased with time of exposure and as the inflammatory condition developed and elastosis gradually became evident during the postirradiation phase. The authors concluded that the relative effects of elastogenesis by UV and elastolysis by inflammatory cellular infiltrate will determine the extent of the observable elastosis.

Hairless mice have proven to be a more convenient model system in which to test UV-irradiation effects, and in the past decade many studies have shown that the dermal alterations are similar to those found in humans. Most studies have used Skh-I albino mice or the HRS/J strain with Westinghouse FS20 or FS40 sunlamps (peak irradiance at 313 nm) as a source of UVB. UVA sources are generally fluorescent BL blacklights (peak irradiance at 365 nm), UVASUN 3000 lamps, which emit long wavelength UVA from 340 to 400 nm, and xenon arc lamps with a variety of filters to selectively remove short-wavelength UVA. Qualitatively similar results have been obtained; however, variations in dosing schedules make direct comparisons difficult. Skin was thickened (32) and a variety of histological changes were noted (33,34). An increase in elastic fibers occurred continuously throughout the radiation period, eventually reaching a state of elastosis. Elastin fibers were thickened and appeared in focal masses; in the deep dermis this effect became more noticeable during the period following irradiation. There were large increases in acid mucopolysaccharides in the upper third of the dermis in an area where losses of mature collagen and increases in reticulin fibers were observed.

At the electron microscopic level (35), duplication of the basement membrane was observed with the excess material either continuous with or separated from the original membrane. Anchoring fibrils were attached to the duplicated membrane. At relatively modest doses of UVB, the collagen morphology was not drastically altered. Only rarely were fibers seen with indistinct outlines. After higher exposures (33), collagen bundles showed a heterogeneity in size.

The conclusions emerging from these and related studies is that chronically UVB-irradiated skin is hypercellular both with regard to fibroblasts and inflammatory cells. Fibroblasts are primarily of the "secondary" type

(36), committed to the synthesis of elastin, with few "primary" fibroblasts synthesizing collagen except in the subepidermal zone, where there is an increase in the synthesis of reticulin, which has been shown to be type III collagen (37). Inflammatory cells can elaborate collagenases and elastases, which may contribute to the altered appearance of the matrix components. The rate of irradiation and the magnitude of the total dose, together with the time of observation, will determine the exact appearance of the damage.

2. Elastin

Quantitation of the histologically apparent increase in elastin has been accomplished by analysis of desmosine content (38–40). This is an amino acid found in elastin hydrolysates that is derived from the lysine-containing cross-links and is a reliable index of elastin content. After a total dose of 0.68 J/cm^2, radioimmunoassayable desmosine increased from 1.6 to 2.6 ng/mm^2 of skin (38); at greater UVB doses (39,40), the values were greater. Table 1 lists examples of published data, all expressed in the same units. A remarkable consistency is evident. In another study, a dose of about 4.8 J/cm^2 led to an increase in elastin, determined gravimetrically, from 0.86 to 2.00 mg (41). This material gave a hydroxyproline:proline ratio of 0.11 and thus was identified as elastin essentially free of collagen.

Table 1 Desmosine Levels in the Skin of Hairless Mice After UVB Irradiation

Group	UVB dose	Desmosine (ng/mm^2)	Ref.
Control	—	1.6 ± 0.1	38
UVB	0.68 J/cm^2	2.6 ± 0.3**	38
Control	—	3.5 ± 0.6	39
UVB	3.5 J/cm^2	6.7 ± 0.4**	39
12 weeks post-UVB		7.3 ± 0.6**	39
Control, 6 weeks	—	1.8 ± 0.6	40
UVB, 6 weeks	0.72 J/cm^2	2.3 ± 0.7	40
Control, 12 weeks	—	1.9 ± 0.7	40
UVB, 12 weeks	1.44 J/cm^2	2.3 ± 1.1	40
Control, 18 weeks	—	2.5 ± 1.2	40
UVB, 18 weeks	2.16 J/cm^2	5.2 ± 0.5*	40

*$p < 0.05$, **$p < 0.01$ vs. control.
Data were converted to ng/mm^2 for comparison.
Source: Ref. 40.

Mouse elastic fibers are generally sparse and small, which make their visualization difficult by electron microscopy. A recent study utilized a tannic acid–uranyl acetate staining procedure to render the fibers electron dense (19). During the early stages of chronic irradiation with UVB, fibers increased in number and were mostly covered with matrix. Damaged fibers were evident and, significantly, were of a type not observed in older nonirradiated animals. With further irradiation the extent of damage increased to the point of photodegradation, with increasing amounts of amorphous material and disappearance of microfibrils.

3. *Collagen*

Direct comparisons of various studies on the effects of UVB are complicated by a multiplicity of irradiation protocols used, different strains of mice (or other animals), light sources, and the definition of a minimal erythemic dose (MED). Early observations on the effects of UV irradiation on collagen recorded a loss of the insoluble fraction but no change in the various soluble fractions (30). Later histological studies described loss of staining in the upper dermis and attributed this to some kind of damage which altered the dye-binding properties of the collagen (33,34). In the last few years newer analytical methodology has been used to more precisely define the effects on collagen. After a total exposure of 0.68 J/cm^2 of UVB (12 doses over 4 weeks), no changes in collagen content, as measured by hydroxyproline, were found (38), although significant increases in elastin had occurred. Interestingly, prolyl hydroxylase activity decreased in this experiment as a result of irradiation with UVA, but not with UVB.

With specific antisera, Berrebi et al. (42) demonstrated a decrease in type I and an increase in type III collagen following irradiation of hairless mice with a xenon arc solar simulator. Changes in elastin during this period were consistent with a significant degree of photodamage. Similar alterations in the ratio of type I and III collagens have been characterized. Thus Plastow et al. (43,44) irradiated Skh-I mice with about 0.5 MED (0.062 J/cm^2) of UVB, 5 days per week, for up to 36 weeks. Collagen was analyzed in cyanogen bromide digests of whole dorsal skin by polyacrylamide gel electrophoresis and the amount of type III expressed as a percentage of the total collagen. Similar samples were processed from ventral (nonirradiated) skin to serve as controls. With increasing radiation dose, the ratio of dorsal to ventral type III collagen increased, which was interpreted as an absolute increase in type III collagen induced by irradiation. The effect appeared to be partially reversible since there was a tendency for these ratios to return to preirradiation values when the UVB was discontinued, which may be a reflection of the postirradiation repair reported in earlier histological studies (33). This study confirmed an earlier finding of increased type III collagen (by

SDS PAGE analysis of cyanogen bromide digests) in chronic elastosis in human subjects (45) and lent support for this animal model.

A different method of analysis was used in a study of Schwartz et al. (46). Animals were irradiated with UVB in a protocol which gave 1.65 J/cm^2 at 5 weeks and 4.8 J/cm^2 after 10 weeks exposure. Pepsin-soluble extracts were analyzed for types I and III collagen by PAGE under reducing conditions. No difference was found in the percentage of type III collagen between control animals (19.6 $\pm$ 1.2) and those irradiated for 10 weeks (21.3 $\pm$ 1.0). Radioimmunoassay demonstrated a 35% loss of amino propeptide of type III collagen after 5 weeks and an 80% loss after 10 weeks of irradiation. Immunoblotting revealed the presence of lower molecular weight species, consistent with the cleavage of the aminopropeptide from type III collagen. Electron microscopy showed that this material no longer was present as a coating on the surface of the collagen fibers.

In a recent biochemical study, Kligman et al. irradiated Skh-I mice with approximately 4 MED (0.06 J/cm^2) UVB, three times per week for up to 24 weeks (48). Collagen was isolated, at 4-weekly intervals, by acid extraction, pepsin treatment, and salt precipitation and then analyzed for types I and III by interrupted PAGE. Total collagen (per unit skin area) in irradiated animals increased relative to control up to 16 weeks then declined to control levels at 20 weeks. The ratio of type III to total collagen was significantly increased only at 8 weeks ($p < 0.05$) and at 24 weeks ($p < 0.005$).

In a similar irradiation protocol, with doses of 0.12 J/cm^2 per week, Chatterjee et al. (40) demonstrated that there were no significant changes in skin collagen content (as measured by hydroxyproline content) when expressed per unit area; a slight reduction was observed ($p < 0.01$), however, in irradiated animals after 24 weeks of UVB (2.88 J/cm^2) when the data weree expressed in terms of unit wet weight.

Another animal model has recently been introduced into the area of photoaging. The miniature pig has a skin structure similar in many ways to humans with respect to epidermal thickness and was used by researchers from L'Oreal in a recent study in which collagen changes were investigated subsequent to UVB irradiation (49). By immunofluorescence techniques, it was demonstrated that collagen type I decreased while type III increased after UVB irradiation.

How can these discrepancies be reconciled? Differences in methodology can be a source of disagreement. Thus, while total collagen analysis by hydroxyproline determination showed either no change with UVB-irradiation (38,40) or a steady increase for 16 weeks followed by a decrease (48), immunofluorescence measurements in mouse and minipig systems showed decreases in type I collagen and increases in type III collagen (42,49). The ratio of type III to type I + type III or the percentage of type III has been measured in several laboratories with various results. Following UVB irradiation, this ratio either 1) increased when the analysis was done by PAGE with extracts of whole skin (43,44) or when assessed by immunofluorescence (42,49), or 2) did not change when the analysis was done by PAGE on collagen isolated by pepsin treatment and salt fractionation (46,48). In a recent study (50), an 86% increase in the level of mRNA and a 52% increment in the protein for type III collagen following UVB irradiation was found; there was no change in type I mRNA. Thus two quite different methodologies gave similar results.

Differences in irradiation protocols, coupled with a lack of definitive studies on the effects of the frequency and dose magnitude, make direct comparisons impossible. Whenever practical, it has been suggested that tissue samples by exchanged for comparison and that techniques be standardized (51).

It must be remembered that UVB irradiation induces an inflammatory condition, the severity of which depends upon, among other things, the dosing regimen. End-stage actinic damage is characterized by massive destruction of the extracellular matrix by digestive enzymes secreted by cells of the inflammatory infiltrate. At earlier stages there must be a point when the degradation begins to exceed the capacity of the fibroblasts to synthesize new collagen, which would explain, for example, the sudden downturn in collagen synthesis at 20 weeks after a sustained increase observed in one study (48).

4. Other Matrix Components

Glycosaminoglycans have been quantitated and expressed in terms of the uronic acid content in two studies. Over the course of 24 weeks, uronic acid content did not change significantly either in the control or irradiated animals when the results were expressed per unit weight of skin (40); however, when expressed in terms of skin area there were significant increments in the irradiated animals. At 6, 12, 18, and 24 weeks at 0.12 J/cm^2 of UVB per week, the percentage increases were 62, 66, 64, and 72, respectively. Control values showed a tendency to increase with time. These observations were in agreement with the histological analysis of this study and of others (33,34, 49).

Using differential salt extraction, a separate study (41) measured the changes in hyaluronic acid, chondroitin sulfate, and heparin after a total dose of 4.8 J/cm^2 of UVB over 10 weeks. All three components were significantly increased.

Elastase and collagenase activity, which would be expected to have some role in the turnover of the extracellular matrix, responded to UVB irradiation in a manner similar to that of their substrates elastin and collagen (40). On a unit area of skin basis, elastase increased, whereas collagenase remained constant. It should be mentioned that there are inherent uncertainties in the measurements of these two enzymes. Elastase activity is ''elastaselike'' activity measured against a synthetic substrate, which may not mirror the activity to which tissue elastin is susceptible; collagenase activity is influenced by the presence of endogenous inhibitors, which may be regulated independently from collagenase.

It has been pointed out (40) that many of the changes accompanying UVB irradiation merely reflect the changes in skin thickness when the data are expressed per unit skin area. Since collagen changes, per unit area, during the first 16 weeks of UVB irradiation were not observed (48), and actually decreased at the end of 24 weeks, it can be concluded that, at least in this study, the relative amount of collagen was decreased, and this decrement may be of sufficient magnitude to cause major alterations in the mechanical properties of the skin.

B. UVA

Early attempts to produce elastosis were generally unsuccessful. The studies used blacklights as a source of UVA and, to eliminate traces of radiation of wavelengths below 320 nm, window glass (52) or mylar (53) filters were used. Even after doses of up to 5400 J/cm^2 there was no histological evidence of the elastotic changes that resembled those that were produced by UVB in parallel experiments. More recent studies have been able to demonstrate some changes in elastin. In general the changes were small, taking the form of slight increases in fine fibers extending deeper into the dermis than was the case for UVB-induced changes (54), and some areas devoid of elastic fibers in the vicinity of damaged collagen were observed. Xenon arc sources of UVA, passed through a WG 345 filter, appeared to give greater elastic fiber hyperplasia than blacklight sources (55); this is probably due to the greater proportion of wavelengths in the 315–340 nm range.

The contribution of a UVB component is undoubtedly the explanation for the elastosis observed in an older study with Ng/$^-$ naked mice (56). The light source was reported to have a relative output of 0.03 W at 280–315 nm versus 8.5 W at 315–380 nm. Over the 8.5-month period of irradiation, the accumulated dose can be calculated to be approximately 9500 J/cm^2 UVA and almost 40 J/cm^2 UVB. Collagen staining was lost from the upper dermis, and elastic fibers were in tangled aggregates. Electron microscopic examination showed disintegrating collagen fibers with a disparity in diameters but also some evidence of new collagen in the form of 30- to 50-nm-diameter fibers. It is not possible to ascribe any of these alterations to UVA in view of the large amount of UVB present in the light source.

The possibility that UVA has a unique effect on elastic fibers has been addressed in two studies with long-term exposures to UVA. In both cases

(57,58) elastic fiber hyperplasia was observed, but of a type different from that produced by UVB. Comparing the results of exposure of up to 10,000 J/cm² of UVA (UVASUN > 340 nm) and 3–4.5 J/cm² UVB (58), Zheng and Kligman (57) reported the following differences: UVA: fibers sparse in upper dermis, hyperplastic in middermis with increased microfibrils lacking the elastin coating; UVB: fibers more numerous in the upper dermis and more degraded than with UVA.

At the histological level UVA causes only minor damage to collagen (54, 55), usually restricted to foci in the upper dermis. No inflammatory filtrate is observed. Biochemically, the changes are more interesting. Kligman and Gebre (59) found that UVA exposure led to an increased resistance of the collagen to pepsin digestion. The percentage of insuluble material was 16–18% in age-matched control animals and increased to 63% after 4000 J/cm² and to 86% after 8000 J/cm² of UVA (UVASUN > 340 nm). When 4000 J/cm² were delivered to another group of animals using a xenon arc solar simulator and a WG 360 filter, which provides wavelengths in the range 335–400 nm, the percentage of pepsin-resistant collagen rose to 90%. The time required to deliver the dose with this radiation source was 60 weeks compared to 25 weeks for the UVASUN source, and the authors concluded that a time factor must be involved to explain the greater collagen insolubility. It was proposed that increased cross-linking could be a likely explanation (59,60). In this regard, the effect of UVA is quite different from UVB. Collagen from normal skin and UVB-irradiated skin is 85–100% solubilized by pepsin.

Collagen content was significantly reduced by UVA in a dose-dependent manner without any change in the degree of hydroxylation. This observation may be related to the finding (38) that the activity of collagen proline hydroxylase was decreased by about 50% after short-term UVA irradiation.

UVA effects on ground substances are less well defined. Increases are invariably found (53–55) particularly in the upper dermis. Hyaluronic acid and sulfated GAGs were only slightly increased (55).

C. Action Spectra

Action spectra have been obtained in hairless mice for a variety of photobiological effects such as erythema/edema (62), pyrimidine dimer formation (63), and carcinogenesis (64). UVB is more effective than UVA. The spectra for erythema/edema were similar for mice and humans.

Three studies were conducted on the wavelength dependence of UV-induced connective tissue changes (64–66). Bissett et al. used a xenon arc and dual grating monochrometer and exposed mice over a range from 280 to 380 nm at intervals of 5–10 nm (64). A wide variety of indices of photodamage was examined, and plots of the responses against wavelength gave action spectra. With the exception of skin sagging, all parameters tested, such as skin thickness, wrinkling, tumor formation, as well as elastosis, collagen damage, and GAG content, showed maximum sensitivity in the UVB region of the spectrum from 290 to 300 nm. Skin sagging response was clearly different and centered at about 340 nm in the UVA region.

Studies by Wulf et al. (65) and Kligman and Sayre (66) achieved essentially the same results for elastosis.

D. Surface Features

Hairless mice develop wrinkling on the dorsum following UVB irradiation (39,54). The pattern takes the form of a series of parallel structures arranged perpendicular to the length of the animal with a spacing of 2–3 mm. It is likely that this pattern is derived from the manner in which the loose skin of the animal lies in folds. Upon irradiation and damage to the substructure, and continual flexure, the folds are "permanently" impressed into the skin; the line pattern does not disappear when the skin is gently stretched so that the wrinkling observed resembles the appearance of photoaged human skin (39). Quantitation of the wrinkling can be achieved by visual grading on live animals (54) or by analysis of skin surface replicas (39), which are examined under low-angle illumination and arbitrary values assigned to the shadowing.

The extent of wrinkling is dose-dependent, but the exact relationship between the irradiation protocol and the final state is not known. In a recent study (67), the development of wrinkling was monitored during two dosing regimens: suberythemal 3 times per week for 20 weeks (total dose 0.72 J/cm^2) and 4 MED reached after 4 weeks and continued for a further 10 weeks (total dose 2.1 J/cm^2). Degree of wrinkling was dose-dependent, but there was a time factor. Thus at 0.5 J/cm^2 total dose, the degree of wrinkling was greater in the animals receiving the radiation at the lower flux level; the onset of wrinkling was earlier with the high-dose regimen. These results draw attention to the necessity for strict following of a protocol in order to obtain reproducible results.

UVA does not cause wrinkling of this type. Rather, the result is a sagging of the skin (54). Changes were progressive during a 31-week period (for a total of 750 J/cm^2) and continued when the radiation was stopped. A loss of elasticity is associated with the sagging phenomenon and can be quantitated in a pinch test (68). The rebound time from a simple skin pinch increased steadily to a value almost three times greater than control after 400 J/cm^2. Animals damaged with UVB showed no such impairment, demonstrating that the effects of UVA and UVB can be clearly separated.

IV. MECHANISMS

A. Prevention

Whatever the mechanism of the effects of UV irradiation, it is not surprising that sunscreen agents afford protection. Studies in the mouse model have adequately demonstrated that concomitant application of sunscreens during chronic irradiation protocols completely prevented the connective tissue damage (33,34,44,55). When damage had already been incurred, the sunscreen stopped further damage and promoted partial reversal of the existing damage (33,34).

Parsol 1789 (4-*t*-butyl-4′-methoxydibenzoylmethane) absorbs significantly in the 320–385 nm region and was shown to prevent effectively the UVA-induced skin sagging (54).

UVB generates reactive oxygen species in skin and a large body of literature describes the potential for damage by these agents (69), which is outside the scope of this review. It is not known to what extent these oxidative reactions contribute to the chronic alterations observed in collagen, elastin, and other components of the extracellular matrix. It has been presumed (39,40,54,55) that the changes in skin surface features and mechanical properties mirror changes in the dermal connective tissue during UVB-induced damage and also in the repair process. Thus a variety of studies have been carried out on the prevention of skin damage by different classes of agents using noninvasive procedures to monitor the development of wrinkling during chronic irradiation with either UVA or UVB. Superoxide-scavenging agents such as α-tocopherol, ascorbic acid, and propyl gallate delayed the onset of wrinkling (70). Histological evaluation showed that collagen damage was significantly reduced. Scavengers of singlet oxygen such as conjugated dienes (71) were effective in reducing UVB-induced wrinkling as were topically applied anti-inflammatory agents such as ibuprofen, naproxen, and hydrocortisone (72). Of these three classes of compounds, only the anti-inflammatories prevented the UVA-induced skin sagging.

β-Carotene is capable of quenching excited triplet states and, to a lesser extent, scavenging free radicals. As a dietary admix it significantly reduced the number of skin tumors in UVB-irradiated hairless mice (73). However, when animals were fed a diet containing 1% β-carotene, there was no effect on the development of elastosis, collagen damage, or increases in ground substance (74); 0.02% 13-*cis*-retinoic acid, which has anti-inflammatory properties, was also ineffective. The irradiation protocol used in this latter study was not identical to that used in other studies (70,71), which may explain the difference in results.

A role for tissue iron in the generation of reactive oxygen species has been proposed (75). Thus chronic UVB releases iron from damaged microvasculature, which can then participate in a variety of reactions, one of which may be the generation of very damaging hydroxy radicals in the presence of hydrogen peroxide (76). Topical application of iron chelators, such as 1,10-phenanthroline and 2,2′-dipyridylamine, significantly delayed the onset of wrinkling, whereas nonchelating analogs were ineffective in photoprotection.

B. Reversal

Some of the effects of chronic UVB exposure can be reversed, particularly in the hairless mouse model system. This "repair" was characterized by a band of normal-looking collagen immediately below the epidermis, which appeared to have displaced the elastotic fibers, forcing them into a compressed band in the lower dermis (33,34). Retinoic acid was found to accelerate this process (77). The same dermal repair phenomenon was shown to occur with 13-*cis*-retinoic acid in a study which also showed that the UVB-induced wrinkling was completely effaced (39). There is now good biochemical evidence for collagen synthesis induced by retinoic acids as determined 1) at the level of gene transcription of both type I and III (78), 2) by increases in collagen content detected by radioimmunoassay of the aminopropeptide of type III (79), and 3) from stimulated incorporation of radioactive proline into collagen hydroxyproline in organ cultures (80). A recent report (81) has proposed that the retinoic acid effect on collagen gene transcription is mediated by transforming growth factor-β (TGF-β) produced in epidermal keratinocytes. This cytokine could diffuse into the dermis and stimulate fibroblasts to produce collagen. Moreover, the effect is dependent upon prior damage to the tissue, since specific stimulation of collagen gene transcription does not take place in age-matched control animals.

Although the collagen appears to be normalized, the elastotic material is not eliminated, rather, it is translocated to the lower dermis where it remains, at least for the duration of the experiment. Chemical analysis of desmosine during repair showed that there was no change in the tissue level (39) after retinoic acid treatment.

Essentially the same picture emerges from recent studies in humans (82), with a significant discrepancy in the extent of the effect. Topical retinoic acid treatment led to improvement of fine wrinkling, which was probably the result of epidermal thickening. Dermal changes were modest, but electron microscopy showed that there were oriented collagen bundles in the papillary dermis and a significant increase in anchoring fibrils (83). It is likely that the discrepant degree of repair in human skin compared to the findings in the mouse model reflect the difference in life span of the two species.

C. Induction of Damage

A state of chronic inflammation in some degree seems to be associated with UVA- or UVB-induced damage. Infiltrating inflammatory cells contain the degradative enzymes collagenase and elastase to account for some of the changes observed in these matrix components. Whether the small amount of radiation that reaches the dermis and acts directly to initiate inflammation is adequate is not known. Other mechanisms could contribute. Keratinocytes are known to be capable of producing a large number of cytokines and mediators in response to UVB irradiation. Interleukin 1 (IL-1) and TNF-α, for example, are found in the granular and spinous layers of the epidermis; after UVB irradiation (84), amounts are markedly increased with immunostaining observed in the basal cells. These two factors have profound effects on the metabolism of connective tissue components (85).

Many other factors undoubtedly have roles in the photodamage/repair process, and it would be difficult to single out any as principal players. Under the influence of chronic UV irradiation, it is possible that an array of mediators is produced that tilts the balance of anabolic/catabolic effects in such a way as to result in damage. Conversely, retinoid-induced repair, for example, can be visualized as a change in the direction of anabolic balance.

REFERENCES

1. Council on Scientific Affairs. Council Report: Harmful effects of ultraviolet radiation. J Am Med Assoc 1989; 262:380–4.
2. MacKie RM, Elwood JM, Hawk JLM. Links between exposure to ultraviolet radiation and skin cancer. JR Coll Physicians Lond 1987; 21:91–6.
3. Kopf AW, Rigel DS, Friedman RJ. The rising incidence and mortality rate of malignant melanoma. J Dermatol Surg Oncol 1982; 8:760–1.
4. McCarthy WH, Shaw HM. Skin cancer in Australia. Med J Aust 1989; 150:469–70.
5. Kligman LH, Kligman AM. Photoaging. In: Fitzpatrick TB, Eisen AZ, Wolff K, Freedberg IM, Austen KF, ed. Dermatology in general medicine. Vol 1. 3d ed. New York: McGraw-Hill, 1986:1470–5.
6. Gilchrest BL. Skin and aging processes. Boca Raton, FL: CRC Press, Inc., 1984: 97–116.
7. Lavker RM, Kligman AM. Chronic heliodermatitis: a morphological evaluation of chronic actinic dermal damage with emphasis on the role of mast cells. J Invest Dermatol 1988; 90:325–30.
8. O'Brien JP, Regan W. A study of elastic tissue and actinic radiation in "aging," temporal arthritis, polymyalgia rheumatica and atherosclerosis. J Am Acad Dermatol 1991; 24:765–76.
9. Eggset G, Kavli G, Krokan H, Volden G. Ultraviolet and visible light penetration of epidermis. Photobiochem Photophys 1984; 8:163–7.

10. Bruls WAG, Slaper H, van der Leun JC, Berrens L. Transmission of human epidermis and stratum corneum as a function of thickness in the ultraviolet and visible wavelengths. Photochem Photobiol 1984; 40:485–94.
11. Everett MA, Yeargers E, Sayre RM, Olson RL. Penetration of epidermis by ultraviolet rays. Photochem Photobiol 1966; 5:533–42.
12. Kligman AM. Early destructive effect of sunlight on human skin. J Am Med Assoc 1969; 210:2377–80.
13. Mitchell RE. Chronic solar elastosis: a light and electron microscopic study of the dermis. J Invest Dermatol 1967; 48:203–20.
14. Kligman AM. Perspectives and problems in cutaneous gerontology. J Invest Dermatol 1979; 73:39–46.
15. Smith JG, Davidson EA, Sams WM, Clark RD. Alterations in human dermal connective tissue with age and chronic sun damage. J Invest Dermatol 1962; 39: 347–50.
16. Lavker RM. Structural alterations in exposed and unexposed aged skin. J Invest Dermatol 1979; 73:59–66.
17. Braverman IM, Fonferko E. Studies in cutaneous aging. I. The elastic fiber network. J Invest Dermatol 1982; 78:434–43.
18. Braverman IM, Fonferko E. Studies in cutaneous aging. II. The microvasculature. J Invest Dermatol 1982; 78:444–8.
19. Hirose R, Kligman LH. An ultrastructural study of ultraviolet-induced elastic fiber damage in hairless mouse skin. J Invest Dermatol 1988; 90:697–702.
20. Chen VL, Fleischmajer R, Schwartz E, Palia M, Timpl R. Immunochemistry of elastotic material in sun-damaged skin. J Invest Dermatol 1986; 87:334–7.
21. Mera SL, Lovell CR, Jones RR, Davies JD. Elastic fibres in normal and sun-damaged skin: an immunohistochemical study. Brit J Dermatol 1987; 117:21–7.
22. Yamauchi M, Woodley DT, Mechanic GL. Aging and crosslinking of skin collagen. Biochem Biophys Res Commun 1988; 152:898–903.
23. Yamauchi M, Prisayanh P, Haque Z, Woodley DT. Collagen cross-linking in sun-exposed and unexposed sites of aged human skin. J Invest Dermatol 1991; 97:938–41.
24. Kumakiri M, Hashimoto K, Willis I. Biologic changes due to long-wave ultraviolet irradiation on human skin: ultrastructural study. J Invest Dermatol 1977, 69:392–400.
25. Zelickson AS, Mottaz JH, Zelickson BD, Muller SA. Elastic tissue changes in skin following PUVA therapy. J Am Acad Dermatol 1980; 3:186–92.
26. Gilchrest BA, Soter NA, Hawk JLM, Barr RM, Black AK, Hensby CN, Mallet AI, Greaves MW, Parrish JA. Histological changes associated with ultraviolet-A-induced erythema in normal human skin. J Am Acad Dermatol 1983; 9:213–19.
27. Wright ET, Shellow WVR. The histopathology of wrinkles. J Soc Cosmet Chem 1973; 24:81–5.
28. Lavker RM, Kwong F, Kligman AM. Changes in skin surface patterns with age. J Gerontol 1980; 35:348–54.
29. Kligman AM, Zheng P, Lavker RM. The anatomy and pathogenesis of wrinkles. Br J Dermatol 1985; 113:37–42.

30. Sams WM, Smith JG, Burk PG. The experimental production of elastosis with ultraviolet light. J Invest Dermatol 1964; 43:467–71.
31. Nakamura K, Johnson WC. Ultraviolet light induced connective changes in rat skin: a histopathologic and histochemical study. J Invest Dermatol 1968; 51: 253–58.
32. Vogel HG, Alpermann HG, Futterer E. Prevention of changes after UV-irradiation by sunscreen products in skin of hairless mice. Arch Dermatol Res 1981; 270:421–8.
33. Kligman LH, Akin FJ, Kligman AM. Prevention of ultraviolet damage to the dermis of hairless mice by sunscreens. J Invest Dermatol 1982; 78:181–9.
34. Kligman LH, Akin FJ, Kligman AM. Sunscreens promote the repair of ultraviolet radiation-induced dermal damage. J Invest Dermatol 1983; 81:98–102.
35. Feldman D, Bryce GF, Shapiro SS. Ultrastructural effects of UVB radiation and subsequent retinoic acid treatment on the skin of hairless mice. J Cutan Pathol 1991; 18:46–55.
36. Kewley MA, Williams G, Stevens FS. Studies of elastic tissue formation in the developing bovine ligamentum nuchae. J Pathol 1978; 124:95–101.
37. Fleischmajer R, Gay S, Perlish JS, Cessarini JP. Immunoelectron microscopy of type III collagen in normal and scleroderma skin. J Invest Dermatol 1980; 75:189–91.
38. Johnston KJ, Oikarinen AI, Lowe NJ, Clark JG, Uitto J. Ultraviolet radiation-induced connective tissue changes in the skin of hairless mice. J Invest Dermatol 1984; 82:587–90.
39. Bryce GF, Bogdan NJ, Brown CC. Retinoic acids promote the repair of the dermal damage and the effacement of wrinkles in the UVB-irradiated hairless mouse. J Invest Dermatol 1988; 91:175–80.
40. Chatterjee R, Benzinger MJ, Ritter JL, Bissett DL. Chronic ultraviolet B radiation-induced biochemical changes in the skin of hairless mice. Photochem Photobiol 1990; 51:91–7.
41. Schwartz E. Connective tissue alterations in the skin of ultraviolet irradiated hairless mice. J Invest Dermatol 1988; 91:158–61.
42. Berrebi C, Fiat F, Fortanier A, Horiszny D, Rouers-Pele D, Vila T. J Invest Dermatol 1987; 89:319.
43. Plastow SR, Lovell CR, Young AR. UVB-induced collagen changes in the skin of the hairless albino mouse. J Invest Dermatol 1987; 88:145–8.
44. Plastow SR, Harrison JA, Young AR. Early changes in dermal collagen of mice exposed to chronic UVB irradiation and the effects of a UVB sunscreen. J Invest Dermatol 1988; 91:590–2.
45. Lovell CR, Plastow SR, Russell-Jones R, Thomas J. Collagen in actinic elastosis. J Invest Dermatol 1984; 82:566.
46. Schwartz E, Cruickshank FA, Perlish JS, Fleischmajer R. Alterations in dermal collagen in ultraviolet irradiated hairless mice. J Invest Dermatol 1989; 93: 142–6.
47. Fleischmajer R, Perlish JS, Timpl R. Collagen fibrillogenesis in human skin. Ann NY Acad Sci 1985; 460:246–57.

48. Kligman LH, Gebre M, Alper R, Kefalides NA. Collagen metabolism in ultraviolet irradiated hairless mouse skin and its correlation to histochemical observations. J Invest Dermatol 1989; 93:210–14.
49. Fourtanier A, Berrebi C. Miniature pig as an animal model to study photoaging. Photochem Photobiol 1989; 50:771–84.
50. Borchiellini C, Labat-Robert J, Fourtanier A, Kern P. UVA and UVB induced changes in collagen and fibronectin biosynthesis in the skin of hairless mice. Abstract, international symposium, Collagen and skin: from basic research to the clinic, Reims, France, September 12–13, 1991.
51. Kligman LH. Reply to letter to the editor. J Invest Dermatol 1990; 94:828–9.
52. Poulsen JT, Staberg B, Wulf HC, Brodthagen H. Dermal elastosis in hairless mice after UV-B and UV-A applied simultaneously, separately or sequentially. Br J Dermatol 1984; 110:531–8.
53. Pfau RG, Hood AF, Morison WL. Photoageing: the role of UVB, solar-simulated UVB, visible and psoralen UVA radiation. Br J Dermatol 1986; 114:319–27.
54. Bissett DL, Hannon DP, Orr TV. An animal model of solar-aged skin: histological, physical, and visible changes in UV-irradiated hairless mouse skin. Photochem Photobiol 1987; 46:367–78.
55. Kligman LH, Akin FJ, Kligman AM. The contributions of UVA and UVB to connective tissue damage in hairless mice. J Invest Dermatol 1985; 84:272–6.
56. Berger H, Tsambaos D, Mahrle G. Experimental elastosis induced by chronic ultraviolet exposure. Arch Dermatol Res 1890; 269:39–49.
57. Zheng PS, Kligman LH. UVA radiation-induced ultrastructural changes in hairless mouse skin: a contrast to UVB-induced damage. J Invest Dermatol 1991; 96:584.
58. Fourtanier A, Labat-Robert J, Berrebi C, Kern P. Effect of chronic UVA irradiation on hairless mouse skin. J Invest Dermatol 1989; 92:429.
59. Kligman LH, Gebre M. Biochemical changes in hairless mouse skin collagen after chronic exposure to ultraviolet-A radiation. Photochem Photobiol 1991; 54:233–7.
60. Trautinger F, Trenz A, Raff M, Kokoschka EM. Influence of UV radiation on dermal collagen content in hairless mice. Arch Dermatol Res 1989; 281:144.
61. Cole CA, Davies RE, Forbes PD, D'Aloisio LC. Comparison of action spectra for acute cutaneous response to ultraviolet radiation: man and albino hairless mouse. Photochem Photobiol 1983; 37:623–31.
62. Ley RD, Peak MJ, Lyon LL. Induction of pyrimidine dimers in epidermal DNA of hairless mice by UVB: an action spectrum. J Invest Dermatol 1983; 80:188–91.
63. Cole CA, Forbes PD, Davies RE. An action spectrum for UV photocarcinogenesis. Photochem Photobiol 1986; 43:275–84.
64. Bissett DL, Hannon DP, Orr TV. Wavelength dependence of histological, physical, and visible changes in chronically UV-irradiated hairless mouse skin. Photochem Photobiol 1989; 50:763–69.
65. Wulf HC, Poulsen T, Davies RE, Urbach F. Narrow-band UV radiation and induction of dermal elastosis and skin cancer. Photodermatol 1989; 6:44–51.

66. Kligman LH, Sayre RM. An action spectrum for ultraviolet induced elastosis in hairless mice: quantification of elastosis by image analysis. Photochem Photobiol 1991; 53:237–42.
67. Kiss I, Chen S, Tramposch KM. The effect of high and low ultraviolet-B dose exposure on the degree of hairless mouse skin wrinkling. Photochem Photobiol 1991; 53:109–12.
68. Bryce GF, Bogdan NJ. The effect of UVA irradiation on the mechanical properties of the skin of hairless mice. J Invest Dermatol 1991; 96:584.
69. Black HS. Potential involvement of free radical reactions in ultraviolet light-mediated cutaneous damage. Photochem Photobiol 1987; 46:213–21.
70. Bissett DL, Chatterjee R, Hannon DP. Photoprotective effect of superoxide-scavenging antioxidants against ultraviolet radiation-induced chronic skin damage in the hairless mouse. Photodermatol Photoimmunol Photomed 1990; 7: 56–62.
71. Bissett DL, Majeti S, Fu J-JL, McBride JF, Wyder WE. Photoprotective effect of topically applied hexadienes against ultraviolet radiation-induced chronic skin damage in the hairless mouse. Photodermatol Photoimmunol Photomed 1990; 7:63–7.
72. Bissett DL, Chatterjee R, Hannon DP. Photoprotective effect of topical anti-inflammatory agents against ultraviolet radiation-induced chronic skin damage in the hairless mouse. Photodermatol Photoimmunol Photomed 1990; 7:153–8.
73. Mathews-Roth MM, Krinsky NI. Carotenoids affect development of UV-B induced skin cancer. Photochem Photobiol 1987; 46:507–9.
74. Kligman LH, Mathews-Roth MM. Dietary β-carotene and 13-*cis*-retinoic acid are not effective in preventing some features of UVB-induced dermal damage in hairless mice. Photochem Photobiol 1990; 51:733–5.
75. Bissett DL, Chatterjee R, Hannon DP. Chronic ultraviolet radiation-induced increase in skin iron and the photoprotective effect of topically applied iron chelators. Photochem Photobiol 1991; 54:215–23.
76. Puppo A, Halliwell B. Formation of hydroxyl radicals from hydrogen peroxide in the presence of iron. Biochem J 1988; 249:185–90.
77. Kligman LH, Chen HD, Kligman AM. Topical retinoic acid enhances the repair of ultraviolet damaged dermal connective tissue. Conn Tiss Res 1984; 12:139–50.
78. Kim HJ, Bogdan NJ, Brown CC, Bryce GF. Effect of topically applied retinoic acids on collagen gene expression in the UVB-irradiated hairless mouse: a possible mechanism for retinoic acid-induced repair of dermal damage. J Invest Dermatol 1989; 92:458.
79. Schwartz E, Cruickshank FA, Mezick JA, Kligman LH. Topical all-trans-retinoic acid stimulates collagen synthesis in vivo. J Invest Dermatol 1991; 96:975–8.
80. Chen S, Kiss I, Tramposch KM. Effect of all-*trans*-retinoic acid on dermal collagen synthesis in UVB-irradiated and non-irradiated hairless mice. J Invest Dermatol 1990; 94:513.
81. Kim HJ, Bogdan NJ, D'Agostaro LJ, Gold LI, Bryce GF. The effect of topical retinoic acids on the levels of collagen MRNAS during the repair of UVB-induced

dermal damage in the hairless mouse and the possible role of TFG-β as a mediator. J Invest Dermatol 1992; 98:359–363.
82. Weiss JS, Ellis CN, Headington JT. Topical tretinoin improves photoaged skin. J Am Med Assoc 1988; 258:527–32.
83. Ellis CN, Weiss JS, Hamilton TA, Headington JT, Zelickson AS, Voorhees JJ. Sustained improvement with prolonged topical tretinoin (retinoic acid) for photodamaged skin. J Am Acad Dermatol 1990; 23:629–37.
84. Oxholm A, Oxholm P, Staberg B, Bendtzen K. Immunohistological detection of interleukin 1-like molecules and tumor necrosis factor in human epidermis before and after UVB-irradiation. Br J Dermatol 1988; 118:369–76.
85. Duncan MR, Berman B. Differential regulation of collagen, glycosaminoglycan, fibronectin, and collagenase activity production in cultured human adult dermal fibroblasts by interleukin 1-alpha and beta and tumor necrosis factor-alpha and beta. J Invest Dermatol 1989; 92:699–706.

8

Antioxidant Protection Against Ultraviolet Light-Induced Skin Damage

Edward Pelle, Kenneth Marenus, and Daniel Maes
Estee Lauder Research Laboratories, Melville, New York

I. INTRODUCTION

Lifestyle changes in the latter half of the twentieth century, such as an increase in outdoor leisure activities, and also the misperception that suntanned skin is synonomous with good health have contributed to a rise in actinic damage in human skin. Currently, however, due to the sudden increase in the incidence of photocarcinogenesis and the long-term effects of overexposure referred to as photoaging, increased consumer awareness has led to the development of more effective photoprotective agents.

Beginning with work by Pathak and Stratton (1), the presence of free radicals in skin, after exposure to ultraviolet (UV) light, became established. Later, Meffert et al. (2) found the presence of thiobarbituric acid–reacting substances (TBARS) in UV-exposed epidermis, which indicated the possible involvement of oxygen free radicals in photodamaged skin. Thus, intervention by antioxidants in the form of topically applied vehicles became, in addition to UV absorbers, another approach for protecting skin. Although free radicals have been implicated as a cause of aging (3,4), this chapter discusses the differences between aging and photoaging and how this latter form of skin damage may be caused by free radicals. Mechanisms of antioxidant action in preventing this damage are also reviewed, followed by a summary of some of our laboratory's efforts to more clearly define the relationship between UV-induced oxidative stress in skin and its protection by antioxidants.

II. PHOTOAGING IS DISTINCT FROM AGING

Two kinds of aging are recognized: intrinsic and extrinsic. Intrinsic aging occurs inevitably as an organism progresses in chronological age, whereas extrinsic aging results from environmental effects such as UV exposure. UV exposure–related aging has been shown to be distinct from intrinsic or premature aging and is caused by chronic solar exposure leading to actinic damage in skin. Studies by Kligman (5) and Gilchrest (6) have clearly differentiated these two phenomena. In addition, the difference between exposed and unexposed skin, even when examined on the same subject, has been shown to be significant (7).

The action spectra for photodamage are generally designated as UVA (400–320 nm) and UVB (320–280 nm). Although the dermis is situated beneath the epidermis, it is still susceptible to damage, particularly to the longer wavelengths (Fig. 1). As demonstrated by Kligman (8), UVA penetrates deeper into the dermis than UVB and is responsible for some of the resultant damage. In the dermis, sun-damaged skin is characterized by a thickening of the

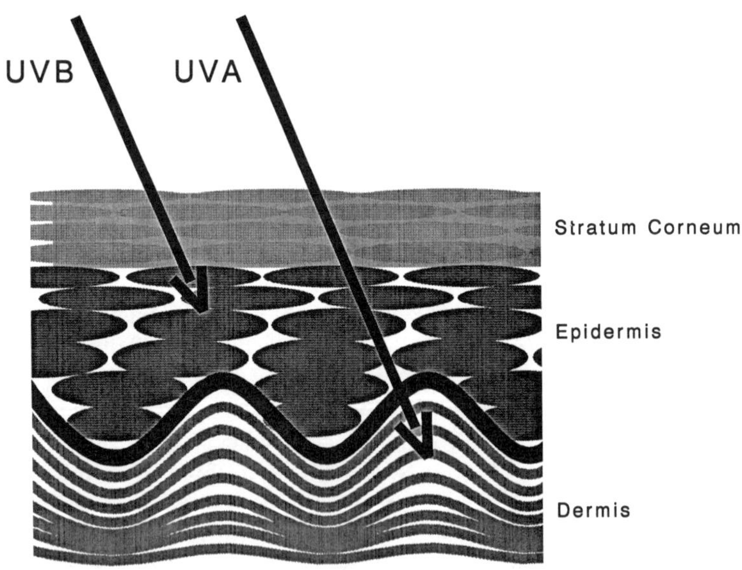

Figure 1 The depths of penetration by ultraviolet light. UVB penetrates into the epidermis, while UVA penetrates deeper into the dermis.

elastin matrix, which appears tangled and disorganized (9). Although there is an apparent hyperplasia (10), pockets devoid of elastin persist and have been observed at the ultrastructural level by Braverman and Fonferko (11). These workers also noted differences but additionally found sun-damaged skin to contain elements of both photoaged and chronologically aged skin.

Collagen is also significantly affected by UV and free radicals. Here a decrease in collagen bundles (12) with a concomitant increase in soluble collagen (13) have been detected. This increase was found to represent new collagen synthesis after hairless mice had been subjected to long-term UVB exposure but protected with sunscreens (14). This indicated not only that damage could be inhibited but also that repair could be stimulated if damage were limited. Glycosaminoglycans similarly increase (15), as do the number of fibroblasts that exhibit active metabolism (16). Another difference is an increase in membrane-like material around microvessels, which leads to a thickened microvasculature (17). Furthermore, partially degranulated mast cells have also been found to infiltrate the dermis (18) leading to chronic inflammation.

In epidermis, aged skin is generally observed to be thinned, whereas photoaged skin is coarse and pigmented (19). Melanin content increases (and therefore melanin radicals), Langerhans cells decrease (20), and keratinocyte cell cultures exhibit lower plating efficiencies (21). UVB absorption by DNA chromophores also leads to thymine-thymine (TT) dimerization followed by neoplastic growth if the cell's repair capacity is deficient. Thus, although the acute effects of UV such as erythema are present, many long-term changes are observed in skin relative to UV overexposure, such as wrinkles and loss of elasticity (22). Nevertheless, these changes can be minimized not only by the application of UV absorbers but also by antioxidants.

III. PHOTOPROTECTION BY ANTIOXIDANTS

As outlined in other parts of this volume, antioxidant defense mechanisms are present in cutaneous tissue. However, the requirements for protection may be greater in skin depending on the amount of ultraviolet insult. Work done by Danno et al. (23) described the role of oxygen intermediates and antioxidants in epidermal cells after UV irradiation by showing that when mice were irradiated and then injected intravenously with superoxide dismutase (SOD), the number of sunburn cells decreased. Although superoxide anions are not considered very reactive, it is the dismutation reaction that produces hydrogen peroxide followed by the Fenton reaction, which then generates hydroxyl radicals, which are believed to account for much of the observed cellular damage (Fig. 2).

According to Ogura et al. (24), hydroxyl radicals were clearly implicated by ESR measurements to be directly responsible for lipid radical formation

$$O_2^{\cdot-} + O_2^{\cdot-} + 2H^+ \rightarrow O_2 + H_2O_2$$

$$H_2O_2 + Fe^{2+} \rightarrow OH^{\cdot} + OH^- + Fe^{3+}$$

Figure 2 Superoxide radicals react to form hydrogen peroxide followed by reaction with reduced iron to form hydroxyl radicals.

in irradiated epidermal homogenates. This experiment agreed with studies by Fuchs et al. (25), who previously had evaluated the cutaneous antioxidant response to these UVB-generated oxygen free radicals and found no significant reductions in superoxide dismutase but did find decreases in the levels of tocopherol, ubiquinol, and catalase. Similar results, though less dramatic, were found when UVA was used in a follow-up study (26). Glutathione, a reductive type antioxidant and a cofactor for glutathione peroxidase, also decreased in the above experiments and, as demonstrated by Hanada et al. (27), correlated with the production of UV-damaged cells. Recently, active oxygen species have also been shown to be responsible for UV-induced decreases in Langerhans cells but only after the depletion of endogenous superoxide dismutase (28).

Furthermore, oxygen intermediates cause membrane damage in the form of lipid peroxidation (29), which then allows other free radicals and response modifiers such as calcium to leak out and cause further havoc. Erythema is another source of free radical–mediated damage, which is controllable by antioxidants. For example, during periods of inflammation, surveillance cells are recruited which contain neutrophils in the infiltrate. Neutrophils are involved in the formation of oxygen free radicals, such as superoxide anions ($O_2^{\cdot-}$) and hydroxyl radicals ($OH^{\cdot}$) (30), and can ingress into the dermal matrix, thereby causing tissue damage at the inflammatory site. Also during normal metabolism, oxygen free radicals form as a consequence of the univalent reduction of oxygen in the electron transport chain or from some other endogenous source. Thus, increased metabolism by fibroblasts after UV treatment may lead to a rise in the spectrum of free radical damage.

Other recent developments in this area include descriptions of some well-characterized antioxidant interactions, like the regeneration of tocopheroxy radicals by ascorbic acid, which has now been demonstrated in UV-irradiated skin (31). Another study carried out by Maeda et al. (32) assessed the long-term effects of UVA on mice and were able to correlate changes in collagen with that of antioxidant levels in skin. Finally, DNA can react with hydroxyl radicals to form OH adducts (33) or with hydroperoxides (34), leading to damaged DNA unless the cell can deactivate these reactive species. In summary, all of these findings clearly establish a causal link between photodamage and its prevention by antioxidants (Table 1).

Table 1 Effects of UV Light on Skin and Free Radical Associated Changes (FRAC)

Location	Target	FRAC
Stratum corneum	Ceramides	?
	Cholesterol	?
Epidermis	Melanin	Yes
	Membrane lipids	Yes
	Sunburn cell formation	Yes
	Cell plating efficiency	?
	Inflammation	Yes
	DNA damage	Yes
	Langerhans cell depletion	Yes
Dermis	Elastosis	Yes
	Collagen destruction	Yes
	Fibroblast increase	Yes
	Glycosaminoglycans	?
	Inflammatory filtrate	Yes
	Thickened microvasculature	?

IV. SUMMARY OF EXPERIMENTAL RESULTS

Aside from their ability to scavenge and reduce free radicals, antioxidants must meet certain other requirements if they are to be used in a topically applied product. First, they must be stable at room temperature and maintain activity in product vehicles. Second, they must be small, lipophilic enough, or packaged well enough in liposomal vehicles to penetrate the stratum corneum and migrate into the epidermis where much of the experimentation described above has been done. This does not mean that the stratum corneum is neglected as a target site, because it too contains unsaturated fatty acyl side chains in its ceramides and unsaturated double bonds in its cholesterol. Together these two molecules comprise approximately 70% of the dry lipid weight of stratum corneum (35), both of which are susceptible to oxidation. UV has also been found to peroxidize squalene on the surface of human skin (36). Hence, effective product development entails stability, a multitargeted approach based on various skin sites, and a mix of antioxidants to deactivate the range of radicals generated. This section describes some of our laboratory's work on stability and antioxidant activity at the biochemical, cellular, and tissue levels.

A. Stability

Tocopheryl acetate is a common ingredient and is esterified for easier formulation. After application, it becomes activated by esterases in skin, which cleave the acyl moiety to produce the topcopherol form. Its stability has been studied in our laboratory by assaying for its continued presence in products by high performance liquid chromatography (HPLC). Aliquots from finished products were removed at various time points and analyzed with a C18, reverse-phase column using a mobile phase consisting of methanol/water (90:10). Tocopheryl acetate was detected by its UV absorption at 286 nm. As can be seen from Figure 3, tocopheryl acetate maintained its integrity for more than one year. Further work assessing photodestruction of antioxidants by sunlight is another area of active research in our laboratory.

B. Biochemical

In order to quickly screen antioxidants and compare their relative antioxidant potential, a test system was developed that assessed lipophilic antioxidants in membranes of unilamellar phosphatidylcholine liposomes or hydro-

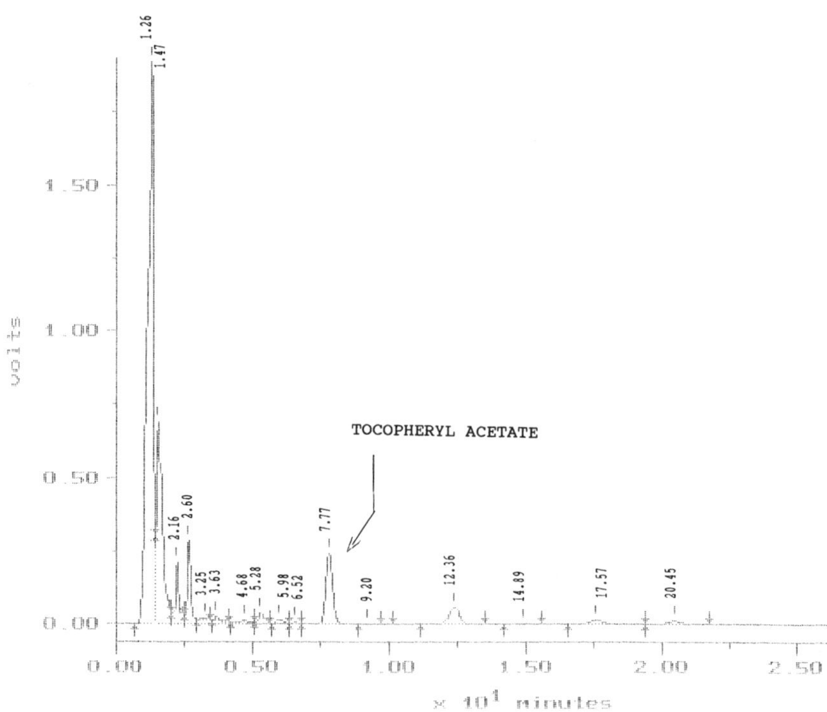

Figure 3 HPLC chromatogram of 12-month sample containing 95% of starting amount of tocopheryl acetate. Samples were made 20% in methanol and 50 μl injected into a 30-cm, ODS column. Calculations were made by comparing peak areas with those of standards.

philic antioxidants in the aqueous phase of liposomal suspensions before and after UV irradiation (37,38). In this system, TBARS were induced by either UVA or UVB. Routinely, however, UVC (280–220 nm) was used because the same results were found for any type of UV except that samples could be irradiated much faster with UVC due to its higher energy. UV absorption by the antioxidant itself had little bearing on the final result. This was illustrated by the fact that tocopheryl linoleate (TL), which has nearly the same UV absorption spectrum as tocopherol, was inactive in this assay system, whereas an equivalent amount of tocopherol at 5 mol% inhibited TBARS production by 58%. TBARS inhibition by antioxidants was shown to be UV dose-and antioxidant concentration-dependent (Fig. 4). Thus, free radicals could be induced by UV and consequent lipid peroxidation inhibited by various antioxidants in a comparative manner (Table 2). The validity of this model was further reinforced by measuring the reduction of linoleate and linolenate side chains of phosphatidylcholine after irradiation and also by determining tocopheroxy radical recycling by ascorbic acid. Nevertheless, this assay is not inclusive for all antioxidants since some well-characterized antioxidants like glutathione were ineffective whereas peroxyl radical scavengers were very effective. Rather, these data provide a selective window as to what may occur in vivo relative to peroxyl radicals and how they are involved in the complex mechanisms of oxidative stress in skin.

Superoxide anion scavengers have been tested for activity in our laboratory with conventional assays. Cytochrome *c* reduction (39) and pyrogallol oxidation (40) are routinely used. The activity of hydroxyl radical scavengers is determined using the conversion of dimethyl sulfoxide to methanesulfinic acid (41).

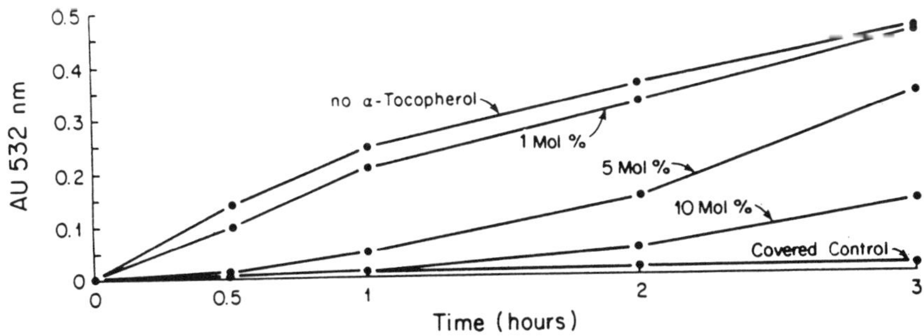

Figure 4 Inhibition of UVC-induced (2.8 mW/cm²) TBARS in the presence of increasing amounts of alpha-tocopherol over time. TBARS were measured by their absorbance maxima at 532 nm.

Table 2 Comparative Antioxidant Efficacy[a]

Antioxidant	Average % inhibition
Lipophilic antioxidant (concentration = 5 mol%)	
Nordihydroguaiaretic acid (NDGA)	100.0
Butylated hydroxyanisole (BHA)	96.0
Catechin	89.8
Butylated hydroxytoluene (BHT)	76.4
alpha-Tocopherol	58.0
Chlorogenic acid	16.4
Hydrophilic antioxidant (concentration = 5 mM)	
Uric acid	61.1
Cysteine	52.7
Ascorbic acid	50.2
Dithiothreitol	20.2
Glutathione	0
Superoxide dismutase (500 units)	0

[a]At 2 hours UVC exposure.

C. Cellular

Human keratinocytes in cell culture and their response to UV radiation is another way in which antioxidant activity is evaluated in our laboratory. Since oxygen free radicals are utilized in the biosynthesis of inflammatory mediators and also in their deployment in tissue (42), controlling these processes should alleviate photodamage in skin. Initiation of the arachidonic acid (AA) cascade following UV trauma is the rate-limiting step in production of mediators like prostaglandins. As described by De Leo et al. (43), UV-stimulated phospholipase A_2 releases arachidonic acid from membrane-bound phospholipids. Our work focused on extending this finding in order to assay for antioxidant efficacy (44). Briefly, tritiated AA was incubated with cells prior to irradiation. After either UVA, UVB, or UVC treatment, tritium was detected in media by liquid scintillation counting. When a mix of antioxidants was included either before or just after exposure, decreased levels of tritium were detected. As shown in Figure 5, antioxidants appear to inhibit this reaction.

D. Tissue Studies

Topical application of antioxidants has been shown to yield significant reductions in photodamaged skin. Roschupkin et al. (45) demonstrated this with a single high dose of UV, and work by Bisset et al. (46) determined photoprotection against long-term effects. This latter report evaluated gross

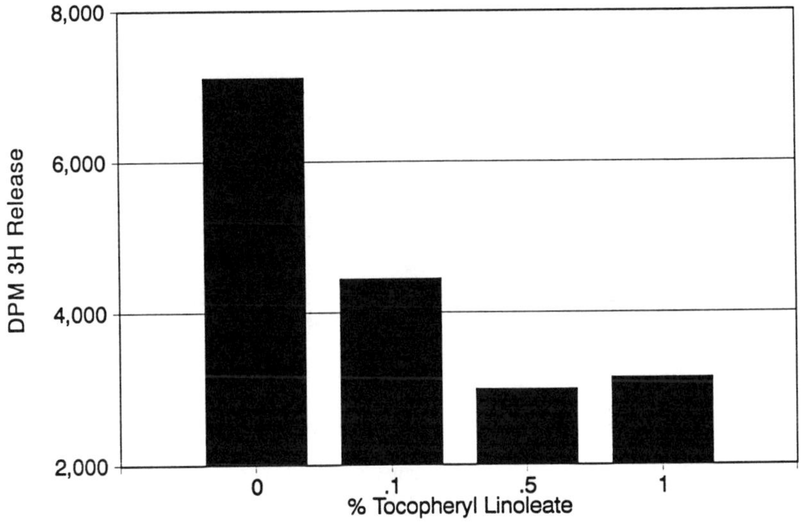

Figure 5 Tritium release from AA-labeled human keratinocytes after UVB irradiation (100 mJ/cm²). Increasing amounts of tocopheryl linoleate inhibited release into cell media.

morphologic changes and agreed with earlier work by Black et al. (47), who found inhibition of UV-induced tumorigenesis by antioxidants. Our approach was to measure photoprotection with more sensitive indicators over a shorter period of time (48). Over a 12-week period, one group was exposed to suberythemal doses of UVB (ca. 150 mJ/cm²), while another group, exposed in the same manner, were treated with a mix of antioxidants. In addition, two control groups receiving either no treatment or only application of product vehicle were similarly tested. At various time points, tissues were obtained and measured for the presence of UV-damaged epidermal cells, ultrastructural changes, and TT dimers. Figure 6 illustrates the effectiveness of antioxidant treatment in preventing damage to epidermal cells. Also, minimal cytoplasmic disorganization of viable cells was observed with transmission electron microscopy. Lastly, the antioxidant-treated group had fewer TT dimers as determined by sensitivity to T4 endonuclease V followed by agarose gel electrophoresis size separation and visualization by ethidium bromide staining.

E. Clinical Studies

In a preliminary set of experiments, it was observed that copper diisopropyl salicylate (Cu-DIPS), a superoxide dismutase mimic, significantly reduced

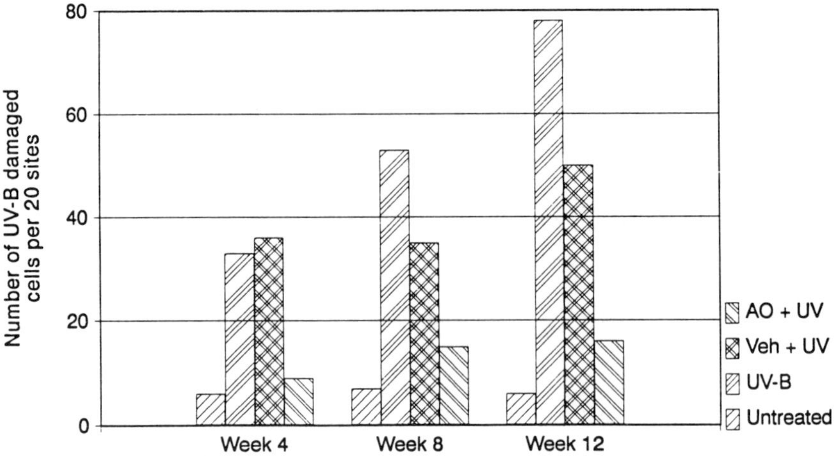

Figure 6 Chronic UVB exposure increased the number of damaged epidermal cells as determined by the presence of pyknotic nuclei, shrunken cytoplasm, and sunburn cells. Antioxidant mix: NDGA, BHT, ascorbic acid, TL, and SOD.

UV-induced erythema, which agreed with Sorenson (49), who had first used this molecule to control inflammatory episodes in arthritics. This finding was followed by a clinical study (50) in which subjects were treated with both a sunscreen and the same sunscreen spiked with 0.3 mM Cu-DIPS. After application, subjects were then irradiated with erythemogenic doses of UVB and graded visually for redness by an independent panel. This experiment was repeated with two other sunscreens containing different concentrations of UV absorbers. Our results from this study showed that in each case, sunscreens containing Cu-DIPS inhibited erythema more than the corresponding unspiked sunscreen.

Another study (51) was carried out which evaluated immediate pigment darkening (IPD) in UVA-irradiated human skin. Since the work by Pathak and Stratton implies photooxidation of melanin because of free radical generation by UVB, IPD may follow a similar pathway. Hence, antioxidants were used to inhibit this effect in a test consisting of ten panelists with Type III to V skin. After determining the minimum IPD dose requirement for each panelist, an antioxidant mix was applied and the exposure repeated until an IPD response could be visually determined. The ratio of IPD dose in protected skin to IPD dose in unprotected skin was used as a measure for determining protection factors. As a result, the average ratio for vehicle-only treated skin was 1.68 (± 0.17) as compared to the average ratio for antioxidant-treated skin which was 2.07 (± 0.21); significantly showing protection by antioxidants.

V. SUMMARY

Over the last several years, much progress has been made in delineating differences between aging and photoaging skin and how free radicals play an integral part in facilitating photodamage. Instrumental to this increased understanding has been the study of antioxidant behavior during oxidative stress and its role in photoprotection. Yet, despite intensive research, the mechanisms of oxidative damage to biomolecules remain somewhat controversial and idiopathic. Nevertheless, research in this field has clearly demonstrated the beneficial effects of topical antioxidant usage. By replacing antioxidants, which decrease upon acute or chronic UV exposure, significant UV protection can be achieved.

Further, new ways are being found to make better use of antioxidants. Indeed, antioxidants may mediate photoprotection in ways different from the commonly accepted modes of action. For example, although retinoids have antioxidant activity (52), they may also provide protection from UV penetration by increasing epidermal thickness. Thus, by combining new discoveries such as these with the findings outlined in this review, we are challenged by the exciting prospect of learning more about UV damage to skin and developing more effective photoprotective products.

ACKNOWLEDGMENTS

The authors gratefully acknowledge the work of other members of our laboratory (in alphabetical order): Kasman, K., Mammone, T., McKeever, M., Muizzuddin, N., Oppenheimer, J., Rothstein, S., and Sparacio, R. (nee: Colletti).

REFERENCES

1. Pathak MA, Stratton K. Free radicals in human skin before and after exposure to light. Arch Biochem Biophys 1968; 123:468–76.
2. Meffert H, Diezel W, Sonnichsen N. Stable lipid peroxidation products in human skin: detection, ultraviolet light-induced increase, pathogenic importance. Experientia 1976; 32:1397–8.
3. Harman D. Free radical theory of aging: the "free radical" diseases. Age 1984; 7:111–31.
4. Cutler RG. Aging and oxygen radicals. In: Taylor AE, Matalon S, Ward P, ed. Clinical Monograph Series. Bethesda, MD: Amer. Physiol. Soc., 1986: 251–85.
5. Kligman LH. Photoaging: manifestations, prevention, and treatment. Dermatol Clinics 1986; 4:517–28.
6. Gilchrest BA. Dermatoheliosis (sun-induced aging). In: Skin and aging processes. Boca Raton, FL: CRC Press, 1989:97–116.
7. Lavker RM. Structural alterations in exposed and unexposed aged skin. J Invest Dermatol 1979; 73:59–66.

8. Kligman LH, Akin FJ, Kligman AM. The contributions of UVA and UVB to connective tissue damage in hairless mice. J Invest Dermatol 1985; 84:272–6.
9. Kligman AM. Early destructive effect of sunlight on human skin. JAMA 1969; 210:2377–80.
10. Sams, Jr, WM, Smith, Jr, JG. Alterations in human dermal fibrous connective tissue with age and chronic sun damage. In: Montagna W, ed. Advances in Biology of Skin. Vol VI. Oxford: Pergamon Press, 1964:199–210.
11. Braverman IM, Fonferko E. Studies in cutaneous aging: I. The elastic fiber network. J Invest Dermatol 1982; 78:434–43.
12. Lavker RM, Zheng P, Dong G. Aged skin: a study by light, transmission, and scanning electron microscopy. J Invest Dermatol 1987; 88:44s–51s.
13. Smith, Jr, JG, Davidson EA, Sams, Jr, WM, Clark RD. Alterations in human dermal connective tissue with age and chronic sun damage. J Invest Dermatol 1962; 39:347–50.
14. Kligman LH, Akin FJ, Kligman AM. Sunscreens promote repair of ultraviolet radiation-induced dermal damage. J Invest Dermatol 1983; 81:98–102.
15. Sams, Jr, WM, Smith, Jr, JG. The histochemistry of chronically sun damaged skin. J Invest Dermatol 1961; 37:447–53.
16. Kligman LH, Akin FJ, Kligman AM. Prevention of ultraviolet damage to the dermis of hairless mice by sunscreens. J Invest Dermatol 1982; 78:181–9.
17. Braverman IM, Fonferko E. Studies in cutaneous aging: II. The microvasculature. J Invest Dermatol 1982; 78:444–8.
18. Lavker RM, Kligman AM. Chronic heliodermatitis: a morphologic evaluation of chronic actinic dermal damage with emphasis on the role of mast cells. *J. Invest. Dermatol. 90*, 325–330.
19. Kligman LH. Skin changes in photoaging: characteristics, prevention, and repair. In: Balin AK, Kligman AM, ed. Aging and the skin. New York: Raven Press, 1989:331–46.
20. Gilchrest BA, Murphy GF, Soter NA. Effect of chronologic aging and ultraviolet irradiation on Langerhans cells in human epidermis. J Invest Dermatol 1982; 79:85–8.
21. Gilchrest BA. Relationship between actinic damage and chronologic aging in keratinocyte cultures of human skin. J Invest Dermatol 1979; 72:219–23.
22. Cerimele D, Serri F. Clinical and histological alterations of human skin from sunlight. In: Galo U, Santamaria L, ed. Research progress in organic, biological, and medicinal chemistry. Vol 3. New York: American Elsevier, 1972:623–39.
23. Danno K, Horio T, Takigawa M, Imamura S. Role of oxygen intermediates in UV-induced epidermal cell injury. J Invest Dermatol 1984; 83:166–8.
24. Ogura R, Sugiyama M, Nishi J, Haramaki N. Mechanism of lipid radical formation following exposure of epidermal homogenate to ultraviolet light. J Invest Dermatol 1991; 97:1044–7.
25. Fuchs J, Huflejt ME, Rothfuss LM, Wilson DS, Carcamo G, Packer L. Impairment of enzymic and nonenzymic antioxidants in skin by UVB irradiation. J Invest Dermatol 1989; 93:769–73.
26. Fuchs J, Huflejt ME, Rothfuss LM, Wilson DS, Carcamo G, Packer L. Acute effects of near ultraviolet and visible light on the cutaneous antioxidant defense system. Photochem Photobiol 1989; 50:739–44.

27. Hanada K, Gange RW, Connor MJ. Effect of glutathione depletion on sunburn cell formation in the hairless mouse. J Invest Dermatol 1991; 96:838–40.
28. Horio T, Okamoto H. Oxygen intermediates are involved in ultraviolet radiation-induced damage of Langerhans cells. J Invest Dermatol 1987; 88:699–702.
29. Dixit R, Mukhtar H, Bickers DR. Studies on the role of reactive oxygen species in mediating lipid peroxide formation in epidermal microsomes of rat skin. J Invest Dermatol 1983; 81:369–75.
30. Babior BM. Oxidants from phagocytes: agents of defense and destruction. Blood 1984; 64:959–66.
31. Packer L, Kagan VE, Witt EH, Fuchs J, Goldman R, Scita G. Light-induced generation of chromanoxyl radicals and their recycling in skin: protective or damaging effect of vitamin E? J Invest Dermatol 1991; 96:551 (117A).
32. Maeda K, Naganuma M, Fukuda M. Effects of chronic exposure ultraviolet-A including 2% ultraviolet-B on free radical reduction systems in hairless mice. Photochem Photobiol 1991; 54:737–40.
33. Jovanovic SV, Simic MG. The DNA guanyl radical: kinetics and mechanisms of generation and repair. Biochim Biophys Acta 1989; 1008:39–44.
34. Roti Roti JL, Cerutti PA. Gamma ray-induced thymine damage in mammalian cells. Int J Rad Biol 1974; 25:413–17.
35. Yardley HJ. Epidermal lipids. In: Goldsmith LA, ed. Biochemistry and physiology of the skin. Oxford: Oxford University Press, 1983:363–81.
36. Ohkido M, Yoshino K, Matsuo I. Lipid peroxide of human skin. In: Current problems in dermatology. Vol. 10. Tokyo: University of Tokyo Press, 1981: 269–78.
37. Pelle E, Maes D, Padulo GA, Kim E-K, Smith WP. In vitro model to assess alpha-tocopherol efficacy. Ann NY Acad Sci 1989; 570:491–4.
38. Pelle E, Maes D, Padulo GA, Kim E-K, Smith WP. An in vitro model to test relative antioxidant potential: ultraviolet-induced lipid peroxidation in liposomes. Arch Biochem Biophys 1990; 283:234–40.
39. McCord JM, Fridovich I. Superoxide dismutase. J Biol Chem 1969; 244:6049–55.
40. Marklund S, Marklund G. Involvement of the superoxide anion radical in the autoxidation of pyrogallol and a convenient assay for superoxide dismutase. Eur J Biochem 1974; 47:469–74.
41. Babbs CF, Steiner MG. Detection and quantitation of hydroxyl radical using dimethyl sulfoxide as molecular probe. In: Packer L, Glazer AN, ed. Methods in enzymology. Vol. 186 (Pt. B). San Diego, CA: Academic Press, 1990:137–47.
42. Pryor WA, ed. Free radicals in biology. Vol VI. San Diego, CA: Academic Press, 1984.
43. De Leo VA, Hanson D, Weinstein IB, Harber LC. Ultraviolet radiation stimulates the release of arachidonic acid from mammalian cells in culture. Photochem Photobiol 1985; 41:51–6.
44. Mammone T, Jadlos S, Muizzuddin N, Marenus K, Maes D, Smith WP. Inhibition of UV-B induced arachidonic acid release in vitro as a model for evaluation of anti-inflammatory ingredients. J Soc Cosmet Chem 1989; 40:245A.
45. Roshchupkin DI, Pistsov MY, Potapenko AY. Inhibition of ultraviolet light-induced erythema by antioxidants. Arch Dermatol Res 1979; 266:91–4.

46. Bissett DL, Hillebrand GG, Hannon DP. The hairless mouse as a model of skin photoaging: its use to evaluate photoprotective materials. Photodermatology 1989; 6:228-33.
47. Black HS, Lenger WA, Gerguis J, Thornby JI. Relation of antioxidants and level of dietary lipid to epidermal lipid peroxidation and ultraviolet carcinogenesis. Cancer Res 1985; 45:6254-9.
48. Marenus K, Muizzuddin N, Kasman K, Colletti R, Meiselman N, Maes D, Smith WP. The use of antioxidants in providing protection from chronic suberythemal UV-B exposure. Proc Intl Fed Soc Cosmet Chem 1990; 16:24-34.
49. Sorenson JRJ. Copper chelates as possible active forms of the antiarthritic agents. J Med Chem 1976; 19:135-48.
50. Pelle E, Marenus K. Panel study conducted at Harrison Research Labs, Inc., Study No.: 87-7009RF, 1987.
51. Muizzuddin N, Marenus K, Maes D. Assessment of UV-A protection by sunscreen ingredients and products. American Academy of Dermatology Meeting, Dec. 1991, Abstract.
52. Hiramatsu M, Packer L. Antioxidant activity of retinoids. In: Packer L, ed. Methods in enzymology. Vol. 190 (Pt. B). San Diego, CA: Academic Press, 1990:273-80.

9

Reactive Oxidants in Photosensitized Skin Inflammations

Narendranath S. Ranadive and I. Aravind Menon
University of Toronto, Toronto, Ontario, Canada

I. INTRODUCTION

Phototoxic cutaneous reactions are commonly observed in skin exposed to large doses of light. Similar effects are seen with moderate doses if photosensitizers are present in the skin. These phototoxic reactions are usually manifested as immediate or delayed erythema followed by hyperpigmentation and peeling or scaling of the skin. In the early phase of moderate sunburn reactions, edema with dermal infiltrate consisting mainly of neutrophils has also been observed (1). Phototoxic reactions seem to have complex mechanisms, and the response under experimental conditions seems to depend on a number of factors such as the wavelength of incident radiation, duration of exposure, type of skin color, penetration of radiation, endogenous or exogenous molecules present at the site, chemical nature of the reactive species generated by the absorption of incident light, and the site of localization of a given photosensitizer.

Chronic exposure for a long period of time results in a number of changes in skin such as wrinkling, atrophy, altered pigmented macule, dilation of superficial blood vessels in skin, and actinic keratoses. Chronic sun exposure has been shown to alter functional activity of cellular components such as a change in melanin-forming activity in basal melanocytes or impaired transfer of melanin from melanocytes to keratinocytes (2) and alterations in fibroblast function, which may be related to connective tissue damage as seen in actinic elastosis (3). The involvement of various cytokines generated and released at the site have also been implicated in the initial damage (4). The loss of functional activity of Langerhans cells may impair the immunologic mechanisms associated with normal physiological defense processes (5,6). Sunburn reactions are characterized by immediate transient erythema followed by variable latent periods during which the skin appears normal; the subsequent erythema and edema are relatively prolonged. Inflammatory response is generally followed by epidermal hyperproliferation and hyperpigmentation. Histologic and biochemical evidence of injury to the surrounding tissue and of the repair process in the skin sites is detectable for several days (7).

The phototoxic reactions induced by photosensitizers can easily be demonstrated by applying the test agent to the skin and then exposing the skin to the appropriate light source. These reactions in most cases are evaluated in terms of the degree of erythema and edema produced (8,9). Some studies on the inflammations induced by ultraviolet radiation (UVR) and photosensitizers such as anthracene and porphyrins have been conducted with respect to vascular permeability and hyperemia using ^{125}I-labeled albumin and ^{51}Cr-labeled erythrocytes in mouse and guinea pig skin (10–13). These studies demonstrate the usefulness of the radioactive tracer technique in the study of the dynamics of phototoxic reactions. Recent investigations on the quantitation of phototoxic reactions have thrown some light on the mediators involved and the kinetics of the reactions (14–17).

II. BIOLOGICAL SIGNIFICANCE OF REACTIVE SPECIES

There are three main types of photochemical reactions induced by photoactivated sensitizers (18). Type I reactions proceed by generation of free radicals either through the charge transfer to O_2 to form superoxide anion (O_2^-) or through the interaction with suitable substrate such as unsaturated fatty acids:

$$S \text{—}h\nu \rightarrow {}^1\Delta S \rightarrow {}^3S$$

$${}^3S + RH \rightarrow R\cdot + H^+ + S\cdot^-$$

$$R\cdot + O_2 \rightarrow R\text{-}OO\cdot \quad \text{and} \quad S\cdot^- + O_2 \rightarrow O_2\cdot^- + S$$

Type II reactions are mediated by generation of singlet oxygen:

$$S \text{—}h\nu \rightarrow {}^1\Delta S \rightarrow {}^3S + O_2 \rightarrow {}^1\Delta O_2 + S$$

$${}^1\Delta O_2 + RH \rightarrow R\text{-}OO\cdot + O_2$$

$$R\text{-}OO\cdot + H^+ \rightarrow ROOH$$

The reactions of the excited photosensitizer directly with substrates such as proteins, lipids, and DNA to form adducts may be classified as Type III reactions (19).

Reactive oxygen species and free radicals act as mediators in a number of biological disorders. Because of the short life time of singlet oxygen ($^1O_2$), the effects of this species are restricted to the cellular location where it is formed (20,21). On the other hand, both O_2^- and H_2O_2 can survive for relatively longer periods, unless of course they are enzymically or chemically scavenged, and can exert the oxidative damage at distal locations. These effects often constitute a combination of oxidative reactions on several cellular components. Oxidative damage can produce lipid peroxidation, DNA cross-linking, protein cross-linking, etc. (22,23). These can induce a variety of conditions such as stroke, cataract, edema, aging, arthritis, inflammation, and cancer.

III. MECHANISTIC APPROACH TO THE EVALUATION OF PHOTOTOXIC REACTIONS

A number of in vitro experiments have demonstrated the cytotoxic effects of reactive oxygen species and other free radicals. These studies have been carried out using tissue specimens, cell suspensions, subcellular organelles, and cellular components. These systems provide simplified models for investigating the involvement of specific chemical species as well as for identifying the particular changes produced in the cellular components. These studies have provided indications as to the possible effects that may be produced in vivo.

A. In Vitro Studies

The in vitro studies (24,25) consist of cytotoxic reactions on the cell surface, or the study of the nature of reactive species involved in a given cellular damage and the cellular components involved, as well as effects of reactive species on the following cell functions: 1) activation of cells for release of biological mediators, 2) expression of cell surface molecules involved in cell communication, and 3) intracellular reactions with various organelles and macromolecules.

B. In Vivo Studies

The in vivo studies (26) consist of cutaneous inflammatory response induced by the following phototoxic reactions:

1. Gross anatomic observations, such as edema, erythema, scaling, etc.
2. Histology of the lesion

3. Quantitative evaluation of cellular infiltrate (neutrophils, macrophages, platelets, lymphocytes etc.), plasma fluid, hemorrhage
4. Kinetics of the biological events at the site of reaction
5. Biochemical analysis of the lesion; quantitation of enzymes of free radical scavenging systems, phospholipases, oxygenases, ornithine decarboxylase, and various products of oxidative processes

IV. ROLE OF MELANINS IN CUTANEOUS INFLAMMATORY REACTIONS

A. Physical and Chemical Properties of Melanins

Melanins are complex heteropolymers of a variety of colors—black, brown, red, etc.—occurring in a wide range of animals and plants (27). In the human body, melanins are present most conspicuously in the skin, hair, and eyes. These melanins are considered to be two chemically different but biogenetically interrelated types of pigments: eumelanin (black-brown in color and insoluble in dilute alkali) and pheomelanin (yellow to reddish-brown and soluble in dilute alkali). It is doubtful whether "pure" eumelanins and pheomelanins (i.e., "dopa melanin" and "cysteinyldopa melanin") commonly occur in nature. Many natural melanins are intermediates between pure dopa melanin and cysteinyldopa melanin (28–30).

Melanins catalyze several oxidation-reduction reactions involving reversible oxidation and reduction of melanins (31). Melanins contain free radicals, which can be detected by electron spin resonance (ESR) spectroscopy (32); ESR spectroscopy has been used to distinguish between eumelanin and pheomelanin (29,30).

B. Photoprotection by Melanins

Melanins absorb the entire range of UV and visible radiations. Hence it acts as an effective sunscreen (33,34). The early concept regarding the protective action of melanin was that it acts as a sunscreen. While this possibility has not been contradicted, the photochemical reactivity of melanins may have a role in the protective effect of melanins. It is possible that melanins may act as scavengers of reactive species formed during irradiation (35). Melanin protects against UV damage to DNA (36,37). The peroxidation of lipids induced by UVR alone and by UVR with pheomelanin was inhibited by eumelanin (I. A. Menon et al., unpublished results). Therefore it seems that the influence of melanins on the biological effects of UVR would be rather complex. While eumelanin as well as pheomelanin, both of which are present in skin (38), would absorb UVR and thus provide protection against the radiation, pheomelanin would also act as a photosensitizer. The overall effects would be the sum total of these effects. The exact effects produced in a given biological condition would depend upon a variety of factors, such as the type and quantity of melanin present, the spectrum of radiation, the presence of scavenging enzymes, e.g., catalase, superoxide dismutase (SOD), and peroxidase, and antioxidants, e.g., sulfhydryl compounds, ascorbic acid, etc.

C. Photosensitization by Melanins

UV irradiation affects the oxidation-reduction properties of eumelanin and pheomelanin differently (39). It is possible that the differences in the chemical structures of eumelanin and pheomelanin may result in differences in the number and excitability of the oxidizing and reducing groups as well as the free radicals in these melanins (40).

Irradiation of red hair melanin (a source of pheomelanin) (41) and dopa melanin (42) produces superoxide and H_2O_2. Irradiation of red hair melanin produces more superoxide than the irradiation of black hair melanin (a source of eumelanin) (43). Similarly the irradiation of red hair melanin induces the oxidation of NADH to a larger extent in comparison to the irradiation of black hair melanin (43). It is conceivable that other oxidative and reduction reactions by pheomelanin may be enhanced by UV irradiation more than in the case of eumelanin.

Until recently it was thought that, with respect to melanin, the quality of melanin was the major factor involved in protection against the deleterious chronic and acute effects of UVR, and that the susceptibility to UVR was due to the lesser amounts of melanin in the skin. More recent reports (16, 44–48) have indicated that not only the *total quantity* of melanin but also the *types* of melanin may be responsible for susceptibility to UVR. It has been definitely established that Celtic individuals have pheomelanin in the hair, and some reports indicate that the skin of these individuals may also have pheomelanin.

Irradiation of several types of cells, e.g., Ehrlich ascites carcinoma (EAC) cells, cultured human melanoma cells, and rat peritoneal mast cells, produces severe cellular damage, which may be observed by several techniques. Among the various photobiological effects of pheomelanin in vitro, the following may be cited: 1) cell lysis as observed by the leakage of ^{51}Cr from labeled cells, release of intracellular lactic dehydrogenase and transplantability of the cells (45), 2) release of histamine from mast cells (49), 3) lesions in intracellular DNA (unpublished results), 4) peroxidation of cellular unsaturated fatty acids (46,48,50), and 5) damage of the ultrastructure of cells (51). In all of the above cases, irradiation of the cells in the presence of pheomelanin produced more damage than irradiation with eumelanin. The presence of sulfhydryl compounds such as cysteine and glutathione increases the formation of H_2O_2 and also enhances the photobiological effects. The mechanism of these effects appears to be complex and would depend upon the composition of the media and the irradiation source. The available evidence points to the role of O_2^-, H_2O_2, free radicals produced during the irradiation of the melanin, and probably hydroxyl radical ($^{\cdot}OH$). The photochemical and photobiological properties of eumelanin and pheomelanin have been discussed in a recent review (47).

The phototoxic inflammatory response in the skin induced by the reactive species generated by melanin was studied using a rabbit model (16). It was observed that when low concentrations of pheomelanin were injected into the skin and the skin was exposed to UV-visible light, an increase in vascular permeability (iVP) as well as an accumulation of polymorphonuclear leukocytes (PMNs) were produced. On the other hand, there was no effect in the dark. At higher concentrations of pheomelanin there was a slight increase in vascular permeability and accumulation of polymorphonuclear leukocyte (aPMN) in the dark. On irradiation both these parameters were increased markedly (16).

Although melanin in the skin is present mostly within the melanocytes, mainly located in epidermal region, while mast cells are found in the dermal and subepidermal layers, the relatively long-lived reactive intermediates such as H_2O_2 and lipid hydroperoxides formed by the action of free radicals can diffuse to the site of target cells. Melanin is also present in skin in diffuse state as "melanin dust" (52). The question of the location of melanin within the melanosomes or as melanin dust with respect to the target cells involved in the effects of UV radiation is difficult to answer directly. However, in vitro studies lend themselves to some relevant considerations. Irradiation of melanin, especially in the presence of naturally occurring sulfhydryl compounds such as cysteine, produces H_2O_2, which is relatively stable and could diffuse across a number of cell layers (46). The irradiation of melanin in the presence of free fatty acids or cellular lipid components has been shown to produce lipid peroxidation (48,50,53). Some of the intermediates in lipid peroxidation such as hydroperoxides are fairly stable (23). Thus, although the reactive free radicals in melanin produced by irradiation are short-lived, it is conceivable that these free radicals could react with several biomolecules producing reactive compounds which are relatively long-lived and could permeate to other sites in the skin.

V. PORPHYRIN-INDUCED SKIN REACTIONS

The inflammatory reactions in skin induced by bacterial endotoxins and immune complexes have been studied using radiolabeled markers (54–56).

The radioactive tracers have been successfully used in quantitation and in study of the kinetics of the changes occurring in vascular permeability, accumulation of PMNs, hyperemia, thrombus formation, and hemorrhage. These studies have helped in the understanding of the mediators involved and the sequential mechanism in elaboration of the inflammatory response. The application of similar techniques in the study of phototoxic reactions has proved to be fruitful.

Guinea pigs systemically sensitized to protoporphyrin when exposed to light of wavelength 400–410 nm showed erythema, crusting, and delayed growth of hair at exposed sites. The histological examination showed dermal edema, dilation of blood vessels, and infiltration of polymorphonuclear and mononuclear cells. The blood samples taken one hour after irradiation showed dose-dependent inhibition of hemolytic activity of the complement. The erythrocytes from these samples had increased susceptibility to hemolysis induced by 400–410 nm light in vitro (57). Further studies with guinea pigs treated with cobra venom factor (depletion of complement components C3 and C5) and cyclophosphamide (depletion of PMN) showed that the cutaneous reactions induced by demethylchlorotetracycline followed by UVA irradiation are inhibited in treated animals. These studies suggest the participation of complement and PMNs in phototoxic inflammations (58). Moreover hematoporphyrin-evoked cutaneous phototoxicity has been shown to be mediated through mast cell degranulation, probably by the activation of the complement system, and the delayed response is partly due to PMNs (12). It is possible that the primary target may be mast cells, which upon degranulation induce vascular leakage at the sites resulting in local complement activation. This would create local inflammatory response at the site of irradiation.

Recent studies have shown that porphyrins plus light produce profound effects on the eicosanoid metabolism of mast cells, endothelial cells, macrophages, and epidermal and dermal homogenates, and induce activation of factor XII dependent pathways. In vitro, prostaglandin D_2 was generated when mast cells were exposed to protoporphyrin and radiation. Exposure of cultured human umbilical vein endothelial cells or human omental microvascular endothelial cells to uroporphyrin or protoporphyrin and radiation resulted in generation of $PGF_{2\alpha}$. Activities of epidermal eicosanoid-metabolizing enzymes in protoporphyric mice were suppressed following in vivo exposure to radiation, whereas activities of the dermal enzymes were enhanced 18 hours after radiation. These effects of porphyrins on skin seem to contribute to the development of porphyrin-induced cutaneous phototoxicity (59,60).

VI. CUTANEOUS REACTIONS INDUCED BY XANTHENE DYES

Xanthene dyes have been shown to induce phototoxic reactions as judged by the immediate wheal and flare followed by moderate to intense erythema and edema after 24 hours in scarified human skin (8). Similar observations were made by Morikawa et al. (9) in animals. A rabbit model was developed for the quantitation of inflammation at the sites of free radical generation and to study the sequence of events involved in the inflammatory response using Rose Bengal (RB), a xanthene dye, plus light. This model is being used to study the phototoxicity of various compounds, the mediators involved, and the mechanisms by which final injury is produced (14–17).

The studies showed dose response between the amount of RB injected at the site and the intensity of the reactions observed. As little as 0.01 nmol of RB produced demonstrable iVP and aPMN. These reactions also showed dependence on the period of irradiation. A clear inflammatory response was seen after 90 minutes of irradiation under the experimental conditions described (14).

Immediate phototoxic reactions induced by other xanthine dyes such as erythrosine, eosine Y, and fluorescein have been studied in rabbit skin (61). Increase in vascular permeability and accumulation of PMNs showed a similar pattern as that of RB; however, erythrosine was about 10 times and eosine Y about 40 times less effective than RB. Fluorescein was ineffective under similar conditions. The phototoxicity paralleled photosensitizing efficiencies of the triplet quantum yields for RB, erythrosine, eosine Y, and fluorescein, which are 0.75, 0.63, 0.57, and 0.03, respectively (62).

A. Effects of Scavengers of Reactive Oxygen Species

The inhibition of these reactions by β-carotene indicates that the observed cutaneous reactions were predominantly mediated by $^1O_2$. Moreover, SOD and catalase did not inhibit the RB-induced iVP and aPMN. This strongly suggests that the reactive species involved in the observed photic reactions was $^1O_2$. Although RB is known to generate predominantly $^1O_2$, the other possibilities cannot be completely ruled out, since it has been shown that under certain conditions β-carotene scavenges oxy-radicals, lipid peroxides, and the triplet states of the photosensitizers (63,64).

B. Role of Mast Cells in Photoinduced Reactions

Cutaneous phototoxic reactions to various photosensitizers such as coal tar, benoxaprofen, pyrenes, acridines, etc. occur within a few minutes after exposure to light and are characterized by burning and itching with development of wheal and flare formation (65). The reactions seem to be mediated by the release of inflammatory agents from dermal mast cells and generation of arachidonic acid metabolites (66,67). These reactions are distinguished from photoallergic contact dermatitis, which occurs late and resembles cell-mediated delayed hypersensitivity. In vitro studies have shown that irradiation of mast cells in the presence of optimal concentrations of porphyrins produced cell lysis; on the other hand, the suboptimal doses did not destroy the cell viability but rendered cells unresponsive to the secretory release of histamine by compound 48/80 (68–70). It has also been shown that the release of mediators from mast cells stimulated by acridine and light was correlated with the loss of cell viability, while the release induced by pyrene and light was much higher than the loss of cell integrity (71). Similarly, mast cells irradiated in the presence of pheomelanin at lower concentrations released higher amounts of histamine than cell lysis as determined by ^{51}Cr release (49). It has been shown that isolated mast cells can be stimulated for histamine release by a system consisting of H_2O_2, eosinophil peroxidase, and halide (72, 73). This histamine release is of noncytotoxic nature at low concentrations of peroxidase, and the cells undergo cytotoxic reactions only at high concentrations of peroxidase. It has also been shown that H_2O_2 added exogenously as well as that generated by the xanthine oxidase system activates mast cells for noncytotoxic histamine release (74). The type of damage to the cell membrane is related to both the concentration and the reactive chemical species generated (75). It is possible that the primary target for the reactive species generated during irradiation of the skin may be different, and the agents released or generated from these targets may act as degranulating or lytic substances on mast cells.

The iVP due to RB-induced phototoxic inflammation was inhibited up to 70% by chlorpheniramine maleate, an antihistaminic drug. This inhibition seems to be due to the antagonistic effect of the drug since in in vitro experiments, histamine and ^{51}Cr release from the isolated mast cells produced by light in the presence of RB was not affected by this drug. Thus the release of histamine, presumably from mast cells, at the site of the phototoxic reaction seems to be the predominant cause of iVP. However, the direct effect of the reactive species on the microvasculature may be a contributing factor. The results of the study on RB-induced phototoxic reactions in rabbit show that the photo-induced reactive species, mainly $^1O_2$, produce skin lesions by degranulation of mast cells as well as by recruiting other cellular components such as neutrophils and activating humoral factors, especially complement (14).

C. Generation of Tolerant State

The RB-induced increase in vascular permeability and PMN migration is of an immediate nature, i.e., occurs within the first 3 hours of irradiation. The reaction subsides gradually after 24 hours if the initial insult is not too severe.

When such moderate insult is repeated, the skin becomes tolerant to subsequent exposure to light in the presence of RB. The second exposure rendered the skin sites unresponsive (desensitized) to the third challenge 24 hours later. This unresponsive state is temporary, i.e., the desensitized sites are fully recovered in 72 hours. Since the iVP observed in phototoxic reactions, as mentioned above, seems to be predominantly due to the release of histamine at the site of injury, it was thought that the decrease in response after repeated exposure to light may be due to the destruction or inactivation of mast cells at the sites. This possibility was investigated by studying the response of photoinduced "desensitized sites" to the agents known either to activate mast cells or to stimulate aPMN by chemotaxis (15).

D. Factors Involved in Generation of Tolerant State

When compound 48/80, an agent known to degranulate mast cells, was injected into sites that were tolerant to RB plus light, iVP was not observed. This observation suggests that either the mast cells were depleted from the sites or the mast cells in the sites were rendered unresponsive to compound 48/80 by previous exposure to light (15). The development of unresponsive state of mast cells to various secretory stimuli and cross-desensitization to these stimuli is known (76–78). The secretory process of the histamine release has also been shown to be desensitized by previous exposure of the cells to photo-induced reactive species (69,71).

It was further observed that the sites desensitized to RB plus light were unresponsive to exogenously injected histamine. This suggests not only that mast cell depletion is involved in the observed desensitization but also that the component(s) required for the elaboration of the action of histamine are partially abrogated. It may be speculated on the basis of the work on the contractile activity of guinea pig *taeniae coli* induced by erythrosine and light (79) that the regulation of Ca-fluxes may be impaired. It is also possible that the receptors for histamine action (H1) may have been damaged by the reactive species.

When zymosan-activated serum (ZAS) was injected into desensitized sites, neither iVP nor aPMN occurred (15). This suggests not only that the ability of mast cells to respond to C5a des arg (anaphylatoxin in ZAS) is abrogated in the desensitized sites but also that the migration of PMNs in response to this factor is inhibited at the desensitized sites. Since migration of PMNs to extravascular spaces involves adherence of these cells to vascular endothelium, the loss of such property, probably due to the lack of expression of endothelial leukocyte adhesion molecule (ELAM-1) at irradiated sites, may cause the decrease in PMN migration when ZAS is injected into these sites.

It is also possible that changes in matrix due to cross-linking of the ground substance induced by reactive species may hinder the migration of PMNs.

E. Temporary Tolerance to Solar Urticaria

It is known that temporary tolerance to solar urticaria can be established by repeated exposure of the skin to subthreshold dose of radiation (80,81). These observations have been used by Ramsay (82) in successful treatment of solar urticarias using an actinic lamp (320–400 nm). Since mast cell degranulation was shown to be associated with solar urticaria (83), it was presumed that tolerance may be related to either a depletion of mast cells or an increase in threshold of the stimulus required for the degranulation of mast cells (84,85). This explanation is, however, not satisfactory since antihistaminic drugs had no effect on the minimal radiation dose required to induce uritcarial response (81,85). Moreover, the skin of urticaria patients made tolerant by exposure to light did not show an appreciable decrease in mast cell density or a change in appearance. It was also observed that the tolerant skin produced inflammatory response when challenged with mast cell degranulating agents, compound 48/80 (82), or codeine (84). It was therefore suggested that some other mediators and cell types may be contributing to this reaction. Recent studies on the induction of tolerance to solar "allergic urticaria" (IgE-mediated) suggest that the mast cell activation is blocked due to unavailability of allergen-binding sites of IgE (86). Solar urticaria independent of IgE has a mast cell component, although other cell types are also involved. It is possible that when the initial reactions are minimal, subtle changes seem to occur at the reaction sites. These changes are manifested as the loss of normal pathophysiological response to the known mediators such as histamine and anaphylatoxin and conceivably to the various cytokines. The downregulation of the expression of adhesion molecules such as ELAM-1 and/or ICAM-1 involved in migration and accumulation of PMNs may be a contributing factor in tolerance (15). Psoralen-UVA (PUVA)-induced phototoxicity also manifests the tolerant state, which is independent of mast cells; here again the inhibition of PMN migration seems to be the cause of tolerance. This aspect will be discussed later.

VII. PHOTOSENSITIZED REACTIONS BY PSORALENS

Psoralens can be excited to the triplet state when irradiated with UVA and can modify membrane lipids and proteins by either oxygen-dependent or oxygen-independent pathways. It is well established that the topical application of 8-methoxy psoralen (8-MOP) followed by UVA radiation produces delayed erythema (24–48 hours postirradiation) accompanied by edema and sometimes vesiculation. Skin phototoxicity has been suggested to be due to the oxidation of unsaturated fatty acids in the membrane by free radical mechanism (87). Both type I reactions involving substrate oxidation by radicals and type II reactions involving the formation of $^1O_2$ in psoralen photosensitization have been reported (88,89). Covalent bonding between furocoumarins and membrane components by type III reaction and consequent intracellular events in $^1O_2$-independent reactions have been discussed recently by Dall'Acqua and Martelli (22). It should also be noted that the adduct formation between DNA and various furocoumarins has been extensively studied (90). Many characteristics regarding the inflammatory response developed by PUVA treatment have been described. The roles of psoralen-DNA adduct and the reactive oxygen species have been implicated in these reactions (91).

A. Phototoxic Reactions to Topically Applied Psoralens

It has been claimed that there is a correlation between $^1O_2$ generation and erythema formation by 5-methoxy psoralen, 8-MOP, trimethyl psoralen, and angelicin (92). However, the direct relationship between the reactive oxygen species and erythema is unlikely in view of the fact that the reactive oxygen species generated in skin sites injected with RB produced inflammatory reactions manifested by iVP and aPMN within the first 3 hours of irradiation without immediate or delayed erythema (14). On the other hand, the reactions induced by 8-MOP are distinct from those produced by $^1O_2$-generating dyes such as RB and erythrosine in that the former reactions reach full intensity 24 hours after irradiation and are marked by intense erythema, which persists for at least 72 hours, while the later reactions occur within the first 3 hours of irradiation without erythema and completely subside within 12 hours. Although both reactions manifest iVP and aPMN, 8-MOP reactions are quantitatively more severe and are of substantially longer duration than the RB-induced reactions. As expected, topical application of 8-MOP consistently evoked inflammatory reactions 24 hours after irradiation, which depended upon 8-MOP concentration and radiation dose (17). Moreover, it has been reported that there is no statistically significant protection by carotenoid against PUVA-induced erythema in human skin (93).

The fact that topically applied 8-MOP did not produce an immediate reaction suggests that the sites of reactive oxygen formation (if 8-MOP reactions are related to reactive oxygen) were such that the cells responsible for the immediate reactions were not accessible. In this regard, it may be recalled that the mast cell is an important cell type involved in the immediate reactions mediated by RB photosensitization. On the other hand, the development of phototoxic reactions at the sites where 8-MOP was topically applied was delayed and largely independent of mast cells (Table 1). It may be suggested from these observations that the major reaction mechanism in 8-MOP–induced phototoxicity is not mediated by $^1O_2$, but is due to the reaction between 8-MOP in the triplet state generated by light and cellular constituents. However, putative initial reaction, whether with DNA, proteins, or lipids, is still a matter of conjecture (19).

B. Effect of Depletion of Mast Cells

In order to see whether the delayed iVP and aPMN seen with 8-MOP were related to mast cells, the sites were depleted of mast cells by repeated injections of compound 48/80, an agent known to degranulate mast cells. 8-MOP was topically applied to the sites thus prepared and then exposed to light. The results (Table 2) clearly show that the removal of mast cells had very little effect on iVP and aPMN, suggesting that the mast cells play a minor role in 8-MOP–induced phototoxic reactions. In this regard it is interesting to note that the blockade of H1 receptor by Terfenadine had no effect on the intensity of erythema by UVB or UVC as measured with a reflectance instrument at 4, 8, and 24 hours after irradiation, suggesting that histamine,

Table 1 Effect of Depletion of Mast Cells on 8-MOP Induced Phototoxic Reactions in Rabbit Skin

	Plasma (μL)		PMN × 10^3	
Treatment	Dark	Light	Dark	Light
Blank sites	14.4 ± 0.7	15.4 ± 1.6	48.5 ± 7.4	50.6 ± 7.1
8-MOP[a]	25.5 ± 1.9	190.6 ± 16.0	84.4 ± 16.0	295.7 ± 18.8
8-MOP in mast cell–depleted sites[b]	26.2 ± 0.7	169.3 ± 8.5	86.6 ± 9.9	220.1 ± 12.8
Compound 48/80[3] in virgin sites	42.6 ± 1.7			
Comp. 48/80 in mast cell–depleted sites[c]	14.4 ± 0.5			

[a]8-Methoxypsoralen (0.1% cream) plus light treatment (day 4) given 24 h before sacrificing rabbit.
[b]The skin sites were depleted of mast cells by injecting 10, 20, and 30 μg of compound 48/80 on day 1, 2, and 3, respectively.
[c]10 μg Compound 48/80 injected 2 h before sacrificing rabbit.

Table 2 Effect of 4-Bromophenacyl bromide[a] on Phototoxic Reactions Induced by 8-MOP in Rabbit Skin

Treatment	μl Plasma (iVP)	Inhib. (%)	PMN × 10^3 (aPMN)	Inhib. (%)	Erythema
8-MOP[b]	85.6	0	75	0	+ + +
8-MOP + BPAB[c] 0.05%	26.5	69.0	41	54.3	–
8-MOP + BPAB[c] 0.10%	17.2	79.9	48	36.0	–
8-MOP + BPAB[c] 0.50%	16.3	81.0	38	49.3	–

[a]4-Bromophenacyl bromide in cream applied 24 h prior to irradiation.
[b]8-Methoxypsoralen; 0.1% (W/W) in cream base applied 30 min before irradiation.
[c]BPAB = 4-Bromophenacyl bromide concentrations (W/W) in cream base.

acting on the H1 receptor, is not a significant mediator of UVB or UVC erythema (94).

C. Modulation of Phospholipase A_2

The modulation of inflammatory response to topical 8-MOP plus light by 4-bromophenacyl bromide (BPAB), a phospholipase A_2 inhibitor, has recently been studied. As seen in Table 2, the 8-MOP plus light–induced iVP was inhibited and erythema was completely abolished. These results suggest that phospholipase A_2 activation is an obligatory step in the 8-MOP–induced cutaneous inflammatory response. It is interesting to note that recent in vitro studies show that keratinocytes, when irradiated by UVA (330–400 nm), release arachidonic acid and cyclooxygenase metabolites and increase the synthesis of phosphoryl choline. It has been suggested that irradiation of keratinocytes by UVA activates the cellular phospholipase A_2 and phospholipase C. Furthermore, UVB (400–1600 J/cm^2) stimulated a dose-dependent release of [^{3}H]-choline from choline-labeled phospholipids of human keratinocytes within minutes of irradiation. The released [^{3}H]-choline was predominantly in the form of glycerophosphorylcholine. Examination of the label remaining in membranes of cells after irradiation by HPLC revealed that the origin of the released [^{3}H]-choline was the membrane phosphatidylcholine/lysophosphatidylcholine. It seems that UVB stimulates phospholipase A and a lysophospholipase (95,96).

D. The Effect of Indomethacin, a Cyclooxygenase Inhibitor

It is generally believed that indomethacin decreases UVB-induced erythema by inhibition of prostaglandin (PG) synthesis (97,98). On the other hand, PUVA-induced erythema seems to be resistant to indomethacin (99). Schwarz et al. (100) observed the inhibition of PUVA-induced erythema; however, part of this inhibition was shown to be related to photoabsorption. Recent study of Imokawa and Tejima (101), however, clearly shows that the erythemal response in guinea pig induced by PUVA is indeed due to the activation of PG synthesis. Investigations on the effect of UVB irradiation on mature and less mature guinea pig keratinocytes showed that the amounts of PGs produced decreased in the following order: $PGF_{2\alpha} > PGE_2 > PGE_1 >$ 6-keto-$PGF_{1\alpha} > (TX)B_2$. These results suggest that the cyclooxygenase is being activated by UV irradiation (102).

Studies on the modulation of inflammatory response in rabbit skin induced by topical application of 8-MOP and UV-visible radiation by topically applied indomethacin, a known inhibitor of cyclooxygenase activity, revealed an interesting phenomenon. A low concentration of indomethacin (less than 5%) inhibited 8-MOP-induced iVP and erythema, but aPMN response was enhanced. On the other hand, high concentrations (7.5–10%) of indomethacin inhibited erythema, iVP, and aPMN (Table 3). These results suggest that indomethacin at low concentrations in vivo inhibits cyclooxygenase activity, while at higher concentrations it may inhibit the phototoxic reactions through its action on phospholipase A_2 activity.

E. Inhibition of 5-Lipoxygenase

Preliminary studies on the effect of 5-lipoxygenase inhibitor (MK886) on 8-MOP-induced phototoxicity showed that iVP and aPMN were completely inhibited, while there was no effect on erythemal response (Kumar et al.,

Table 3 Effect of Indomethacin Treatment[a] on Phototoxic Reactions Induced by 8-MOP in Rabbit Skin

Treatment	ul Plasma (iVP)	Inhib. (%)	PMN × 10^3 (aPMN)	Inhib. (%)	Erythema
8-MOP[b]	123.0	0	312	0	+ + + +
8-MOP + Indo[c] 0.45%	98.0	20.3	643	106.1[d]	+ +
8-MOP + Indo[c] 1.0%	82.4	33.0	483	57.4[d]	+ ±
8-MOP + Indo[c] 2.5%	73.3	40.4	562	80.0[d]	+ ±
8-MOP + Indo[c] 5.0%	27.4	77.7	322	3.3[d]	+
8-MOP + Indo[c] 7.5%	9.2	92.5	12	96.2	−
8-MOP + Indo[c] 10.0%	1.0	99.2	0	100.0	−

[a]Indomethacin cream was applied topically 24 h prior to irradiation.
[b]8-Methoxypsoralen 0.1% in cream base applied 30 min before irradiation.
[c]Indo; Indomethacin concentrations (W/W) in cream base.
[d]Percent enhancement.

unpublished results). Since MK886 has been shown to be a potent inhibitor of 5-lipoxygenase (103), these findings suggest that the arachidonic acid metabolites, mainly leukotrienes generated by 5-lipoxygenase, are involved in 8-MOP–induced iVP and aPMN.

F. Biochemical Parameters in Phototoxic Reactions

Wheeler et al. (104) found that although there is correlation between edema formation, induction of ornithine decarboxylase (ODC), and depletion of glutathione, there was no increase in oxidized glutathione in 8-MOP–UVA treated mice. Furthermore, they could not induce similar reactions in mice with 3-carbmethoxy-psoralen ($^1O_2$ producer). The depletion of glutathione and induction of ODC in the skin as a result of UVR or by 12-O-tetradecanoyl-phorbol-13-acetate (TPA) application has implications in tumor promotion (105,106). However, it is not clear whether decrease in glutathione and induction of ODC has any relation to the inflammatory response, namely, delayed erythema and edema, seen in 8-MOP–UVA treated mice. It is also not explainable how the decrease in glutathione would induce delayed edema, which seems to be related to iVP through the release of lysosomal mediators from emigrated PMNs. The generation of arachidonic acid metabolites by both cyclooxygenase and lipoxygenase pathways seems to be an important factor in this phototoxicity. This presumption is based on the observation that the topically applied indomethacin at low concentrations inhibits 8-MOP–induced iVP and erythema, while at higher concentrations the aPMN is also inhibited. Furthermore, we have recently observed that phospholipase A_2 inhibitor BPAB inhibits 8-MOP–induced phototoxicity and that the topical application of MK886, an inhibitor of lipoxygenase (103), inhibited 8-MOP–induced iVP and aPMN but not erythema.

The skin sites to which 8-MOP cream was applied when repeatedly exposed to light for 3 consecutive days (24 hours apart) produced minimal reaction on the fourth day as far as iVP and aPMN are concerned. If 8-MOP cream was reapplied to the repeatedly exposed sites on the third day prior to irradiation, the iVP and aPMN were significantly less than the virgin sites, although erythemal response remained the same (17). It seems the 8-MOP sites, when repeatedly exposed to light, become desensitized. This phenomenon of desensitization has been observed previously with RB-sensitized sites and has been discussed earlier. Since the desensitization to RB-induced photoreactions is manifested by inability to produce iVP and aPMN, similar absence of iVP and aPMN response in repeatedly irradiated 8-MOP sites suggests that there may be a common component related to the desensitized state. Moreover we have also found that the skin sites can be cross-desensitized to RB and 8-MOP (unpublished data).

G. Pathogenesis of Inflammatory Reactions

The repeated exposure of RB- or 8-MOP–treated sites to light render these sites unresponsive to the subsequent challenge. The sites thus prepared did not respond with aPMN when challenged with ZAS or with iVP when injected with compound 48/80 or histamine. The failure of PMNs to migrate into photodesensitized sites when ZAS was injected is suspected to be due to the loss of adherence of PMNs to endothelial cells. This adhesion is accomplished by the expression of ELAM-1 (107). It is known that the endothelial cell cultures stimulated with bacterial endotoxin and cytokines display the adhesion of PMNs by expressing cell surface glycoprotein, ELAM-1. Whether repeated free radical attack on these cells abrogates the ability to elaborate adhesion property is not known, although some circumstantial evidence exists. In this regard it may be noted that recent studies on intercellular adhesion molecule-1 (ICAM-1) expressed by keratinocytes as a response to interferon gamma (IFN-γ) and tumor necrosis factor (TNF) is modulated when these cells are subjected to UVR (108). It was found that the IFN-γ–induced ICAM-1 expression on cultured human keratinocytes is inhibited if these cells are exposed to strong UVR 24 hours prior to treatment with INF. However, if these cells are allowed to recover for 48 to 90 hours, the ICAM-1 expression is augmented. The studies of Krutmann et al. (109) further show that sublethal doses of UVR inhibit the ICAM-1 expression induced by TNFα, TNFβ, and IFN-γ in human keratinocytes in dose-dependent fashion. Human melanocytes are also capable of expressing ICAM-1 as a response to tumor necrosis factor (TNFα), interleukin-1 (IL-1) and IFN-γ (110); however, its modulation by UVR is not yet demonstrated.

ICAM-1, ELAM-1, and vascular cell adhesion molecule-1 (VCAM-1) are cytokine-regulated cell surface molecules involved in leukocyte adhesion. In experimental cutaneous inflammation in human volunteers, it was found that after two minimal erythema doses (MED) of UVB, ELAM-1 expression on vascular endothelium, and PMN infiltration were first observed at 6 hours and became maximal at 24 hours following UVB treatment. VCAM-1 was minimally expressed in control skin, and no induction of VCAM-1 was seen following challenge with either 2 or 8 MED UVB. UVB irradiation did not induce expression of ICAM-1 on basal keratinocytes. Thus expression of ELAM-1 seems to play an important role in phototoxic reactions (111).

It may be speculated that the 8-MOP photosensitized reactions are initiated by yet unrecognized putative step(s) involving activation of phospholipase A_2 that generates lipid mediators through the activation of arachidonic acid metabolism. The elaboration of IL-1 and TNFα by traumatized cells would result in their interaction with the endothelial cells to express ELAM-1. The lipoxygenase pathway is known to produce various leukotrienes, including LTB4, a chemoattractant for PMN. The observation that erythemal response to 8-MOP is independent of iVP and aPMN and that MK886, an inhibitor of 5-lipoxygenase, inhibits iVP and aPMN but not erythema suggests that the generation of prostaglandins by cyclooxygenase may be predominantly involved in hyperemia. This would also explain why erythema induced by 8-MOP is not affected although iVP and aPMN are inhibited at the sites made unresponsive by repeated irradiation of RB-injected sites. It should also be noted that erythemal response by guinea pig to PUVA is found to be independent of complement (112).

VIII. UVB-INDUCED CUTANEOUS REACTIONS

It has been postulated that UVB triggers a cutaneous inflammatory response by directly inducing epidermal keratinocytes to elaborate specific pro-inflammatory cytokines and adhesion molecules. The consequences are activation of dermal microvascular endothelial cells and selective accumulation of specific mononuclear cells in the dermis and epidermis. Thus, keratinocytes may act as "signal transducers," capable of converting exogenous stimuli into the production of cytokines, adhesion molecules, and chemotactic factors (acting in an autocrine and paracrine fashion) responsible for initiation of variety of cutaneous inflammations (113).

Preliminary studies with UVB show that a dose of 3–4 J/cm^2 induces phototoxic reaction 12–24 hours after irradiation and is manifested by erythema, iVP, and aPMN. Thus, in certain respects these reactions seem to have components similar to those of PUVA-induced phototoxicity.

The febrile response induced by an acute dose of UVB in rabbits peaks at 5 hours and is accompanied by a prolonged increase in serum IL-1 activity and by significant alterations in the serum divalent cation levels. The irradiated rabbits become euthermic in 24 hours, although the serum IL-1 levels remain elevated. These findings further suggest the role of IL-1 in the pathogenesis of phototoxicity after an acute dose of UV irradiation (114).

UVB irradiation inhibits induction of systemic immunity at high doses and local reactions at low doses in animals and is associated with the appearance of transferable T-suppressor lymphocytes. Immunosuppressed patients show increased rates of skin cancer, suggesting immunologic involvement in the regulation of skin cancer development. UVB alters circulating lymphocyte subtypes and contact hypersensitivity responses in the humans. In vitro irradiation of keratinocytes with UV alters their ability to secrete the specific cytokines such as IL-1, and these cells are induced to produce immunosuppressive factors. The exposure of epidermal cells and Langerhans cells to UVR also alters their ability to present some of the antigens (115). Since immunosuppressed patients have increased rates of skin cancer, it has been suggested that immunologic mechanisms may be involved in the regulation of cutaneous oncogenesis in humans (116).

Recent studies have shown that the exposure of keratinocytes to psoralen plus UVA elaborated factor(s) depressed delayed-type hypersensitivity (DTH) in mice. This effect was independent of the ability of given psoralens to produce mono- or bifunctional adducts. Interestingly, the exposure of these cells to lower doses of UVA released a factor that inhibits contact hypersensitivity but not DTH. These studies again suggest that the keratinocyte-derived factor(s) are involved in systemic immune suppression and that the type of factor released is related to the dose of psoralen plus UVA (117).

Immunosuppressive effects of low levels of UVB radiation on cutaneous immune responses have been attributed to deleterious effects of UVB radiation on epidermal Langerhans cells. On the basis of in vitro studies on the effect of UVB irradiation on Langerhans cell cultures, Tang and Udey (118) suggested that the inhibitory effects of low doses of UVB radiation on Langerhans cell accessory function may result from prevention of upregulation of ICAM-1 expression. Murine and human keratinocytes produce an IL-1–like factor that appears to be similar if not identical to monocyte-derived IL-1. UVB irradiation had a significant effect on murine keratinocyte IL-1 alpha expression. High-dose UV (300 mJ/cm^2) inhibited IL-1 alpha expression at 4, 8, 24, and 48 hours post-UV, whereas a lower dose of UV (100 mJ/cm^2) inhibited it at 4 and 8 hours post-UV, but induced its expression at 24 and 48 hours post-UV (119).

IX. ULTRAVIOLET RADIATION AND CANCER

It is generally recognized that UVR is a major factor responsible for many of the common cutaneous cancers, i.e., squamous cell carcinoma, basal cell carcinoma, and malignant melanoma (120–123). Moreover, clinical evidence indicates that UVR-induced carcinogenesis and/or aging are more prevalent and severe among people with red hair and "Celtic-type" skin than among people with black hair and dark skin (124). It has been shown that UV-visible irradiation of pheomelanin produces considerably larger amounts of reactive oxygen species and melanin free radicals than irradiation of eumelanin (40, 43). There is substantial evidence to suggest that the cellular pro-oxidant state caused by increased concentration of reactive oxygen species and organic peroxides can promote initiated cells to neoplastic growth (125). Since pheomelanin on irradiation can bring about pro-oxidant state, it is possible that the greater sensitivity of Celtic-type skin to UVR may be related to the generation of excessive amounts of free radicals.

The incidences of basal cell epithelioma and squamous cell carcinoma, the two most common skin cancers, seem to be related to factors such as total sunlight exposure during lifetime, intensity and duration of UVR in sunlight, and genetic predisposition. The fact that the incidence of squamous cell carcinoma in patients who received renal transplants and received long-term immunosuppressant therapy is much higher than in the normal population suggests that immunological mechanisms may be involved in human skin cancers related to exposure to sunlight. It is observed that patients with squamous or basal cell carcinoma have fewer T lymphocytes in circulation. There is also some correlation between the size of the tumor and the degree of lymphocyte infiltrations. Although these observations by themselves are not substantive, recent studies in experimental models of UVR-induced carcinogenesis in mice show: 1) UVR-induced tumors are mostly nontransplantable but become transplantable when host mice are either irradiated or given immunosuppressive agents, 2) the suppressor cells can be induced in mice by UVR and suppressor activity can be transferred to normal mice by the transfer of the lymphoid cells, and 3) the generation of cytotoxic memory cells can be inhibited by the suppressor T lymphocytes or the factor elaborated by these cells (126).

It is widely believed that cancer is caused by the alteration of cellular DNA resulting from chromosomal aberration or mutations. The various aspects of UVR-induced DNA damage and repair have been studied extensively (127). Near UVR is known to produce reactive oxygen species; moreover, DNA damage (single strand breaks) by such agents has been documented (128). It has also been shown that phagocyte-generated oxygen free radicals induce mutations (129). A number of findings suggest that in chronic inflammatory states, free radicals generated from phagocytes may produce mutagenic effects. Thus the phagocyte-mediated mutations may be a link between chronic inflammatory processes and malignancy.

ACKNOWLEDGMENTS

We wish to thank Dr. Janak Kumar for unpublished experimental data. We would like to thank Hansa Menon for assistance in the preparation of the manuscript. We acknowledge the grants from the Ontario Ministry of Labour and Medical Research Council.

REFERENCES

1. Epstein JH. The pathological effects of light on the skin. In: Pryor WA, ed. Free radicals in biology. Vol 3. New York: Academic Press, 1977:219–49.
2. Olson RL, Nordquist J, Everett MA. The role of epidermal lysosomes in melanin physiology. Br J Dermatol 1970; 83:189–199.
3. Johnson BE, Daniels F, Magnus IA. Response of human skin to ultraviolet light. In: Giese AC, ed. Photophysiology. Vol 4. New York: Academic Press, 1968: 139–202.
4. Kupper TS, Horowitz M, Birchall N, Mizutami H, Coleman D, McGuire J. Hematopoietic, lymphopoietic and proinflammatory cytokines produced by human and murine keratinocytes. Ann NY Acad Sci 1988; 548:262–70.

5. Aberer W, Schuler G, Stengl G, Honigsmann H, Wolff K. Ultraviolet light depletes surface markers of Langerhans cells. J Invest Dermatol 1981; 76:202–10.
6. Baadsgaard O. In vivo ultraviolet irradiation of human skin results in profound perturbation of the immune system. Arch Dermatol 1991; 127:99–109.
7. Hawk JLM, Parrish JA. Responses of normal skin to ultraviolet radiation. In: Regan JD, Parrish JA, eds. The science of photomedicine. New York: Plenum Press, 1983:219–;60.
8. Kaidbey KH, Kligman AM. Identification of topical photosensitizing agents in humans. J Invest Dermatol 1978; 70:149–51.
9. Morikawa F, Nakayama Y, Fukuda M, Hamano M, Yokoyama Y, Nagura T, Ishihara M, Toda K. Techniques for evaluation of phototoxicity and photoallergy in laboratory animals. In: Fitzpatrick TB, Pathak MA, Harber LC, Seiji M, Kukita A, eds. Sunlight and man. Tokyo: University of Tokyo Press, 1974: 529–57.
10. Argenbright WW, Forbes PD, Stewart GJ. Quantitation of phototoxic hyperemia and permeability to protein: I. Inhibition by histamine (H1 and H2) receptor antagonist in mouse skin. J Invest Dermatol 1980; 75:417–20.
11. Kamide RI, Gigli I, Lim HW. Participation of mast cells and complement in the immediate phase of hematoporphyrin-induced phototoxicity. J Invest Dermatol 1984; 82:485–90.
12. Lim HW, Young L, Hagan M, Gigli I. Delayed phase of hematoporphyrin-induced phototoxicity: modulation by complement, leukocytes and antihistamines. J Invest Dermatol 1985; 84:114–17.
13. Woodward DF, Raval P, Owen DAA. Reevaluation of the effect of nonsteroidal anti-inflammatory agents on UV-induced cutaneous inflammation. Agents Actions 1981; 11:711–17.
14. Ranadive NS, Menon IA, Shirwadkar S, Persad S. Quantitation of cutaneous inflammations induced by reactive species generated by UV-visible irradiation of rose bengal. Inflammation 1989; 13:483–94.
15. Ranadive NS, Kumar JR, Menon IA. Desensitization of rabbit skin by repeated exposure to UV-visible light of sites injected with rose bengal. Inflammation 1990; 14:223 37.
16. Menon IA, Ranadive NS, Shirwadkar S, Persad S, Haberman HF. Role of melanins and drugs in cutaneous photosensitivity. In: Moreno G, Pottier RH, Truscott TG, eds. Photosensitization: molecular, cellular and medical aspects. New York: Springer-Verlag, 1988:259–62.
17. Kumar JR, Ranadive NS, Menon IA, Haberman HF. Photoinduced cutaneous inflammatory response by psoralens. J Photochem Photobiol 1992; 14:125–37.
18. Laustriat G. Molecular mechanism of photosensitization. Biochimie 1986; 68: 771–78.
19. Beijersbergen van Henegouwen GMJ, Wijn ET, Schoonderwoerd SA, Dall'Acqua F. Method for the determination of in vivo irreversible binding of 8-methoxypsoralen (8-MOP) to epidermal lipids, proteins and DNA/RNA of rats after PUVA treatment. J Photochem Photobiol B:Biol 1989; 3:631–35.
20. Krinsky NI. Biology and photobiology of singlet oxygen. In: Bors W, Saran M, Tait D, eds. Oxygen radicals in chemistry and biology. Berlin: Walter de Gruyter, 1984:455–64.

21. Valenzeno DP. Photomodification of biological membranes with emphasis on singlet oxygen mechanisms. Photochem Photobiol 1987; 46:147–60.
22. Dall'Acqua F, Martelli P. Photosensitizing action of furocoumarins on membrane components and consequent intracellular events. J Photochem Photobiol B: Biol 1991; 8:235–54.
23. Girotti AW. Photodynamic lipid peroxidation in biological systems. Photochem Photobiol 1990; 51:497–509.
24. Kornhauser A, Wamer W, Lambert LA. Light induced dermal toxicity; effects on cellular and molecular level. In: Marzulli FN, Maibach HI, eds. Dermatotoxicology. New York: Hemisphere Pub. Corporation, 1991:527–69.
25. Girotti AW. Mechanisms of photosensitization. Photochem Photobiol 1983; 38:745–51.
26. Bouclier M, Henby CN, Cavey D. Animal models used in cutaneous pharmacology. In: Galli CL, Henby CN, Marinowich M, eds. Skin pharmacology and toxicology, recent advances; series A: life sciences. Vol 181. New York: Plenum Press, 1990:99–120.
27. Prota G. Recent advances in the chemistry of melanogenesis in mammals. J Invest Dermatol 1980; 75:122–7.
28. Prota G, Rorsman H, Rosengren AM, Rosengren E. Phaeomelanin pigments from a human melanoma. Experientia 1976; 32:970–1.
29. Sealy RC, Hyde JS, Felix CC, Menon IA, Prota G, Swartz HM, Persad SD, Haberman HF. Novel free radical in synthetic and natural pheomelanins: distinction between dopa melanins and cysteinyldopa melanins by ESR spectroscopy. Proc Natl Acad Sci USA 1982; 79:2885–9.
30. Sealy RC, Hyde JS, Felix CC, Menon IA, Prota G. Eumelanin and pheomelanin: characterization by electron spin resonance spectroscopy. Science 1982; 217:545–7.
31. Gan EV, Haberman HF, Menon IA. Electron transfer properties of melanin. Arch Biochem Biophys 1976; 173:666–72.
32. Sealy RC, Felix CC, Hyde JS, Swarz HM. Structure and reactivity of melanins: influence of free radicals and metal ions. In: Pryor WA, ed. Free radicals in biology. Vol 4. New York: Academic Press, 1980:210–59.
33. Morison WL. What is the function of melanin? Arch Dermatol 1985; 121:1160–3.
34. Kollias N, Sayre R, Zeise L, Chekedel M. Photoprotection by melanin. J Photochem Photobiol B: Biol 1991; 9:135–60.
35. Korytowski W, Kalyanaraman B, Menon IA, Sarna T, Sealy RC. Reaction of superoxide anions with melanins: electron spin resonance and spin trapping studies. Biochim Biophys Acta 1986; 882:145–53.
36. Ishikawa T, Kodama K, Matsumoto J, Takayama S. Photoprotective role of epidermal melanin granules against ultraviolet damage and DNA repair in guinea pig skin. Cancer Res 1984; 44:5195–9.
37. Hill HZ, Huselton C, Pilas B, Hill GJ. Ability of melanins to protect against the radiolysis of thymine and thymidine. Pigment Cell Res 1987; 1:81–6.
38. Thody A, Higgins E, Wakamatsu K, Ito S, Burchill S, Marks J. Pheomelanin as well as eumelanin is present in human epidermis. J Invest Dermatol 1991; 97: 340–4.

39. Menon IA, Persad S, Haberman HF, Kurian CJ. A comparative study of the physical and chemical properties of melanins isolated from human black and red hair. J Invest Dermatol 1983; 80:202–6.
40. Sarna T, Menon IA, Sealy RC. Photoinduced oxygen consumption in melanin systems. II. Action spectra and quantum yields for pheomelanins. Photochem Photobiol 1984; 39:805–9.
41. Chedekel MR, Agin PP, Sayre RM. Photochemistry of pheomelanin: action spectrum for superoxide production. Photochem Photobiol 1980; 31:553–5.
42. Felix CC, Hyde JS, Sarna T, Sealy RC. Melanin photoreactions in aerated media: electron spin resonance evidence for production of superoxide and hydrogen peroxide. Biochem Biophys Res Commun 1978; 84:335–41.
43. Persad S, Menon IA, Haberman HF. Comparison of the effects of UV-visible irradiation of melanins and melanin-hematoporphyrin complexes from human black and red hair. Photochem Photobiol 1983; 37:63–8.
44. Menon IA, Persad S, Ranadive NS, Haberman HF. Formation of superoxide and cell damage during UV-visible irradiation of melanin. In: Bors W, Saran M, Tait D, eds. Oxygen radicals in chemistry and biology. Berlin: Walter de Gruyter, 1984:673–9.
45. Menon IA, Persad S, Ranadive NS, Haberman HF. Effects of ultravioletvisible irradiation in the presence of melanin isolated from human black or red hair upon Ehrlich ascites carcinoma cells. Cancer Res 1983; 43:3165–9.
46. Menon IA, Persad S, Ranadive NS, Haberman HF. Role of superoxide and hydrogen peroxide in cell lysis during irradiation in vitro of Ehrlich ascites carcinoma cells in the presence of melanin. Can J Biochem Cell Biol 1985; 63:278–83.
47. Ranadive NS, Menon IA. Role of reactive oxygen species and free radicals from melanins in photoinduced cutaneous inflammations. Pathol Immunopathol Res 1986; 5:118–39.
48. Menon IA, Persad S, Ranadive NS, Haberman HF. Photobiological effects of eumelanin and pheomelanin. In: Bagnara J, Klaus SN, Paul E, Schartl M, eds. Pigment cell 1985: biological, molecular and clinical aspects of pigmentation. Tokyo: University of Tokyo Press, 1985:77–85.
49. Ranadive NS, Shirwadkar S, Persad S, Menon IA. Effect of melanin induced free radicals on isolated rat peritoneal mast cells. J Invest Dermatol 1986; 86: 303–7.
50. Ezzahir A. The influence of melanins on the photoperoxidation of lipids. J Photochem Photobiol B: Biol 1989; 3:341–9.
51. Lea PJ, Pawlowski A, Persad S, Menon IA, Haberman HF. Ultrastructural changes produced in Ehrlich ascites carcinoma cells by ultraviolet-visible radiation in the presence of melanins. Pigment Cell 1988; 1:402–9.
52. Klein LE, Nordlund JJ. Genetic basis of pigmentation and its disorders. Int J Dermatol 1981; 20:621–31.
53. Dixit R, Mukhtar H, Bickers DR. Studies on the role of reactive oxygen species in mediating lipid peroxidation in epidermal microsomes of rat skin. J Invest Dermatol 1983; 81:369–75.

54. Kopaniak MM, Movat HZ. Kinetics of acute inflammation induced by E. coli in rabbits. II. The effect of hyperimmunization, complement depletion and depletion of leukocytes. Am J Pathol 1983; 110:13–29.
55. Cybulski MI, Movat HZ. The application of ^{51}Cr-labeled PMN-leukocytes in quantitating PMN kinetics in systemic and local inflammatory mediated processes: effect of endotoxin, complement and interleukin I. Surv Synth Pathol Res 1983; 1:208–28.
56. Crawford JP, Movat HZ, Ranadive NS, Hay JB. Pathway to inflammation induced by immune complexes: development of Arthus reaction. Fed Proc 1982; 41:2583–7.
57. Lim HW, Gigli I. Role of complement in porphyrin-induced photosensitivity. J Invest Dermatol 1981; 76:4–9.
58. Lim HW, Gigli I. Complement-derived peptides in phototoxic reactions. In: Forbes PD, Davies RE, eds. Experimental and clinical photoimmunology. Cleveland: CRC Press, 1983:81–93.
59. Lim H. Effects of porphyrins on skin. Ciba Found Symp 1989; 146:148–58.
60. Lim H. Mechanisms of phototoxicity in porphyria cutanea tarda and erythropoietic protoporphyria. Immunol Ser 1989; 46:671–85.
61. Kumar JR, Ranadive NS, Menon IA. Photo-induced inflammatory response by erythrosine. J Toxicol Cut Ocular Toxicol 1991; 10:233–44.
62. Gandin EY, Lion Y, Van der Vorst A. Quantum yield of singlet oxygen production by xanthene derivatives. Photochem Photobiol 1983; 37:271–8.
63. Barton GW, Ingold KU. Beta-carotene: an unusual type of lipid antioxidant. Science 1984; 224:569–73.
64. Krinsky NI, Deneke SM. Interaction of oxygen and oxy-radicals with carotinoids. J Natl Cancer Inst 1982; 69:205–10.
65. Harber LC, Kochevar IE, Shalita AR. Mechanism of photosensitization to drugs in humans. In: Regan JD, Parrish JA, eds. The science of photomedicine. New York: Plenum Press, 1982:323–47.
66. Kaidbey KH, Nonaka S. Action spectrum for anthracene-induced photosensitization of human skin. Photochem Photobiol 1984; 29:375–80.
67. Ruzika T, Walter JF, Printz MP. Changes in arachidonic acid metabolism in UV-irradiated hairless mouse skin. J Invest Dermatol 1982; 81:300–3.
68. Ranadive NS, Menon IA, Haberman HF. Photolysis of mast cells in the presence of protoporphyrin. Acta Derm Venereol Stockh 1979; 59:493–7.
69. Ortner MJ, Abhold RH, Chigenell CF. Effect of protoporphyrin on histamine secretion by rat peritoneal cells: dual phototoxic reaction. Photochem Photobiol 1981; 33:355–60.
70. Abhold RH, Leid RW, Magnusson JA, Hegreberg GA. Inhibition of rat peritoneal mast cell secretion by protoprphyrin and incandescent light. Int Arch Allergy Appl Immunol 1985; 76:126–32.
71. Gendimenico GJ, Kochevar IE. Degranulation of mast cells and inhibition of the response to secretory agents by phototoxic compounds and ultraviolet radiation. Toxicol Appl Pharmacol 1984; 76:376–82.
72. Henderson WR, Chi EY, Klebanoff SJ. Eosinophil-peroxidase-induced mast cell secretion. J Exp Med 1980; 152:265–79.

73. Stendahl O, Molin L, Lindroth M. Granulocyte mediated release of histamine from mast cells. Effect of myeloperoxidase and its inhibition by anti-inflammatory sulfone compounds. Int Arch Allergy Appl Immunol 1983; 70:277–84.
74. Ohmori H, Yamamato I, Agaki M, Tasaka K. Properties of hydrogen peroxide-induced histamine release from rat mast cells. Biochem Pharmacol 1980; 29: 741–5.
75. Menon IA, Shirwadkar S, Ranadive NS. Nature of the oxygen species generated by xanthine oxidase involved in secretory histamine release from mast cells. Biochem Cell Biol 1989; 67:397–403.
76. Fox PC, Basciano LK, Siraganian RP. Mouse mast cell activation and desensitization for immune aggregate-induced histamine release. J Immunol 1982; 129: 309–19.
77. Morrison DC. Receptor modulation and mast cell secretion. J Invest Dermatol 1978; 71:85–91.
78. Ranadive NS, Ruben DH. Mechanism of histamine release from rat mast cells by compound 48/80. Comparison with the release induced by cationic protein. Int Arch Allergy Appl Immunol 1973; 44:745–58.
79. Matthews EK, Mesler DE. Photodynamic effect of erythrosine on smooth muscle cells of guinea-pig taenia coli. Br J Pharmacol 1984; 83:555–66.
80. Rubin L, Beal PL, Rothmans S. A method of protection for patients with solar urticaria. J Invest Dermatol 1947; 8:189.
81. Sams WM, Epstein JH, Winkelmann RK. Solar urticaria: investigations on pathogenetic mechanisms. Arch Dermatol 1969; 101:520–3.
82. Ramsay CA. Solar urticaria treatment by inducing tolerance to artificial radiation and natural light. Arch Dermatol 1977; 113:1222–5.
83. Hawk JL M, Eady RAJ, Challoner AVJ, Kobza-Black A, Keahey TM, Greaves MW. Elevated blood histamine levels and mast cell degranulation in solar urticaria. Br J Clin Pharmacol 1980; 9:183.
84. Keahey TM, Lavker RM, Kaidbey K, Atkins PC, Zweiman B. Studies on the mechanism of clinical tolerance in solar urticaria. Br J Dermatol 1984; 110:327–38.
85. Parrish JA, Jaenicke KF, Morison WL, Momtaz K, Shea C. Solar urticaria: treatment with PUVA and mediator inhibitors. Br J Dermatol 1982; 106:575–80.
86. Leenutaphong V, Holzle E, Plewig G. Solar urticaria: studies on mechanisms of tolerance. Br J Dermatol 1990; 122:601–6.
87. Potapenko AY. New trends in photobiology. Mechanisms of photodynamic effects of furocoumarins. J Photochem Photobiol B: Biol 1991; 9:1–33.
88. Blan QA, Grossweiner LI. Singlet oxygen generation by furocoumarins: effect of DNA and liposomes. Photochem Photobiol 1987; 45:177–83.
89. Pathak MA. Mechanism of psoralen photosensitization reactions. Natl Cancer Inst Monogr 1984; 66:41–6.
90. Caffieri S, Lucchini V, Rodighiero P, Miolo G, Dall'Acqua F. 3,4 and 4′,5′-Photocycloadducts between 4′-methylangelicin and thymine from DNA. Photochem Photobiol 1988; 48:573–8.

91. Kittler L, Lober G. Photoreactions of furocoumarins with nucleic acids, proteins, membranes and their subunits. Studia Biophys 1988; 124:97–114.
92. De Mol NJ, Beijersbergen van Henegouwen GMJ. Relation between some photobiological properties of furocoumarins and their extent of singlet oxygen production. Photochem Photobiol 1981; 33:815–19.
93. Wolf C, Steines A, Honigsmann D. Oral carotenoid protects human skin against UV erythema, psoralen phototoxicity and UV-induced DNA damage. J Invest Dermatol 1988; 90:55–7.
94. Farr P, Diffey B, Humphreys F. A quantitative study of the effect of terfenadine on cutaneous erythema induced by UVB and UVC radiation. J Invest Dermatol 1986; 87:771–4.
95. Hanson D, DeLeo V. Long-wave ultraviolet light induces phospholipase activation in cultured human epidermal keratinocytes. J Invest Dermatol 1990; 95:158–64.
96. DeLeo V, Scheide S, Meshulam J, Hanson D, Cardullo A. Ultraviolet radiation alters choline phosopholipid metabolism in human keratinocytes. J Invest Dermatol 1988; 91:303–8.
97. Kaidbey KH, Kurban AK. The influence of corticosteroids and topical indomethacin on sunburn erythema. J Invest Dermatol 1976; 66:153–6.
98. Snyder DK, Eaglstein W. Topical indomethacin and sunscreen. Br J Dermatol 1974; 90:91–3.
99. Morison WL, Paul BS, Parrish JA. The effect of indomethacin on longwavelength ultraviolet induced delayed erythema. J Invest Dermatol 1977; 68:120–33.
100. Schwarz T, Gschnait F, Greiter F. Photoprotective effect of topical indomethacin—an experimental study. Dermatologica 1985; 171:450–8.
101. Imokawa G, Tejima T. A possible role of prostaglandin in PUVA induced inflammation—implication by organ cultured skin. J Invest Dermatol 1989; 92:296–300.
102. Karmali R, Safai B. Ultraviolet-evoked prostaglandin biosynthesis in varying stages of keratinocyte differentiation in guinea pig skin. Prostaglandins Leukot Med 1984; 15:277–86.
103. Rouzer CA, Ford-Hutchinson AW, Morton HE, Gillard JW. MK886, a potent and specific leukotriene biosynthesis inhibitor blocks and reverses the membrane association of 5-lipoxygenase in ionophore-challenged leukocytes. J Biol Chem 1990; 265:1436–42.
104. Wheeler LA, Aswad A, Connor MJ, Lowe N. Depletion of glutathine and the induction of inflammation by 8-methoxypsoralen plus UVA radiation. J Invest Dermatol 1986; 87:658–62.
105. Lowe NJ, Connor MJ, Breeding JH, Russell DH. Epidermal polyamine profiles after chronic ultraviolet irradiation. Carcinogenesis 1983; 4:671–4.
106. Perchellet JP, Owen MD, Posey TD, Orten DK. Schneider BA. Inhibitory effect of glutathione levels rising agents on ODC induction and mouse skin tumor promotion by tetradecanoyl phorbol acetate. Carcinogenesis 1985; 6: 567–73.

107. Bevilacqua MP, Stegelin S, Gimbrone M, Seed V. Endothelial leukocyte adhesion molecule I: an inducible receptor for neutrophils related complement regulatory proteins and lectins. Science 1989; 243:1160–2.
108. Norris DA, Lyons MB, Middleton MH, Yohn JJ, Kashihara-Sawami M. Ultraviolet radiation can either suppress or induce expression of intercellular adhesion molecule 1 (ICAM-1) on the surface of cultured human keratinocytes. J Invest Dermatol 1990; 95:132–8.
109. Krutmann J, Kock A, Schauer E, Parlow F, Moller A, Kapp A. Tumor necrosis factor beta and ultraviolet radiation are potent regulators of human keratinocyte ICAM-1 expression. J Invest Dermatol 1990; 95:127–31.
110. Yohn JJ, Critelli M, Lyons MB, Norris DA. Modulation of melanocyte intercellular adhesion molecule-1 by immune cytokines. J Invest Dermatol 1990; 95:235–7.
111. Norris P, Poston R, Thomas D, Thornhill M, Hawk J, Haskard D. The expression of endothelial leukocyte adhesion molecule-1 (ELAM-1), intercellular adhesion molecule-1 (ICAM-1), and vascular cell adhesion molecule-1 (VCAM-1) in experimental cutaneous inflammation: a comparison of ultraviolet B erythema and delayed hypersensitivity. J Invest Dermatol 1991; 96:763–70.
112. Rauterberg AD, Jung EG, Berger R, Rauterberg EW. Phototoxic erythema following PUVA treatment: independence of complement. J Invest Dermatol 1990; 94:144–9.
113. Barker J, Mitra R, Griffiths C, Dixit V, Nickoloff B. Keratinocytes as initiators of inflammation. Lancet 1990; 337:211–14.
114. Ansel J, Luger T, Green I. Fever and increased serum IL-1 activity as a systemic manifestation of acute phototoxicity in New Zealand white rabbits. J Invest Dermatol 1987; 89:32–7.
115. Granstein R. Photoimmunology. Semin Dermatol 1990; 9:16–24.
116. Granstein R. Ultraviolet radiation effects on immunologic function. Reg Immunol 1990; 3:112–19.
117. Aubin F, Krupke ML, Ulrich SE. Activation of keratinocytes with psoralen plus UVA radiation induces the release of soluble factors that suppress delayed and contact hypersesntivity. J Invest Dermatol 1991; 97:995–1000.
118. Tang A, Udey M. Inhibition of epidermal Langerhans cell function by low dose ultraviolet B radiation. Ultraviolet B radiation selectively modulates ICAM-1 (CD54) expression by murine Langerhans cells. J Immunol 1991; 146: 3347–55.
119. Ansel J, Luger T, Lowry D, Perry P, Roop D, Mountz J. The expression and modulation of IL-1 alpha in murine keratinocytes. J Immunol 1988; 140:2274–8.
120. Epstein JH. Photomedicine. In: Smith KC, ed. The science of photobiology. New York: Plenum Press, 1989:155–92.
121. Gallagher RP, Elwood JM, Yang CP. Is chronic sunlight exposure important in accounting for increase in melanoma incidence? Internatl J Cancer 1989; 44:813–15.
122. Henriksen T, Dahlback A, Larsen SHH, Moan J. Ultraviolet-radiation and skin cancer—effect of an ozone layer depletion. Photochem Photobiol 1990; 51:579–82.

123. Urbach F. Potential effects of altered solar ultraviolet radiation on human skin cancer. Photochem Photobiol 1989; 50:507–14.
124. McGovern VJ. Etiology of melanomas. In: Milton GW, ed. Malignant melanoma of the skin and mucous membrane. New York: Churchill Livingstone, 1977:1–25.
125. Cerutti PA. Prooxidant state and tumor promotion. Science 1985; 227:375–81.
126. Morison WL, Parrish JA. Photoimmunology. In: Regan JD, Parrish JA, eds. The science of photomedicine. New York: Plenum Press, 1983:293–320.
127. Harm W. Biological effects of ultraviolet radiation. Cambridge: Cambridge University Press, 1980.
128. Tyrell RM. A common pathway for protection of bacteria against solar UVA and oxidizing agent H_2O_2. Mutation Res 1985; 145:129–36.
129. Weitzman SA, Gordon LI. Inflammation and cancer; role of phagocyte generated oxidants in carcinogenesis. Blood 1990; 76:655–63.

10

Solar Ultraviolet Effects on Mammalian Cell DNA

Meyrick J. Peak and Jennifer G. Peak
Argonne National Laboratory, Argonne, Illinois

Irene E. Kochevar
Massachusetts General Hospital and Harvard Medical School, Boston, Massachusetts

I. INTRODUCTION

Solar photons in the ultraviolet (UV) range are of prime importance in the etiology of human skin cancers (basal and squamous cell carcinomas), and they contribute to a deadly skin cancer, malignant melanoma. Skin is exposed to photons of the terrestrial solar spectrum, including wavelengths from 290 nm through the longer UV, visible, and infrared wavelengths. The regions of this spectrum that are most important for human photobiology are the UVB (290–320 nm) and UVA (320 nm to visible light) ranges (1). A barely detectable number of 290-nm photons reaches the earth's surface because of the attenuating effects of the stratospheric ozone layer, but the flux increases markedly with increasing wavelength in the UVA region. These photons can penetrate into the viable layers of skin, with increased penetration by longer-wavelength photons, so that at 365 nm about 30% of the incident energy penetrates through the stratum corneum into the viable epidermis (2).

The carcinogenic effects of UVC and UVB photons have been known for decades (3), and light-induced lesions in DNA are believed to be the initiating events. This is consistent with the high probability that UVB photons will be absorbed by DNA (4), resulting in the formation of photoproducts, especially pyrimidine dimers and adducts. Recent evidence supports an important role for photoreactivable dimers in UV carcinogenesis in the South American opossum (the marsupial *Monodelphis domestica*), which was selected as an experimental target because its cells have high photolyase activity (5).

The UVB region may become even more important in terms of increased cancer risk if an environmental increase in UVB flux takes place due to stratospheric ozone depletion (6). Ozone levels over Antarctica have reached record low levels, and detectable erosion of the ozone layer (at faster than predicted rates) over temperate latitudes in summer has now been measured (7,8).

UVA photons, long considered biologically harmless, have recently been shown to be weakly carcinogenic in experiments in which shorter-wavelength (UVB and UVC) photons were carefully eliminated (9,10). Because of the low absorptivity of DNA in this wavelength region, it is highly likely that photons of wavelengths longer than 320 nm are absorbed by non-DNA chromophores that participate in UVA-induced carcinogenesis (11).

The observation of UVB and UVA carcinogenesis has motivated the study of interactions between these photons and DNA. One strategy is to irradiate cultured cells with monochromatic photons, which are emitted from a high-pressure mercury arc lamp. An important advantage of using monochromatic rather than broad-band sources is that the latter emit a mixture of photons that may have very different effects. Furthermore, a minor component at the short-wavelength end of a broad-band source may contribute overwhelmingly to the biological action, because UVB is more biologically efficient than UVA by 2–4 orders of magnitude (especially in the case of effects on DNA). Thus, the effect from the longer-wavelength component may be obscured by that from the smaller number of shorter-wavelength photons (12).

A practical problem in experimentation with UVA photons is the low irradiances available from current sources, which results in long exposure times. This is often a special problem with monochromatic sources. For instance, at 365 nm, the F_{37} (fluence resulting in 37% survival of the population) of human P3 teratocarcinoma cells is more than 1 MJ m^{-2} (13), which translates to about 30 minutes of exposure at the fluence rate available from a 2500-W, high-pressure mercury source coupled to a monochromator. Yet, DNA repair processes may be very rapid. For instance, the time to repair half of the single-strand breaks produced by x-rays or 365-nm photons in human epithelioid P3 cells is 2.5 minutes (14,15), so all breaks produced by 365-nm photons will be concomitantly repaired by the cells during the exposure period unless the repair is inhibited. One tactic is to keep the cells cold (0.5 °C) or on ice during exposures, which slows the metabolic repair processes to beneath detectable levels (14).

II. MUTAGENESIS

The carcinogenic effects of solar ultraviolet radiation, particularly the UVB component, have been amply demonstrated in experiments with animals (3).

These studies used broadband sources, and there are intractable problems involved in conducting such experiments with the monochromatic sources currently available. However, in vitro experiments can be used to obtain action spectra for mammalian cell mutagenicity that provide information about the potential genotoxicity of photons of different energies. Thus, insights into carcinogenicity may be obtained from in vitro experiments. Solar mutagenicity studies have used both broad-band and monochromatic sources. The most natural broad-band source is the sun itself, and this was used by Hsie to expose Chinese hamster ovary (CHO) cells in summer (July) at Oak Ridge, Tennessee (16). The frequency of production of thioguanine (TG)-resistant mutants (defined as resistance to genetic change at the hyoxanthine-guanine phosphoribosyl transferase locus) was eight mutants per surviving cell per minute of exposure to solar radiation, although the fluence rate was not given. When we exposed human P3 cells to a solar simulator, seven TG-resistant mutants per minute exposure at 1 kW m^{-2} were detected (unpublished data). Other studies have used a variety of broad-band sources (from fluorescent light fixtures to black light) to produce mutations in mammalian cells (reviewed in Ref. 17). However, in the case of sources containing predominantly UVA photons, the possibility that the mutagenic response was due to a small component of short-wavelength photons was not stringently excluded. Studies using monochromatic photons in conjunction with careful elimination of short-wavelength photons have unequivocally shown that UVA photons of wavelength as long as 365 nm are mutagenic (13,18,19). A weak positive dose response has been demonstrated even at 405 nm (13).

All the action spectra described thus far show a close correlation with the absorption spectrum of DNA in the UVB (and UVC) region, with a corresponding steep fall-off in efficiency in the UVB region (13,18–21); thus the probability of obtaining a mutation matches the probability of the absorption of a photon by DNA, as illustrated by Figure 1. The quantum efficiency for mutation in the UVA region is orders of magnitude lower than that in the UVB region, but it is still well above the level predicted by DNA absorption, so (as for other DNA effects described below) these genotoxic effects must be mediated by non-DNA sensitizers. When mutagenesis studies are compared, it is clear that results from different laboratories in which different cell types were used are quantitatively quite varied in the UVA region. For instance, Tyrrell (21) found no mutants (TG or ouabain resistance) produced in human lymphoblastoid cells by any UVA photons, whereas Wells and Han (19) found significant numbers produced by 334-nm, but not 365-nm photons. In contrast, Jones et al. (13) found no mutants at 334 nm, but obtained strongly positive results at 365 nm in human P3 epithelioid cells. Enninga et al., using human diploid foreskin fibroblasts (18), measured about 20 times greater frequency of mutants at 365 nm than that described by Jones et al. This constitutes further evidence for an indirect photosensitization mechanism, because the explanation for the variation must lie in qualitative

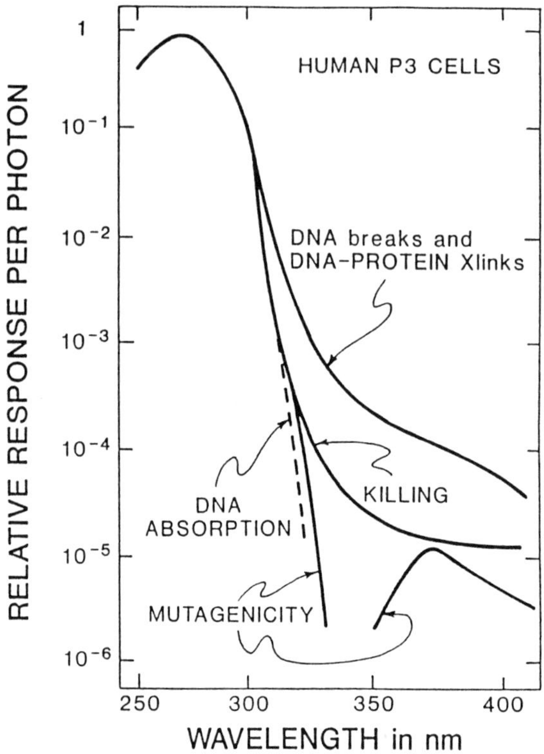

Figure 1 Action spectra for changes produced in human epithelioid cells. DNA absorption (----) uncertain for wavelengths longer than 320 nm. (From Ref. 17.)

or quantitative differences in the sensitizers that act as primary chromophores for photochemical reactions in the target cells (see Sec. VI).

III. DNA CHANGES RESULTING FROM SOLAR UV PHOTONS

Both the types of DNA lesions and the chemical mechanisms (see below) whereby they are produced depend markedly on wavelength. Pyrimidine photoproducts, comprising cyclobutane-type dimers, and (6-4) pyrimidine photoadducts are produced in DNA in high yields by the direct absorption of UVB photons by DNA. Supporting this is the close correlation of the absorption characteristics of DNA between 254 and 320 nm and the action spectra for the induction of these lesions (18,20,22,23). Although detectable, the yields of pyrimidine photoproducts induced by UVA photons are low by comparison, especially at longer wavelengths such as 365 nm. At 405 nm, pyrimidine dimers are not detectable (24). Cleaver et al. (25) discussed the relative biological significance of cyclobutane and (6-4) pyrimidine-pyrimidone photoproducts in human cells, based on experiments with a revertant xeroderma mutant.

Mitchell et al. (26) described other pyrimidine photoproducts caused by UVB photons in SV40 DNA irradiated in vivo and in vitro. For instance, yields of *Escherichia coli* endonuclease sites (primarily cytosine hydrates) were 1–2% of the frequency of pyrimidine cyclobutane dimers. Possible causes for the discrepancies in the previous literature regarding thymine glycol yields were suggested. Further work is necessary to determine the significance of these and other minor lesions such as purine-pyrimidine photoadducts (27) in tumorigenesis and the induction of skin cancers by sunlight.

UVA photons produce DNA lesions more characteristic of ionizing radiation: single-strand (24,28–32) and double-strand (33) DNA breaks and DNA-to-protein crosslinks (DPCs) (34). In the specific case of single-strand breaks induced by 405-nm photons in human P3 epithelioid cells, about half the breaks assayed by the alkaline elution technique are alkali-labile sites, as revealed by the convex shape of the elution profiles (35). Such profiles were not observed for other cell types, such as Chinese hamster ovary cells (36) exposed to 405-nm photons, and the nature of the alkali-labile sites produced in the human cells remains enigmatic.

Kleiman et al. (37) described the induction of DNA-DNA interstrand crosslinks in bovine lens cells exposed to a UVB broad-band source. However, it did not appear that proteinase K was used in these experiments, so that DPCs and DNA-DNA crosslinks were not distinguished. We have observed a barely detectable yield of DNA-DNA crosslinks after UVB and UVA exposures (i.e., crosslinks remaining after exposure to proteinase K).

A compilation of the initial yields (with cells meticulously kept cold throughout exposure and thereafter until lysis) of single-strand breaks and DPCs (Table 1) induced by various energies of radiation was presented by Peak and Peak (38,39), and the yield of double-strand breaks has been described (33). This comparison on the basis of the number of lesions produced at the lethal fluence clearly showed that cyclobutane dimers are the predominant lesions in the UVB region, but UVA and visible photons are more effective than UVB at producing single-strand breaks and DPCs. An even more striking observation is that the initial yields of single-strand breaks and DPCs produced by UVA radiation are more than an order of magnitude greater than the initial yields produced by ionizing radiation when analyzed on this basis (39). This implies that if DNA breaks and DPCs are critical lesions, those inflicted by ionizations may be of a more deleterious type than those produced by UVA photons. One way to study the severity of DNA lesions induced by radiation is to follow their repair within the cell.

Table 1 Relative Yields of Lesions Induced by Ionizing and Nonionizing Radiation in Human P3 Cells

Radiation	D_o	Lesions/Cell·D_o[a] DPCs	Single-strand breaks
X-rays	100 cGy	200	840
254 nm	10 J/m^2	80	140
290 nm	31 J/m^2	130	420
313 nm	3950 J/m^2	153	430
334 nm	476,000 J/m^2	1560	3400
365 nm	1,500,000 J/m^2	1590	9000
405 nm	3,590,000 J/m^2	1360	3500
434 nm	>4,200,000 J/m^2	>270	>2200

[a]For DNA content of a cell equal to 3×10^{12} daltons.
Source: Ref. 39.

Finally, little information is as yet available on the induction of such lesions as thymine glycols, urea residues, or apurinic sites by UVA photons in mammalian cells (40). Early reports indicating that thymine glycols are formed in high yields by 313-nm photons (41) were not confirmed by more recent studies (6,42).

IV. CELLULAR REPAIR OF DNA LESIONS PRODUCED BY UVA PHOTONS

Two major pathways exist for removal of the pyrimidine photoproducts that are produced in high yields by UVB radiation, namely photoreactivation and the excision repair processes. These repair systems have been studied for decades, with much of the early work using bacteria and yeast (see Ref. 43 for overview). Many recent studies have reported on the repair of the pyrimidine dimers (44) and (6-4) adducts in mammalian systems. Some studies have indicated that repair of (6-4) photoproducts is more efficient than dimer repair. The repair of pyrimidine dimers, (6-4) photoproducts, Dewar isomers, and DNA breaks in human fibroblasts has been described (45–47). There have also been several in vivo studies. The repair of pyrimidine dimers in mammalian skin exposed to artificial solar radiation was described by Strickland and Creasey (48), who used hamsters, and others have measured repair of dimers in human skin in situ (49–51). In the marsupial *Monodelphis domestica*, which possesses photolyase enzyme activity, both excision repair and concomitant photoreactivation of these two types of lesions were reported by Mitchell et al. (46).

Far fewer studies have been carried out on the repair of the types of lesions characteristically produced by UVA radiation, such as DNA breaks and DPCs. For single-strand breaks, as is the case with the breaks produced by ionizing radiation, indications are that normal mammalian cells have efficient enzyme systems to seal these lesions (30,32); both of these studies used human lymphocytes. More recently, Peak et al. (14) described complete rejoining of the breaks produced by 365-nm monochromatic UVA in human epithelioid cells, but in xeroderma pigmentosum cells, a fraction of the breaks were unrepairable. Churchill et al. (36,52) reported complete repair of breaks by wild-type CHO cells, AA8, whereas the radiation-sensitive mutant, EM9, showed less efficient DNA repair after exposure to 365-nm photons and, to a lesser extent, to 405-nm near-violet and 434-nm blue-light photons. These studies demonstrate that break repair is important for cells exposed to UVA and visible light photons and that unrepaired DNA strand breaks might account for observed cellular sensitivity to these radiations. It should be noted that although strand breaks are apparently all rapidly rejoined by normal human cells, the assays used measure only DNA size, so no information is currently available about the fidelity of the repair or the prevalance of misrepair in different cell types.

Only one published study has analyzed the cellular repair of DPCs produced by solar photons: Rosenstein et al. (53) followed crosslink processing in various frog cell lines exposed to a solar simulator. The lesions increased during repair incubation, then were completely removed by 24 hours. Clearly, repair responses for this lesion are very different from the response to DNA breakage. In unpublished experiments, we have confirmed an increase in DPCs in epithelioid cells exposed to a potent crosslinking regimen—405-nm photons in the presence of protoporphyrin 1X—during repair incubation, followed by partial elimination (25% remaining) of the lesions after 24 hours of repair incubation (Fig. 2). More work is necessary to further elucidate this pathway.

At each wavelength studied, double-strand breaks were only detected at fluences greatly exceeding the F_{37} value (33). Any repair studies would thus involve a population of thoroughly dead cells. Analysis of the biological significance of these particular events will have to await the development of more sensitive assays.

V. MECHANISMS OF DNA DAMAGE

Damage to DNA by UVB photons results from photochemical reactions that are initiated by direct absorption of the photons by the DNA bases. The

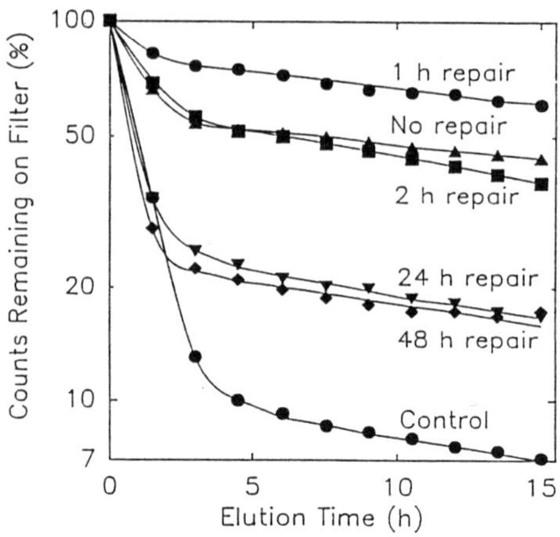

Figure 2 Analysis by alkaline DNA filter elution of repair of DPC induced by 405-nm photons plus protoporphyrin IX. The extrapolation of the slow-eluting DNA to the ordinate gives a direct function of the DPC remaining.

mechanisms for formation of a (6-4) pyrimidine photoadduct involves the singlet state of pyrimidine; the formation of cyclobutylpyrimidine dimers mainly occurs from the triplet state of the pyrimidines. These direct mechanisms for UVB action are supported by the well-substantiated observation that the absorption spectrum for DNA correlates closely with the action spectrum for DNA damage in this region (Fig. 1), as recently reviewed by Coohill et al. (17).

DNA is virtually transparent to UVA photons (11). Thus, in the UVA region, the mechanisms for inflicting DNA damage in cells differ completely from those in the UVB region, requiring the absorption of photons by endogenous photosensitizer molecules. These mechanisms involve photosensitized oxidations, since oxygen is required for DNA changes caused by UVA photons in cells. DNA damage is also inflicted in the presence of added sensitizer (54). The requirement for oxygen in the formation of most (if not all) DPCs and strand breaks caused by 365-nm photons in human cells was shown by Peak and coworkers, evidence that these UVA photons operate by means of photodynamic action (28,34).

Photosensitized oxidations can be either Type I or Type II reactions, as recently defined clearly by Foote (55). Both reactions use oxygen, thus the observation of oxygen dependence does not distinguish between them. The first step in these reactions is the absorption of UVA photons by a sensitizer, which consequently reaches an excited state. The involvement of a non-DNA sensitizer as the primary chromophore in UVA-DNA interactions is supported by action spectroscopy, which reveals that more DNA-damaging events occur than are justified by the absorption of the photons by DNA (39). Compounds suspected of being sensitizers in mammalian cells include riboflavin, porphyrins, quinones, reduced nicotinamide cofactors, and bilirubin. Evidence supporting a role for bilirubin in UVA action has been described (56). Peak et al. (57), using an in vitro model DNA system, demonstrated that DNA breakage by 334-nm UVA was enhanced in the presence of a series of naturally occurring nucleotide cofactors and nucleic acid components, such as NADH and thiolated bases. The excited sensitizer subsequently reacts either with the substrate or solvent (Type I) or with oxygen (Type II). The initial products of Type I reactions are radicals or radical ions, whereas Type II reactions produce reactive oxygen species such as singlet molecular oxygen or superoxide anion (Fig. 3).

Which particular reactive oxygen species are responsible for the DNA damage caused by exposure of cells to UVA photons is still not clear, but there is some suggestive evidence. Irradiation of human cells with UVA in an aqueous D_2O environment enhances the rate of accumulation of DPCs and DNA strand breaks (28,34). This is indirect evidence for the production of singlet molecular oxygen (a Type II reaction) because the decay of singlet oxygen is slowed by more than an order of magnitude in D_2O compared with H_2O, providing that the lifetime of singlet oxygen is limited by its interaction with the solvent (58). Experiments in which singlet oxygen was generated outside of the sample and allowed to diffuse to the DNA (59) demonstrated that singlet oxygen itself has the energy to break covalent bonds in DNA (60). In other studies, putative sensitizers were irradiated with UVA photons of various energies, resulting in the generation of superoxide anion (61,62). While superoxide anion itself does not react with DNA, its dismutation product, H_2O_2, in the presence of a suitable transition metal such as Fe^{2+} will generate the hydroxyl radical, the species thought to be responsible for 75% of DNA breaks caused by ionizing radiation (63). Thus DNA damage may involve a series of reactions, the sum of which is the metal-catalyzed Haber-Weiss reaction. A recent report that UVA photons cause the release of Fe^{3+} from ferritin (64) suggests that the metal ion for the Haber-Weiss reaction may also result from UVA exposure (see Fig. 3). Another species that may initiate DNA breaks is the ferryl radical, as suggested by Linn and coworkers (65).

Isolated transforming DNA has been shown to be protected against the inactivating effects of 334-nm UVA by catalase, superoxide dismutase, and mannitol, which is consistent with the model that the DNA damage is the result of photodynamic reactions involving superoxide anion, H_2O_2, and hydroxyl radicals (66). Recent evidence using an isolated DNA model strongly supports a role for the hydroxyl radical in strand cleavage caused by 365-nm

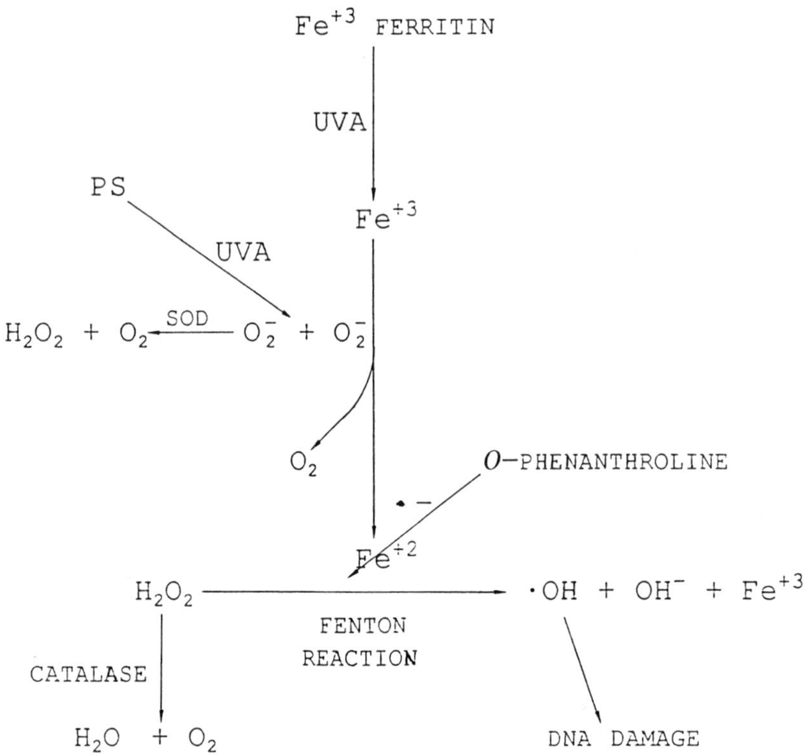

Figure 3 Scheme showing possible pathways from the absorption of the UVA photons by endogenous species to formation of DNA-damaging hydroxyl radicals. UVA photons are absorbed by a photosensitizer (PS) that transfers an electron to dissolved oxygen to form superoxide anion (O_2^-). UVA photons may also cause release of ferric ions (Fe^{+3}) from cellular protein. Superoxide dismutase (SOD) produces H_2O_2 from O_2^- and reduces Fe^{+3} to Fe^{+2}. The H_2O_2 and Fe^{+2} enter into the Fenton reaction to produce ·OH. This reaction can be inhibited by *o*-phenanthroline, which complexes Fe^{+2}.

photons (67), because the relative rates for quenching of the breakage of DNA by 365-nm UVA by acetate, formate, mannitol, and azide mimicked their rate constants for quenching of hydroxyl radical by these agents (68). The same result was obtained for gamma-ray exposures, where breakage is known to be mediated by the hydroxyl radical. A putative scheme describing the interconversion of reactive oxygen species is shown in Figure 3.

Although the in vitro results are suggestive, there is no definitive evidence as to which reactive species are involved in the production of the DNA damage in mammalian cells exposed to UVA radiation, and some recent evidence appears to be contradictory. H_2O_2, long suspected to be an important molecule in UVA effects, was shown to be produced in suspensions of cells exposed to UVA photons (14). However, the H_2O_2 yield was not sufficient to account for the measured DNA damage, a conclusion reached also by Tyrrell and Pidoux (69), who favored singlet oxygen as the mediator of DNA damage caused by UVA. In experiments using xeroderma pigmentosum cells, the breakage caused by H_2O_2 exposure was markedly inhibited in the presence of the metal chelator *o*-phenanthroline, but this compound had no effect on the breakage resulting from exposure to UVA photons at 365 nm (14). In contrast to the in vitro experiments described above, this implies that the metal-catalyzed Haber-Weiss reaction, which leads to production of hydroxyl radicals, is not the main pathway leading to DNA breaks in these cells caused by 365-nm photons.

It is clear that insufficient work has been carried out to describe definitively the chemical events that occur between the delivery of UVA photons to the surface of the cell and the ultimate DNA changes that they cause. The concept to be borne in mind is that the type of cell, the sensitizers involved, the sensitizer concentrations, and the particular energy of the UVA photon all affect the breakage at DNA bonds and the mechanism by which that breakage occurs.

VI. PROTECTIVE MECHANISMS

Halliwell and Gutteridge (70) and Tyrrell et al. (71,72) recently provided thorough reviews of the cellular defense pathways against the oxidative stress experienced by cells and tissues exposed to UVA photons; this section provides a brief overview. Defense mechanisms fall into three general types: repair of damage (see Sec. IV), constitutive antioxidant molecules and enzymes, and inducible UVA-stress genes that lead to protection. Constitutive antioxidant molecules (73) include a variety of cellular compounds that have been invoked as important agents for scavenging radicals, such as ascorbic acid, carotenoids, and sulfhydryl-containing compounds. Of prime importance in mammalian cells is glutathione.

Tyrrell and Pidoux (69,74) demonstrated a marked sensitization to both UVA and UVB radiation in cultured human skin fibroblasts whose glutathione levels had been depleted by the drug buthione-[S,R]-sulfoximine (BSO). As many as 80% of the lethal events caused by exposure of the glutathione-depleted cell population were not measured in control fibroblasts.

In our laboratory, Dudek (manuscript in preparation) has demonstrated that human epithelioid P3 cells enriched with metallothionein, a naturally occurring, cysteine-rich peptide, incur fewer DNA breaks and are also protected against the lethal effects of 365-nm UVA. A dose modification factor of 1.2 was measured at the D_{37} level, and this was elevated to 2.0 when the cells were in a glutathione-depleted state. It is clear that a combination of glutathione and metallothionein could be of major significance in protection against the oxidation events driven by UVA photons.

Catalase and superoxide dismutase have both been investigated as putative antioxidant enzymes that protect against UVA radiation. Recently, Peak and coworkers compared the responses to 365-nm UVA of a pair of Chinese hamster ovary (CHO) cells, one rich in catalase and the other having normal catalase levels. There was no difference in cell killing or in yields of DNA strand breaks after UVA exposures (75). Another study demonstrated that P3 epithelioid cells, which lack detectable catalase, do not incur more DNA breaks after exposure to 365-nm UVA than do catalase-containing xeroderma pigmentosum cells (14). Although less data is available on the protective role of superoxide dismutase, it also appears not to be of significance as a protective agent against UVA radiation. In our laboratory, we saw no difference in 365-nm–induced DNA breaks in mouse L cells having different levels of the enzyme (unpublished data). However, we have not yet been able to obtain mammalian cells completely deficient in the dismutase activity. It should be noted, however, that small changes in catalase (but not superoxide dismutase) activities have been reported in skin exposed to UVA (76). Also, Maeda et al. (77) reported significant elevation of superoxide dismutase in the skin of hairless mice chronically exposed to a UVA photon source spiked with 2% UVB.

The most significant finding in the area of induced-enzyme synthesis comes from the reports of Tyrrell and coworkers (78–81). These authors have shown that induction of heme oxygenase occurs as a general response to oxidants (H_2O_2, menadione, the tumor promoter 12-*O*-tetradecanoylphorbol-13-acetate, or gamma rays) including UVA radiation, or to agents that lower glutathione concentration, such as BSO, arsenite, iodoacetamide, cadmium, or diamide. This response was observed in a wide range of cells, including keratinocytes, lung and colon fibroblasts, and cells from rat, hamster, monkey, and marsupials, and also in whole animal tissues such as kidney and liver (although not all of these were tested with UVA). No response was seen to UVC or UVB, consistent with the reduced or lack of oxidative effects of those photons. The induction of heme oxidase is now considered to be a general response to any oxidative stress to which cells may be exposed. It can be inferred from these experimental observations that, apart from its catalytic role in the formation of biliverdin from hemoglobin, heme oxidase has a protective function against oxidative attack, especially since the induction of this enzyme was observed in cells that are not involved in hemoglobin catabolism.

The nature of the protection afforded by heme oxygenase remains conjectural. The induction of heme oxygenase will result in a dramatic decrease in intracellular heme levels, which could have several protective effects. For instance, the iron-catalyzed (including heme) Fenton reaction that produces hydroxyl radical from peroxides will be reduced. Further, there will be an increase in the product biliverdin, a known antioxidant that may itself be a protective compound.

VII. SUMMARY

It is clear that there is a large potential for a spectrum of DNA changes to occur in the skin of organisms exposed to solar UV photons, from pyrimidine photoproducts resulting from direct UVB radiation to oxidation events resulting from the UVA component. Action spectroscopy (82) reveals that UVB photons are much more efficient at producing DNA damage than are UVA photons, but the greater number of UVA photons in the environment, coupled with its greater penetration into skin, may offset this effect. Tyrrell and Pidoux (83) present a detailed computation and analysis of action spectra, which shows that in human keratinocytes and fibroblasts, between 20 and 60% (depending upon the solar zenith angle) of the toxic effects of solar radiation is due to the UVA component. An important consequence of chronic occurrence of these different types of DNA damage in human skin is tumor formation, the cancer of greatest incidence in humans. It is interesting to note that solar radiation consists of UVB radiations thought to initiate carcinogenesis (84), as well as oxidizing UVA radiations, since oxidations are thought to act as tumor promoters (85).

ACKNOWLEDGMENTS

This work was supported in part by the U.S. Department of Energy, Office of Energy Research, Office of Health and Environmental Research, under Contract #W-31-109-ENG-38, and in part by PHS by grants R01-34492 and RO1-37848 awarded by the National Cancer Institute, U.S. Department of Health and Human Services.

REFERENCES

1. Johnson FS, Mo I, Green AES. Average latitudinal variation in ultraviolet radiation at the earth's surface. Photochem Photobiol 1976; 23:179–88.
2. Bruls WAG, Slaper H, van der Leun JC, Berrens L. Transmission of human epidermis and stratum corneum as a function of thickness in the ultraviolet and visible wavelengths. Photochem Photobiol 1984; 40:485–94.
3. Blum HF. Carcinogenesis by ultraviolet light. Princeton: Princeton University Press, 1959.

4. Setlow RB. The wavelengths of sunlight effective in producing skin cancer: a theoretical analysis. Proc Natl Acad Sci USA 1974; 71:3363–6.
5. Ley RD, Applegate LA, Fry RJM, Sanchez AB. Photoreactivation of ultraviolet radiation induced skin and eye tumors in *Monodelphis domestica*. Cancer Res 1991; 51:6539–42.
6. Kelfkens G, de Gruijl FR, van der Leun JC. Ozone depletion and increase in annual carcinogenic ultraviolet dose. Photochem Photobiol 1990; 52:819–23.
7. Appenzeller T. Ozone loss hits us where we live. Science 1991; 254:645.
8. Kerr RA. Ozone destruction worsens. Science 1991; 252:204.
9. Sterenborg HJCM, van der Leun JC. Tumorigenesis by a long wavelength UV-A source. Photochem Photobiol 1990; 51:325–30.
10. Kligman LH, Crosby MJ, Miller SA, Hitchins VM, Beer JZ. In: Urbach F, ed. Biological effects of UV-A radiation. Overland Park, KS: Valdenmar Publishing Co. In press.
11. Sutherland JC, Griffin KP. Absorption spectrum of DNA for wavelengths longer than 320 nm. Radiat Res 1981; 86:399–409.
12. Peak MJ, Peak JG. Molecular photobiology of UVA. In: Urbach F, Gange RW, eds. The biological effects of UVA radiations. New York: Plenum Press, 1986: 42–56.
13. Jones CA, Huberman E, Cunningham, ML, Peak MJ. Mutagenesis and cytotoxicity in human epithelial cells by far- and near-ultraviolet radiations: action spectra. Radiat Res 1987; 110:244–54.
14. Peak JG, Pilas B, Dudek EJ, Peak MJ. DNA breaks caused by monochromatic 365-nm radiation or hydrogen peroxide and their repair in human epithelioid and xeroderma pigmentosum cells. Photochem Photobiol 1991; 54:197–203.
15. Peak MJ, Peak JG, Carnes BA, Chang-Liu CM, Hill CK. DNA damage and repair in rodent and human cells after exposure to JANUS fission spectrum neutrons: a minor fraction of single-strand breaks as revealed by alkaline elution are refractory to repair. Int J Radiat Biol 1989; 55:761–72.
16. Hsie AW, Li AP, Machanoff R. A fluence-response study of lethality and mutagenicity of white, black and blue fluorescent light, sunlamp and sunlight irradiation in Chinese hamster ovary cells. Mutat Res 1977; 45:333–42.
17. Coohill TP, Peak MJ, Peak JG. The effects of the ultraviolet wavelengths of radiation present in sunlight on human cells in vitro. Photochem Photobiol 1987; 46:1043–50.
18. Enninga IC, Groenendijk RTL, Filon AR, van Zeeland AA, Simmons JWIM. The wavelength dependence of u.v.-induced pyrimidine dimer formation, cell killing and mutation induction in human diploid skin fibroblasts. Carcinogenesis 1986; 7:1829–36.
19. Wells RL, Han A. Action spectra for killing and mutation of Chinese hamster ovary cells exposed to mid- and near-ultraviolet monochromatic radiation. Mutat Res 1984; 129:251–8.
20. Doniger J, Jacobson ED, Krell K, DiPaolo JA. Ultraviolet light action spectra for neoplastic transformation and lethality of Syrian hamster embryo cells correlate with spectrum for pyrimidine dimer formation in cellular DNA. Proc Natl Acad Sci USA 1981; 78:2378–82.

21. Tyrrell RM. Mutagenic action of monochromatic UV radiation in the solar range on human cells. Mutat Res 1984; 129:103–10.
22. Jacobson ED, Krell K. Ultraviolet wavelength regions implicated in toxic and mutagenic effects of broad-spectrum radiation from fluorescent lamps in L5178Y mouse lymphoma cells. Mutat Res 1979; 62:535–8.
23. Chan GL, Peak MJ, Peak JG, Haseltine WA. Action spectrum for the formation of endonuclease-sensitive sites and (6–4) photoproducts induced in DNA fragment by ultraviolet radiation. Int J Radiat Biol 1986; 50:641–8.
24. Han A, Peak MJ, Peak JG. Induction of DNA-protein cross-linking in Chinese hamster ovary cells by monochromatic 365 and 405 nm ultraviolet light. Photochem Photobiol 1984; 39:343–8.
25. Cleaver JE, Cortes F, Karentz D, Lutze LH, Morgan WF, Player AN, Vuksanovic L, Mitchell DM. The relative biological importance of cyclobutane and (6-4) pyrimidine-pyrimidone dimer photoproducts in human cells: evidence from a xeroderma pigmentosum revertant. Photochem Photobiol 1988; 48:41–9.
26. Mitchell DL, Jen J, Cleaver JE. Relative induction of cyclobutane dimers and cytosine photohydrates in DNA irradiated in vitro and in vivo with ultraviolet-C and ultraviolet-B light. Photochem Photobiol 1991; 54:741–6.
27. Bose SN, Davies RJH, Sethi SK, McCloskey JA. Formation of an adenine-thymine photoadduct in the deoxydinucleoside monophosphate d(TpA) and in DNA. Science 1983; 220:723–5.
28. Peak MJ, Peak JG, Carnes BA. Induction of direct and indirect single-strand breaks in human cell DNA by far- and near-ultraviolet radiations: action spectrum and mechanisms. Photochem Photobiol 1987; 45:381–7.
29. Hirshi M, Netrawali MS, Remsen JF, Cerutti PA. Formation of DNA single-strand breaks by near-ultraviolet and γ-rays in normal and Bloom's syndrome skin fibroblasts. Cancer Res 1981; 41:2003–7.
30. Holmberg M, Almassy Zs, Langerberg M, Niejahr B. The repair of strand breaks in human lymphocytes exposed to near UV-radiation (UVA) and far UV-radiation (UVC). Photochem Photobiol 1985; 41:437–44.
31. Rosenstein BS, Ducore JM. Induction of DNA strand breaks in normal human fibroblasts exposed to monochromatic ultraviolet and visible wavelengths in the 260–546 nm range. Photochem Photobiol 1983; 38:51–5.
32. Roza L, van der Schans GP, Lohman PHM. The induction and repair of DNA damage and its influence on cell death in primary human fibroblasts exposed to UV-A or UV-C radiation. Mutat Res 1985; 146:89–98.
33. Peak JG, Peak MJ. Ultraviolet light induces double-strand breaks in DNA of cultured human P3 cells as measured by neutral filter elution. Photochem Photobiol 1990; 52:387–93.
34. Peak MJ, Peak JG, Jones CA. Different (direct and indirect) mechanisms for the induction of DNA-protein crosslinks in human cells by far- and near-ultraviolet radiations (290 and 405 nm). Photochem Photobiol 1985; 42:141–6.
35. Peak MJ, Peak JG. Convex curvatures of alkaline elution profiles of DNA from human cells irradiated with 405 nm UVA: evidence for induction of slowly developing alkali-labile sites. Photochem Photobiol 1989; 50:379–83.

36. Churchill ME, Peak JG, Peak MJ. Repair of near-visible- and blue-light-induced DNA single-strand breaks by the CHO cell lines AA8 and EM9. Photochem Photobiol 1991; 54:639–44.
37. Kleiman NJ, Wang R-W, Spector A. Hydrogen peroxide-induced DNA damage in bovine lens epithelial cells. Mutat Res 1990; 240:35–45.
38. Peak MJ, Peak JG. DNA-to-protein crosslinks and backbone breaks caused by far- and near-ultraviolet, and visible radiations in mammalian cells. In: Simic MG, Grossman L, Upton AC, eds. Mechanisms of DNA damage and repair: implications for carcinogenesis and risk assessment. New York: Plenum Press, 1986: 193–202.
39. Peak MJ, Peak JG. Comparison of initial yields of DNA-to-protein crosslinks induced in cultured human cells by x-rays, gamma rays, 0.85-MeV neutrons, far- and near-ultraviolet light, and blue light. Mutat Res 1991; 246:187–96.
40. Leadon SA. Production and repair of DNA damage in mammalian cells. Health Physics 1990; 59:15–22.
41. Hariharan PV, Cerutti PA. Formation of products of the 5,6-dihydroxydihydrothymine type by ultraviolet light in HeLa cells. Biochem 1977; 16:2791–5.
42. Demple B, Linn S. DNA *N*-glycosylases and UV repair. Nature 1980; 287:203–8.
43. Setlow RB. Twenty five years of DNA repair. In: Castellani A, ed. DNA damage and repair. New York: Plenum Press, 1987: 1–10.
44. Hoeijmakers JHJ, van Duin M, Koken M, Yasui A, Jaspers NGJ, Westerveld A, Bootsma D. Isolation and characterization of genes involved in mammalian excision repair. In: Castellani A, ed. DNA damage and repair. New York: Plenum Press, 1987: 27–36.
45. Mitchell DL, Humphrey RM, Adair GM, Thompson LH, Clarkson JM. Repair of (6-4) photoproducts correlates with split-dose recovery in UV-irradiated normal and hypersensitive rodent cells. Mutat Res 1988; 193:53–63.
46. Mitchell DL, Applegate LA, Nairn RS, Ley RD. Photoreactivation of cyclobutane dimers and (6-4) photoproducts in the epidermis of the marsupial, *Monodelphis domestica*. Photochem Photobiol 1990; 51:653–8.
47. Mitchell DL, Haipek CA, Clarkson JM. (6–4) Photoproducts are removed from the DNA of UV-irradiated mammalian cells more efficiently than cyclobutane pyrimidine dimers. Mutat Res 1985: 143:109–12.
48. Strickland PT, Creasey JS. Induction and repair of cyclobuta-dithymidine photoproducts in hamster skin by solar-simulated UV radiation. J Photochem Photobiol 1989; 3:17–24.
49. D'Ambrosio SM, Whetstone JW, Slazinski L, Lowney E. Photorepair of pyrimidine dimers in human skin in vivo. Photochem Photobiol 1981; 34:461–4.
50. Sutherland BM, Gange RW, Freeman SE, Sutherland JC. 1987. DNA damage and repair in human skin in situ. In: Castellani A, ed. DNA damage and repair. New York: Plenum Press, 1987: 157–68.
51. Freeman SE. Variations in excision repair of UVB-induced pyrimidine dimers in DNA of human skin in situ. J Invest Dermatol 1988; 90:814–17.
52. Churchill ME, Peak JG, Peak MJ. Correlation between cell survival and DNA single-strand breaks repair proficiency in the Chinese hamster ovary cells lines AA8 and EM9 irradiated with 365-nm UVA radiation. Photochem Photobiol 1990; 53:229–36.

53. Rosenstein BS, Lai L-W, Ducore JM, Rosenstein RB. DNA-protein crosslinking in normal and solar UV-sensitive ICR2A frog cell lines exposed to solar UV-radiation. Mutat Res 1989; 217:219–26.
54. Kochevar IE, Dunn DA. Photochemistry involving the nucleic acids. In: Munson H, ed. Bioorganic photochemistry. Vol. 1. New York: Wiley-Interscience, 1990: 273–315.
55. Foote CS. Definition of Type 1 and Type 11 photosensitized oxidation. Photochem Photobiol 1991; 54:659.
56. Rosenstein BS, Ducore JM, Cummings SW. The mechanism of bilirubin photosensitized DNA strand breakage in human cells exposed to phototherapy light. Mutat Res 1983; 112:397–406.
57. Peak JG, Peak MJ, MacCoss M. DNA breakage caused by 334-nm ultraviolet light is enhanced by naturally occurring nucleic acid components and nucleotide coenzymes. Photochem Photobiol 1984; 39:713–16.
58. Merkel PB, Nilsson R, Kearns DR. Deuterium effects on singlet oxygen lifetimes in solutions: a new test of singlet oxygen reactions. J Am Chem Soc 1972; 94:1030–2.
59. Midden WR, Wang SY. Singlet oxygen generation for solution kinetics: clean and simple. J Am Chem Soc 1983; 105:4129–35.
60. Blazek ER, Peak JG, Peak MJ. Singlet oxygen induces frank strand breaks as well as alkali- and piperidine-labile sites in supercoiled plasmid DNA. Photochem Photobiol 1989; 49:607–13.
61. Cunningham ML, Krinsky NI, Giovanazzi SM, Peak MJ. Superoxide anion is generated from cellular metabolites by solar radiation and its components. J Free Rad Biol Med 1985; 1:381–5.
62. Cunningham ML, Peak JL, Peak MJ. Single-strand DNA breaks in rodent and human cells produced by superoxide anion or its reduction products. Mutat Res 1987; 184:217–22.
63. Hall EJ. Radiobiology for the radiobiologist. 2nd ed. New York: Harper and Row, 1978.
64. Aubailly M, Santus R, Salmon S. Ferrous ion release from ferritin by ultraviolet-A radiations. Photochem Photobiol 1991; 54:769–73.
65. Imlay JA, Chin SM, Linn S. Toxic damage by hydrogen peroxide through the Fenton reaction in vivo and in vitro. Science 1988; 240:640–2.
66. Peak MJ, Ito A, Foote CS, Peak JG. Photosensitized inactivation of DNA by monochromatic 334-nm radiation in the presence of 2-thiouracil: genetic activity and backbone breaks. Photochem Photobiol 1988; 47:809–13.
67. Peak MJ, Peak JG. Hydroxyl radical quenching agents afford similar protection against DNA breakage caused by 365-nm UVA or by gamma radiation. Photochem Photobiol 1990; 51:649–52.
68. Farhataziz, Ross AB. Selected specific rate of reaction of transients from water in aqueous solutions. III. Hydroxyl radical and perhydroxyl radical and their radical ions. Nat Std Ref Data Syst 1977; 59:1–113.
69. Tyrrell RM, Pidoux M. Endogenous glutathione protects human skin fibroblasts against the cytotoxic action of UVB, UVA, and near-visible radiations. Photochem Photobiol 1986; 44:561–4.
70. Halliwell B, Gutteridge JMC. Free radicals in biology and medicine. Oxford: Clarendon Press, 1989: 86–179.

71. Tyrrell RM. UVA (320-380 nm) radiation as an oxidative stress. In: Seis H, ed. Oxidative stress: oxidants and antioxidants. New York: Academic Press, 1991: 57–83.
72. Tyrrell RM, Keyse SM, Moraes EC. Cellular defense against UVA (320-380 nm) and UVB (290-320 nm) radiations. In: Ricklis E, ed. Photobiology: the science and its applications. New York: Plenum Press, 1991:861–71.
73. Black HS. Potential involvement of free radical reactions in ultraviolet light-mediated cutaneous damage. Photochem Photobiol 1987; 46:213–21.
74. Tyrrell RM, Pidoux M. Correlation between endogenous glutathione content and sensitivity of cultured human skin cells to radiation at defined wavelengths in the solar ultraviolet range. Photochem Photobiol 1988; 47:405–12.
75. Peak MJ, Jones CA, Sedita BA, Dudek EJ, Spitz DR, Peak JG. Evidence that hydrogen peroxide generated by 365-nm UVA radiation is not important in mammalian cell killing. Radiat Res 1990; 123:220–3.
76. Fuchs J, Hufleit MW, Rothfoss LM, Wilson DS, Garcano G, Packer L. Acute effects of near ultraviolet and visible light on the cutaneous antioxidant defense system. Photochem Photobiol 1989; 50:739–44.
77. Maeda K, Naganuma M, Fukuda M. Effects of chronic exposure of ultraviolet-A including 2% ultraviolet-B on free radical reduction systems in hairless mice. Photochem Photobiol 1991; 54:737–40.
78. Keyse SM, Tyrrell RM, Both near ultraviolet radiation and the oxidizing agent hydrogen peroxide induce a 32-kDa stress protein in normal human skin fibroblasts. J Biol Chem 1987; 262:14821–5.
79. Keyse SM, Tyrrell RM. Induction of the heme oxygenase gene in human skin fibroblasts by hydrogen peroxide and UVA (365 nm) radiation: evidence for the involvement of the hydroxide radical. Carcinogenesis 1990; 11:787–91.
80. Keyse SM, Applegate LA, Tromvoukis Y, Tyrrell RM. Oxidant stress leads to transcriptional activation of the human heme oxygenase in cultured skin fibroblasts. Mol Cell Biol 1990; 10:4967–9.
81. Applegate LE, Luscher P, Tyrrell RM. Induction of heme oxygenases: a general response to oxidant stress in cultured mammalian cells. Cancer Res 1991; 51: 974–8.
82. Coohill TP. Action spectra again? Photochem Photobiol 1991; 54:859–70.
83. Tyrrell RM, Pidoux M. Action spectra for human skin cells: estimates of the relative cytotoxicity of the middle ultraviolet, near ultraviolet and violet regions of sunlight on epidermal keratinocytes. Cancer Res 1987; 47:1825–9.
84. Hart RW, Setlow RB, Woodhead AD. Evidence that pyrimidine dimers in DNA can give rise to tumors. Proc Natl Acad Sci USA 1977; 74:5574–8.
85. Cerutti PA. Prooxidant states and tumor promotion. Science 1985; 227:375–81.

11

Oxidative Stress, Cytokines, and Lymphocyte Activation

Marie-Anne Gougerot-Pocidalo
Hôpital Bichat, Paris, France

Jean-Pierre Revillard
University of Lyon, Lyon, France

I. INTRODUCTION

Oxidative stress can be defined as molecular alterations generated by oxidant molecules. The most important biological oxidants are superoxide, hydrogen peroxide, hydroxyl radical, hypochlorous acid, heme-associated ferryl species, and water-soluble and lipid-soluble peroxyl radicals (1–3). Among these biological oxidants, free radicals, i.e., molecules with an unpaired electron, are particularly reactive. Extensive studies with biological material in vitro have clearly shown that reactive free radicals are able to produce chemical modifications of, and damage to, proteins, lipids, carbohydrates, and nucleotides (2,4). Highly reactive free radicals produced in biological systems are essentially confined to their immediate vicinity as a direct consequence of their extreme chemical reactivity. It has been calculated, for instance, that hydroxyl radical $OH^{\cdot}$ will have a half-life of less than a nanosecond and a diffusion of less than 100 nm (4). Furthermore, each cell is equipped with highly potent antioxidant mechanisms, which involve primarily glutathione (GSH) and the selenium-dependent enzyme GSH-peroxidase in the cytoplasm, and alpha-tocopherol in the membranes. Many natural and synthetic antioxidant compounds have been evaluated (5).

Oxidative tissue injury may be induced by a wide variety of physical or chemical aggressions. These include, for instance, ischemia-perfusion, ultra x or γ irradiation, thermal injury, and exposure to hyperbaric or normobaric oxygen, to nitrous oxide, or to a variety of chemicals including carcinogenic and mutagenic products. The immune system uses reactive oxygen species produced by phagocytic cells as a major defense mechanism against microorganisms. However, the same reactive oxygen species contribute to tissue lesions associated with inflammatory reactions (4,6,7).

Since the initial descriptions of immunosuppression induced by x or γ irradiation, it was believed that, depending on its duration and intensity, any kind of oxidative stress could be deleterious to the lymphoid tissues. Indeed several experimental models of in vivo exposure to normobaric or hyperbaric oxygen were reported, demonstrating both cell depletion in the lymphoid system and functional impairment of the remaining lymphocytes. Knowing that all living organisms are exposed to repeated oxidative stress, it was assumed that an altered reduction/oxidation (redox) balance might play a role in the progressive immunodeficiency associated with ageing, along with an increased risk of tumors and atherosclerosis, and that dietary supplementation with antioxidant compounds could help to delay the occurrence of that immunodeficiency. More recently the role of oxidative stress was discussed in the pathogenesis of viral infection (8,9), especially in the progression of human immunodeficiency virus (HIV) infection towards AIDS (10,11). However, any direct association between oxidative stress and immunodeficiency is merely an oversimplification of what appears now to be a quite complex and still partly unknown set of molecular interactions. Indeed oxidative reactions would seem to be required for the production of some cytokines and/or involved in their biological effects as well as in T-lymphocyte activation.

In this brief review we summarize current knowledge and controversies on the mechanisms whereby oxidative stress and natural or therapeutical antioxidants including thiol compounds can modulate inflammatory cytokine synthesis and activity and how they control T-cell activation. We then discuss some membrane and intracellular signaling systems that may represent targets for oxidative stress and present some indirect evidence for the contribution of an altered redox balance in HIV replication in vitro and in the development of immunodeficiency during the course of HIV infection. It should be kept in mind that despite accumulation of a wide array of experimental data over the last 20 years, it is not yet possible to present a unique and simplified pathway that would allow to predict the effect of antioxidant therapy on the immune response.

II. IMMUNOSUPPRESSION INDUCED BY OXIDATIVE STRESS

A. In Vivo Experiments

Exposure of mice to normobaric oxygen was reported to induce the following sequential alterations: 1) decreased numbers of antibody-secreting cells after immunization with sheep erythrocytes, a T_H-dependent antigen, after 24 hours of hyperoxia; 2) decreased responses to TNP-LPS, a T-independent antigen and diminution of contact type hypersensitivity reactions to the hapten oxazolone; 3) decreased proliferative responses of spleen lymphocytes to concanavalin A demonstrable after only 24 hours of hyperoxia, whereas proliferative responses to phytohemagglutinin and to lipopolysaccharide were decreased only after 96 hours of hyperoxia (12). Furthermore, a marked depletion of cellularity was observed in the thymus spleen and lymph nodes of these animals, whether adrenalectomized or not (13). Comparable data were reported in the rat (14). Further studies indicated that the depressed proliferative response to concanvalin A in mice exposed to normobaric oxygen was not due to the presence of suppressor cells in the spleen and would be restored by addition of intact peritoneal cells in the cultures or of 2-mercaptoethanol (15), reduced glutathione, L-cystine or L-cysteine, but not selenomethionine (16). The magnitude of hyporesponsiveness to concanavalin A was correlated with the decrease of cellular sulfhydryl groups (SH). Oxygen scavengers such as superoxide dismutase, catalase, mannitol, and vitamin E did not protect against the decrease in SH groups and responsiveness to concanavalin A (17). Similarly, a single injection of diethyldithiocarbamate before exposure to normobaric oxygen was reported to prevent lymphoid depletion and functional immunological alterations (18).

These data clearly demonstrated that oxidative stress in vivo induced first functional alterations of T_H cells and, only after prolonged exposure, defects of delayed type hypersensitivity and antibody secreting cells. As regards macrophage phagocytic activity, the data were controversial (19–22), suggesting that those cells were less susceptible than T lymphocytes to hyperoxia. Other studies showed that in mice, rats, and guinea pigs, hyperbaric oxygen resulted in the suppression of tuberculin reaction, adjuvant disease, allograft rejection, and allergic encephalomyelitis (23) and prevented the development of antibody responses and autoimmune symptoms of New Zealand black and MRL/*lps* mice (24).

B. In Vitro Experiments

Analysis of oxidative stress in lymphocyte cultures has been performed using different experimental protocols, including addition of hydrogen peroxide (50–200 μM), hypoxanthine and xanthine oxidase, or glucose and glucose oxidase, exposure to normobaric 95% O_2 atmosphere, or to UV-irradiated culture medium, addition of polymorphonuclear neutrophils stimulated by opsonized zymosan or phorbol esters in the presence of exogenous superoxide dismutase or catalase to determine the relative contribution of superoxide and hydrogen peroxide. Several parameters of T-cell activation have been investigated, including proliferative responses to vegetal lectins, allogeneic stimulator cells, cell cycle analysis, interleukin-2 (IL-2) production, and expression of CD25, the α chain of the IL-2 receptor. Other studies focused on natural killer cell or cytotoxic T-cell activities. The main findings can be summarized as follows: depending on experimental conditions, in vitro oxidative stress was reported to induce cell killing as assessed by membrane permeability to ethidium bromide (25) or alteration in the cytoskeleton as assessed by capping induced by concanavalin A (26) and a wide array of functional alterations including impaired cytotoxic activity (27,28) and impaired proliferative responses to the lectins concanavalin A, pokeweed mitogen, and phytohemagglutinin (29–31). Indirect evidence suggested that hydrogen peroxide but not superoxide or singlet oxygen was responsible for these impaired proliferative responses inasmuch as quite similar effects were achieved by direct addition of hydrogen peroxide and no protection was afforded by superoxide dismutase (25,28,30–32). The role of macrophages as a main source of free radicals was suggested (31,32), and the contribution of both myeloperoxidase from polymorphonuclear neutrophils and halides was established (27). The amplifying activity of ferrous ions through the Fenton reaction was shown indirectly by the protective effect of iron-chelating agents like desferrioxamine (25,33). Finally, strong indirect evidence was provided for a role of lipid peroxidation and generation of suppressive lipidic derivatives by macrophages, with possible candidates being prostaglandin E_2 and leukotriene B_4 (31,34–37). Further evidence for a role of lipid peroxidation arose from the observation that the lipid antioxidants butylated hydroxyanisole, butylated hydroxytoluene, and *n*-propylgallate blocked the lymphocytotoxic effect of hydrogen peroxide (25). A remarkable feature of impaired T-cell proliferative responses secondary to oxidative stress was the contrast between unimpaired IL-2 secretion (25,29) and reduced expression of transferrin receptor and IL-2 α chain receptor (29). These data suggested a blockade in the G1 phase of the cell cycle. Another mechanism suggested by Carson and coworkers (38) implied a later block in the cell cycle induced by accumulation of poly(ADP)-ribose synthesized in lymphocytes with unrepaired strand breaks.

III. ROLE OF OXIDATIVE STRESS AND LIPOXYGENASE PATHWAY IN T-CELL ACTIVATION

T-cell activation is triggered by a molecular interaction between the T-cell receptor (TcR) CD3 complex and a peptide associated with a major histocompatibility complex–encoded molecule on the antigen presenting cell, a superantigen (e.g., staphylococcal enterotoxins), a lectin, or antibodies to TcR/CD3 epitopes. Full activation requires a second signal delivered by the antigen-presenting cell, the nature of which is still a matter of debate. This signal may be provided by cognate interaction of cell surface adhesion molecules (e.g., CD2/LFA3, LFA-1/ICAM1, and ICAM-2) (39) or by cytokines such as cell-associated or secreted interleukin-1 (40). The initial metabolic events that control T-cell activation are the activation of a phospholipase C, possibly through coupling by a G protein, resulting in hydrolysis of phosphatidyl inositol 4,5-biphosphate to yield diacylglycerol and inositol 1,4,5-triphosphate. These second messengers appear to mediate the mobilization of intracellular and extracellular Ca^{2+} and activation of protein kinase C (PKC). Direct stimulation of these second messengers by a calcium ionophore in association with a phorbol ester results in T-cell activation and proliferation (40). In parallel, TcR cross-linking activates different protein tyrosine kinases as, for instance, the $p59^{fyn}$ associated with CD3 and the $p56^{lck}$ associated with CD4 and CD8, whereas the phosphotyrosine phosphatase located in the intracytoplasmic portion of the CD45 molecule induces the dephosphorylation of several phosphotyrosine substrates including $p56^{lck}$ itself (41). Subsequently, intracellular signals induce a cascade of sequential gene expression events. A large number of genes (roughly 200) are expressed over a period of a few days, resulting in a profound change of the T cell. Among those genes the IL-2 gene has become a model of study, and a great number of studies have focused on transcription factors that interact with *cis*-acting elements associated with a transcriptional enhancer in the 300 bp region directly upstream of the IL-2 promoter (reviewed in Ref. 42).

Several studies suggest the implication of oxidative processes in T-cell responses. The data rely essentially on the use of antioxidants (43–52). In several reports the involvement of oxidative products has been related to the arachidonic acid cascade, since it was demonstrated that inhibition of the lipoxygenase pathway blocked lymphocyte responses (44,46,48–50,52). It was suggested that inhibition of lipid peroxidation selectively decreased the events depending on PKC activation, namely, expression of the light chain of IL-2 receptor and phorbol ester–induced pH changes (52). In agreement with these data, it was shown that lipid oxidation products decreased the K_a for PKC activation (53). Further insight into the role of oxidation in T-cell activation was given by the demonstration that oxidative modification of the regulatory domain of membrane bound and/or catalytically active form of PKC resulted in a calcium and phospholipid-independent activation of the enzyme (54). In turn, further oxidation of the molecule resulted in the formation of an oxidatively modified inactive PKC more susceptible that the intact enzyme to calpain-mediated proteolysis. These mechanisms could account for both the activation of PKC by phorbol esters that generate free radicals and for the downregulation of this enzyme upon prolonged exposure of the cells to phorbol esters (54).

IV. A ROLE FOR THIOL COMPOUNDS IN T-CELL ACTIVATION

It has been known for a long time that addition of 2-mercaptoethanol to murine lymphoid cell cultures was required to achieve optimal primary in vitro antibody response, cytotoxic activity, and proliferative responses to T-cell mitogens (55–57). The mechanisms of action of 2-mercaptoethanol may involve several pathways, (58) but the best-documented end result is the formation of a mixed disulfide with cysteine (59) and the increase in intracellular reduced glutathione levels (60,61). During lymphocyte activation the level of glutathione declines, but the presence of 2-mercaptoethanol or various thiol-reducing agents partially prevents this decline (62). The role of intracellular glutathione was further demonstrated by the inhibition of T-cell mitogenic and cytotoxic responses induced by buthionine sulfoximine, a selective inhibitor of glutathione synthesis by inhibition of the enzyme γ-glutamylcysteine synthetase, and by 2-cyclo-hexene-1-one, which decreases intracellular levels of glutathione. These compounds inhibit mitogen-induced T-cell proliferation (63,64) and cytotoxic activities of natural killer cells and cytotoxic T lymphocytes (65). The inhibition of DNA synthesis was also demonstrated in T cells activated by the CD2 pathway, which is accessory cell independent (66). Furthermore, the proliferative capacity of T cells stimulated by lectins could be correlated with their intracellular glutathione content (67–69). Both decrease in parallel in ageing and autoimmune-prone mice (70).

Even more direct evidence was recently provided by experiments of cell sorting on the basis of glutathione content measured by using monochlorobimane in flow cytometry (71). Early reports suggested that glutathione was involved in initial stages of T-cell activation because 2-mercaptoethanol was effective only if added during the first hours of culture. However, more recent data support the view that defect in glutathione does not affect cytokine secretion (e.g., IL-2), which occurs during the G_1 phase of the cell cycle, but rather control IL-2–dependent events (binding, internalization, and progression to the S phase) (72–75). These recent data are in complete agreement with the above-reported effects of exogenous oxidative stress such as culture in normobaric oxygen. Finally, initial experiments with murine cell cultures showed that the positive effect of 2-mercaptoethanol could be replaced by macrophages. Activated macrophages release glutathione and other sulfhydryl compounds in the extracellular medium (76–78) and could therefore provide T lymphocytes at their contact with cysteine, thereby contributing to the control of the oxidant mechanisms triggered during the activation of these cells (79).

V. OXIDATIVE STRESS AND INFLAMMATORY CYTOKINES

The role of free radicals and cytokines (IL-1, IL-6, IL-8, and tumour necrosis factor [TNF]) has been extensively investigated in inflammation, but it is only recently that a direct relationship between oxidation and cytokine synthesis or cytokine activities on their target cells could be established. Recombinant human TNFα was reported to induce the generation of reactive oxygen species by neutrophils (80–82) at concentrations as low as 0.5 ng/ml. The effect was neutralized by TNF-specific antibodies but not by polymyxin B, thus excluding the possible role of contaminant lipopolysaccharide. TNFα also increased the oxidative burst of neutrophils in response to zymosan, but not to glucan, by increasing the expression of surface CR3 complement receptors (83). The sensitivity of several tumor cell lines to the cytotoxic effect of TNFα was negatively correlated with the intracellular GSH levels (84). It was enhanced by pretreatment with DL-buthionine-(S,R)-sulfoximine, an inhibitor of GSH synthesis, or by administration of diethylmaleate, a compound that irreversibly binds GSH and neutralizes its antioxidant activity. These data suggest that reactive oxygen species may represent the major intracellular effector mechanism of the toxicity of TNF. Conversely, target cells exposed to exogenous oxidative stress (e.g., H_2O_2) become more resistant to TNF and express lower levels of TNF receptors (85). Furthermore, in vitro exposure to TNF selectively enhances MnSOD activity, but not CuZnSOD or catalase activities, in endothelial cells. Such a negative feedback control would appear necessary to limit the extension of the oxidative stress generated by TNFα (86). The enzyme MnSOD is known to be inducible by oxidative stress in various cell types (87). Oxidative stress has been reported by several authors to increase the synthesis of inflammatory cytokines. In fact, hydrogen peroxide and periodate were reported to enhance lipopolysaccharide-induced TNF production by monocytes, whereas antioxidants like butylated hydroxyanisole markedly suppressed TNF synthesis (88). In a similar fashion, exposure to ionizing radiation was reported to increase the expression of TNFα mRNA in several cell lines (89) and UV irradiation to stimulate the synthesis and release of TNFα by keratinocytes (90). More recently quite comparable data were reported with IL-1 and IL-6. Ultraviolet light stimulates the synthesis of IL-6 by keratinocytes (91). Human monocytes cultured in 95% normobaric oxygen synthesized increased amounts of biologically active IL-1 whether stimulated or not by lipopolysaccharide or opsonized zymosan (92). Exposure of monocytes to free radicals generated by polymorphonuclear neutrophils stimulated synthesis and release of IL-1. The effect could be attributed to superoxide anion but not to hydrogen peroxide (93). The generation of Il-1 activity was amplified by preincubation with catalase but suppressed by preincubation with superoxide dismutase or vitamin E (93).

A common mechanism accounting for the stimulation of IL-1, IL-6, and TNFα synthesis under oxidative stress is likely to involve the binding to regulatory elements of enhancer factors that are generated by activation of protein kinase C (PKC), including the nuclear factor κB as discussed below. Supporting this hypothesis, inhibitors of PKC were reported to prevent IL-1 synthesis by monocytes cultured in normobaric oxygen (92). Furthermore two regulatory elements, NF-κB and the activator protein-1 (AP-1) binding sites, are located in 5′ of the IL-6 promoter upstream and downstream of the recently identified IL-1 responsive elements (reviewed in Ref. 94). The AP-1 transcription factor is formed by the association of products of the cellular oncogenes c-*jun* and c-*fos*. In cell exposed to oxidative stress generated by the xanthine/xanthine oxidase system c-*fos* mRNA is detected after 15 minutes of exposure, peaking at 30 minutes, and c-*jun* mRNA is detected after 30 minutes and peaked at 120 minutes (95). The expression of c-*fos* was inhibited by superoxide dismutase but not by catalase, indicating a role for superoxide anion, whereas hydrogen peroxide alone did not induce c-*fos* mRNA expression in this system. Finally the expression of c-*fos* was partly inhibited by two PKC inhibitors and completely blocked by 2-aminopurine, a broad-spectrum protein kinase inhibitor, and by chelation of calcium (95). In another study it was shown that the formation of AP-1 by dimerization of the Fos and Jun proteins was modulated by redox of a single conserved cysteine residue in the DNA-binding domain of the two proteins (96). Furthermore, in the same study a nuclear protein was identified that reduced Fos and Jun and stimulated their DNA-binding activity. Altogether the results suggest that transcriptional activity mediated by AP-1 binding factors may be regulated by a redox mechanism.

VI. OXIDATE STRESS NF-κB AND HIV INFECTION

Nuclear factor κB (NF-κB) was initially identified as a regulatory protein that binds the κB enhancer element located in the intron between the J and C genes of the κ immunoglobulin light chain (97). Identical or closely related elements were subsequently discovered upstream of a great number of cellular genes, especially those of cytokines (IL-2, IL-3, IL-4, IL-6, GM-CSF, IFN-β, IFN-γ, TNAα) the α chain (CD25) of IL-2 receptor, the protooncogene c-*fos*, β_2-microglobulin, and class I genes of the major histocompatibility complex. The κB sequence is also present in the long terminal repeat of HIV and in the cytomegalovirus genome. The protein NF-κB is a heterodimer made of two polypeptides, p50 and p65, that are expressed constitutively in the cytoplasm and associated with a 68-kd inhibitor I-κB to form a biologically inactive complex. Phosphorylation of I-κB induces the dissociation of the complex and the translocation of NF-κB to the nucleus, thus permitting the binding of p50 and p65 to their specific κB sequences. Several studies on NF-κB have used the long terminal repeat of the HIV genome linked to a reporter gene to investigate the signals involved in the activation of this transcription factor. Activation of gene expression in such models is usually well correlated with induction of viral replication in chronically infected cell lines and in acutely infected peripheral blood mononuclear cells. Conversely, the data obtained with these models cannot be extended to the regulation of expression of cellular genes (e.g., IL-2 and IL-2 receptor) associated with κB elements, because several additional regulatory elements (e.g., AP-1, oct-1, NFAT-1) control that expression in a coordinated manner.

Phosphorylation of I-κB and subsequent translocation and binding of NF-κB can be induced by TNFα, TNFβ, IL-1 (98), or by direct stimulation of PKC by phorbol esters. It is also triggered by bacterial lipopolysaccharides, T-cell mitogens, antibodies directed against the TcR/CD3 complex, and calcium ionophores. Induction by TNFα is not (exclusively) mediated by PKC activation because TNFα can induce NF-κB binding in cell lines that do not show increased binding in response to phorbol ester (98). Intracellular thiol levels play a key role in regulating NF-κB activation. Stimulation by TNFα and/or phorbol esters activates NF-κB and decreases intracellular thiols while N-acetyl–L-cysteine prevents this thiol decrease and blocks the activation of NF-κB (99). Binding of NF-κB can be inhibited by treatment with the oxidizing agent diamide, and this effect is reversed by 2-mercaptoethanol (100). Schreck and coworkers recently reported the activation of NF-κB by treatment of $CD4^+$ T-cell line with hydrogen peroxide, double-stranded RNA, calcium ionophore, lectins, lipopolysaccharide, phorbol esters, and ionomycin (101). Whatever the activating agent, NF-κB activation could be prevented by addition of N-acetyl–L-cysteine, suggesting that generation of reactive oxygen intermediates represented the common pathway utilized by these various stimulating agents to activate NF-κB. More recently, further evidence for a control of the translocation of NF-κB by the constitutive redox status of lymphocytes and monocytes was reported by N. Israel and coworkers (submitted manuscript). These authors used phenolic, lipid-soluble, chain-breaking antioxidants (butylated hydroxyanisole, nordihydroquairetic acid, and vitamin E) to show that peroxyl radical scavenging in unstimulated or TNFα- or phorbol ester–stimulated T- and monocytic cell lines led to the blockade of NF-κB–dependent activation functions. The antioxidative effect of butylated hydroxianisole was accompanied by an increase in thiol but not glutathione content in stimulated or unstimulated T cells. Furthermore the secretion of TNFα was abolished in presence of butylated hydroxyanisole. Altogether the data suggest that a basal redox equilibrium tending towards oxidation is a prerequisite for full activation of protein kinase C–dependent pathways regulating the activity of the NF-κB–dependent genes, but there is no need to postulate a role for reactive oxygen species induction in the mechanism of action of NF-κB activators.

In agreement with these data, several reports using either chronically HIV-infected monocytic cell lines or acutely infected peripheral blood T cells or T-cell lines demonstrated the stimulation of in vitro infection by hydrogen peroxide (102) and its suppression by ascorbate (103), N-acetyl-L-cysteine (104,105), glutathione and glutathione ester (105,106), and 2-mercaptoethanol (106). Further indirect evidence for a positive effect of oxygen radical intermediates on HIV replication came from the demonstration of the enhancing effect of ultraviolet light, mitomycin (107), and hyperthermia (108) on HIV gene expression.

If HIV replication in vitro is controlled by the redox status of the infected cell, is there any indirect evidence for increased oxidative stress and/or altered antioxidant defense mechanisms in HIV-infected patients? If present, are such abnormalities more pronounced during the progression of the disease towards full-blown AIDS than in asymptomatic or pauci-symptomatic HIV-infected subjects? Positive answers to these questions could provide a rationale for setting up clinical trials of antioxidant agents as adjuvant therapy in association with antiretroviral drugs in HIV infection. A great number of concordant data suggest indeed that increased lipid peroxidation (109,110) and decreased serum glutathione (111) and thiol levels (112) can be demonstrated in HIV-infected patients and that these alterations are increased in full-blown AIDS. However, further clinical studies are urgently needed to ascertain the role of malnutrition, opportunistic infections, tumors, and drugs in these abnormalities and to determine the precise cellular redox status in these patients (10,11,113).

VII. CONCLUSIONS

The extremely complex relationship between oxidative stress, cytokines, and regulation of T-cell activation is progressively emerging into a coherent yet fairly incomplete picture. Apparent discrepancies among experimental data can often be accounted for by a unidirectional interpretation, which tends to associate, in an oversimplified causal relationship, oxidative stress with immunodepression and antioxidant therapies with immunostimulation. In fact the production of free radicals is a physiological process that is likely to be critical for the synthesis of mediators such as prostaglandins and leukotrienes and for intracellular signals that control gene expression in activated cells. These free radicals are characterized by an extremely short half-life and small diffusion space, and any cell is equipped with highly efficient antioxidant devices both in the lipidic membrane (vitamin E) and in water solution in the cytoplasm (glutathione). Glutathione synthesis must be adjusted to the rate of consumption (oxidation) determined by the oxidative stress. Its rate of synthesis depends on the source of extracellular glutamate cysteine or other thiol compounds and methionine, on the transport of these amino acids to the intracellular compartment, and on enzymatic activities. The enzymes glutathione peroxidase and glutathione reductase are critical in maintaining a stable redox status of the cell, which can be evaluated by determination of cellular reduced glutathione.

T-cell activation and subsequent proliferation as well as monocyte activation with synthesis of Il-1, IL-6, and TNFα seem to be critically dependent on oxidative processes inasmuch as they are blocked by antioxidants. One of the fairly well-documented intracellular signaling cascades which seems to depend on reactive oxygen species includes activation of PKC and phosphorylation of I-κB resulting in the translocation of NF-κB. Other pathways triggered by free radicals are suggested and remain to be deciphered (e.g., TNFα, c-*jun*, and c-*fos* products). Of note, a limited level of oxidative stress is required to trigger T-lymphocyte and monocyte activation, whatever the nature of the mitogen. Furthermore the whole sequential program of gene expression leading to complete T-cell activation and cell proliferation can be triggered by sodium periodate (0.1–1 mM) or exposure to neuraminidase and glucose oxidase.

Sustained oxidative stress and/or defective antioxidative mechanisms may lead to immunodeficiency through multiple pathways. The outcome depends within each target cell on the balance between: 1) the intensity and duration of the oxidizing aggression and 2) the efficacy of antioxidants in the aqueous and lipid phases and that of DNA repair mechanisms. Lymphoid cells may either undergo necrosis with multiple single-strand DNA breaks or remain viable but with altered responses to antigenic stimulation, characterized by impaired proliferation in contrast to preserved secretion of IL-2 (and possibly other cytokines as IFN-γ). Nonlymphoid cells including monocytes/macrophages are more resistant to oxidative injury than lymphocytes, and oxidative stress triggers an inflammatory reaction dominated by the synthesis of IL-1 and TNFα and by the production of arachidonic acid derivatives such as prostaglandins and leukotrienes, which in turn may contribute to the diminution of T-cell activation.

Antioxidant therapy (scavenger compounds and/or inducers of glutathione synthesis) may protect the immune system in a situation of oxidizing stress (e.g., x or γ irradiation, exposure to hyperbaric or normobaric oxygen). However, in vitro experiments suggest that antioxidants could exert an immunosuppressive effect in absence of oxidative stress, but experimental demonstration of such an activity in vivo remains unconclusive.

Finally, the cumulative effects of repeated oxidative stress should be considered in the light of cell life span and renewal. In this respect oxidative destruction of neutrophils that may occur during phagocytosis remains harmless in view of the very short half-life of these cells, as long as hematopoiesis is not impaired. Conversely, destruction of T cells, if occurring in adults after thymic involution, may result in progressive reduction of the T-cell repertoire. B cells, on the other hand, are produced by the bone marrow throughout life, and even if they seem to be more susceptible to oxidation in vitro, the B-cell compartment in the body is relatively resistant when compared with that of T cells.

REFERENCES

1. Halliwell B, Gutteridge JMC. Free radicals in biology and medicine. 2nd ed. Oxford, UK: Clarendon Press, 1989.
2. Halliwell B, Hoult JRS, Blake DR. Oxidants, inflammation and antiinflammatory drugs. FASEB J 1988; 2:2867–73.
3. Halliwell B, Gutteridge JMC. Oxygen free radicals and iron in relation to biology and medicine: some problems and concepts. Arch Biochem Biophys 1986; 246:501–14.
4. Slater TF. Free-radical mechanisms in tissue injury. Biochem J 1984; 222:1–15.
5. Halliwell B. How to characterize a biological antioxidant. Free Rad Res Comms 1990; 9:1–32.
6. Beaman L, Beaman BL. The role of oxygen and its derivatives in microbial pathogenesis and host defense. Ann Rev Microbiol 1984; 38:27–48.
7. Rice-Evans C, Halliwell B (eds). Free radicals: methodology concepts. London: Richelieu Press, 1988.
8. Oda T, Akaike T, Hamamoto T, Suzuki F, Hirano T, Mada H. Oxygen radicals in influenza-induced pathogenesis and treatment with pyran polymer-conjugated SOD. Science (Wash DC) 1989; 244:974–6.
9. Akaike T, Ando M, Oda T, Doi T, Ijiri S, Araki S, Maeda H. Dependence on O_2^- generation by xanthine oxidase of pathogenesis of influenza virus infection in mice. J Clin Invest 1990; 85:739–45.
10. Revillard JP. Oxidative stress and HIV infection. In: Retroviruses of human A.I.D.S. and related animal diseases. Colloque des Cent Gardes (Girard M, Valette L, eds). Fondation Mérieux, 1991:299–305.
11. Halliwell B, Cross CE. Reactive oxygen species, antioxidants, and acquired immunodeficiency syndrome—sense or speculation? Arch Intern Med 1991; 151: 29–31.
12. Levacher-Place M, Gougerot-Pocidalo MA, Rouveix B, Kraus L, Pocidalo JJ. Characterization of immunological depression in mice exposed to normobaric oxygen. Clin Exp Immunol 1983; 54:580–6.
13. Gougerot-Pocidalo MA, Jacquet B, Pocidalo JJ. Mise en évidence de l'effet dèpresseur de l'oxygène normobare sur le système lymphoide de la souris. C R Acad Sc, Paris (série III) 1982; 294:925–7.

14. Rouveix B, Jacquet B, Gougerot-Pocidalo MA, Levacher M, Pocidalo JJ. Lymphoid tissue depression induced by normobaric oxygen. Its prevention by endotoxin in rat. Int J Immunopharmacol 1982; 4:363 (abst).
15. Gougerot-Pocidalo M.-A., Fay M, Pocidalo JJ. In vivo normobaric oxygen exposure depresses spleen cell in vitro Con A response. Effects of 2-mercaptoethanol and peritoneal cells. Clin Exp Immunol 1984; 58:428–35.
16. Kraus L, Gougerot-Pocidalo MA, Lacombe P, Pocidalo JJ. Depression of Con A proliferative response of immune cells by in vitro hyperoxic exposure—protective effects of thiol compounds. Int J Immunopharmacol 1985; 7:753–60.
17. Gougerot-Pocidalo M.-A., Fay M, Roche Y, Lacombe P, Marquetty C. Immune oxidative injury induced in mice exposed to normobaric O_2: effects of thiol compounds on the splenic cell sulfhydryl content and con A proliferative response. J Immunol 1985; 135:2045–51.
18. Lacombe P, Carré I, Fay M, Pocidalo JJ. In vitro O_2-induced depression of T and B lymphocyte activation is reversed by diethyldithiocarbamate (DDC) treatment. Immunol Lett 1988; 18:99–108.
19. Fisher AB, Diamond S, Mellen S. Effect of O_2 exposure on metabolism of the rabbit alveolar macrophage. J Appl Physiol 1974; 37:341–5.
20. Jacquet B, Gougerot-Pocidalo MA. Functional activities of alveolar macrophages in rat exposed to hyperoxia (normobaric O_2). Ann Immunol (Inst Pasteur) 1983; 134C:93–104.
21. Murphey SA, Hyams JS, Fisher AB, Root RK. Effects of oxygen exposure on the in vitro functions of pulmonary alveolar macrophages. J Clin Invest 1975; 56:503–11.
22. Rister M. Effect of hyperoxia on phagocytosis. Blut 1982; 45:157–66.
23. Warren J, Sacksteder MR, Thuning CA. Oxygen immunosuppression: modification of experimental allergic encephalomyelitis in rodents. J Immunol 1978; 121:315–20.
24. Saito K, Tanaka Y, Ota T, Eto S, Yamashita Y. Suppressive effect of hyperbaric oxygenation on immune responses of normal and autoimmune mice. Clin Exp Immunol 1991; 86:322–7.
25. Allan IM, Lunec J, Salmon M, Bacon PA. Reactive oxygen species selectively deplete normal T lymphocytes via a hydroxyl radical dependent mechanism. Scand J Immunol 1987; 26:47–53.
26. Kraut EH, Sagone AL. The effect of oxidant injury on the lymphocyte membrane and functions. J Lab Clin Med 1981; 98:697–703.
27. El-Hag A, Clark RA. Immunosuppression by activated human neutrophils. Dependence on the myeloperoxidase system. J Immunol 1987; 139:2406–13.
28. Grever MR, Thompson VN, Balcerzak SP, Sagone AL. The effect of oxidant stress on human lymphocyte cytotoxicity. Blood 1980; 56:284–8.
29. Gougerot-Pocidalo M-A, Fay M, Roche Y, Chollet-Martin S. Mechanisms by which oxidative injury inhibits the proliferative response of human lymphocytes to PHA. Effect of the thiol compound 2-mercaptoethanol. Immunology 1988; 64:281–8.
30. Freed BM, Rapoport R, Lempert N. Inhibition of early events in the human T-lymphocyte response to mitogens and alloantigens by hydrogen peroxide. Arch Surg 1987; 122:99–104.

31. Metzger ZVI, Hoffeld JT, Oppenheim JJ. Macrophage-mediated suppression. I. Evidence for participation of both hydrogen peroxide and prostaglandins in suppression of murine lymphocyte proliferation. J Immunol 1980; 124:983–8.
32. Sagone AL, Husney R, Guter H, Clark L. Effect of catalase on the proliferation of human lymphocytes to phorbol myristate acetate. J Immunol 1984; 133: 1488–94.
33. Hoepelman IM, Bezemer WA, van Doornmalen E, Verhoef J, Marx JJM. Lipid peroxidation of human granulocytes (PMN) and monocytes by iron complexes. Br J Haemat 1989; 72:584-8.
34. Goodman MG, Weigle WO. Modulation of lymphocyte activation: inhibition by an oxidation product of arachidonic acid. J Immunol 1980; 125:593–600.
35. Gualde N, Rabinovitch H, Fredon M, Rigaud M. Effects of 15-hydroperoxy-eicosatetraenoic acid on human lymphocyte-sheep erythrocyte rosette formation and response to concanavalin A associated with HLA system. Eur J Immunol 1982; 12:773–7.
36. Payan DG, Goetzl EJ. Specific suppression of human T lymphocyte activation by leukotriene B_4. J Immunol 1983; 131:551–3.
37. Gutteridge JMC. The role of superoxide and hydroxyl radicals in phospholipid peroxidation catalyzed by iron salts. FEBS Lett 1982; 150:454–8.
38. Carson DA, Seto S, Wasson B. Lymphocyte dysfunction after DNA damage by toxic oxygen species. A model of immunodeficiency. J Exp Med 1986; 163:746–51.
39. Springer TA. Adhesion receptors of the immune system. Nature 1990; 346:425–34.
40. Cambier JC (ed). Ligands, receptors and signals transduction in regulation of lymphocyte function. Washington, DC: American Society for Microbiology, 1990.
41. Altman A, Coggeshall KM, Mustelin T. Molecular events mediating T cell activation. Adv Immunol 1990; 48:227–360.
42. Riegel JS, Corthesy B, Flanagan WM, Crabtree GR. Regulation of the interleukin-2 gene. In: Kishimoto T, ed. Interleukins: molecular biology and immunology, Basel: Karger, 1992:266–98.
43. Chaudri G, Clark IA, Hunt NH, Cowden WB, Ceredig R. Effect of antioxidants on primary alloantigen-induced T cell activation and proliferation. J Immunol 1986; 137:2646–52.
44. Farrar WL, Humes JL. The role of arachidonic acid metabolism in the activity of interleukin-1 and 2. J Immunol 1985; 135:1153–9.
45. Fidelus RK. The generation of oxygen radicals: a positive signal for lymphocyte activation. Cell Immunol 1988; 113:175–82.
46. Kato K, Yoshikaza Y, Murota S. Contribution of lipoxigenase metabolites to IL-2 production in the early phase of lymphocyte activation. Prostagland Leukotr Med 1988; 22:31.
47. Novogrosky A, Ravid A, Rubin AL, Stenzel KH. Hydroxl radical scavengers inhibit lymphocyte mitogenesis. Proc Natl Acad Sci USA 1982; 79:1171–4.
48. Schultz RM, Altom MG. Modulation of interleukin-1 activity on murine thymocytes by various inhibitors of arachidonic acid oxygenation. Immunopharmacol Immunotoxicol 1988; 10:21–34.

49. Dornand J, Sekkat C, Mani JC, Gerber M. Lipoxigenase inhibitors suppress IL2 synthesis: relationship with $[Ca2^+]_i$ rise and the events dependent on protein kinase C activation. Immunol Lett 1987; 16:101–6.
50. Gerber M, Ball D, Michel F, Crastes de Paulet A. Mechanism of enhancing effect of irradiation on production of IL2. Immunol Lett 1985; 9:279–83.
51. Sekkat C, Dornand J, Gerber M. Oxidative phenomena are implicated in human T cell stimulation. Immunology 1988; 63:431–7.
52. Dornand J, Gerber M. Inhibition of murine T-cell responses by antioxidants: the targets of lipo-oxygenase pathway inhibitors. Immunology 1989; 68:384–91.
53. O'Brian CA, Ward NE, Bernard Weinstein I, Bull AW, Marnett LJ. Activation of rat brain protein kinase C by lipid oxidation products. Biochem Biophys Res Commun 1988; 155:1374–80.
54. Gopalakrishna R, Anderson WB. Ca^{2+-} and phospholipid-independent activation of protein kinase C by selective oxidative modification of the regulatory domain. Proc Natl Acad Sci USA 1989; 86:6758–62.
55. Broome JD, Jeng MW. Promotion of replication in lymphoid cells by specific thiols and disulfides in vitro. Effects on mouse lymphoma cells in comparison with splenic lymphocytes. J Exp Med 1973; 138:574–92.
56. Ohmori H, Yamamoto I. Activation of murine lymphocytes by 2-mercaptoethanol and related thiol compounds and its mechanism. I. Relationship between mitogenic activities and augmenting effects on antibody synthesis in vitro. Immunopharmacology 1981; 3:333–45.
57. Noelle RJ, Lawrence DA. Modulation of T-cell functions. I. Effect of 2-mercaptoethanol and macrophages on T-cell proliferation. Cell Immunol 1980; 50: 416–31.
58. Noelle RJ, Lawrence DA. Modulation of T-cell functions. II. Chemical basis for the involvement of cell surface thiol-reactive sites in control of T-cell proliferation. Cell Immunol 1981; 60:453–69.
59. Ohmori H, Yamamoto I. Mechanism of augmentation of the antibody response in vitro by 2-mercaptoethanol in murine lymphocytes. II. A major role of the mixed disulfide between 2-mercaptoethanol and cysteine. Cell Immunol 1983; 79:173–8.
60. Hoffeld JT. Agents which block membrane lipid peroxidation enhance mouse spleen cell immune activities in vitro: relationship to the enhancing activity of 2-mercaptoethanol. Eur J Immunol 1981; 11:371–6.
61. Ohmori H, Yamamoto I. Mechanism of augmentation of the antibody response in vitro by 2-mercaptoethanol in murine lymphocytes. I. 2-mercaptoethanol-induced stimulation of the uptake of cysteine, an essential amino acid. J Exp Med 1982; 155:1277–90.
62. Zmuda J, Friedenson B. Changes in intracellular glutathione levels in stimulated and unstimulated lymphocytes in the presence of 2-mercaptoethanol or cysteine. J Immunol 1983; 130:362–4.
63. Fischman CM, Udey MC, Kurtz M, Wedner J. Inhibition of lectin-induced lymphocyte activation by 2-cyclohexene-1-one:decreased intracellular glutathione inhibits an early event in the activation sequence. J Immunol 1981; 127:2257–62.
64. Hamilos DL, Wedner HJ. The role of glutathione in lymphocyte activation. I. Comparison of inhibitory effects of buthionine sulfoximine and 2-cyclohexene-1-one by nuclear size transformation. J Immunol 1985; 135:2740–7.

65. MacDermott RP, Bertovich MJ, Bragdon MJ, Nash GS, Leusch MS, Wedner HJ. Inhibition of cell-mediated cytotoxicity by 2-cyclohexene-1-one: evidence for a role for glutathione and/or glutathione-protein interactions in cytolysis. Immunology 1986; 57:521–6.
66. Suthanthiran M, Anderson ME, Sharma VK, Meister A. Glutathione regulates activation-dependent DNA synthesis in highly purified normal human T lymphocytes stimulated via the CD2 and CD3 antigens. Proc Natl Acad Sci USA 1990; 9:3343–7.
67. Noelle RJ, Lawrence DA. Determination of glutathione in lymphocytes and possible association of redox state and proliferative capacity of lymphocytes. Biochem J 1981; 198:571–9.
68. Lacombe P, Kraus L, Fay M, Pocidalo JJ. Lymphocyte glutathione status in relation to their con A proliferative response. FEBS Lett 1985; 191:227–30.
69. Fidelus RK, Tsan MF. Enhancement of intracellular glutathione promotes lymphocyte activation by mitogen. Cell Immunol 1986; 97:155–63.
70. Fidelus RK, Tsan MF. Glutathione and lymphocyte activation: a function of ageing and auto-immune disease. Immunology 1987; 61:503–8.
71. Kavanagh TJ, Grossmann A, Jaecks EP, Jinneman JC, Eaton DL, Martin GM, Rabinovitch PS. Proliferative capacity of human peripheral blood lymphocytes sorted on the basis of glutathione content. J Cell Physiol 1990; 145:472–80.
72. Dröge W, Pottmeyer-Gerber C, Schmidt H, Nick S. Glutathione augments the activation of cytotoxic T lymphocytes in vivo. Immunobiology 1986; 172:151–6.
73. Sugama K, Namba Y, Hatanaka M, Hanaoka M. 2-Mercaptoethanol acts at the restricted stage of interleukin-2 dependent lymphocyte proliferation. J Cell Physiol 1987; 131:23–8.
74. Liang CM, Lee N, Cattell D, Liang SM. Glutathione regulates interleukin-2 activity on cytotoxic T-cells. J Biol Chem 1989; 264:13519–23.
75. Messina JP, Lawrence DA. Cell cycle progression of glutathione-depleted human peripheral blood mononuclear cells is inhibited at S phase. J Immunol 1989; 143:1974–81.
76. Rouzer CA, Scott WA, Griffith OW, Hamill AL, Cohn ZA. Glutathione metabolism in resting and phagocytizing peritoneal macrophages. J Biol Chem 1982; 257:2002–8.
77. Toohey JI. Macrophages and methylthio groups in lymphocyte proliferation. J Supramol Struc Cell Biochem 1981; 17:11–25.
78. Jokay I, Kelemenics K, Földes I. Sulfhydryl groups generated by macrophages into the culture medium. Immunol Lett 1988; 17:217–22.
79. Gmünder H, Eck HP, Benninghoff B, Roth S, Dröge W. Macrophages regulate intracellular glutathione levels of lymphocytes. Evidence for an immunoregulatory role of cysteine. Cell Immunol 1990; 129:32–46.
80. Klebanoff SJ, Vadas MA, Harlan JM, Sparks LH, Gamble JR, Agosti JM, Waltersdorph AM. Stimulation of neutrophils by tumor necrosis factor. J Immunol 1986; 136:4220–5.
81. Larrick JW, Graham D, Toy K, Lin LS, Senyk H, Fendly BM. Recombinant tumor necrosis factor causes activation of human granulocytes. Blood 1987; 69:640–4.

82. Nathan CF. Neutrophil activation on biological surfaces. Massive secretion of hydrogen peroxide in response to products of macrophages and lymphocytes. J Clin Invest 1987; 80:1550–60.
83. Steadman R, Petersen MM, Topley N, Williams D, Matthews N, Spur B, Williams JD. Differential augmentation by recombinant human tumor necrosis factor of neutrophil responses to particulate zymosan and glucan. J Immunol 1990; 144:2712–8.
84. Zimmerman RJ, Marafino BJ, Chan A, Landre P, Winkelhake JL. The role of oxidant injury in tumor cell sensitivity to recombinant human tumor necrosis factor in vivo. J Immunol 1989; 142:1405–9.
85. Baud L, Affres H, Perez J, Ardaillou R. Reduction in tumor necrosis factor binding and cytotoxicity by hydrogen peroxide. J Immunol 1990; 145:556–60.
86. Shaffer JB, Treanor CP, Del Vecchio PJ. Expression of bovine and mouse endothelial cell antioxidant enzymes following TNF-α exposure. Free Rad Biol Med 1990; 8:497–502.
87. Wrong GH, Elwell JH, Oberley LW, Goeddel DV. Manganous superoxide dismutase is essential for cellular resistance to cytotoxicity of tumor necrosis factor. Cell 1989; 58:923–31.
88. Chaudri G, Clark IA. Reactive oxygen species facilitate the in vitro and in vivo lipopolysaccharide-induced release of tumor necrosis factor. J Immunol 1989; 143:1290–4.
89. Hallahan DE, Spriggs DR, Beckett MA, Kufe DW, Weichselbaum RR. Increased tumor necrosis factor α mRNA after cellular exposure to ionizing radiation. Proc Natl Acad Sci USA 1989; 86:10104–7.
90. Köck A, Schwarz T, Kirnbauer R, Urbanski A, Perry P, Ansel JC, Luger TA. Human keratinocytes are a source for tumor necrosis factor α: evidence for synthesis and release upon stimulation with endotoxin or ultraviolet light. J Exp Med 1990; 172:1609–14.
91. Urbanski A, Schwarz T, Neuner P, Krutmann J, Kirnbauer R, Köch A, Luger TA. Ultraviolet light induces increased circulating interleukin-6 in humans. J Invest Dermatol 1990; 94:808–11.
92. Gougerot-Pocidalo M A, Roche Y, Fay M, Perianin A, Bailly S. Oxidative injury amplifies interleukin-1-like activity produced by human monocytes. Int J Immunopharmacol 1989; 11:961–9.
93. Kasama T, Kobayashi K, Fukushima T, Tabata M, Ohno I, Negishi M, Ide H, Takahashi T, Niwa Y. Production of interleukin 1-like factor from human peripheral blood monocytes and polymorphonuclear leukocytes by superoxide anion: the role of interleukin 1 and reactive oxygen species in inflammed sites. Clin Immunol Immunopath 1989; 53:439–48.
94. Akira S, Isshiki H, Nakajima T, Kinoshita S, Nishio Y, Hashimot S, Natsuka S, Kishimoto T. A nuclear factor for the IL-6 gene (NF-IL6). In: Kishimoto T, ed. Interleukins: molecular biology and immunology, Basel: Karger, 1992:299–322.
95. Maki A, Berezesky IK, Fargnoli J, Holbrook NJ, Trump BJ. Role of $[Ca^{2+}]_i$ in induction of c-*fos*, c-*jun*, and c-*myc* mRNA in rat PTE after oxidative stress. FASEB J 1992; 6:919–24.

96. Abate C, Patel L, Rauscher FJ, Curran T. Redox regulation of Fos and Jun DNA-binding activity in vitro. Science 1990; 249:1157–61.
97. Leonardo MJ, Baltimore D. NF-κB: a pleiotropic mediator of inductible and tissue-specific gene control. Cell 1989; 58:227–9.
98. Osborn L, Kunkel S, Nabel GJ. Tumor necrosis factor α and interleukin 1 stimulate the human immunodeficiency virus enhancer by activation of the nuclear factor κB. Proc Natl Acad Sci USA 1989; 86:2336–40.
99. Staal FJT, Roederer M, Herzenberg LA, Herzenberg LA. Intracellular thiols regulate activation of nuclear factor κB and transcription of human immunodeficiency virus. Proc Natl Acad Sci USA 1990; 87:9943–7.
100. Toledano, MB, Leonard WJ. Modulation of transcription factor NF-κB binding activity by oxidation-reduction in vitro. Proc Natl Acad Sci USA 1991; 88:4328–32.
101. Schreck R, Rieber P, Baeuerle PA. Reactive oxygen intermediates as apparently widely used messengers in the activation of the NF-κB transcription factor and HIV-1. EMBO J 1991; 10:2247–57.
102. Legrand-Poels S, Vaira D, Pincemail J, van de Vorst A, Piette J. Activation of human immunodeficiency virus type 1 by oxidative stress. AIDS Res Hum Retrov 1990; 6:1389–97.
103. Harakeh S, Jariwalla RJ, Pauling L. Suppression of human immunodeficiency virus replication by ascorbate in chronically and acutely infected cells. Proc Natl Acad Sci USA 1990; 87:7245–9.
104. Roederer M, Staal FJT, Raju PA, Ela SW, Herzenberg LA, Herzenberg LA. Cytokine-stimulated human immunodeficiency virus replication is inhibited by *N*-acetyl-L-cysteine. Proc Natl Acad Sci USA 1990; 87:4884–8.
105. Kalebic T, Kinter A, Poli G, Anderson ME, Meister A, Fauci AS. Suppression of human immunodeficiency virus expression in chronically infected monocytic cells by glutathione, glutathion ester, and N-acetylcysteine. Proc Natl Acad Sci USA 1991; 88:986–90.
106. Wu J, Levy EM, Black PH. 2-mercaptoethanol and n-acetylcysteine enhance T cell colony formation in AIDS and ARC. Clin Exp Immunol 1989; 77:7–10.
107. Valerie K, Delers A, Bruck C, Thiriart C, Rosenberg H, Debouck C, Rosenberg M. Activation of human immunodeficiency virus type 1 by DNA damage in human cells. Nature 1988; 333:78–81.
108. Stanley SK, Bressler PB, Poli G, Fauci AS. Heat shock induction of HIV production from chronically infected promonocytic and T cell lines. J Immunol 1990; 145:1120–6.
109. Sonnerborg A, Carlin G, Akerlund B, Jarstrand C. Increased production of malondialdehyde in patients with HIV infection. Scand J Infect Dis 1988; 20: 287–90.
110. Revillard JP, Vincent CMA, Favier AE, Richard MJ, Zittoun M, Kazatchkine MD. Lipid peroxidation in human immunodeficiency virus (HIV) infection. J AIDS 1992; 5:637–8.
111. Bühl R, Holroyd KJ, Mastrangeli A, Cantin AM, Jaffre HA, Wells FB, Saltini C, Crystal RG. Systemic glutathione deficiency in symptom-free HIV-seropositive individuals. Lancet 1989; ii:1294–8.

112. Eck HP, Gmunder H, Hartmann M, Petzoldt D, Daniel V, Dröge W. Low concentrations of acid-soluble thiol (cysteine) in the blood plasma of HIV-1 infected patients. Biol Chem 1989; 370:101–8.
113. Dröge W, Eck HP, Gmunder H, Mihm S. Requirement for prooxidant and antioxidant states in T cell mediated immune responses—relevance for the pathogenetic mechanism of AIDS. Klin Wochenscrift 1991; 69:21–2.

12

Oxidative Stress in Skin Chemical Carcinogenesis

Rajesh Agarwal and Hasan Mukhtar
Case Western Reserve University, Cleveland, Ohio

I. INTRODUCTION

A. Background

It is becoming increasingly clear that most human tumors develop from the progeny of a single cell through multiple sequential alterations in growth regulatory genes (1). Nonmelanoma skin cancer is the most common neoplasm occurring in the human population, which is suggested to be directly related to the skin's continuing exposure to the environment (2). It has been estimated that of all the new cancers diagnosed annually in the Western countries, one third of them develop in the skin (2). The role of environmental agents, both physical and chemical, in causation of the human skin cancers is gaining increasing acceptance due to their continuous rise in number (2). While solar radiation is the major cause of skin cancers (3), it is appreciated that significant levels of cancer-causing chemical agents such as polluted air, automobile exhaust, and tobacco smoke are also present in the environment, contributing significantly to the growing incidences of cutaneous neoplasms in human population (2,4). The risk factors for human skin cancer include environmental pollutants, occupational exposure to coal tar, pitch, creosote, arsenic compounds, radium, and excessive exposure to the sun (2).

Considerable evidences support the importance of chemical agents as causative factors for carcinogenesis (2,4). Several experimental studies have demonstrated the tumorigenic potential of a large number of chemical substances, which include molecules of diverse chemical classes, organic and inorganic, and natural products as well as those synthesized in the laboratory or produced by the industry (2,4). Polycyclic aromatic hydrocarbons (PAHs) are the major class of skin chemical carcinogens generated as products of incomplete combustion of fossil fuels and other organic matter (5). The potential role of PAHs in the causation of human cancer is strongly suggested by their environmental occurrence and exceptionally high carcinogenic potency in experimental animals (2,4,5).

B. General Concepts of Multistage Skin Carcinogenesis

For almost half a century, mouse skin has been used as the convenient model to study the mechanism of carcinogenesis and modulation of sequential steps involved in this process (2,4–6). As depicted in Figure 1, it is recognized that chemical carcinogenesis in murine skin and possibly in human skin and other tissues is a stepwise process of at least three distinct stages known as initiation, promotion, and progression (2,6–9). The most common benign tumors induced by chemicals are papillomas and keratoacanthomas, whereas common malignancies are squamous cell carcinomas, basal cell carcinomas, and melanomas (2,6–9).

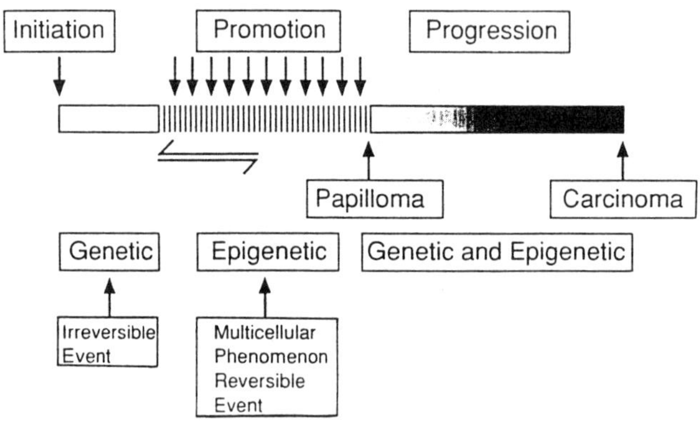

Figure 1 Multistep cutaneous carcinogenesis showing initiation, promotion and progression.

Several lines of evidence have suggested that the induction of cancer is a multistep process requiring both initiating and promoting substances for the development of a malignant tumor (2,6–9). In murine skin, tumors are generally induced by a single topical application of the initiating agent followed by repeated topical applications of the promoting agent on the initiated skin (2). This is known as the two-stage carcinogenesis protocol (2,6–9). The initiation stage is essentially an irreversible step in which genetic changes occur in gene(s) controlling differentiation (2,6–9). The promotion stage leads to the formation of visible lesions, which is a reversible step, at least in early stages, and it is believed that clonal expansion of the initiated cell population occurs during this stage of carcinogenesis (2,6–9). 7,12-Dimethylbenz(a)anthracene (DMBA), a synthetic PAH, is a conventional tumor-initiating agent employed in most of the studies on murine skin carcinogenesis (2,6–10). The initiation phase of cutaneous chemical carcinogenesis in a two-stage carcinogenesis protocol can be achieved by a single topical application of subthreshold doses of a carcinogen such as DMBA. This subthreshold dose of carcinogen does not produce tumors over the life span of mice. Promotion with multiple applications of croton oil or a phorbol ester such as 12-O-tetradecanoylphorbol-13-acetate (TPA), however, leads to the development of benign papillomas after a short latency period (2,6–10). The majority of the benign papillomas developed by initiation-promotion protocol regress, whereas only a few of them develop to carcinomas. It has been shown that a lapse of even 1 year between the application of initiator and the beginning of promoter treatment produces tumor response similar to that observed when promotion is started only 1 week after initiation with the carcinogen (2,6–11). A single topical application of a large dose of chemical carcinogen such as DMBA on mouse skin can also result in cutaneous tumor induction, this is often called as complete carcinogenesis protocol (2,6). The latency period in this protocol, however, becomes longer, and the number of papillomas produced is lower than in initiation-promotion protocol. Also, the repeated applications of subthreshold doses of a carcinogen induce cutaneous tumors, but the number of carcinomas developed with this protocol is much higher than that developed when DMBA is used as the initiating agent and TPA is employed as a promoting agent (2,6–11).

Tumor promotion in mouse skin is also subdivided into two steps known as stage I and stage II (12). Following initiation of mice with DMBA, stage I of promotion can be achieved by limited skin applications of TPA, A23187, hydrogen peroxide, 4-O-methyl TPA, or by skin wounding (13). Subsequently, stage II of promotion can be achieved by repeated skin applications of TPA or the weak complete tumor promoter mezerein (13). Tumor promoters are usually nonmutagenic and noncarcinogenic agents and application of them alone to normal mouse skin does not produce any tumors. Although the phorbol esters are the most potent of the mouse skin tumor promoters, a variety of other compounds have also been shown to possess skin tumor-promoting activity (14,15). Among them are a number of free radical–generating compounds, such as benzoyl peroxide (BPO), lauroyl peroxide, and decanoyl peroxide (15). Out of various free radical–generating compounds, BPO is widely used in industry as a polymerization initiator (16), as a bleaching agent for flour and cheese (17), and as an additive in cosmetics and pharmaceuticals, especially those employed for the treatment of acne vulgaris (17,18). BPO is an effective tumor promoter when applied repetitively after a single subcarcinogenic dose of DMBA in a two-stage carcinogenesis protocol (14). Although BPO does not directly activate protein kinase C (PKC), it has been shown to be effective in the induction of ornithine decarboxylase (ODC) and epidermal hyperplasia and in increasing the number of dark basal keratinocytes (19,20), the known markers of tumor promotion in skin (2,6).

The majority of the benign papillomas formed by skin application of initiating and promoting agents regress spontaneously, whereas a few of them convert to carcinomas. This is known as tumor progression, the mechanism of which is incompletely understood (21,22). Tumor progression is an irreversible step and probably requires additional insult to the cellular genome of the papilloma cell (21,22). It has been shown that the rate of malignant conversion of benign skin papillomas is increased by topical application of genotoxic agents such as urethane, *N*-methyl-*N'*-nitro-*N*-nitrosoguanidine and 4-nitroquinoline-*N*-oxide (23,24). Based on these observations, it was suggested that malignant conversion may result from further genetic changes in the papilloma cells. Furthermore, the observed ineffectiveness of TPA in enhancing malignant conversion was assumed to be due to its being a non-mutagen. In contrast to this suggestion, the studies by Slaga and coworkers showed that BPO can also enhance the malignant conversion of benign lesions at least in DMBA-initiated mouse skin (25,26). Athar et al. (27) have shown that in addition to BPO, other free radical–generating compounds may also accelerate the conversion of benign papillomas to malignant carcinomas.

C. Oxidative Stress

While oxygen is critical for all living aerobic organisms, paradoxically, hyperoxia or the inappropriate metabolism of oxygen is a well-defined and investigated cause of toxicity in biological systems (28). The toxic effects of oxygen are not due to molecular oxygen per se but are related with the reactive metabolites generated as the by-products of normal oxidative metabolism (28,29). Several studies in experimental animals have shown that reactive oxygen species (ROS) including superoxide anion, hydrogen peroxide, hydroxyl radical, and singlet oxygen, which are generated by enzymatic and nonenzymatic reactions from molecular oxygen, produce physiological, biochemical, and pathological changes in the cellular systems (30). Cellular injury caused by reactive oxygen involves the generation of superoxide anion and hydrogen peroxide, both of which can directly damage cells (30); the most reactive and most important mediator of such damage, however, appears to be hydroxyl radical. Excessive production of these species occurs in certain disease states or as a result of toxic insult by selected foreign compounds and/or drugs (30,31). Besides, some ROS are generated as the by-products of certain endogenous enzyme reactions. It is important to emphasize at this point that tissues possess a variety of systems to prevent and/or minimize oxidative injury (30–34). The balance between cellular production and catabolism of these oxidants is critical for the maintenance of tissue homeostasis (28,29,35,36). In general, it is appreciated that in a particular cell only a limited amount of protective enzymes and chemicals are synthesized to protect the cell under normal physiological conditions (37). Therefore, if an oxidative stress above that which normally occurs in a cell is applied, it results in either enhanced protective mechanism or in oxidative damage, depending on the magnitude of the oxidative stress and the rate of its application in that cell (37).

There is substantial evidence to implicate the involvement of free radicals, particularly those derived from molecular oxygen in all three stages of skin chemical carcinogenesis (37–41). Such an assumption is further supported by the studies suggesting that antioxidants that scavenge free radicals directly or interfere with the generation of free radical–mediated events lead to the inhibition of neoplasia induction (39,40). While the antioxidant studies are beyond the scope of this review, several extensive reviews on this topic have been published recently (39,40,42). This chapter, however, provides a comprehensive review of current evidence regarding the possible role of oxidative stress in all three stages of cutaneous chemical carcinogenesis.

II. ROLE IN SKIN TUMOR INITIATION

Since the essential obligatory step in tumor initiation is the covalent binding of the carcinogen to macromolecules, especially to DNA in the cell (2,6,43), and since free radicals are known to cause DNA damage (44), they are thought to play an important role in this process. Also, transition metal ions such as iron have been shown to be involved in the generation of superoxide anion–dependent reactive metabolites and may therefore influence tumor initiation (45).

Most carcinogens, especially those of the PAH class, are chemically inert and require metabolic activation primarily by the oxidative reactions catalyzed by specific enzymes to exert its neoplastic effects (46–49). The first step in the oxidative metabolism of PAH skin carcinogens is catalyzed by the cytochrome P-450 isozyme 1A1 (CYP1A1) (48–50). The reactive metabolites capable of binding and initiating the carcinogenic response are known as "ultimate" carcinogens, and their immediate metabolic precursors are called "proximate" carcinogens (46–49). The process of metabolic activation of PAH, especially benzo(*a*)pyrene (BP), has been extensively studied using liver and skin preparations (48,49). These studies have suggested that oxygenation is the initial step in producing proximate and/or ultimate carcinogenic metabolites (48,49). While most studies on the enzymatic activation of chemical carcinogens, especially BP, have focused on two enzymatic pathways, i.e., monooxygenation by CYP1A1 followed by hydrolysis of the epoxides by epoxide hydrolase (48,49), it has become increasingly clear that activation of cellular peroxidases, including the prostaglandin H (PGH) synthase, also plays an important role in the metabolism of many carcinogens, including PAH-type skin carcinogens (51). The model horseradish peroxidase/H_2O_2 system has also been found to metabolize BP by one-electron oxidation (52). The metabolic activation of PAHs via different oxygenation processes leading to the formation of proximate and/or ultimate cariogenic metabolites and/or noncarcinogenic metabolites and their role in skin tumor initiation are described in following sections. BP and/or DMBA are discussed as prototypes since most of the skin tumor initiation studies have been performed with these two carcinogenic PAHs (2,6).

A. Metabolism of PAHs by CYP1A1 and Epoxide Hydrolase

The most conventional pathway of PAH metabolism, such as BP, is catalyzed by CYP1A1 and epoxide hydrolase leading to the formation of three classes of products: phenols, dihydrodiols, and quinones. Phenols and dihydrodiols are formed by an initial electrophilic attack of an enzyme-generated oxygen atom (53–55). The same pathway of activation of BP has been postulated in the formation of quinones; though the putative 6-hydroxy BP precursor has not yet been isolated (56,57), the formation of quinones presumably proceeds by autoxidation of 6-hydroxy BP (57). A number of reports suggest that the first step in the formation of quinones does not involve the typical attack of the electrophilic active oxygen to yield 6-hydroxy BP, but instead consists of the loss of one electron from BP to produce a radical cation (58,59). CYP1A1-catalyzed metabolic activation is a predominant pathway for PAHs in which bay-region diol epoxides are the most important, if not exclusive, pathway of activation (60). The most important metabolite of BP formed by the CYP1A1 monooxygenase system is the proximate carcinogen BP-7,8-diol, an ultimate product of hydrolysis of BP-7,8-epoxide by epoxide hydrolase (48,49). This compound is further epoxidized by enzyme-generated oxygen atoms and leads to the formation of BP-7,8-diol-9,10-epoxide (BPDE). Based on the stereospecificity of BP-7,8-diol, four stereoisomeric forms of BPDE, namely, *anti-cis-*, *anti-trans-*, *syn-cis-*, and *syn-trans-*BPDE, have been identified (48). However, extensive studies from several laboratories have identified *anti-trans-BPDE (7β*,8α-(+)-dihydroxy-9α-10α-epoxy-BP) as the "ultimate" carcinogen of the BP, which binds with DNA and initiates tumorigenesis (48). The formation of putative "proximate" and "ultimate" carcinogens from BP via CYP1A1 and epoxide hydrolase pathway is illustrated in Figure 2. Similarly, all other carcinogenic PAHs contain a phenanthrene structure, which forms a so-called "bay region" (60). It has been suggested that the dihydrodiol epoxides with the epoxy group in this region exhibit the highest biological activity of all isomeric compounds because of steric hindrance and increased chemical reactivity (60). Potent skin carcinogen and most frequently used tumor initiator in animal experiments, DMBA, is also metabolized by the pathway quite similar to that of BP (2 and references therein). In case of DMBA, DMBA-3,4-diol is the "proximate" carcinogen and DMBA-3,4-diol-1,2-epoxide is the "ultimate" carcinogen of the parent compound (2,6,10 and references therein).

B. Metabolism of PAHs by Peroxyl Free Radicals

As reviewed in detail by Marnett (51), there has been an increased interest in studying the involvement of free radicals in skin carcinogenesis (37,39–41, 51) due to the fact that free radicals are formed in biological systems and contribute substantially to physiological and pathological processes. Recent studies linking free radicals and cancer induction have focused on intermediates of oxygen reduction such as superoxide anion and the hydroxyl radical (40). The increased generation of ROS by leukocytes during skin inflammation has been correlated with the metabolic activation of PAHs in mouse skin (61). The oxidative stress produced by polymorphonuclear leukocytes (PMNs), which are activated by pretreatment of skin with TPA 16 hours before and immediately prior to initiation, is also reported to enhance the metabolic activation of BP-7,8-diol to reactive metabolite via myeloperoxidase (MPO)-dependent pathways (62). This study further supports that the oxidative stress induced by metabolically stimulated leukocytes infiltrating the sites of inflammation in mouse skin could directly or indirectly enhance the tumor-initiating effect of skin carcinogens. Stereochemical analysis of

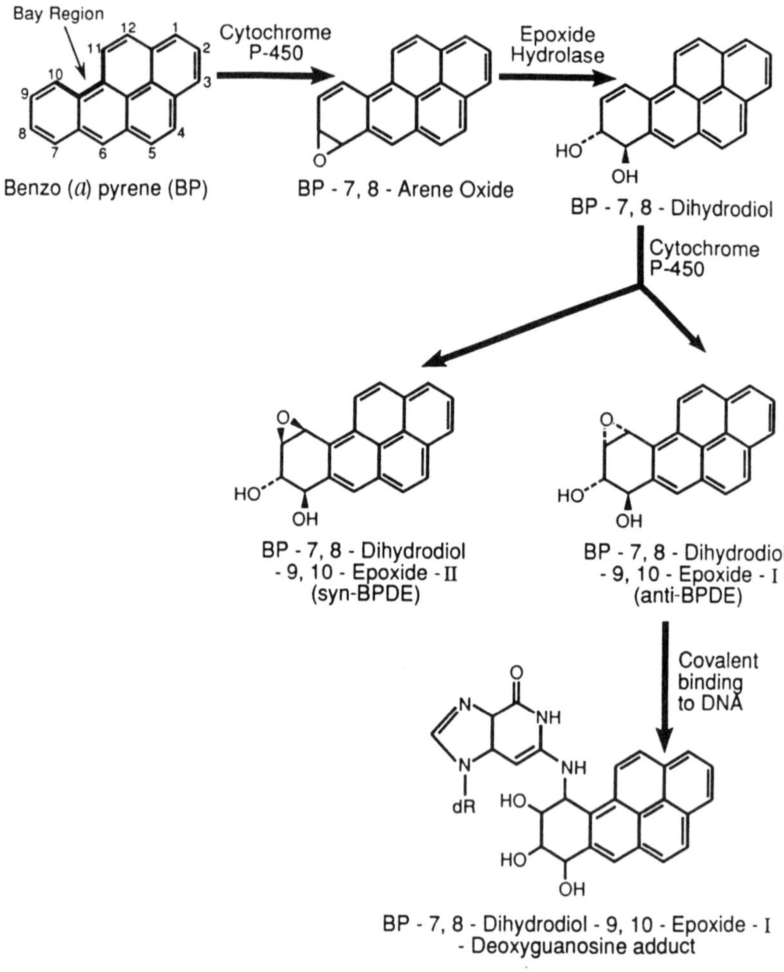

Figure 2 Cytochrome P-450 isozyme 1A1 and epoxide hydrolase–dependent metabolic activation of benzo(*a*)pyrene to ultimate carcinogenic metabolite, *anti*-BPDE, and its subsequent binding to DNA.

BP-tetrols have shown that primarily *anti*-diolepoxides are formed from (±)-BP-7,8-diol by TPA-stimulated PMNs with the *anti/syn* ratio of 6 and by MPO-hydrogen peroxide system with the ratio of >5; addition of azide, a MPO inhibitor, however, decreases the tetrol formation from BP-diol by both the systems (63). Based on these studies, it was suggested that BP-7,8-diol can be epoxidized via a peroxyl radical or a ferryl oxygen transfer–mediated reaction (63). Another family of oxygen-centered free radicals that received less attention in the past, but has been found equally important in recent years as a causative factor in neoplasia induction including skin tumor initiation, is the peroxyl radical (51), which is defined in detail in the following sections.

1. Oxygen-Centered Free Radicals: Peroxyl Radicals

Molecular oxygen, when catalyzed via one-electron reduction by redox cycling or by PMNs, produces superoxide anions, which exist in equilibrium with a protonated form perhydroxyl radical (64). Perhydroxyl radicals with the H atom replaced by an organic group are called peroxyl radicals (65), which are formed in biological systems primarily as a result of autoxidation such as lipid peroxidation (66,67). As compared to other oxygen-centered free radicals, peroxyl radicals are stable species capable of diffusing to remote cellular locations. The organic analogs of hydroxyl radicals are called alkoxyl radicals (68). The major pathway of their formation is metal-catalyzed one-electron reduction of organic hydroperoxides. Lipid hydroperoxides, the initial products of lipid peroxidation, are the major organic hydroperoxides in animal tissues (69). The predominant fate of carbon-centered radicals is the coupling with molecular oxygen to produce peroxyl radicals; the formation of oxygen-centered free radicals in a cell either by the reduction of molecular oxygen or by autoxidation results in the generation of peroxyl radicals (65–67,70). The stability of peroxyl radicals suggests that they can diffuse to cellular loci remote from the site of their generation and, therefore, are considered as prime mediators of radical-induced pathology (65–67,70).

2. Peroxyl Radical–Mediated Hydroperoxide-Dependent Metabolic Oxidation of PAHs

The interest in the involvement of peroxyl radicals in skin tumor initiation was a result of the seminal observation from the laboratory of Marnett in which cooxidation of PAHs during prostaglandin (PG) biosynthesis was described (51). PGH synthase, a membrane-bound enzyme widely distributed in mammalian tissue, has also been shown to oxygenate polyunsaturated fatty acids such as arachidonic acid to bicyclic peroxides, which are subsequently converted to PGs, thromboxanes, and prostacyclin (71). In addition, it also has peroxidase activity that catalyzes the reduction of hydroperoxy endoperoxide to hydroxy endoperoxide by a reducing agent (72). Several laboratories have shown that chemical carcinogens that require metabolic activation are cooxidized to reactive metabolites by this hydroperoxide-dependent reaction (73,74). For example, PAH diols, aflatoxin B_1, and aromatic amines are cooxidized to mutagenic derivatives during prostaglandin biosynthesis (74–78). As PGH synthase is structurally and functionally unrelated to cytochrome P-450, arachidonic acid–dependent cooxidation has been suggested as an alternate pathway to classical NADPH-dependent oxidation for carcinogen activation (73–79). This assumption is further supported by the study showing that arachidonic acid treatment prior to that of DMBA stimulates tumor initiation in skin and that the simultaneous administration of PGH synthase inhibitor, indomethacin, abolishes this enhanced effect (51,74).

BP has been shown to be cooxidized during PG biosynthesis to BP-quinones, derived from 6-hydroxy-BP (80), which is unique among BP phenols as it is not formed via the intermediate aromatic epoxide (81), whereas BP-7,8-diol is cooxidized by PGH synthase to the *anti*-BPDE (82,83). The correlation studied between peroxyl radicals and PAH-diol-epoxide formation further suggests that the peroxyl radicals formed during arachidonic acid metabolism may also be oxidants that epoxidize the DMBA-3,4-diol to the ultimate carcinogenic metabolite, DMBA-3,4-diol-1,2-epoxide (51,74). Recent studies in epidermis have established that the major source of peroxyl radicals for BP-7,8-diol epoxidation is generated from lipid peroxidation instead of the enzymes of arachidonic acid oxygenation (84). This observation is further supported by the fact that indomethacin does not inhibit the oxidation of BP-7,8-diol to ultimate carcinogenic diol-epoxide (84). The importance of PAH-diol-epoxide as the ultimate carcinogenic form, and the fact that through this pathway PAH-diol is epoxidized, whereas PAH is not, has been of considerable interest to workers studying the pathways of carcinogen cooxidation. The kinetics of cytochrome P-450–dependent monooxygenase induction, DNA adduct formation, and skin tumor initiation suggest that in intact epidermis, peroxyl radical–mediated epoxidation of PAH-diol to diol-epoxide may be an effective alternative pathway. Such as assumption is based on the fact that the induction of CYP1A1 is not absolutely required for the metabolic activation of low tumor-initiating doses of carcinogenic PAHs in mouse skin (85). Hematin, the prosthetic group of PGH synthase, has also been shown to catalyze BP-7,8-diol epoxidation by unsaturated fatty acid hydroperoxides (86–88).

3. *Stereoselectivity of Peroxyl Radical–Dependent Epoxidation of BP-7,8-Diol*

The stereoselectivity of epoxidation of BP-7,8-diol by peroxyl radicals is found to be different from CYP1A1 (51,74,89). Oxygen inserted by peroxyl radicals occurs on the same side of the molecule as the 8-hydroxyl group, whereas in the case of CYP1A1, it is determined primarily by the orientation of the pyrene moiety with respect to the active site of the enzyme (89, 90). (−)-BP-7,8-diol, generated in vitro from BP, is epoxidized by both peroxyl radicals and CYP1A1 to the (+)-enantiomer of the *anti*-BPDE, whereas (+)-BP-7,8-diol is epoxidized by peroxyl radicals mainly to the

(–)-enantiomer of the *anti*-BPDE but by CYP1A1 to the (+)-enantiomer of the *syn*-BPDE (51,74). The distinctive stereochemistry of the epoxidation of (+)-BP-7,8-diol by peroxyl radicals and CYP1A1 has provided a useful method to differentiate the two pathways of epoxidation in vitro and potentially in vivo (51,74). Using mouse skin homogenate, cell culture system, freshly isolated mouse epidermal cells, and control and β-naphthoflavone (β-NF)-treated skin from female CD-1 mice, Marnett and coworkers (51), and Eling and co-workers (74) have also established that in the skin of untreated animals where the levels of CYP1A1 (and other cytochrome P-450 isozymes) are undetectable or very low, the conversion of BP-7,8-diol to BP-7,8-diol-9,10-epoxide may be carried out preferentially by peroxyl radical-dependent pathways. However, in CYP1A1 inducer-treated animals, the P-450-dependent pathway predominates (91). Similarly, Melikian et al. (92) have also shown that topical application of radiolabeled (+)-BP-7,8-diol to mouse skin results in the formation of (–)-*anti*- and (+)-*syn*-BPDE in the ratio of 4 to 1, respectively. While the pretreatment with indomethacin produced no effect on this ratio, pretreatment with β-NF resulted in twice the yield of (+)-*syn*-BPDE in the first half-hour, which also diminished with the time (92). On the contrary, co-treatment with catechol showed decreased formation of (–)-*anti*-BPDE as well as lipid peroxidation (92). Based on these studies, it was suggested that a peroxyl radical-mediated epoxidation is involved in the oxidative metabolism of (+)-BP-7,8-diol in mouse skin in vivo (92).

C. Metabolism of PAHs by Lipoxygenase

Based on the above-mentioned oxygen-centered free radical studies, the question arises whether other peroxidation processes operating in tissues, such as the skin, that do not involve hematin but also produce fatty acid-derived radicals are able to cooxygenate BP or BP-7,8-diol. Several indirect observations have suggested that lipoxygenase, which seems to dominate over PGH synthase in certain tissues like human bronchus and human skin, not only represents the main source of lipid peroxyl radicals but also supports cooxygenation (93–95).

Lipoxygenases (EC 1.12.11.12) are nonheme iron dioxygenases widely distributed in plants and animals. In mammals, these enzymes are found in cells such as platelets (96) and PMNs (97) and in organs such as the brain, lung, spleen (98), and skin (99). The lipoxygenase activity represents an important part of the unsaturated fatty acids such as arachidonic acid or linoleic acid cascade leading to the formation of hydroperoxyeicosatetraenoic acid (HPETE), leukotrienes, and lipoxins (100). By a mechanism analogous to the hematin system, fatty acid hydroperoxides have been shown to be metabolized by lipoxygenase to alkoxyl radicals (101,102), and during this process oxidation of chemical carcinogens can occur (74).

Byczkowski and Kulkarni (103), using radioactive (+)-*trans*-BP-7,8-diol and very pure soybean lipoxygenase and linoleic acid, have shown the formation of *anti-trans*-BP-7,8,9,10-tetrol, the hydrolyzed product of the ultimate carcinogenic metabolite *anti*-BPDE. The epoxidation was reported to be dependent on the enzyme concentration and inhibited by nordihydroguaiaretic acid (NDGA), a selective inhibitor of lipoxygenase. This study was the first direct evidence showing the ability of lipoxygenase to catalyze the epoxidation of BP-7,8-diol to the ultimate carcinogenic metabolite. Subsequently, Eling and colleagues (104) also showed the epoxidation of BP-7,8-diol by 5- and 15-lipoxygenase in the presence of arachidonic acid or 15-HPETE and found that *anti*-BPDE was the major metabolite formed by this pathway, further supporting the earlier observation. Though these studies have shown the lipoxygenase-catalyzed metabolism of BP-7,8-diol to *anti*-BPDE, the role of mammalian skin lipoxygenase in the metabolism of carcinogenic compounds came recently by the studies from our laboratory (105–107). In a series of studies we have shown that the incubation of radioactive BP with epidermal cytosol prepared from human, mouse, or rat epidermis in the presence of arachidonic acid results in the metabolism of BP, however, the major metabolites identified on HPLC were BP-quinones (105–107). Under similar experimental conditions, the incubation of radioactive BP-7,8-diol resulted in the formation of *anti-trans*-BP-7,8,9,10-tetrol, the hydrolyzed product of ultimate carcinogenic metabolite of BP, as the major metabolite (105–107). These metabolic reactions were inhibited significantly by the selective inhibitors of lipoxygenase, whereas indomethacin showed no such inhibition; these observations supported that the bioactivation of BP and BP-7,8-diol was mediated by mammalian epidermal lipoxygenase-dependent metabolism of arachidonic acid (105–107). Further studies showed that the incubations carried under anaerobic conditions significantly decreased the metabolism of BP and BP-7,8-diol, suggesting that the oxygen inserted in the BP-metabolite formation is derived from the atmosphere (105–107). When arachidonic acid was replaced with other fatty acids, in identical incubation conditions only unsaturated fatty acids such as γ-linoleic acid and 15-HPETE, but not saturated fatty acid such as oleic acid, supported the metabolism of BP and BP-7,8-diol to the same metabolites (105–107). While the results of these studies suggest that in rodent and human epidermal cytosol, lipoxygenase-dependent metabolism of arachidonic acid plays an important role in the formation of noncarcinogenic and carcinogenic metabolites of BP and BP-7,8-diol, in vivo studies utilizing animal experiments are still to be conducted to further establish the role of this pathway in skin tumor initiation.

D. DNA Binding of PAH Metabolites Formed by Various Metabolic Pathways

Due to their electrophilic nature, the ultimate carcinogens such as the bay-region dihydrodiol epoxides react spontaneously with all kinds of nucleophilic groups of cellular macromolecules such as protein and specifically DNA, forming covalent bonds (2,6). Such reactions are thought to be essential for the neoplastic transformation of the target cell (2,6). Considering the irreversibility of tumor formation, at least in the initiation stage, the close correlation between mutagenic and carcinogenic efficacy, and the high incidence of cancers in patients with defective DNA repair (e.g., in xeroderma pigmentosum), DNA (mainly the hydroxyl and amino groups of the nucleobases) is generally assumed to be the critical target molecule for ultimate carcinogens (2). As reviewed by Cavalieri and Rogan (108), the PAH-metabolites form by either of the metabolic pathways described above have a great tendency to bind to cellular macromolecules, especially to DNA. Such covalent binding is the ultimate process, which results the neoplastic transformation of the target cell (2,6). The DNA binding of carcinogens usually shows a high degree of specificity. For example, the ultimate carcinogenic metabolite of BP has been found to be almost exclusively bound to the 2-amino group of guanine residues in epidermal DNA, whereas the metabolites of N-acetoxy-2-acetylamino-fluorene prefer the C-8 position of guanine (2). The nature of various DNA adducts from carcinogenic and noncarcinogenic enantiomers of BPDE was studied by Cheng et al. (109), who suggested that a *trans* product with deoxyguanosine is the major product formed with either enantiomer of the *anti*-dihydrodiol epoxide. The CYP1A1-dependent metabolism of BP leading to *anti*-BPDE formation and its covalent binding to DNA resulting in tumor initiation is well established (48, 49). However, peroxyl radical–dependent epoxidation of BP forming *anti*-BPDE followed by covalent binding to DNA has only been shown recently by Marnett and colleagues (91,110,111) and others (92). Using epidermis from CD-1 mice, they showed that topical application of (+)-BP-7,8-diol results in similar amounts of (−)-*anti*-BPDE-deoxyguanosine (dGuo) and (+)-*syn*-BPDE-dGuo formation, whereas in β-NF (a conventional inducer of CYP1A1)–pretreated animals, (+)-*syn*-BPDE-dGuo was the major product. Based on this study, it has been suggested that in mouse skin two independent pathways of metabolic activation occur in vivo. In another study, Hughes and Phillips (112) showed the DNA-binding ability of different dibenzopyrenes and BP when applied topically either separately or in combination to murine skin. The majority of skin DNA adducts were obtained within 21 days of treatment, however, low levels of adducts persisted for at least 3 months (112). Lecoq et al. (113) studied the metabolic activation of dibenz(*a,h*)anthracene, related diol and *anti*- and *syn*-diol-epoxide when applied topically to the shaved skin of C57B1/CB1 mice in terms of epidermal DNA binding of metabolites. The results of this study suggested that most of the skin DNA adducts formed in PAH-treated mice are derived from its diol metabolite, whereas fewer were due to conversion of this diol to the bay-region *anti*-3,4-diol-1,2-oxide.

III. ROLE IN SKIN TUMOR PROMOTION

The role of oxidative stress, more specifically of ROS and free radicals and free radical–generating compounds, in skin tumor promotion has been studied and reviewed extensively (39–41). Lines of evidence reported in this regard (39–41) report that 1) lipoxygenase and cyclooxygenase-catalyzed metabolism of arachidonic acid induces skin tumor promotion and inflammation, 2) in general, tumor promoters increase the generation and decrease the degradation of ROS, 3) certain organic peroxides and free radical–generating systems exert tumor-promoting activities and mimic or enhance certain molecular events related to tumor promotion, 4) tumor promoters are known modulators of antioxidant defense mechanism, 5) depletion of physiological antioxidants such as α-tocopherol, selenium, and glutathione increases tumor induction, and treatment with compounds capable of augmenting the cysteine pool such as *N*-acetyl-cysteine provides protection against phorbol ester–mediated tumor promotion in mouse skin, 6) various antioxidants and free radical scavengers inhibit the biochemical and biological effects of tumor promoters, and 7) as reviewed by Perchellet and Perchellet (40), since tumor promoters are not known to be mutagenic and do not bind to DNA, their effects on aneuploidy, chromosomal aberration, sister chromatid exchange, DNA strand breakage, and gene amplification have been thought to be mediated through ROS generated within the phorbol ester–treated epidermal cells and/or released by activated phagocytic cells infiltrating the skin during inflammation (40). In the following sections of this chapter, emphasis has been placed on defining the role of oxidative stress on skin tumor promotion via different mechanisms involving molecular oxygen, ROS, free radical–generating systems, and antioxidant enzyme systems. The role of antioxidant, however, is beyond the scope of this review.

A. Lipoxygenase- and Cyclooxygenase-Catalyzed Metabolism of Arachidonic Acid in Skin Tumor Promotion

The topical application of phorbol ester TPA to mouse skin or its treatment in certain epidermal cells is known to result in a number of biochemical alterations, changes in cellular functions, and histological changes leading to skin tumor promotion (114). Among these, the ones that best correlate with skin tumor-promoting effects include epidermal hyperplasia, inflammation, and increase in the number of dark basal keratinocytes (2,6). It is, however, difficult to establish which of these effects or many other effects of phorbol ester tumor promoters are obligatory or sufficient components of the tumor-promotion process. The induction of inflammation in skin mediated by TPA is believed to be governed by lipoxygenase- and cyclooxygenase-catalyzed metabolites of arachidonic acid, specifically hydroxyeicosatetraenoic acid (HETE) and PGs, respectively (114). Such an assumption is strongly supported by the studies showing effects of inhibitors of various parts of these metabolic pathways on the inhibition of skin tumor promotion (39,40,42,114,115). These inhibitory studies support the involvement of arachidonic acid metabolism pathways in skin tumor promotion; readers further interested in these studies are referred to several recent reviews in this area of research (40,42,115).

Lipoxygenases, a family of structurally and functionally related enzymes which are ubiquitous in their distribution and catalyze the dioxygenation of polyenolic fatty acids carrying one or more 1,4-*cis,cis*-pentadiene systems, e.g., arachidonic acid (116–118), have also been detected in the skin of various mammalian species (119–123). Arachidonic acid is released from membrane phospholipids by phospholipase activity in the skin by a number of stimuli, e.g., phorbol ester tumor promoters (124). As reviewed by several workers (114,116–118,125,126), depending on their nature, different lipoxygenases catalyze stereospecific incorporation of molecular oxygen at the carbon 5,8,9,11,12, or 15 positions of arachidonic acid, resulting in the formation of HPETEs. These HPETEs are further metabolized into various biologically active metabolites; one of these formed by hydroperoxidase is HETE (114,116–118,125,126). Lipoxygenase-catalyzed metabolites of arachidonic acid have been considered to play an important role in many physiological and pathophysiological events, including inflammation and growth regulation; the presence of these metabolites in the mammalian epidermis has been reported by several laboratories (124–130). One member of the lipoxygenase family, 12-lipoxygenase, is characterized in the epidermis of mice (119,120), rats (121), guinea pigs (122), humans (121,123), and in human and murine keratinocytes (131). The 12-HETE metabolite of arachidonic acid formed by catalytic action of 12-lipoxygenase (125,126) is detected in appreciable amounts in psoriatic epidermis as compared to normal tissue (132). In addition, 12-HETE is also characterized in the homogenates from normal human (133) and mouse (127) epidermis. In contrast to 12-lipoxygenase, only very little 5-lipoxygenase activity has been observed in epidermis from mice (131) and humans (134), which is shown to generate 5-HETE and leukotriene B_4 (125,126). In addition to these studies, 15-lipoxygenase activity has been reported in human keratinocytes under in vitro conditions

(121,135), as well as in the cytosolic fraction of epidermis from CD-1 mice (119,136). The presence of 8-lipoxygenase in epidermis has been reported by Gschwendt et al. in NMRI mice (137,138). Fischer et al. (139) reported 5-, 8-, 12-, and 15-lipoxygenases in the epidermis from SSIN and C57BL/6J mice.

In addition to studies showing the predominant low levels of 5-, 8-, 12- and 15-lipoxygenases in different mammalian skin samples and epidermal cell culture systems, in recent years several studies have shown that treatment of mammalian skin or cell-free extract with TPA results in the induction of 8-lipoxygenase, identified by the generation of 8-HETE (137–140). Marks and coworkers showed that when treated with phorbol ester tumor promoters such as TPA, 8-HETE is generated in cell-free preparations of mouse epidermis in vivo, but not in extracts of normal epidermis (137,138). In another study, Fischer et al. (139) showed that single topical application of TPA to the skin of inbred SENCAR (SSIN) and C57BL/6J mice result in elevated levels of 8-lipoxygenase only in SSIN mice within 24 hours, whereas in C57BL/6J mice elevated levels are observed after multiple TPA treatments. Recently, Furstenberger et al. (140) have characterized in more detailed fashion the 8-lipoxygenase activity induced by TPA in NMRI mouse skin in vivo. The results of this study showed that the enzyme was inducible maximally at 24 hours after TPA treatment, strictly dependent on the age of the mice and TPA dose, and was inhibited by cycloheximide (140). In SENCAR mouse, we have also studied the effect of single and multiple topical applications of TPA on the induction of 8-lipoxygenase activity. As analyzed by 8-HETE formation on HPLC, single TPA application resulted in considerable induction of 8-lipoxygenase, which otherwise was undetectable in the epidermis of untreated or acetone-treated control animals. When TPA was applied twice or thrice at an interval of 72 hours, 1.5- and 2.8-fold increases in the 8-lipoxygenase activity over that observed with single application were evident. The results of several studies by other laboratories as well as by our present findings suggest that 8-lipoxygenase–catalyzed arachidonic acid metabolites play an important role in skin tumor promotion, at least in the case of phorbol ester–type skin tumor promoters (126,139,140). As discussed by Furstenberger et al. (140), the 8-lipoxygenase gene belongs to the group of phorbol ester–responsive genes thought to be regulated via a protein kinase C–dependent pathway.

Cyclooxygenases, the microsomal enzymes, also play an important role in cutaneous inflammation, cell proliferation, and skin tumor promotion (114,115,125,126). As reviewed by several laboratories (125,1;26), arachidonic acid, released predominantly from skin phosphatidylcholine by the activation of a skin phospholipase A_2 due to a variety of mechanical, chemical, or hormonal stimuli, undergoes oxidative metabolism via the cyclooxygenase pathway resulting in the formation of PGs. The biosynthesis of PGs is well studied in skin preparations from several mammalian species, including humans (141–149). As reported by Furstenberger et al. (150), adult NMRI mouse epidermis in vivo oxidatively metabolizes arachidonic acid via cyclooxygenase pathway leading to the formation of PGE_2, $PGF_{2\alpha}$, and PGD_2 as determined by radioimmunoassay and HPLC. In the skin, several studies have shown the involvement of PGs in inflammatory processes (151, 152), epidermal growth in response to wound repair (153), and in proliferative skin disease psoriasis (154,155). PGE_2, the major PG product (126), is also reported in fresh human keratinocytes obtained from trypsinization of human epidermis (156). Increased synthesis of PGE_2 and $PGF_{2\alpha}$ has been shown in murine keratinocytes in culture obtained from neonatal as well as adult animals and treated with TPA (157,158).

While PGs have not been shown to be involved in cell proliferation in normal epidermis, they do play an important role in the induction of hyperplasia induced by TPA (150,159). In addition to dominant inflammatory effects, TPA application also produces epidermal hyperproliferation, including ODC induction (160), associated with epidermal hyperplasia (161). Since in addition to these effects, TPA finally promotes the development of tumors in initiated skin (2), the role of cyclooxygenase pathway in tumor promotion is being studied by several laboratories (126). Using CD-1 and NMRI mice it was reported that prior application of indomethacin, a cyclooxygenase inhibitor, to that of TPA results in dose-dependent inhibition of tumor development of the skin (150,162); on the contrary, an inverse effect was observed in SENCAR mice—indomethacin was found to induce the tumor development (163). Further studies to examine this apparent discrepancy showed that in the case of SENCAR mice, lipoxygenase-catalyzed metabolism of arachidonic acid and its role in tumor promotion was a predominant pathway (164) and the treatment of SENCAR mouse keratinocytes with indomethacin results in increased generation of 12-HPETE (164). The results of this study suggest that while indomethacin inhibits the cyclooxygenase pathway, the lipoxygenase pathway starts working in a more profound manner. In addition to these studies, several laboratories have shown the role of cyclooxygenase pathway in skin tumor promotion utilizing various inhibitors of cyclooxygenase and investigating the role of different PG treatments on skin tumor promotion (126). Furstenberger and Marks (126) have also studied and reviewed the role of PGs in the first and second stages of tumor promotion in mouse skin.

B. Reactive Oxygen Species and Skin Tumor Promotion

The role of ROS in skin tumor promotion has been studied by several laboratories using cell culture system, epidermal homogenate, and in vivo animal models (39,40,165,166). Goldstein et al. (167,168) showed that phorbol ester tumor promoters stimulate superoxide anion production in PMNs and that the inhibitors of tumor promotion such as protease inhibitors, retinoids and anti-inflammatory steroids inhibit the generation of this ROS (167,168). As studied and reviewed by Fischer (165), the alternate method of measuring ROS generation in TPA-treated PMNs is by measuring photoemission and chemiluminescence (CL)—a method associated with the generation and energy dissipation of superoxide anion and singlet oxygen (169). In further studies, Fischer showed the generation of ROS in mouse epidermal cells from newborn SENCAR mice and treated with TPA. The results of this study showed that CL response was dependent on TPA dose and cell number (165). This study was extended with a series of phorbol ester tumor promoters with varying promotion effect; however, TPA was found to be most effective among all of them (165). In several other studies, based on inhibition, the production of ROS has also been studied in tumor promoter–treated assay systems (165).

The role of hydrogen peroxide in TPA-caused tumor promotion came from the study by DeChatelet et al. (170), showing that TPA rapidly induces the cellular production of hydrogen peroxide. Several lines of evidence later correlated the release of hydrogen peroxide in response to TPA. Phorbol ester tumor promoters with varying tumor promotion ability were found to induce the generation of hydrogen peroxide by macrophages during a 1-hour incubation time (171); TPA was also found to stimulate hydrogen peroxide production in mouse skin in vivo after its application (172,173). Perchellet et al. (174,175) also studied the production of hydrogen peroxide following TPA and other skin tumor promoter treatment and its role in skin tumor promotion. The results of these studies showed that TPA rapidly stimulates hydrogen peroxide generation in intact epidermal cells as well as in cell-free systems incubated in the absence or presence of enzymic and nonenzymic systems that produce ROS (174,175). These results suggested that TPA increases the levels of peroxidation and cellular damage in mouse epidermis via the stimulation of prooxidant activities of various endogenous sources of ROS (40). In another study, TPA-caused induction of xanthine oxidase, a superoxide anion–generating system that mimics the effect of TPA on ODC induction (176) and promotion of cell transformation (177), has also been shown in murine epidermis (178). Recently, utilizing an in vivo model, ROS have also been shown to induce skin inflammation (179). Since DNA strand break, sister chromatid exchange, chromosomal aberrations, aneuploidy, and gene amplification are thought to be mediated through ROS (40), the generation of ROS mediated by tumor promoters can be directly connected with their tumor-promoting ability.

In a series of studies utilizing cell culture system, generation of ROS and their role in tumor promotion have also been studied. Similar to TPA, the xanthine-xanthine oxidase system has been shown to induce ODC activity in primary murine keratinocytes, an effect strongly correlated with tumor promotion (180). The role of ROS in tumor promotion has also been demonstrated in murine JB6 epidermal cells (181,182), mouse embryo C3H/10T1/2/C18 fibroblasts (183–186), and F_1 offspring of SSIN × C57BL/6J mice (187). While the generation and inhibition of degradation of ROS by phorbol ester skin tumor promoters has been correlated with their ability to cause DNA damage and chromosomal aberrations, etc., resulting in tumor promotion (40), the relevance of these events to a specific stage of tumor promotion still awaits more studies.

C. Free Radical-Generating Compound—Organic Peroxides and Tumor Promotion

The role of free radical–generating organic peroxides in tumor promotion was implicated by the seminal observation of Slaga et al. (14) that BPO promotes the development of tumors in DMBA-initiated SENCAR mouse skin; BPO, however, possesses neither tumor-initiating nor complete carcinogenic activity (188). Although BPO does not directly activate PKC, it has been shown to be effective in the induction of ODC and epidermal hyperplasia and in increasing the number of dark basal keratinocytes (189,190), the known markers of tumor promotion in skin (2,6). BPO has also been shown to result in the transformation of JB6 mouse epidermal cell line (191), induction of ODC activity in cultured epithelial cells (192), and DNA strand breaks in several types of cell lines including keratinocytes (192–194). The tumor-promoting activity of BPO was ascribed due to its free radical–generating ability since the antioxidant BHA and BHT inhibited the tumor-promoting activity of this compound (14). In further studies, several other diacylperoxides and hydroperoxides such as lauroyl peroxide, decanoyl peroxide, dicumyl peroxide, cumene hydroperoxide, 2,2-azobis-(2-amidinopropane), *tert*-butyl hydroperoxide, methylethylketone peroxide, and *tert*-butyl hydroperoxybenzoate were also identified as very active or weak tumor promoters in murine skin (195). Using isolated mouse keratinocytes (195) and human carcinoma skin keratinocytes (196), it has been shown that all the skin tumor promoter organic peroxides generate free radicals, which are involved in their tumor-promoting action. Recently, Slaga and coworkers (15) have also reported the tumor-promoting effects of some additional organic peroxides utilizing short-term markers of skin tumor promotion.

D. Antioxidant Enzymes and Tumor Promotion

As reviewed by Sun (39), the antioxidant enzyme system contains superoxide dismutase (SOD), catalase (CAT), glutathione peroxidase (GPx), glutathione reductase (GR), and glucose-6-phosphate dehydrogenase (G6PD). The basic functions of these antioxidant enzymes are to protect the normal cellular physiology of the tissue against the toxic ROS generated during normal metabolism or under oxidative stress (37–41). While detailed biochemistry of these enzymes is beyond the scope of this review, in brief, 1) SOD dismutates two molecules of superoxide anion, resulting in the formation of hydrogen peroxide and molecular oxygen, 2) CAT catalyzes the degradation of hydrogen peroxide into water and molecular oxygen, 3) GPx catalyzes the oxidation of reduced glutathione (GSH) to oxidized glutathione (GSSG) at the expense of hydrogen peroxide, 4) GR catalyzes the reaction GSSG + NADPH + $H^+ \longrightarrow$ 2GSH + $NADP^+$, and 5) G6PD catalyzes the reaction G-6-P + $NADP^+ \longrightarrow$ 6-phosphoglucolactone + NADPH + H^+ (40). It has been suggested that these enzymes exert their protective effects against cell damage by oxygen and/or its metabolites via enzymatic and nonenzymatic mechanisms (37–41). While SOD, CAT, and GPx have been considered primary antioxidant enzymes due to their involvement in the direct elimination of ROS, GR and G6PD are secondary antioxidant enzymes that participate in the detoxification of ROS by decreasing peroxide levels or by maintaining a steady supply of metabolic intermediates such as glutathione and NADPH for the primary antioxidant enzymes (37–41). The nonenzymatic physiological antioxidants include sulfhydryl compounds such as reduced glutathione and thiols, NADPH, ascorbate, α-tocopherol, selenium, β-carotene, bilirubin, etc.; their depletion causes an increase in tumor induction (37–41, 197).

The role of antioxidative enzymes in skin tumor promotion has been investigated by several laboratories (40). It has been shown that the specific activities of epidermal SOD and CAT are decreased significantly by a single or multiple treatment of murine skin with tumor promoters such as TPA (198,199). Perchellet and coworkers examined the in vivo and in vitro effects of TPA on GSH:GSSG ratio and GPx activity in epidermal homogenate prepared from mouse treated topically with TPA and in murine epidermal suspensions added with TPA (40,200–204). The results of these studies showed an approximate 20–40% decrease in Gpx activity and considerable reduction in GSH:GSSG ratio (40,200–204). Recently, Reiners et al. (205,206) examined the levels of several antioxidant/prooxidant enzymes in murine skin following topical application of TPA and throughout the cancer development. The results of these studies showed a considerable to significant decrease in SOD, CAT, and GPx activities and a significant increase in xanthine oxidase activity in TPA-treated mouse skin, papillomas and carcinomas induced by two-stage carcinogenesis protocol, and skin adjacent to the papillomas when compared to controls (205). Based on these observations, it was concluded that TPA regulates the changes in the activities of antioxidant enzymes involved in ROS metabolism characteristically associated with papilloma and carcinoma phenotype (205).

IV. ROLE IN MALIGNANT CONVERSION

Increasing amounts of direct and indirect evidence suggest that oxidative stress generated by different systems including ROS and free radical-generating compounds are involved in the malignant conversion of benign papillomas to squamous cell carcinomas, at least in murine skin (37–41). The free radical–generating compound BPO, which is also an oxidant tumor promoter, has been shown to enhance the progression stage of murine skin carcinogenesis (25,26). Our laboratory has also shown that other free radical–generating compounds such as *tert*-butyl hydroperoxybenzoate and 2,2-azobis-(2-amidinopropane) are effective in converting benign papillomas, either induced chemically or by ultraviolet radiation on SENCAR or SKH-1 hairless mouse skin, to malignant carcinomas (27). Since cutaneous chemical carcinogens and tumor promoters generate ROS directly or indirectly, and since various antioxidants effectively inhibit biochemical and biological events associated with tumor initiation, promotion, and/or malignant conversion, it is conceivable that ROS-induced damage in macromolecules may contribute to the transition of epidermal target cells from a preneoplastic state to malignancy (39–41).

V. CONCLUSION

The simplest and most acceptable model of cutaneous carcinogenesis and possibly other sites distinguishes three different stages: initiation, promotion, and progression. There is convincing evidence to suggest that oxidative stress might be involved in all three stages of carcinogenesis, at least in experimental animal tumor bioassay systems. However, relatively little direct human data are available. Evidence also indicates that oxidants, which are generated by multiple pathways in mammalian cellular systems, may be natural carcinogens and contribute in accelerating the formation of premalignant lesions and their conversion to malignancy. It is known that active oxygen released by inflammatory phagocytes can induce mutations and chromosomal aberrations in neighboring target cells (207). These oxidant-induced DNA sequence changes, though rare, may affect the activities of proto-oncogenes and suppressor genes (208), which in turn facilitate cell growth and differentiation. In recent years oxidant stress in mouse epidermal cells has been shown to induce the protooncogenes c-*fos* and c-*myc* (208). The role of induction of poly ADP ribosylation during oxidative stress of epidermal cells is also gaining importance (208). The role of oxidative stress in cutaneous carcinogenesis gained further support from studies showing that antioxidants produce protective effects in experimental studies (39–42). In this chapter, we have attempted to provide evidence that prolonged oxidative stress, i.e., chronic exposure to singlet oxygen, superoxide anion, hydrogen peroxide, and/or hydroxyl radicals, can result in cell injury and plays a role in all three stages of murine skin chemical carcinogenesis. If suitable methodology to measure oxidative stress in human epidermal cells is established, then the role of oxidative stress in human skin carcinogenesis may be better studied and established.

ACKNOWLEDGMENTS

The studies were supported by USPHS grants ES-1900, CA-51802, PO1-CA-48735, P-30-AR-39750, and by research funds from the Department of Veterans Affairs. We thank Ms. Sandra Evans for excellent help in preparing the manuscript.

REFERENCES

1. Pitot HC, Dragan YP. Facts and theories concerning the mechanisms of carcinogenesis. FASEB J 1991; 5:2280–6.
2. Agarwal R, Mukhtar H. Cutaneous chemical carcinogenesis. In: Mukhtar H, ed. Pharmacology of the skin. Boca Raton, FL: CRC Press, 1991:371–87.
3. Matsui MS, DeLeo VA. Longwave ultraviolet radiation and promotion of skin cancer. Cancer Cells 1991; 3:8–12.
4. Boyd JA, Barrett JC. Genetic and cellular basis of multistep carcinogenesis. Pharmacol Ther 1990; 46:469–86.
5. Egan H, Castegnaro M, Kunte H, Bogovski P, Walker EA, eds. Environmental carcinogens selected methods of analysis. Vol. 3. Analysis of polycyclic aromatic hydrocarbons in environmental samples. Lyon: IARC Publ No. 29, 1979.
6. DiGiovanni J. Modification of multistage skin carcinogenesis in mice. In: Ito N, Sugano H, eds. Modification of tumor development in rodents. Vol. 33. Basel: Karger, 1991:192–9.
7. Slaga TJ. Multistage skin tumor promotion and specificity of inhibition. In: Slaga TJ, ed. Mechanisms of tumor promotion. Vol. II. Tumor promotion and skin carcinogenesis. Boca Raton, FL: CRC Press, 1984:189–96.
8. Boutwell RK. Some biological aspects of skin cacinogenesis. Prog Exp Tumor Res 1964; 4:207–49.
9. Yuspa SH. Cutaneous chemical carcinogenesis. J Am Acad Dermatol 1986; 15: 1031–44.
10. Sims P. The metabolic activation of chemical carcinogens. Br Med Bull 1980; 36:11–18.
11. Mukhtar H, Merk HF, Athar M. Skin chemical carcinogenesis. Clin Dermatology 1989; 7:1–10.
12. Furstenberger G, Kinzel V. Schwarz M, Marks F. Partial inversion of the initiation-promotion sequence of multistage tumorigenesis in the skin of NMRI mice. Science 1985; 230:76–8.
13. Cohen SM, Ellwein LB. Genetic errors, cell proliferation and carcinogenesis. Cancer Res 1991; 51:6493–505.

14. Slaga TJ, Klein-Szanto AJP, Triplett LL, Yotti LP, Trosko JE. Skin tumor-promoting activity of benzoyl peroxide, a widely used free radical-generating compound. Science 1981; 213:1023–5.
15. Gimenez-Conti I, Viaje A, Chesner J, Conti C, Slaga TJ. Induction of short-term markers of tumor promotion by organic peroxides. Carcinogenesis 1991; 12:563–9.
16. Synthetic organic chemicals, U.S. production and sales. Washington, DC: US International Trade Commission, 1977.
17. Criteria for a Recommended Standard Occupational Exposure to Benzoyl Peroxide. DHEW/PUB/NIOSH-77/166, 1977.
18. Karasz AB, DeCocco F, Maxstadt JJ. Gas chromatographic measurements of benzoyl peroxide (as benzoic acid) in cheese. J Assoc Off Anal Chem 1974; 57: 706–9.
19. Locniskar M, Belury MA, Cumberland AG, Patrick KE, Fischer SM. The effect of dietary lipid on skin tumor promotion by benzoyl peroxide: comparison of fish, coconut and corn oil. Carcinogenesis 1991; 12:1023–8.
20. Bonfil RD, Momiki S, Conti CJ, Klein-Szanto AJP. Benzoyl peroxide enhances the invasive ability of a mouse epidermal carcinoma cell line. Int J Cancer 1989; 44:165–9.
21. Noweell PC. Mechanisms of tumor progression. Cancer Res 1986; 46:2203–7.
22. Marks F. Furstenberger G. The conversion stage of skin carcinogenesis. Carcinogenesis 1990; 11:2085–92.
23. Hennings H, Shores R, Wenk ML, Spangler EF, Tarone, Yuspa SH. Malignant conversion of mouse skin tumors is increased by tumor initiators and unaffected by tumor promoters. Nature 1983; 304:67–9.
24. O'Connell JF, Klein-Szanto AJP, DiGiovanni DM, Fries JW, Slaga TJ. Malignant progression of mouse skin papillomas treated with ethylnitrosourea, N-methyl-N-nitro-N-nitrosoguanidine, or 12-O-tetradecanoylphorbol-13-acetate. Cancer Lett 1986; 30:269–74.
25. Reiners JJ, Jr, Nesnow S, Slaga TJ. Murine susceptibility to two-stage carcinogenesis is influenced by the agent used for promotion. Carcinogenesis 1984; 5: 301–7.
26. O'Connell JF, Klein-Szanto AJP, DiGiovanni DM, Fries JW, Slaga TJ. Enhanced malignant progression of mouse skin tumors by the free-radical generator benzoyl peroxide. Cancer Res 1986; 46:2863–5.
27. Athar M, Lloyd JR, Bickers DR, Mukhtar H. Malignant conversion of ultraviolet radication and chemically induced mouse skin benign tumors by free radical generating compounds. Carcinogenesis 1989; 10:1841–5.
28. Pryor WA. The role of free radical reactions in biological systems. In: Pryor WA, ed. Free radicals in biology. Vol. 1. New York: Academic Press, 1976: 1–50.
29. Cross CE, Halliwell B, Borish ET, Pryor WA, Ames BN, Saul RL, McCord JM, Harman D. Oxygen radicals and human disease. Ann Intern Med 1987; 107:526–45.
30. Hayaishi O, Imamura S, Miyachi Y, eds. The biological role of reactive oxygen species in skin. New York: Elsevier, 1987.

31. Kalyanaraman B, Sohnle PG. Generation of free radical intermediates from foreign compounds by neutrophil-derived oxidants. J Clin Invest 1985; 75:1618–22.
32. Hsieh GC, Acosta D. Dithranol-induced cytotoxicity in primary cultures of rat epidermal keratinocytes. I. The role of reactive oxygen species. Toxicol Appl Pharmacol 1991; 107:16–26.
33. Farber E. Chemical carcinogenesis, a biologic perspective. Am J Pathol 1982; 2:271–96.
34. Finkelstein E, Rosen GM, Rauckman EJ. Spin trapping of superoxide and hydroxyl radical: practical aspects. Arch Biochem Biophys 1980; 200:1–16.
35. Zimmerman R, Cerutti PA. Active oxygen acts as a promoter of transformation in mouse C3H/10T1/2/C18 fibroblasts. Proc Natl Acad Sci USA 1984; 81:2085–7.
36. Fridovich I. Superoxide radicals: an endogenous toxicants. Ann Rev Pharmacol Toxicol 1983; 23:239–57.
37. Floyd RA. Role of oxygen free radicals in carcinogenesis and brain ischemia. FASEB J 1990; 4:2587–97.
38. Fischer SM, Floyd RA, Copeland ES. Workshop report from the division of research grants, antional institute of health. Oxy radicals in carcinogenesis—a chemical pathology study section workshop. Cancer Res 1988; 48:3882–7.
39. Sun Y. Free radicals, antioxidant enzymes, and carcinogenesis. Free Radical Biol Med 1990; 8:583–99.
40. Perchellet J, Perchellet EM. Antioxidants and multistage carcinogenesis in mouse skin. Free Radical Biol Med 1989; 7:377–408.
41. Athar M, Agarwal R, Bickers DR. Mukhtar H. Role of reactive oxygen species in skin. In: Mukhtar H, ed. Pharmacology of the skin. Boca Raton, FL: CRC Press, 1991:269–79.
42. Fuchs J, Friesleben H-J, Packer L. Antioxidants in the skin. In: Mukhtar H, ed. Pharmacology of the skin. Boca Raton, FL: CRC Press, 1991:249–67.
43. Miller EC, Miller JA. Chemical carcinogenesis: mechanisms and approaches to its control. J Natl Cancer Inst 1971; 47:V.
44. Floyd RA. Free radicals and cancer. New York: Marcel Dekker, 1982.
45. Halliwell B, Gutteridge JMC. Oxygen toxicity, oxygen radicals, transition metals and disease. Biochem J 1984; 219:1–14.
46. Levin W, Wood A, Chang R, Ryan D, Thomas P, Yagi H, Thakkar D, Vyas K, Boyd D, Chu SY, Conney AH, Jerina DM. Oxidative metabolism of polycyclic aromatic hydrocarbons to ultimate carcinogens. Drug Metab Rev 1982; 13:555–80.
47. Miller EC, Miller JA. The metabolism of chemical carcinogens to reactive electrophiles and their possible mechanisms of actions in carcinogenesis. In: Searle CE, ed. Chemical carcinogens (ACS Monog 173). Washington DC: American Chemical Society, 1976:737–62.
48. Conney AH. Induction of microsomal enzymes by foreign chemicals and carcinogenesis by polycyclic aromatic hydrocarbons: GHA Clowes memorial lecture. Cancer Res 1982; 42:4875–917.

49. Mukhtar H, Agarwal R, Bickers DR. Cutaneous metabolism of xenobiotics and steroid hormones. In: Mukhtar H, ed. Pharmacology of the skin. Boca Raton, FL: CRC Press, 1991:89–110.
50. Mukhtar H. Epidermal cytochrome P-450. In: Mukhtar H, ed. Pharmacology of the skin. Boca Raton, FL: CRC Press, 1991:139–47.
51. Marnett LJ. Prostaglandin synthase-mediated metabolism of carcinogens and a potential role for peroxyl radicals as reactive intermediates. Environ Health Perspect 1990; 88:5–12.
52. Cavalieri E, Rogan E. Metabolic activation by one-electron and two-electron oxidation in aromatic hydrocarbon carcinogenesis. In: Woo Y-T, Lai DY, Arcos JC, Argus MF, eds. Chemical induction of cancer. Vol. IIIB. New York: Academic Press, 1985:533–69.
53. Cavalieri E, Rogan E. One-electron oxidation in aromatic hydrocarbon carcinogenesis. In: Harvey RG, ed. ACS Symposium Series No. 283: Polycyclic Hydrocarbons and Carcinogenesis. Washington, DC: American Chemical Society, 1985:289–305.
54. Sims P, Grover PL. Involvment of dihydrodiols and diol epoxides in the metabolic activation of polycyclic hydrocarbons other than benzo(a)pyrene. In: Gelboin HV, Ts'o Pop, eds. Polycyclic hydrocarbons and cancer. New York: Academic Press, 1981:117–81.
55. Mukhtar H, Khan WA. Cutaneous cytochrome P-450. Drug Metab Rev 1989; 20:657–73.
56. Cavalieri E, Roth R, Rogan EG. Metabolic activation of aromatic hydrocarbons by one-electron oxidation in relation to the mechanisms of tumor initiation. In: Freudenthal RI, Jones PW, eds. Polynuclear aromatic hydrocarbons; chemistry, metabolism and carcinogenesis. Vol. I. New York: Raven Press, 1976:181–90.
57. Cremonesi P, Cavalieri EL, Rogan EG. One-electron oxidation of 6-substituted benzo(a)pyrenes by manganic acetate. A model for metabolic activation. J Org Chem 1989; 54:3561–70.
58. Cavalieri EL, Devanesan PD, Rogan EG. Radical cations in the horseradish peroxidase and prostaglandin H synthase mediated metabolism and binding of benzo(a)pyrene to deoxyribonucleic acid. Biochem Pharmacol 1988, 37:2183–7.
59. Cavalieri E, Rogan E, Davanesan P, Cremonesi P. Radical cations as precursors in the metabolic formation of quinones from benzo(a)pyrene and 6-fluorobenzo(a)pyrene: fluoro substitution as a probe for one-electron oxidation in aromatic substrate. Biochem Pharmacol 1988; 37:2173–82.
60. Jerina DM, Sayer JM, Thakker DR, Yagi H. Carcinogenicity of polycyclic aromatic hydrocarbons: the bay-region theory. In: Pullman B, Ts'o Pop, Gelboin H, eds. Carcinogenesis: fundamental mechanisms and environmental effects. Dordrecht: Reidel, 1980:1–12.
61. Trush MA, Seed JL, Kensler TW. Oxidant-dependent metabolic activation of polycyclic aromatic hydrocarbons by phorbol ester-stimulated human polymorphonuclear leukocytes: possible link between inflammation and cancer. Proc Natl Acad Sci USA 1985; 82:5193–8.

62. Kensler TW, Egner PA, Moore KG, Taffe BG, Twerdok LE, Trush MA. Role of inflammatory cells in the metabolic activation of polycyclic aromatic hydrocarbons in mouse skin. Toxicol Appl Pharmacol 1987; 90:337–46.
63. Mallet WG, Randolph Mosebrook D, Trush MA. Activation of (±)-*trans*-7,8-dihydroxy-7,8-dihydrobenzo(a)pyrene to diolepoxide by human polymorphonuclear leukocytes or myeloperoxidase. Carcinogenesis 1991; 12:521–4.
64. Fee JA, Valentine JS. Chemical and physical properties of superoxide. In: Michelson AM, McCord JM, Fridovich I, eds. Superoxide and superoxide dismutates. New York: Academic Press, 1977:19–60.
65. Ingold K. Peroxy radicals. Acc Chem Res 1969; 2:1–9.
66. Porter NA. Mechanisms for the autoxidation of polyunsaturated lipids. Acc Chem Res 1986; 19:262–8.
67. Pryor WA. Oxy-radicals and related species: their formation, life-times, and reactions. Annu Rev Physiol 1986; 48:657–67.
68. Gray P, Williams A. The thermochemistry and reactivity of alkoxyl radicals. Chem Rev 1959; 59:239–328.
69. Sheldon RA, Kochi JK. Metal catalysis in peroxide reactions. In: Sheldon RA, Kochi JK, eds. Metal-catalyzed oxidations of organic compounds. Mechanistic principles and synthetic methodology including biochemical processes. New York: Academic Press, 1981:34–70.
70. Walling C. In: Walling C, ed. Free radicals in solution. New York: John Wiley and Sons, 1957:418–27.
71. Needleman P, Turk J, Jakschik B, Morrison A, Lefkowith JB. Arachidonic acid metabolism. Annu Rev Biochem 1986; 55:69–102.
72. Ohki S, Ogino N, Yamamoto S, Hayashi O. Prostaglandin hydroperxoidase, an integral part of prostaglandin endoperoxide synthetase from bovine vesicular gland microsomes. J Biol Chem 1979; 254:829–36.
73. Eling TE, Krauss RS. Arachidonic acid-dependent metabolism of chemical carcinogens and toxicants. In: Marnett LJ, ed. Arachidonic acid metabolism and tumor initiation. New York: Martinus Nijhoff, 1985:83–124.
74. Eling TE, Thompson DC, Foureman GL, Curtis JF, Hughes MF. Prostaglandin H synthase and xenobiotic oxidation. Ann Rev Pharmacol Toxicol 1990; 30:1–45.
75. Marnett LJ, Reed GA, Dennison DJ. Prostaglandin synthetase-dependent activation of 7,8-dihydroxy-7,8-dihydro-benzo(a)pyrene to mutagenic derivatives. Biochem Biophys Res Commun 1978; 82:210–16.
76. Guthrie J, Robertson IGC, Zeiger E, Boyd JA, Eling TE. Selective activation of dihydrodiol of several polycyclic aromatic hydrocarbons to mutagenic products by prostaglandin synthetase. Cancer Res 1982; 42:1620–3.
77. Battista JR, Marnett LJ. Prostaglandin H synthase-dependent epoxidation of aflatoxin B_1. Carcinogenesis 1985; 6:1227–9.
78. Robertson IGC, Sivarajah K, Eling TE, Zeiger E. Activation of some aromatic amines to mutagenic products by prostaglandin endoperoxide synthetase. Cancer Res 1983; 43:476–80.
79. Marnett LJ, Dix TA, Sachs RJ, Siedlik PH. Oxidations by fatty acid hydroperoxides and prostaglandin synthetase. In: Saumeulsson B, Ramwell P, Pao-

letti R, eds. Advances in prostaglandin and thromboxane research. Vol. II. New York: Raven Press, 1983:79–86.
80. Marnett LJ, Reed GA, Johnson JT. Prostaglandin synthetase-dependent benzo(a)pyrene oxidation: products of the oxidation and inhibition of their formation by antioxidants. Biochem Biophys Res Commun 1977; 79:569–76.
81. Lesko S, Caspary W. Lorentzen R, Ts'o POP. Enzyme formation of 6-oxobenzo(a)pyrene radical in rat liver homogenates from carcinogenic benzo(a)pyrene. Biochemistry 1975; 14:3978–84.
82. Marnett L, Johnson JT, Bienkowski MJ. Arachidonic acid dependent metabolism of 7,8-dihydroxy-7,8-dihydro-benzo(a)pyrene by ram seminal vesicles. FEBS Lett 1979; 106:13–16.
83. Sivarajah K, Mukhtar H, Eling TE. Arachidonic acid dependent metabolism of (±)trans-7,8-dihydroxy-7,8-dihydro-benzo(a)pyrene (BP-7,8-diol) to 7,10/8,9 tetrols. FEBS Lett 1979; 106:17–20.
84. Mallet WG, Mosebrook DR, Trush MA. Activaiton of (±)-*trans*-7,8-dihydroxy-7,8-dihydrobenzo[a]pyrene to diolepoxides by human polymorphonuclear leukocytes or myeloperoxidase. Carcinogenesis 1991; 12:521–4.
85. Marnett LJ. Peroxyl free radicals: potential mediators of tumor initiation and promotion. Carcinogenesis 1987; 8:1365–73.
86. Dix TA, Marnett LJ. Free radical epoxidation of 7,8-dihydroxy-7,8-dihydrobenzo(a)pyrene by hematin and polyunsaturated fatty acid hydroperoxides. J Am Chem Soc 1981; 103:6744–6.
87. Dix TA, Fontana R, Panthani A, Marnett LJ. Hematin-catalyzed epoxidation of 7,8-dihydroxy-7,8-dihydrobenzo(a)pyrene (BP-7,8-diol) by polyunsaturated fatty acid hydroperoxides. J Biol Chem 1985; 260:5358–65.
88. Dix TA, Marnett LJ. Conversion of linoleic acid hydroperoxide to hydroxy, keto, epoxyhydroxy and trihydroxy fatty acids by hematin. J Biol Chem 1985; 260:5351–7.
89. Panthananickal A, Marnett LJ. Arachidonic acid-dependent metabolism of 7,8-dihydroxy-7,8-dihydrobenzo(a)pyrene to polyguanylic acid-binding derivatives. Chem Biol Interactions 1981; 33:239–52.
90. Panthananickal A, Weller P, Marnett LJ. Stereoselectivity of the epoxidation of 7,8-dihydroxy-7,8-dihydrobenzo(a)pyrene by prostaglandin H synthase and cytochrome P-450 determined by the identification of polyguanylic acid adducts. J Biol Chem 1983; 258:4411–18.
91. Pruess-Schwartz D, Nimesheim A, Marnett LJ. Peroxyl radical- and cytochrome P-450-dependent metabolic activation of (+)-7,8-dihydroxy-7,8-dihydrobenzo(a)pyrene in mouse skin in vitro and in vivo. Cancer Res 1989; 49:1732–7.
92. Melikian AA, Bagheri K, Hoffmann D. Oxidation and DNA binding of (+)-7,8-dihydroxy-7,8-dihydrobenzo(a)pyrene in mouse epidermis in vivo and effects of coadministration of catechol. Cancer Res 1990; 50:1795–9.
93. Reed GA, Grafstrom RC, Krauss RS, Autrup H, Eling TE. Prostaglandin H synthase-dependent cooxygenation of (±)-7,8-dihydro-7,8-dihydrobenzo(a)pyrene in hamster trachea and human bronchus explants. Carcinogenesis 1984; 5:955–60.

94. Eling T, Curtis J, Battista J, Marnett LJ. Oxidation of (+)-7,8-dihydroxy-7,8-dihydrobenzo(a)pyrene by mouse keratinocytes: evidence of peroxyl radical- and monooxygenase-dependent metabolism. Carcinogenesis 1986; 7:1957–63.
95. Krauss RS, Eling TE. Arachidonic acid-dependent cooxidation: a potential pathway for the activation of chemical carcinogens in vivo. Biochem Pharmacol 1984; 33:3319–24.
96. Nugteren DJ. Arachidonate lipoxygenase in blood platelets. Biochim Biophys Acta 1975; 380:299–302.
97. Borgeat P, Hamberg M, Samuelsson B. Transformation of arachidonic acid and homo-gamma-linolenic acid by rabbit polymorphonuclear leukocytes. J Biol Chem 1976; 251:7816–20.
98. Malle E, Leis HJ, Karadi I, Kostner GM. Lipoxygenases and hydroperoxy/hydroxy-eicosatetraenoic acid formation. Int J Biochem 1987; 19:1012–22.
99. Henke D, Danilowicz R, Eling T. Arachidonic acid metabolism by isolated epidermal basal and differential keratinocytes from the hairless mouse. Biochim Biophys Acta 1986; 876:271–9.
100. Marnett LJ. Arachidonic acid metabolism and tumor initiation. Boston: Martinus Nijhoff Publishing, 1985.
101. deGroot JJMC, Veldink GA, Vliegenthart JFG, Boldingh J, Wever R, van Gelder BF. Demonstration by EPR spectroscopy of the functional role of iron in soybean lipoxygenase-I. Biochim Biophys Acta 1975; 377:71–9.
102. Mansuy D, Cucurou C, Biatry B, Battioni JP. Soybean lipoxygenase-catalyzed oxidations by linoleic acid hydroperoxide: different reducing substrates and dehydrogenation of phenidone and BW 775C. Biochem Biophys Res Commun 1988; 151:339–46.
103. Byczkowski JZ, Kulkarni AP. Lipoxygenase-catalyzed epoxidation of benzo(a)pyrene-7,8-dihydrodiol. Biochem Biophys Res Commun 1989; 159:1199–205.
104. Hughes MF, Chamulitrat W, Mason RP, Eling TE. Epoxidation of 7,8-dihydroxy-7,8-dihydrobenzo(a)pyrene via a hydroperoxide-dependent mechanism catalyzed by lipoxygenase. Carcinogenesis 1989; 10:2075–80.
105. Agarwal R, Raza H, Mukhtar H. Lipoxygenase-catalyzed metabolism of the skin carcinogen benzo(a)pyrene in rodent and human epidermis. J Invest Dermatol 1990; 94:502.
106. Agarwal R, Mukhtar H. Lipoxygenase catalyzed metabolism of benzo(a)pyrene and benzo(a)pyrene-7,8-diol in human skin. Toxicologist 1991; 11:346.
107. Agarwal R, Mukhtar H. Metabolism of benzo(a)pyrene and benzo(a)pyrene 7,8-dihydrodiol catalyzed by human rodent epidermal lipoxygenase. Proc Am Assoc Cancer Res 1991; 32:117.
108. Cavalieri EL, Rogan EG. Radical cations in aromatic hydrocarbon carcinogenesis. Free Rad Res Comms 1990; 11:77–87.
109. Cheng SC, Hilton BD, Roman JM, Dipple A. DNA adducts from carcinogenic and noncarcinogenic enantiomers of benzo[a]pyrene dihydrodiol epoxide. Chem Res Toxicol 1989; 2:334–40.
110. Reddy AP, PreussSchwartz D, Marnett LJ. Separation of (+)-syn-benzo[a]pyrene and (−)-anti-benzo[a]pyrene dihydrodiol epoxide DNA adducts in P-32-postlabelling analysis. Chem Res Toxicol 1992; 5:19–25.

111. Reddy AP. PruessSchwartz D, Ji CA, Gorycki P, Marnett LJ. P-32-postlabelling analysis of DNA adduction in mouse skin following topical administration of (+)-7,8-dihydroxy-7,8-dihydrobenzo[a]pyrene. Chem Res Toxicol 1992; 5:26–33.
112. Hughes NC, Phillips DH. Covalent binding of dibenzpyrenes and benzo[a]pyrene to DNA: evidence for synergistic and inhibitory interactions when applied in combination to mouse skin. Carcinogenesis 1990; 11:1611–19.
113. Lecoq S, She MN, Hewer A, Grover PL, Platt KL, Oesch F, Phillips DH. The metabolic activation of dibenz[a,h]anthracene in mouse skin examined by ^{32}P-postlabelling: minor contribution of the 3,4-diol 1,2-oxides to DNA binding. Carcinogenesis 1991; 12:1079–83.
114. Fischer SM, Slaga TJ. Arachidonic acid metabolism and tumor promotion. Boston: Martinus Nijhoff Publishing, 1985.
115. Nakadate T. The mechanism of skin tumor promotion caused by phorbol esters: possible involvement of arachidonic acid cascade/lipoxygenase, protein kinase C and calcium/calmodulin systems. Jap J Pharmacol 1989; 49:1–9.
116. Needleman P, Turk J, Jakschik BA, Morrison AR, Lefkowith JB. Arachidonic acid metabolism. Annu Rev Biochem 1986; 55:69–102.
117. Pace-Asciak CR, Smith WL. Enzymes in the biosynthesis and catabolism of the eicosanoids: prostaglandins, thromboxanes, leukotrienes and hydroxy fatty acids. Enzymes 1983; 16:543–603.
118. Ruzicka T, Printz MP. Arachidonic acid metabolism in skin: a review. Rev Physiol Biochem Pharmacol 1984; 100:121–32.
119. Kato R, Kakadate T, Yamamoto S, Sugimura T. Inhibition of 12-O-tetradecanoylphorbol-13-acetate-induced tumor promotion and ornithine decarboxylase activity by quercetin: possible involvement of lipoxygenase inhibition. Carcinogenesis 1983; 4:1301–9.
120. Henke B, Danilowicz R, Eling T. Arachidonic acid metabolism by isolated and differentiated keratinocytes from the hairless mouse. Biochim Biophys Acta 1986; 876:271–9.
121. Nugteren DH, Kivits GAA. Conversion of linoleic acid and arachidonic acid by skin epidermal lipoxygenase. Biochim Biophys Acta 1987; 921:135–41.
122. Ruzicka T, Vitto A, Printz MP. Epidermal arachidonate lipoxygenase. Biochim Biophys Acta 1983; 751:369–74.
123. Hammarstrom S, Lingren JA, Marcelo C, Duell EA, Anderson RF, Voorhees JJ. Arachidonic acid transformations in normal and psoriatic skin. J Invest Dermatol 1979; 73:180–3.
124. Bresnick E, Bailey G, Bonney RJ, Wightman P. Phospholipase activity in skin after application of phorbol esters and 3-methylcholanthrene. Carcinogenesis 1981; 2:1119–22.
125. Ziboh VA. Arachidonic acid metabolism in the skin. In: Fischer SM, Slaga TJ, eds. Arachidonic acid metabolism and tumor promotion. Boston: Martinus Nijhoff Publishing, 1985:5–20.
126. Furstenberger G, Marks F. Prostaglandin, epidermal hyperplasia and skin tumor promotion. In: Fischer SM, Slaga TJ, eds. Arachidonic acid metabolism and tumor promotion. Boston: Martinus Nijhoff Publishing, 1985:49–72.

127. Hughes MA, Brash AR. Investigation of the mechanism of biosynthesis of 8-hydroxyeicosatetraenoic acid in mouse skin. Biochim Biophys Acta 1991; 1081:347–54.
128. Holtzman MJ, Turk J, Pentland A. A regiospecific monoxygenase with novel stereopreference is the major pathway for arachidonic acid oxygenation in isolated epidermal cells. J Clin Invest 1989; 84:1446–53.
129. Rosenbach T, Grabbe J, Moller A, Schwanitz HJ, Czarnetzki BM. Generation of leukotrienes from normal epidermis and their demonstration in cutaneous disease. Br J Dermatol Suppl 1985; 28:157–67.
130. Burrall BA, Cheung M, Chiu A, Goetzl EJ. Enzymatic properties of the 15-lipoxygenase of human cultured keratinocytes. J Invest Dermatol 1988; 91: 294–7.
131. Ziboh VA, Gasebolt TL, Marcelo CL, Voorhees JJ. Biosynthesis of lipoxygenase products by enzyme preparations from murine keratinocytes. J Invest Dermatol 1984; 83:248–51.
132. Hammarstrom S, Hamberg M, Samuelsson B, Duell EA, Strawiski M, Voorhees JJ. Increased concentration of non-esterified arachidonic acid. 12L-hydroxy-5,8,10,12-eicosatetraenic acid, prostaglandin E_2 and prostaglandin $F_{2\alpha}$ in epidermis of psoriasis. Proc Natl Acad Sci USA 1975; 72:5130–2.
133. Woollard PM, Leigh I. Stereochemical analysis of 12-hydroxy-5,8,10,14-eicosatetraenoic acid (12-HETE) from adult human epidermal cells in vitro. Br J Dermatol 1987; 116:437a.
134. Ruzicka T. Eicosanoids and the skin. Boca Raton, FL: CRC Press, 1990.
135. Ruzicka T. Arachidonic acid metabolism in normal skin. In Ruzicka T, ed. Eicosanoids and the skin. Boca Raton, FL: CRC Press, 1990:23–31.
136. Nakadate T, Aizu E, Yamamoto S, Kato R. Some properties of lipoxygenase activities in cytosol and microsomal fractions of mouse epidermal homogenate. Prost Leuk Med 1986; 21:305–19.
137. Gschwendt M, Furstenberger G, Kittstein W, Besemfelder E, Hull WE, Hagedorn H, Opferkuch HJ, Marks F. Generation of the arachidonic acid metabolite 8-HETE by extracts of mouse skin treated with phorbol ester in vivo; identification by ^{1}H-n.m.r. and GC-MS spectroscopy. Carcinogenesis 1986; 7:449–55.
138. Furstenberger G, Gschwendt M, Hagedorn H, Marks F. Modulation of the conversion stage of multistep carcinogenesis in mouse skin by eicosanoids. In: Garaci E, Paoletti R, Santoro MG, eds. Prostaglandins in cancer research. Heidelberg: Springer-Verlag, 1987:48–61.
139. Fischer SM, Baldwin JK, Jasheway DW, Patrick KE, Cameron GS. Phorbol ester induction of 8-lipoxygenase in inbred SENCAR (SSIN) but not C57BL/6J mice correlated with hyperplasia, edema, and oxidant generation but not ornithine decarboxylase induction. Cancer Res 1988; 48:658–64.
140. Furstenberger G, Hagedorn H, Jacobi T, Besemfelder E, Stephan M, Lehmann W, Marks F. Characterization of an 8-lipoxygenase activity induced by the phorbol ester tumor promoter 12-O-tetradecanoylphorbol-13-acetate in mouse skin in vivo. J Biol Chem 1991; 266:15738–45.

141. Ziboh VA, Hsis SL. Prostaglandin E_2: biosynthesis and effects on glucose and lipid metabolism in rat skin. Arch Biochem Biophys 1971; 146:100–9.
142. Van Dorp D. Recent developments in the biosynthesis and the analysis of prostaglandins. NY Acad Sci 1971; 180:181–99.
143. Greaves MW, McDonald-Gibson W. Effect of non-steroid anti-inflammatory drugs on prostaglandin biosynthesis by skin. Br J Dermatol 1973; 88:47–50.
144. Wilkinson DI, Walsh JT. Prostaglandin biosynthesis in the epidermis and dermis of young mouse skin and the effects on calcium and cyclic nucleotides. J Invest Dermatol 1977; 68:210–14.
145. Ruzicka T, Printz MP. Arachidonic acid metabolism in guinea pig skin. Biochim Biophys Acta 1982; 711:391–7.
146. Jonsson DC, Angaard E. Biosynthesis and metabolism of prostaglandin E_2 in human skin. Scan J Clin Lab Invest 1972; 29:289–96.
147. Ziboh VA. Biosynthesis of prostaglandin E_2 in human skin: Subcellular localization and inhibition of unsaturated fatty acids and anti-inflammatory drugs. J Lipid Res 1973; 14:377–84.
148. Ruzicka T. Eicosanoids in inflammatory skin diseases. In Ruzicka T, ed. Eicosanoids and the skin. Boca Raton, FL: CRC Press, 1990:53–65.
149. Furstenberger G, Marks F. The role of eicosanoids in normal, hyperplastic, and neoplastic growth in mouse skin. In Ruzicka T, ed. Eicosanoids and the skin. Boca Raton, FL: CRC Press, 1990:107–24.
150. Furstenberger G, Gross M, Marks F. On the role of prostaglandins in the induction of epidermal proliferation, hyperplasia and tumor promotion in mouse skin. In: Powles TJ, Bockman RS, Honn KV, Ramwell P, eds. Prostaglandins and cancer: First International Conference Prostaglandins and Related Lipids. Vol. 2. New York: Alan R Liss, 1982:239–54.
151. Bonta JL, Parnham MJ. Prostaglandins and chronic inflammation. Biochem Pharmacol 1978; 27:1611–23.
152. Samuelsson B. Leukotrienes: a new class of mediators of immediate hypersensitivity reactions and inflammation. In: Samuelsson B, Paoletti R, Ramwell PW, eds. Adv Prostaglandin, Thromboxane and Leukotriene Res. Vol. 11. New York: Raven Press, 1983:1–13.
153. Anggard E, Jonsson CE. Formation of prostaglandins in the skin following a burn injury. In: Ramwell PW, Pharris BB, eds. Prostaglandins in cellular biology. Vol. 1. New York: Plenum Press, 1972:269–91.
154. Kragballe K, Fisher GJ, Voorhees JJ. Eicosanoids, phosphoinositides, and the pathophysiology of psoriasis. In Ruzicka T, ed. Eicosanoids and the skin. Boca Raton, FL: CRC Press, 1990:67–78.
155. Voorhees JJ. Leukotrienes and other lipoxygenase products in the pathogenesis and therapy of psoriasis and other dermatoses. Arch Dermatol 1983; 119:541–7.
156. Forstrom L, Goldyne ME, Winklemann RK. Prostaglandin production by human epidermal cells in vitro: a model for studying pharmacological inhibition of prostaglandin synthesis. Prostaglandins 1974; 8:107–13.
157. Fischer SM, Slaga TJ. Modulation of prostaglandin synthesis and tumor promotion. In: Proceedings of the International Congress on Prostaglandins and Cancer. New York: Alan R. Liss, 1982:255–64.

158. Galey CI, Ziboh VA, Marcelo CL, Voorhees JJ. Activation of phospholipase A and C in murine keratinocytes by phorbol ester 12-O-tetradecanoylphorbol-13-acetate. J Invest Dermatol 1983; 80:345.
159. Marks F, Furstenberger G, Kownatzki E. Prostaglandin E-mediated mitogenic stimulation of mouse epidermis in vivo by divalent cation ionophore A 23187 and by tumor promoter 12-O-tetradecanoylphorbol-13-acetate. Cancer Res 1981; 41:696–702.
160. O'Brien TG. The induction of ornithine decarboxylase as an early, possibly obligatory event in mouse skin carcinogenesis. Cancer Res 1976; 36:2644–53.
161. Marks F, Bertsch S, Grimm W, Schweizer J. Hyperplastic transformation and tumor promotion in mouse epidermis: possible consequences of disturbance of endogenous mechanisms controlling proliferation and differentiation. In: Slaga TJ, Sivak A, Boutwell RK, eds. Carcinogenesis. Vol. 2. New York: Raven Press, 1978:97–116.
162. Verma AK, Ashendel CL, Boutwell RK. Inhibition by prostaglandin synthesis inhibitors of the induction of epidermal ornithine decarboxylase activity, the accumulation of prostaglandins and tumor promotion caused by TPA. Cancer Res 1980; 40:308–15.
163. Fischer SM, Gleason GL, Mills GD, Slaga TJ. Indomethacin enhancement of TPA tumor promotion in mice. Cancer Lett 1980; 10:343–50.
164. Fischer SM, Slaga TJ. Modulation of prostaglandin synthesis and tumor promotion. In: Powles TJ, Bockman RS, Honn KV. Ramwell P, eds. Prostaglandins and cancer: First International Conference (prostaglandins and related lipids). Vol 2. New York: Alan R. Liss, 1982:255–64.
165. Fischer SM. Arachidonic acid metabolism and tumor promotion. In: Fischer SM, Slaga TJ, eds. Arachidonic acid metabolism and tumor promotion. Boston: Martinus Nijhoff Publishing, 1985:21–48.
166. Cerutti PA. Active oxygen and promotion. In: Fischer SM, Slaga TJ, eds. Arachidonic acid metabolism and tumor promotion. Boston: Martinus Nijhoff Publishing, 1985:131–68.
167. Goldstein BD, Witz G, Amoruso M, Troll W. Protease inhibitors antagonize the activation of polymorphonuclear leukocyte oxygen consumption. Biochem Biophys Res Commun 1979; 88:854–60.
168. Goldstein BD, Witz G, Amoruso M, Stone DS, Troll W. Stimulation of human polymorphonuclear leukocytes superoxide anion radical production by tumor promoters. Cancer Lett 1981; 11:154.
169. Kensler TW, Trush MA. Inhibition of phorbol ester stimulated chemiluminescence in human polymorphonuclear leukocytes by retinoic acid and 5,6 epoxyretinoic acid. Cancer Res 1981; 41:216–22.
170. DeChatelet LR, Shirley PS, Johnston RB. Effects of phorbol myristate acetate on the oxidative metabolism of human polymorphonuclear leukocytes. Blood 1976; 47:545–54.
171. Keisari Y, Flescher E, Geva I. Macrophage oxidative burst and related cytotoxicity. I. Differential activation by tumor-promoting and non-tumor-promoting phorbol esters. Int J Cancer 1984; 34:845–8.

172. Goldstein B, Witz G, Zimmerman J, Gee C. Free radicals and reactive oxygen species in tumor promotion. In: Greenwald RA, Cohen G, eds. Oxy radicals and their scavenger systems. Vol. 2. Cellular and medical aspects. New York: Elsevier, 1983:321–5.
173. Sirak AA, Beavis AJ, Robertson FM. Enhanced hydroperoxide production by peripheral blood leukocytes following exposure of murine epidermis to 12-O-tetradecanoylphorbol-13-acetate. Carcinogenesis 1991; 12:91–5.
174. Perchellet EM, Abney NL, Perchellet JP. Stimulation of hydroperoxide generation in mouse skins treated with tumor-promoting or carcinogenic agents in vivo and in vitro. Cancer Lett 1988; 42:169–77.
175. Perchellet JP, Boutwell RK. Enhancement by 3-isobutyl-1-methylxanthine and cholera toxin of 12-O-tetradecanoyl-phorbol-13-acetate-stimulated cyclic nucleotide levels and ornithine decarboxylase activity in isolated epidermal cells. Cancer Res 1980; 40:2653–60.
176. Fischer SM, Reiners JJ, Pence BC, Aldaz CM, Conti CJ, Morris RJ, O'Connell JF, Rotstein JB, Slaga TJ. Mechanisms of carcinogenesis using mouse skin: the multistage assay revisited. In: Langenbach R, Elmore E, Barrett JC, eds. Tumor promoters: biological approaches for mechanistic studies and assay systems. New York: Raven Press, 1988:11–30.
177. Zimmerman R, Cerutti P. Active oxygen acts as a promoter of transformation in mouse embryo C3H/10T½/C18 fibroblasts. Proc Natl Acad Sci USA 1984: 81:2085–7.
178. Reiners JJ, Pence BC, Barcus MCS, Cantu AR. 12-O-Tetradecanoylphorbol-13-acetate-dependent induction of xanthine dehydrogenase and conversion to xanthine oxidase in murine epidermis. Cancer Res 1987; 47:1775–9.
179. Trenam CW, Dabbagh AJ, Morris CJ, Blake DR. Skin inflammation induced by reactive oxygen species (ROS): an in vivo model. Br J Dermatol 1991; 125: 325–9.
180. Kensler TM, Trush MA. Role of oxygen radicals in tumor promotion. Environ Mutagen 1984; 6:593–616.
181. Nakamura Y, Colburn NH, Gindhardt TD. Role of reactive oxygen in tumor promotion: implication of superoxide anion in promotion of neoplastic transformation in JB6 cells by TPA. Carcinogenesis 1985; 6:229–35.
182. Crawford DR, Amstad PA, YinFoo DD, Cerutti PA. Constitutive and phorbol-myristate-acetate regulated antioxidant defense of mouse epidermal JB6 cells. Mol Carcinog 1989; 2:136–43.
183. Shamberger RJ. Increase in peroxidation in carcinogenesis. J Natl Cancer Inst 1972; 48:1491–1497.
184. Kennedy AR, Troll W, Little JB. Role of free radicals in the initiation and promotion of radiation transformation in vitro. Carcinogenesis 1984; 5:1213–18.
185. MacCann J, Choi E, Yamasaki E, Ames BN. Detection of carcinogens as mutagens in the Salmonella/microsome test: assay of 300 chemicals. Proc Natl Acad Sci USA 1975; 72:5135–9.
186. Pryor WA. Cancer and free radicals. In: Shankel DM, Hartman P, Kada T, Hollaender A, eds. Antimutagenesis and anticarcinogenesis mechanisms. New York: Plenum Press, 1986:45–59.

187. Fischer SM, Baldwin JK, Jasheway DW, Patrick KE. Possible dissociation of the phorbol ester-induced oxidant response and tumor promotion in the F_1 offspring of SSIN X C57BL/6J mice. Carcinogenesis 1987; 8:1521–4.
188. Swauger JE, Dolan PM, Kensler TW. Role of free radicals in tumor promotion and progression by benzoyl peroxide. In Mendelsohn ML, Albertini RJ, eds. Mutation and the environment, Part D, progress in clinical and biological research. Vol. 340. New York: Wiley-Liss, 1990:143–52.
189. Binder RL, Volpenhein ME, Motz AA. Characterization of the induction of ornithine decarboxylase activity by benzoyl peroxide in SENCAR mouse epidermis. Carcinogenesis 1989; 10:2351–7.
190. Klein-Szanto AJP, Slaga TJ. Effects of peroxides on rodent skin: epidermal hyperplasia and tumor promotion. J Invest Dermatol 1982; 79:30–4.
191. Gindhart TD, Srinivas L, Colburn NH. Benzoyl peroxide promotion of JB6 mouse epidermal cells: inhibition by ganglioside GT but not retinoic acid. Carcinogenesis 1985; 6:309–11.
192. Saladino AJ, Willey JC, Lechner JF, Grafstrom RC, LaVeck M, Harris CC. Effects of formaldehyde, acetaldehyde, benzoyl peroxide, and hydrogen peroxide on cultured normal bronchial epithelial cells. Cancer Res 1985; 45:2522–6.
193. Hartley JA, Gibson NW, Kilkenny A, Yuspa SH. Mouse keratinocytes derived from initiated skin or papillomas are resistant to DNA strand breakage by benzoyl peroxide: a possible mechanism for tumor promotion mediated by benzoyl peroxide. Carcinogenesis 1987; 8:1827–30.
194. Birnboim HC. DNA strand breaks in human leukocytes induced by superoxide anion, hydrogen peroxide and tumor promoters are repaired slowly compared to breaks induced by ionizing radiation. Carcinogenesis 1986; 7:1511–17.
195. Taffe BG, Takahashi N, Kensler TW, Mason RP. Generation of free radicals from organic hydroperoxide tumor promoters in isolated mouse keratinocytes. J Biol Chem 1987; 262:12143–9.
196. Athar M, Mukhtar H, Bickers DR, Khan IU, Kalyanaraman B. Evidence for the metabolism of tumor promoter organic hydroperoxides into free radicals by human carcinoma skin keratinocytes: an ESR-spin trapping study. Carcinogenesis 1989; 8:1499–503.
197. Yamamoto S, Nakadate T, Aizu E, Kato R. Anti-tumor promoting action of phthalic acid mono-*n*-butyl ester cupric salt, a biomimetic superoxide dismutase. Cacinogenesis 1990; 11:749–54.
198. Solanki V, Rana RS, Slaga TJ. Diminution of mouse epidermal superoxide dismutase and catalase activities by tumor promoters. Carcinogenesis 1981; 2:1141–6.
199. Kensler TW, Bush DM, Kozumbo WJ. Inhibition of tumor promotion by a biomimetic superoxide dismutase. Science 1983; 221:75–7.
200. Perchellet EM, Maata EC, Abney NL, Perchellet J-P. Effects of diverse intracellular thiol delivery agents on glutathione peroxidase activity, the ratio of reduced/oxidized glutathione, and ornithine decarboxylase induction in isolated mouse epidermal cells treated with 12-O-tetradecanoylphorbol-13-acetate. J Cell Phys 1987; 131:64–73.

201. Perchellet J-P, Perchellet EM, Orten DK, Schneider BA. Inhibition of the effects of 12-O-tetradecanoylphorbol-13-acetate on mouse epidermal glutathione peroxidase and ornithine decarboxylase activities by glutathione level-raising agents and selenium-containing compounds. Cancer Lett 1985; 26:283–93.
202. Perchellet J-P, Abney NL, Thomas RM, Guislain YL, Perchellet EM. Effects of combined treatments with selenium, glutathione and vitamin E on glutathione peroxidase activity, ornithine decarboxylase induction and complete and multistage carcinogenesis in mouse skin. Cancer Res 1987; 47:477–85.
203. Perchellet J-P, Abney NL, Thomas RM, Perchellet EM, Maatta EA. Inhibition of multistage tumor promotion in mouse skin by diethyldithiocarbamate. Cancer Res 1987; 47:6302–9.
204. Perchellet J-P, Owen MD, Posey TD, Orten DK, Schneider BA. Inhibitory effects of glutathione level-raising agents and D-α-tocopherol on ornithine decarboxylase induction and mouse skin tumor promotion by 12-O-tetradecanoylphorbol-13-acetate. Carcinogenesis 1985; 6:567–;73.
205. Reiners JJ, Jr, Thai G, Rupp T, Cantu AR. Assessment of the antioxidant/prooxidant status of murine skin following topical treatment with 12-O-tetradecanoylphorbol-13-acetate and throughout the ontogeny of skin cancer. Part I: quantitation of superoxide dismutase, catalase, glutathione peroxidase and xanthine oxidase. Carcinogenesis 1991; 12:2337–43.
206. Reiners JJ, Jr, Kodari E, Cappel RE, Gilbert HF. Assessment of the antioxidant/prooxidant status of murine skin following topical treatment with 12-O-tetradecanoylphorbol-13-acetate and throughout the ontogeny of skin cancer. Part II: quantitation of glutathione and glutathione disulfide. Carcinogenesis 1991; 12:2345–52.
207. Cerutti PA, Trump BF. Inflammation and oxidative stress in carcinogenesis. Cancer Cell 1991; 3:1–7.
208. Amstad P, Crawford D, Muehlematter D, Zbinden I, Larsson R, Cerutti P. Oxidant stress induces the proto-oncogenes, C-*fos* and C-*myc* in mouse epidermal cells. Bull Cancer 1990; 77:501–2.

13

The Defensive Role of Antioxidants in Skin Carcinogenesis

Homer S. Black
Baylor College of Medicine and Veterans Affairs Medical Center, Houston, Texas

I. INTRODUCTION: THE TARGET, THE PROBLEM

The skin represents one of the largest organ systems of the human body, constituting approximately one twelfth of total body weight. Situated at the interface between the body and its environment and acting as a barrier to harmful effects of exogenous chemicals and ultraviolet radiation (UVR) (1), the skin also provides the foremost target for an expansive array of environmental insults. Epidemiological evidence indicates that exogenous factors play a predominant role in inducing and promoting cancers in humans (2). Indeed, a great proportion of all cancers are thought to be related to lifestyle and environmental factors (3). Thus, it is not surprising that the incidence of skin cancer cases among Americans nearly equals the annual incidence of all other malignancies combined, with new cases reaching approximately a half million annually (4). Whereas it is known that UVR of the sun is the major cause of skin cancers (5), there is experimental evidence that exogenous chemicals may act in concert to enhance skin cancer incidence (6). Nevertheless, if a significant proportion of cancers can be attributed to lifestyle and environmental factors, then a great proportion of those cancers should be preventable. Before preventive strategies can be developed, however, some understanding of the processes of carcinogenesis, oxidative stress, and antioxidant reactions is necessary. This chapter is not a comprehensive review of these topics, rather, the author has sought to provide an overview of each, woven into a cohesive account that will allow the reader to appreciate the potential role of the antioxidant defense system in cancer prevention. Even here, the emphasis is on cutaneous cancer with a bias, based upon its predominance as a risk factor, toward UVR-carcinogenesis.

II. CARCINOGENESIS

Carcinogenesis represents the sum of a complex sequence of biochemical events that ultimately results in the appearance of a tumor. Because of the complexity of these biochemical events, but based upon somewhat discrete physiological responses, carcinogenesis has been factitiously segmented into stages. Operationally these stages are defined as *initiation* and *promotion*. That skin should be the premier model in which great strides toward elucidation and delineation of the carcinogenic continuum into various stages have occurred is not surprising, in that the skin represents a model with obvious advantage for cancer studies (7). However, it is now recognized that the multistage characterization of carcinogenesis also applies to a variety of other tissues and organs (8).

A. Initiation

Tumor formation can be *initiated* by a single or sequential application of a subthreshold dose of a carcinogenic agent (9). An important characteristic of the initiation process is that it is accomplished rapidly, compared to subsequent steps in tumor formation, and thus represents an irreversible step (10). In this regard, the multistage carcinogenesis model complements the multihit theory for the monoclonal origin of cancer. The latter contends that neoplasia arises when stem cells accumulate nonlethal, unrepaired, irreversible defects that can be transmitted to successive generations of daughter cells (11). Tumor initiation is generally regarded as a permanent alteration of the cell genotype, which expresses no immediate neoplastic phenotype. In other words, an initiating event is insufficient for skin tumorigenesis unless amplification of the altered genotype is triggered by a subsequent event such as the action of a tumor promoter.

The probable target of carcinogenic agents is DNA, which, if replicated before repair can occur, fixes the initiating genetic damage (12). Data have recently been summarized that provide evidence that, for carcinogenic polycyclic hydrocarbons at least, the target for the initiating mutation in mouse skin carcinogenesis may be the cellular Ha-*ras* proto-oncogene (13). This oncogene has been isolated from skin papillomas, and introduction of the mutated oncogene can substitute for 7,12-dimethylbenz[*a*]-anthracene (DMBA) initiation. In addition, keratinocytes derived from DMBA-initiated skin contains an activated Ha-*ras* oncogene and is capable of forming squamous papillomas in skin grafts on athymic nude mice.

The Ha-*ras* proto-oncogene may be activated by a single base substitution, the frequency of which depends upon the initiating agent. Careful evaluation has shown that over 90% of DMBA-induced skin tumors have a point mutation at the 2nd nucleotide in codon 61 of this oncogene (14). Other initiators may produce a point mutation at the 2nd nucleotide in codon 12. Mutations in codon 12 of the Ha-*ras* oncogene have been shown to occur at a high frequency in human skin cancers originating on sun-exposed body sites (15). Recent studies have also identified the p53 gene as a probable site of genetic damage from excessive UVR exposure in skin cancer patients (16,17). Regardless of the site of the genotypic damage caused by initiating events, it appears that such events result in alterations in the program controlling terminal differentiation. Initiated epidermal cells are consistently refractive to CA^{2+}-induced terminal differentiation (18).

B. Promotion

Just as mutational change is the most likely basis for initiating action of a carcinogen, the mode of action of the *promotion* phase of carcinogenesis appears to be manifestly epigenetic. Farber (12) describes promotion as "the process whereby an initiated tissue or organ develops focal proliferations, one or more which may act as precursors for subsequent steps in the carcinogenic process." Thus, an essential feature of promotion is the creation of a mitogenic environment that imposes a differential effect on initiated cells with respect to their surrounding neighbors.

Promoters produce a number of biochemical changes in initiated cells. These include induction of protein kinase C, increases in protease activity, prostaglandin release, induction of ornithine decarboxylase (ODC) activity with a subsequent increase in macromolecular synthesis, a decrease in epidermal differentiation, and production of reactive oxygen species (13,19, 20). Based upon this array of responses and employing a number of strong and weak promoters, as well as inhibitors, within specific experimental carcinogenesis protocols, it has been possible to segment promotion into two additional steps: a partially reversible stage I promotion, or conversion, and an irreversible stage II, propagation (10,20,21). By such systematic dissection of the carcinogenic continuum, and the biochemical events comprising each segment, we not only gain insight into the very complex process itself, but acquire an important key to formulation of anticancer strategies.

C. Some Practical Considerations for "Complete" Carcinogens

To this end, it should be emphasized that an important characteristic of experimental carcinogenesis is the length of time (latent period) that elapses from the first exposure to a carcinogen until a neoplasm appears (19). Tumor induction time is greatly dependent upon the exposure dose of carcinogen or promoter. Thus, while an agent may operationally be considered an initiator, that is, it does not produce tumors without the subsequent application of a chemical promoter, it must be remembered that most carcinogenic agents behave in this manner only when applied at concentrations below a predetermined carcinogenic threshold dose. Indeed, most carcinogenic agents are "complete" carcinogens, requiring no promoter. This is certainly true for the carcinogenic polycyclic hydrocarbon DMBA, which has been used extensively in two-stage carcinogenesis studies. In fact, UVR, which is considered a complete carcinogen, has been employed in similar fashion. A single dose of UVR, while not producing tumors during a defined experimental period, nevertheless has been shown to be carcinogenic when followed with repeated application of a noncarcinogenic chemical promoter (22).

This relationship of dose and tumor latent period can be employed to examine the sequence of biochemical responses to a "complete" carcinogen such as UVR without the complication of chemical action of a promoter. By adjusting the dose, and thereby increasing the latent period, predictions can be made regarding the initiation/promotion action of a suspected cocarcinogen or promoter by evaluation of the resulting tumor incidence curves (latency) (11). Clearly, the same rationale applies to evaluation of agents that have anticancer properties. Studies based upon this principle have successfully been employed to delimit the principal anticarcinogenic effect of butylated hydroxytoluene (BHT) to the period in which the initiating incitant (UVR) was administered (23).

III. REACTIVE OXYGEN SPECIES AND OXIDATIVE STRESS

Skin performs its protective function as a barrier to environmental insult in an oxygenated atmosphere. Thus, it is not only the direct target of a number of damaging free radicals, perhaps generated by photochemical reactions induced by UVR, but other endogenous reactions mounted in response to insult may themselves contribute to the level of reactive species with which the skin must deal.

For the most part, endogenous sources of reactive species tend to be restricted to derivatives of molecular oxygen, polyunsaturated fatty acids, sulfhydryl compounds, and quinones and related compounds that can easily transfer single electrons. The major reactive oxygen species, superoxide anion ($O_2^{\overline{\cdot}}$), hydrogen peroxide (H_2O_2), and hydroxyl radical ($\cdot OH$), are derived from the mono-, di-, and trivalent reduction of molecular oxygen, respectively (24).

A. Superoxide Anion

Superoxide anion is the primary radical formed by the univalent reduction of molecular oxygen (Eq. 1) and may result from a variety of biochemical processes (25,26).

$$^{3}O_2 \xrightarrow[+H^+]{+e^-} O_2^{\cdot -} \qquad (1)$$

Normally, endogenous factors that influence radical production in the cell are those that regulate respiration and includes the availability of NAD-linked substrates, succinate, and oxygen. If oxygen is present at concentrations that limit its reduction to water by cytochrome oxidase (27), if the respiratory chain is highly reduced, or if an accumulation of reduced cofactors occurs, superoxide anion formation will be favored. Superoxide anion and hydrogen peroxide production usually accounts for only about 1–2% of the total oxygen consumption under normal reduced conditions (28). However, under physiological stress circumstances in which the aforementioned conditions occur, it is estimated that about 10% of the oxygen consumed in respiratory reactions may be lost as free radicals.

B. Hydroxyl Radical

$O_2^{\cdot -}$ is reduced to H_2O_2 via spontaneous or enzymatic catalyzed dismutation:

$$2O_2^{\cdot -} + 2H^+ \rightarrow H_2O_2 + O_2 \qquad (2)$$

Normally, the subsequent nonenzymatic reaction yielding the highly damaging hydroxy radical ($\cdot OH$) proceeds very slowly unless catalyzed by a heavy ion such as iron. When the latter is present, the reaction progresses very rapidly, even at temperatures below 25°C. This type of reaction is exemplified by the Haber/Weiss cycle, in which the metal ion is first reduced by $O_2^{\cdot -}$ (Eq. 3) and then oxidized by H_2O_2 (Eq. 4) to yield the $\cdot OH$ radical.

$$Fe^{3+} + O_2^{\cdot -} \rightarrow Fe^{2+} + O_2 \qquad (3)$$

$$Fe^{2+} + H_2O_2 \rightarrow Fe^{3+} + OH^- + \cdot OH \qquad (4)$$

C. Singlet Oxygen

Singlet oxygen ($^{1}O_2$), which is not a free radical, but rather an excited state species created by absorption of energy, which shifts an electron to an orbit of higher energy level, is also highly reactive and may produce biological effects similar to other reactive oxygen species. It has been shown, in well-controlled trapping experiments employing purified myeloperoxidase (MLPD), H_2O_2, and halide ion, that $^{1}O_2$ is produced (29):

$$H_2O_2 + Cl^- \xrightarrow{\text{MLPD}} H_2O + OCl^- \quad (5)$$
$$H_2O_2 + OCl^- \rightarrow {}^1O_2 + HCl + OH^- \quad (6)$$

Whether such a reaction occurs widely in biological systems is unclear, but the hypohalous acid produced in the reaction, in the presence of H_2O_2, has been a preferred method for chemical generation of ${}^1O_2$. In principle, ${}^1O_2$ could be formed from reactants of the Haber/Weiss cycle (Eq. 7) or via spontaneous dismutation of $O_2^{\bar{}}$ (Eq. 8):

$$O_2^{\bar{}} + H_2O_2 \rightarrow {}^1O_2 + \cdot OH + OH^- \quad (7)$$
$$2O_2^{\bar{}} + 2H^+ \rightarrow {}^1O_2 + H_2O_2 \quad (8)$$

Photosensitized reactions leading to the formation of ${}^1O_2$ have been extensively studied (30). The most common process, energy transfer, involves interaction of a triplet sensitizer (${}^3S$) with triplet-state (ground-state) oxygen (${}^3O_2$). The ground-state sensitizer (S) may be excited (${}^1S^*$) to the triplet (${}^3S$) by photons ($h\nu$).

$$S + h\nu \rightarrow {}^1S^* \rightarrow {}^3S \quad (9)$$
$${}^3S + O_2 \rightarrow S + {}^1O_2 \quad (10)$$

The pathway(s) involved in a sensitized photooxidation depends on the sensitizer, the substrate, and the reaction conditions.

D. Lipid Peroxidation

Biological membranes are rich in polyunsaturated fatty acids (PUFA) and are therefore vulnerable targets for free radical attack. Of particular importance are those free radical–mediated reactions that result in lipid peroxidation (31). This type of chain-propagated reaction usually proceeds by free radical (R·) attack upon PUFA to yield lipid radicals:

$$R\cdot + PUFA \rightarrow PUFA\cdot + RH \quad (11)$$

The newly formed PUFA radical (PUFA·) then reacts with molecular oxygen to form fatty acid peroxy radicals:

$$PUFA\cdot + O_2 \rightarrow PUFA\text{-}O\text{-}O\cdot \quad (12)$$

The fatty acid peroxy radical, by hydrogen abstraction from adjacent PUFA, yields unsaturated hydroperoxides:

$$PUFA\text{-}O\text{-}O\cdot + PUFA \rightarrow PUFA\cdot + PUFA\text{-}O\text{-}OH \quad (13)$$

Participation of the newly formed PUFA radical in a propagating chain reaction can lead to extensive disruption of the finely ordered architecture and function of the cellular membranes. Whereas lipid peroxidation may occur at late stages in some disease processes (32), it should be remembered that formation of a number of important vasoactive and immunopotentiators are dependent upon PUFA peroxidation (33,34). Lipid peroxidation is certainly an early event associated with cutaneous UVR insult (35).

E. Oxidative Stress

Although oxygen toxicity has long been recognized (36), the concept of "oxidative stress" has evolved from early benchmark studies in which it was observed that, in the presence of O_2, the lethal dose of x-ray irradiation was only about one-third that required to produce an equivalent effect under anoxic conditions (37). This "oxygen effect" was attributed to production of radicals from O_2 and H_2O, which interacted with DNA (38). As noted previously, prooxidants, such as the reactive oxygen species, may be generated from a wide spectrum of exogenous and endogenous reactions. It is also apparent that skin, filling its unique protective niche, faces a continuous prooxidant challenge. If there is a significant increase in radical generation, or if antioxidant defenses are depleted, the delicate balance between insult and defense is disrupted and a state of *oxidative stress* occurs.

A number of mechanisms have been identified that exhibit potential for creating a prooxidant state (39). These include:

1. Hyperbaric oxygen tension
2. Radiation, as noted above, but also including UV and photosensitizing agents
3. Xenobiotic metabolism and Fenton-type reactions
4. Modulation of the cytochrome electron-transport chain
5. Peroxisome proliferators
6. Inhibitors of antioxidant defenses
7. Membrane-active agents
8. Inflammatory agents
9. Dietary factors

It should be noted that many of these potential mechanisms are interdependent. For example, the metabolism of xenobiotic quinoid molecules interfere in the redox cycle and results in the formation of reactive oxygen species. This prooxidant state can then lead to a decrease in reduced NADPH with a subsequent release of Ca^{2+}. Pertubations of Ca^{2+} homeostasis have a pronounced influence on development and differentiation (40). Some xenobiotics can stimulate the biosynthesis of peroxisomes, organelles containing peroxide-generating oxidases (41).

Membrane-active agents, including some known tumor promoters, render membranes more susceptible to autoxidation and stimulate release of arachidonic acid and its metabolites. In the first instance, the lipid peroxyl radical may be the major oxygen-centered radical generated. It is relatively stable and capable of diffusing to cellular loci distant from its site of formation. In addition, clastogenic factors, capable of causing chromosome breaks, are suspected of being derived from membrane lipid peroxidation, perhaps from autoxidation of arachidonic acid (42). These long-lived agents may be generated by ionizing radiation, oxy-radical generating systems, and by certain tumor promoters. As superoxide dismutase prevents formation of clastogenic factors in vitro, involvement of superoxide anion is suspected.

With respect to arachidonic acid metabolism, it is recognized that peroxides and oxygen-centered free radicals generated from both cyclooxygenase and lipoxygenase pathways modulate and regulate prostaglandin synthesis (43–45). Cadenas and Sies (46) have provided supporting evidence that the oxidant formed in the initial peroxidative steps of prostaglandin synthesis results from a dismutation reaction and is likely to be $^1O_2$.

$$2\,PGG_2 \rightarrow 2PGH_2 + {}^1O_2 \tag{14}$$

PGG_2 is believed to react with the shared heme site of the cyclooxygenase/hydroperoxidase enzymes. An oxo-ferryl complex results, which then reacts with a second molecule of PGG_2.

$$\underset{}{Fe^{3+}} + \underset{(PGG_2)}{\text{PG-OOH}} \xrightarrow[\text{hydroperoxidase}]{\text{PG}} Fe{=}O^{3+} + \underset{(PGH_2)}{\text{PG-OH}} \tag{15}$$

$$Fe{=}O^{3+} + \text{PG-OOH} \rightarrow \text{PG-OH} + Fe^{3+} + {}^1O_2 \tag{16}$$

The oxo-ferryl complex is converted back to ground-state trivalent iron with the resultant formation of PGH_2 and $^1O_2$. Thus, not only does this represent an autocatalytic reaction, but the variety of active metabolites produced constitute a local control mechanism for recruitment and activation of inflammatory cells. When activated, these inflammatory cells undergo a rapid respiratory burst, which results in production of the entire range of reactive oxygen species (34).

One can readily see how radical-initiated processes can lead to secondary and tertiary free radicals, with a consequent cascade of damaging effects. Propitiously, cells possess a number of enzymatic and nonenzymatic defense mechanisms to control these reactions.

IV. ANTIOXIDANT DEFENSES

Cellular free radical defense mechanisms can be broadly classified as enzymatic and nonenzymatic quenching reactions. The enzyme system includes superoxide dismutase (SOD; consisting of four distinct metalloproteins), catalase (CAT), glutathione peroxidase (GP), glutathione reductase (GR), and glucose-6-phosphate dehydrogenase (G6PD) (47).

The endogenous metabolism of $O_2^{\cdot-}$, the primary radical resulting from the reduction of O_2, occurs through a variety of biochemical steps. SOD first dismutes $O_2^{\cdot-}$ to H_2O_2 (48):

$$O_2^{\cdot-} \xrightarrow{\text{SOD}} H_2O_2 + O_2 \tag{17}$$

Hydrogen peroxide may then be reduced via CAT or GP. When reduced by GP, a coupled reaction requiring glutathione (GSH), GR, and NADPH occurs. G6PD restores the redox potential by reacting with NADP to form 6-phosphoglucolactone (6PGL) and NADPH.

Recently a membrane-associated thioredoxin reductase (TR) has been described, which, in the presence of NADPH, acts as a free radical trap at the external surface of the epidermis (49). Although some concerns have been raised regarding methods employed in assessing its activity (50), TR apparently plays a much more important role in epidermal free radical defense than previously recognized. Figure 1 depicts an integrated view of enzymatic free radical defense mechanisms.

There are many low molecular weight agents that serve as free radical quenchers. Compounds capable of inhibiting free radical reactions are commonly defined as antioxidants (see Fig. 1). The mechanisms of action vary from agent to agent. For example, some directly reduce the more reactive free radical species to less reactive states. Tocopherol is effective in termination of free radical–propagated chain reactions and plays a particularly important role in maintaining the integrity of biological membranes against peroxidative damage. Mannitol is an effective hydroxyl radical quencher (24), whereas certain carotenoids are capable of quenching excited triplet states, scavenging singlet oxygen, as well as inhibiting lipid peroxidation (23).

The type of process in which antioxidants act as free radical scavengers is exemplified by the reaction of ascorbic acid (AAH_2) with the stable free radical diphenylpicrylhydrazyl (DPPH·). Ascorbic acid undergoes oxidation to yield monodehydroascorbic acid radical (AAH·), which reacts with a second radical of DPPH to form dehydroascorbic acid (AA). One molecule of ascorbic acid effectively scavenges two DPPH radicals (51).

$$\text{DPPH}\cdot + AAH_2 \rightarrow DPPH_2 + \text{AAH}\cdot \tag{18}$$
$$\text{DPPH}\cdot + \text{AAH}\cdot \rightarrow DPPH_2 + \text{AA} \tag{19}$$

Antioxidant-mediated termination of chain propagation may proceed via reactions similar to those depicted with lipid peroxidation. It is generally believed that the lipid peroxy radical is the most likely site at which an antioxidant intervenes:

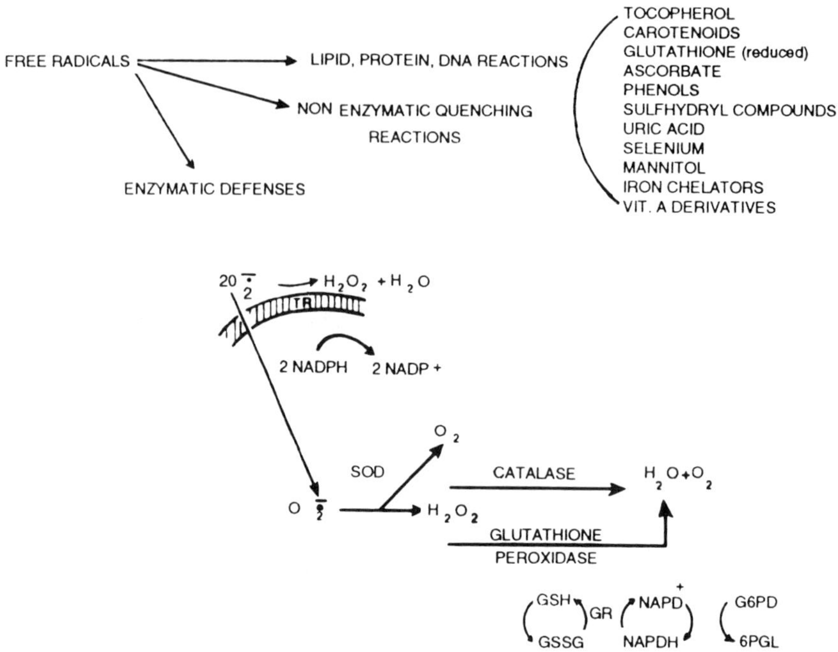

Figure 1 A simplified schema of free radical defense mechanisms. A number of agents are listed that participate in nonenzymatic antioxidant reactions and that exhibit anticarcinogenesis properties. Enzymatic defenses include membrane-bound thioredoxin reductase (TR) and endogenous superoxide dismutase (SOD), both involved in the dismutation of superoxide anion ($O_2^{\overline{\cdot}}$) to hydrogen peroxide (H_2O_2). Hydrogen peroxide may be reduced by either catalase or glutathione peroxidase.

$$\text{PUFA-O-O}\cdot + AH_2 + \text{PUFA} \rightarrow \text{PUFA-O-OH} + AH\cdot + \text{PUFA} \quad (20)$$

The PUFA of this reaction would, in the absence of an antioxidant (AH_2), have reacted with the fatty acid peroxyl radical (PUFA-O-O·) to yield another lipid radical (PUFA·). In the presence of AH_2 the PUFA is spared and the antioxidant reacts with the peroxy radical instead, to yield an antioxidant radical (AH·).

Antioxidant/free radical reactions are not "fail-safe" processes with respect to damage control. Indeed, some antioxidants, under certain conditions, act as prooxidants. Ascorbic and uric acids may actually promote autoxidations by reducing oxygen activators such as transition metals or quinones, both of which can lead to free radical formation. For example, ascorbic acid, in the presence of oxygen and a metal complex, can undergo a redox-induced homolysis leading to the production of perhydroxyl radical or superoxide anion (52). In addition to potential prooxidant action, termination of chain propagation, as depicted above, is contingent upon non-participation of the antioxidant radical (AH·) in further peroxidative reactions.

Ideally, the antioxidant radical would be destroyed by harmless side reactions or restored to an active antioxidant state. Evidence has recently been obtained that establishes a link between enzymatic and antioxidant quenching reactions that achieve this end. McCay et al. (53) report that tocopherol (T) (vitamin E) functions as an electron shuttle in the membrane by quenching reactive radicals, resulting in the formation of the relatively stable tocopheroxy radical (T·), which is then rapidly quenched by a membrane-bound, heat-labile protein ("free radical reductase," or FRR) utilizing GSH as a source of electrons. Such a reaction might be depicted as:

$$T + R\cdot \longrightarrow T\cdot + RH \qquad (21)$$

$$T\cdot \xrightarrow[GSH]{FRR} T \qquad (22)$$

NADPH

GSSG

$NADP^+$

As the authors note, a recycling function such as depicted would be a very efficient mechanism for protecting membranes against oxidative stress, limited only by the reaction rate of FRR and the enzymatic capacity of the cytosol to maintain GSH in the reduced state.

In summary, the major intracellular defense mechanisms are:

1. Enzymatic removal of O_2^- and H_2O_2 by the detoxifying enzymes SOD, CAT, GP, GR, and G6PD
2. Experimentally decreasing the availability of metal catalysts that participate in Fenton reactions yielding ·OH
3. Lipid- and water-soluble antioxidants such as vitamin E, ascorbic acid, and GSH

V. OXIDATIVE STRESS AND CARCINOGENESIS

A. Initiation

As previously noted, DNA is the probable target for carcinogenic agents, and a substantial body of evidence exists that implicates oxidative stress as a significant contributor to the various stages of carcinogenesis (54). Reactive oxygen species are directly mutagenic, an effect mediated via the formation of oxidized DNA bases (55). Thus, Pryor (56) has suggested several mechanisms by which free radicals may be involved in chemically initiated carcinogenesis.

Whereas many polycyclic aromatic hydrocarbons (PAHs) are "activated" by the cytochrome P-450 system, entailing even-electron processes and not free radicals, some PAHs may be oxidized by radicals produced from the activity of prostaglandin synthetase. The resulting epoxide intermediate, in the case of benzo(*a*)pyrene, binds to DNA via nucleophilic attack on the epoxide ring. In this case, only the activation step of the procarcinogen involves radical reactions.

Molecular binding to DNA usually involves positively charged ionic or uncharged electrophilic species that attack and bind by two-electron mechanisms. Generally, the reactive oxygen species discussed above are electrophilic, but not sufficiently so to bind to the double bonds of DNA in competition with abstracting hydrogen atoms from the constituent deoxyribose. However, ·OH is the most electrophilic radical of the oxy-radicals to which DNA is exposed and is able to penetrate the shield of easily abstracted hydrogen atoms in the deoxyribose to reach the DNA bases. Further, ·OH is able to react via both hydrogen atom addition and abstraction, producing thymine glycol and strand nicks, respectively. Although it might be argued that ·OH is so reactive it would be destroyed before reaching DNA targets, a major part of toxicity of H_2O_2 to *E. coli* has been attributed to DNA damage mediated by a Fenton reaction that generates ·OH from H_2O_2, DNA-bound iron, and a constant source of reducing equivalents, e.g., NADH (57). As a result of ·OH-induced strand breakage, poly (ADP-ribose) synthetase is activated, which, in turn, depletes nicotinamide nucleotides. With the simultaneous demand for NADPH, GSH is oxidized by the action of GP on H_2O_2. The resulting reduced concentrations of NADH, NADPH, GSH, and ATP not only stress the antioxidant defense but, combined with effects of reactive species on CA^{2+}-sequestering mechanisms, may lead to rises in cytosolic transition metals that then result in a cascade of radical reactions. Ames (55) has measured the endogenous background level of oxidant-induced DNA damage in vivo and estimates that the total number of all types of oxidative hits on DNA per cell per day in humans may exceed 10^4. Endogenous levels of oxidative damage in rat liver nuclear DNA were determined as one per 130,000 bases. It is obvious that oxidative stress, as mediated through any of the aforementioned mechanisms, could result in the type of genotypic alterations associated with initiation of carcinogenesis. In at least one case, direct evidence has been obtained that demonstrates that a reactive oxygen species (H_2O_2) initiates carcinogenesis (58).

B. Promotion

Tumor promotion involves the modulation of gene expression related to growth and differentiation of initiated cells (39). Perhaps the most convincing evidence for participation of oxidative stress in carcinogenesis comes from studies of chemical tumor promoters and are based on the following points (47,54,59).

1. *Tumor promoters are capable of generating free radicals. Reactive oxygen-generating systems mimic some of the biochemical actions of tumor promoters.* Most tumor promoters induce an inflammatory response where polymorphonuclear cells (PMNs) infiltrate the dermis. Repine et al. (60) demonstrated that the tumor promoter phorbol-12-myristate-13-acetate (PMA) elicited an oxidative burst from PMNs with the subsequent production of reactive oxygen species. Another promoter, 12-O-tetradecanoylphorbol-13-acetate (TPA), is capable of recruiting PMNs to the skin. Oxidants generated by TPA-recruited and stimulated PMNs can also activate penultimate PAHs to potentially genotoxic metabolites in vivo (61). Whereas these studies further define a role for inflammation in carcinogenesis, the direct formation of oxy-radicals has also been demonstrated to occur with TPA-treated epidermal cells (62,63).

2. *Tumor promoters evoke a rapid and sustained decrease in antioxidant defenses.* Decreased epidermal SOD activity occurs within 3 hours after TPA treatment, with maximal effect seen at 17 hours post-treatment (64). Likewise, a similar effect upon CAT activity occurs. GP activity increases over control values for several days after TPA treatment (65). GR activity declines with a concomitant increase in oxidized glutathione levels (66).

3. *Free radical–generating compounds, some commonly applied to the skin in medications (59), exhibit tumor promoter activity.* Benzoyl peroxide, decanoyl peroxide, and cumene hydroperoxide are all effective in the promotion stage of skin carcinogenesis. The latter, when incubated with primary mouse keratinocytes, is capable of producing alkyl radicals, which may be involved in hydrogen abstraction or addition reactions (67). Free radical–generating systems such as xanthine/xanthine oxidase or neutrophils are capable of transforming and promoting cultured mouse fibroblasts, as does TPA (68,69).

4. *Tumor promotion is inhibited by antioxidant enzyme–mimicking agents and natural and synthetic antioxidants.* A number of phenolic antioxidants have been found to inhibit carcinogenesis induced by a substantial variety of chemical carcinogens (70). These include butylated hydroxyanisole (BHA), butylated hydroxytoluene (BHT), and ethoxyquin. Selenium salts and retinoids are also effective inhibitors (70,71) of carcinogenesis, as is the SOD biomimetic, CuDIPS [copper(II)-(3,5 diisopropyl-salicylate)] (72). Antioxidants and antioxidant enzymes inhibit a wide range of transformation and promotion events in in vitro systems (47).

VI. PHOTOCARCINOGENESIS AND ANTIOXIDANTS

Although free radical–mediated reactions have the potential for playing a causal role in a number of UVR-induced pathophysiological conditions, including carcinogenesis, just as with chemically induced carcinogenesis such a relationship is founded upon indirect evidence (73). As with chemical carcinogenesis, implication of free radical involvement in the etiology of photocarcinogenesis remains speculative and is based principally upon: 1) the ability of UVR to generate free radicals, 2) the capacity of UVR-generated free radicals to damage DNA, 3) the ability of UVR to diminish antioxidant defenses, and 4) the ability of antioxidants to modulate photocarcinogenesis.

Unlike x-irradiation, studies employing short UV wavelengths, primarily 254 nm, led to the general conclusion that biological responses induced by UVR were independent of whether the irradiation was administered in the presence or absence of O_2 (74). This represented a major and important divergence from the oxygen effect upon x-irradiation in which oxy-radicals were known to react with DNA. However, UVR was shown to produce free radicals in skin, although the identity (oxygen or carbon centered) of these radicals were not determined (75,76). Recent studies have demonstrated that long-wave UVR (UVA; >320 nm) causes various types of DNA damage, e.g., cyclobutane-type pyrimidine dimers, strand breaks, and DNA crosslinks (77). The latter two types of damage increase at longer wavelengths. Dimer formation at 365 nm is unaffected by oxygen tension, whereas strand breaks in DNA from *E. coli* are completely prevented under anaerobic conditions at this wavelength. These data would suggest radical species specificity with respect to the type of DNA damage induced by UVR. Indeed Peak et al. (78,79) have shown that strand breakage in human cell DNA is enhanced when irradiated at 365 and 405 nm, but not at 290 nm. Further, strand breakage increased in the presence of deuterium oxide, which favors the transient life of $^1O_2$ and implies that this reactive species is involved. Peak et al. (80) suggested that near UVR in vivo effects most likely involved a type II photosensitization process. Using a mutant bacterial strain requiring a porphyrin precursor and in which porphyrin levels could be controlled, these workers were able to show that near UVR lethality was oxygen dependent and that cell death was probably not due to DNA breaks but involved damage to cell membranes. Carraro and Pathak (81) have extensively studied the nature of in vitro and in vivo photosensitization reactions by porphyrins and psoralens. Porphyrin photosensitization is primarily an oxygen-dependent type II photodynamic reaction targeted to lipid-rich membranes and in which $^1O_2$ is a major intermediate. Psoralens evoke both type I (anoxic) and type II reactions, with the former targeted to DNA.

The myriad of biochemical events resulting from UVR obviously complicate efforts to unravel the etiological steps in photocarcinogenesis. Whereas considerable evidence points to participation of photosensitization reactions in UVR insult, Peak and Peak (80) have demonstrated the similarity of 365 nm UVA and gamma radiation effects on DNA strand breakage. Both irradiations induce strand breaks via a common step that apparently involves ·OH. There seems to be some support for ·OH involvement, as Roza et al. (82) have shown that induction of DNA single-strand breaks in human fibroblasts irradiated at > 330 nm is almost completely prevented when irradiation occurs in the presence of catalase. The complexity, and perhaps specificity, of radical-induced damage upon DNA is exemplified by the fact that catalase did not reduce cell death resulting from >330 nm irradiation. Nevertheless, it is becoming apparent that UVR can induce free radical formation in skin and that these radicals represent an oxidative stress that, if not the major source of genotypic damage associated with photocarcinogenesis, is at least a significant contributor thereto.

On first inspection, the conclusion that biological responses induced by short-wavelength UVR (<320 nm) are independent of oxygen tension might appear to bode ill for the theory that reactive oxygen species are involved in the etiology of photocarcinogenesis. After all, UVB (280–320 nm) is the most efficient waveband for the induction of carcinogenesis. However, aside from experimental conditions (for the most part experimental photocarcinogenesis studies have not employed exclusive UVB radiation sources), one is never exposed solely to UVB—a complementary dose of UVA always accompanies exposure. After all, the separatioin of UVR (UVC notwithstanding) into bands of specific wavelengths for the study of carcinogenesis is as factitious as the division of carcinogenesis into initiation and promotion stages, but admittedly as useful. In fact, it is likely that oxygen-independent UVB effects are directly targeted to genomic DNA, resulting in the unrepaired, nonlethal, permanent genotypic alterations characterizing the initiation stage of chemical carcinogenesis. The tumor-promoting properties of UVA (83), and the oxidative processes so induced, are compatible with the action of chemical promoters. Thus, in concert, UVR is recognized as a "complete" carcinogen.

Although UVB may not mediate its effects directly through oxyradical mechanisms, it is very efficient in producing other free radicals, particularly lipid radicals. Not only do these reactions represent a serious threat to the integrity of the cell, but, perhaps as importantly, they diminish many of the skin's natural antioxidant defenses. The latter effect, in concert with oxyradicals produced by companion UVA irradiation, provides a potential for significant damage via a two-pronged attack.

Because it represents an important avenue of oxidative stress, the effect of UVR on antioxidant defenses have recently received much attention (84). A number of studies have addressed the question of what effect a single, acute UVR insult, both UVB and UVA, exerts upon the antioxidant defense system (85–89). Miyachi et al. (90) have shown that a single exposure to UVB (2.1 mW/cm^2; 305 nm) significantly decreases cutaneous SOD activity, although they acknowledge that chronic exposure elevates skin SOD levels. Liposomal SOD injected i.p. 2 hours prior to UVR caused a significant sparing of cutaneous SOD activity. Fuchs et al. (86) found that upon a single large fluence of UVB (300 J/cm^2; >280 nm), an immediate and significant inhibition of GR and CAT activity occurred. They observed no effects upon GP or SOD activity. Glutathione levels decreased, while glutathione disulfide concentrations increased. Ascorbic acid levels were unchanged but tocopherol and ubiquinol 9 and ubiquinone 9 levels decreased significantly. These studies demonstrate that UVB can alter antioxidant defenses. Physiological effects of such alteration can be observed where, at low to moderate UVB doses, endogenous glutathione provides the epidermis with measurable protection against formation of sunburn cells (91). Miyachi et al. (92) have demonstrated that sunburn cell formation can be inhibited by scavenging oxygen intermediates with SOD, CAT, and mannitol. Generally, the fluence levels employed in single-dose, acute studies have been much greater than those required to induce carcinogenesis in multiple-dose, chronic studies. On the other hand, Bissett et al. (93) have demonstrated a photoprotective effect of $O_2^{\cdot -}$-scavenging agents against UVR-induced chronic skin damage with respect to aging. Nevertheless, a careful assessment of antioxidant defense levels within a chronic carcinogenic protocol is needed. Perhaps the most convincing data for involvement of reactive species in carcinogenesis comes from studies in which antioxidants have been shown to modulate photocarcinogenic expression.

Among the more notable compounds that have demonstrable inhibitory effects upon photocarcinogenesis are certain carotenoids and phenolic antioxidants. The photoprotective function of carotenoid pigments against lethal porphyrin-related photosensitization in mice was shown to be oxygen dependent (94). β-Carotene has been shown to be a very efficient quencher of singlet oxygen (95). Mathews-Roth (96), using a singlet oxygen probe, has demonstrated that β-carotene can quench singlet oxygen–driven photochemical reactions that occur in mouse epidermis. This carotenoid is also capable of quenching other free radical–mediated reactions, including lipid peroxidation (97).

Epstein (98) first demonstrated that β-carotene provided significant protection against photocarcinogenesis in hairless mice. Photoprotection was manifested in both delayed tumor appearance and reduced tumor growth rates. These effects were confirmed in dietary studies and extended to include other carotenoid pigments displaying no vitamin A activity (99,100). Although some carotenoid pigments absorb strongly in the UVB region (101), β-carotene does not. In fact, β-carotene does not function in the skin as an optical filter, and thus its protective action is probably related to its ability to quench specific reactive species and their reactions (102).

Other nonenzymatic antioxidants have been shown to provide protection against a range of UVR insults, including carcinogenesis. Pursuant to a study of the potential involvement of the cholesterol photooxidation product and putative carcinogen, cholesterol epoxide (cholesterol-5α, 6α-epoxide), it was found that formation of this compound was UV-dose dependent (103) and levels increased eightfold in skin of chronically irradiated animals, peaking prior to the appearance of squamous cell carcinomas (104). As cholesterol epoxide is a sensitive manifestation of lipid peroxidation, its identification in UVR-treated animals indicate that peroxidative processes are concomitant, if not causal, events in photocarcinogenesis. Ogura and colleagues (105) have examined UVR-induced lipid peroxidation extensively and recently reported that UVR-induced lipid peroxidation is mediated by radical reactions involving reactive oxygen species. Thus, it seemed reasonable that these reactions might be inhibited by antioxidants. Using a broad-spectrum dietary supplement of water- and lipid-soluble antioxidants (including ascorbic acid, reduced glutathione, tocopherol, and butylated hydroxytoluene [BHT], an inverse relationship between cutaneous antioxidant content and cholesterol epoxide formation was observed (106). These findings suggested possible prophylactic effects of systemic antioxidants not only on the formation of this putative carcinogen, but also on the subsequent pathological conditions that might result from, or concurrently with, its formation. When animals were fed the antioxidant mixture in a UVR-carcinogenesis protocol, significantly fewer actinic lesions and tumors occurred than in non–antioxidant-supplemented controls (107). Antioxidants markedly reduced the severity of the tumors that did appear. Pauling et al. (108) corroborated the observations concerning the inhibitory nature of the antioxidant mixture on photocarcinogenesis. Subsequent studies indicated that BHT was the most active principal of the antioxidant mixture (109–111), although a second study from Pauling's group (112) found that ascorbic acid alone could significantly inhibit photocarcinogenesis.

Studies on the mechanism of BHT photoprotection delimited its major site of action to the stratum corneum (113). Spectral transmission between 280 and 320 nm was approximately 65% greater through nonsupplemented control stratum corneum than that from BHT-supplemented animals. The greater UVR absorption of the BHT-supplemented stratum corneum could not be explained by increased thickness or altered physical dimensions of the tissue. It had previously been shown that BHT accrues in the epidermis at levels that would absorb no more than 1% of incident UVR (114). BHT absorbs maximally at 278 nm. It was proposed that BHT, by virtue of its antioxidant properties, might protect keratin (the major constituent of stratum corneum) sulfhydryl moieties from normal maturation or photooxidation, resulting in fewer disulfide cross-links and thereby altering the optical properties of the stratum corneum. Although this proposed mechanism remains speculative, it does point to a heretofore little-considered mechanism for reduction of photoinduced oxidative stress, i.e., UVR-dose diminution resulting from the action of an antioxidant upon normal tissue maturation/differentiation. Oxidants are known to influence the course of development of a number of tissues (40). Investigators should be aware of this possibility and not dismiss dose diminution simply because the potential photoprotective agent does not absorb UVR in the range being administered, as has often been the case.

Lipid peroxidation exemplifies the type of chain reaction initiated by free radicals, with unsaturated fatty acids being the primary center of free radical attack (31,105). Antioxidants act as free radical scavengers and are able to terminate free radical-initiated chain reactions (see Eq. 20). Thus, as a corollary to antioxidant inhibition of photocarcinogenesis, anything that reduces the host's free radical load would be expected to reduce carcinogenic expression. In fact, this is one of the intervention strategies first proposed by Harman (115,116). Black et al. (117) have verified the existence of such a relationship in photocarcinogenesis. When feeding various levels of polyunsaturated lipid, an approximately linear relationship between lipid level and tumor latent period was observed, with the lowest lipid level exhibiting a significantly longer tumor latent period. In addition, with increased lipid intake the number of tumors per animal increased. Antioxidants added to the diet caused significant increases in tumor latency and decreased tumor multiplicity at the two highest lipid levels tested. Mathews-Roth and Krinsky (118) have also reported that as dietary lipid is increased, the ability of β-carotene to protect against photocarcinogenesis declines. There seems little doubt that by increasing the free radical load, photocarcinogenesis is enhanced.

These studies certainly point to the involvement of lipid peroxidatic reactions in that part of photocarcinogenic expression enhanced by unsaturated lipid, but recent studies with menhaden oil, a polyunsaturated lipid rich in eicosapentaenoic acid, an omega-3 fatty acid, indicate that a much more complex relationship exists between lipids and photocarcinogenesis. This polyunsaturated lipid inhibits photocarcinogenesis and other UV-induced inflammatory responses (119). Eicosapentaneoic acid is known to act as a competitive substrate analog to the cyclooxygenase pathway and could block the formation of a number of vasoactive agents and immunomodulators that are important participants in the inflammatory process. At the same time, the synthesis of various prostaglandins, thromboxanes, and leukotrienes via the cyclooxygenase pathway is initiated by peroxidatic reactions. Indeed, menhaden oil has been shown to markedly lower plasma levels of the pro-inflammatory prostaglandin E_2 (120), the latter known to suppress certain immune responses. Menhaden oil has been shown to significantly influence inflammatory and delayed-type hypersensitive immune responses in UV-irradiated animals (121). It is not without justification to think that free radical–mediated reactions exert some regulatory control over immune functions.

In summary, there exist four major lines of evidence, albeit indirect, that support the thesis that oxidative stress, as manifested by reactive oxygen species and their reactions, is involved in the etiology of photocarcinogenesis. These include:

1. Free radicals are formed in UV-irradiated skin.
2. Conditions that exemplify oxidative stress diminish natural antioxidant defenses.
3. Supplementation with antioxidants inhibit photocarcinogenesis.
4. Conditions that increase the free radical load of the host also enhance photocarcinogenesis.

VII. CONCLUSION

In 1962, Urbach (122) remarked that he suspected that a casual comment by Professor Otto Warburg would be validated, i.e., "Pasteur's 'Life without Air' is the key to the problem of cancer." Warburg (123) recognized the importance of the aberrant nature of oxidative metabolism exhibited by cancer cells, a metabolism that favored anaerobic glycolysis and escaped normal regulatory control. Whether this characteristic of cancer cells is cause or effect remains unknown, but we now know that conditions controlling the rate of cellular respiratory reactions can lead to elevated levels of oxy-radicals and, therefore, oxidative stress. In view of the wealth of knowledge gleaned regarding oxidative stress and its potential role in cancer since the time of Warburg's statement, it would not be inappropriate to allow this author license to reinterpret "life without air" as antioxidant. Indeed, antioxidants may hold a key to the cancer problem!

ACKNOWLEDGMENTS

The author's work has been supported by grants from the National Cancer Institute, Veterans Affairs, and currently from the American Institute for Cancer Research.

REFERENCES

1. Thody AJ, Friedmann PS. Functions of the skin. In: Thody AJ, Friedmann PS, eds. Scientific basis of dermatology. Edinburgh: Churchill Livingstone, 1986: 1–5.

2. Higginson J, Muir CS. The role of epidemiology in elucidating the importance of environmental factors in human cancer. Cancer Detect Prev 1976; 1:79–105.
3. Weisburger JH, Cohen LA, Wynder EL. On the etiology and metabolic epidemiology of the main human cancers. In: Hiatt HH, Watson JD, Winsten JA, eds. Origins of human cancer. New York: Cold Spring Harbor Laboratory, 1977:567–602.
4. Scotto J, Fears T, Lisiecki E, Radosevich S. Incidence of nonmelanoma skin cancer in the United States, 1977–1978, preliminary report. Bethesda, MD: National Cancer Institute, 1980.
5. Mason TJ, McKay FW, Hoover R, Blot WJ, Fraumeni Jr JF. Atlas of cancer mortality for U.S. counties: 1950–1969. Washington, DC: DHEW Publ. No. (NIH) 75-780, 1975.
6. Fry RJM, Ley RD. Ultraviolet radiation carcinogenesis. In: Slaga TJ, ed. Mechanisms of tumor promotion II. Boca Raton, FL: CRC Press, 1984:73–96.
7. Berenblum I. Theoretical and practical aspects of the two-stage mechanism of carcinogenesis. In: Griffin AC, Shaw CR, eds. Carcinogens: identification and mechanisms of action. New York: Raven Press, 1979:25–36.
8. Slaga TJ, ed. Mechanisms of tumor promotion I. Tumor promotion in internal organs. Boca Raton, FL: CRC Press, 1983.
9. Slaga TJ. Mechanisms involved in two-stage carcinogenesis in mouse skin. In: Slaga TJ, ed. Mechanisms of tumor promotion II. Tumor promotion and skin carcinogenesis. Boca Raton, FL: CRC Press, 1984:1–16.
10. Boutwell RK. Some biological aspects of skin carcinogenesis. Progr Exp Tumor Res 1964; 4:207–50.
11. Moolgavkar SH, Knudson Jr AG. Mutation and cancer: a model for human carcinogenesis. J Natl Cancer Inst 1981; 66:1037–52.
12. Farber E. Chemical carcinogenesis: a biologic perspective. Am J Pathol 1982; 106:269–96.
13. Perchellet JP, Perchellet EM. Antioxidants and multistage carcinogenesis in mouse skin. Free Radic Biol Med 1989; 7:377–408.
14. Bailleul B, Brown K, Ramsden M, Fee F, Balmain A. Initiation of skin carcinogenesis can occur by induction of carcinogen—specific point mutations in the Harvey-*ras* gene. In: Slaga TJ, Klein-Szanto AJP, Boutwell RK, Stevenson DE, Spitzer HL, D'Motto B, eds. Skin carcinogenesis. Mechanisms and human relevance. New York: Alan R. Liss, Inc., 1989:137–45.
15. Pierceall WE, Goldberg LH, Tainsky MA, Mukhopadhyay T, Ananthaswamy HN. Ras gene mutation and amplification in human nonmelanoma skin cancers. Mol Carcinog 1991; 4:196–202.
16. Brash DE, Rudolph JA, Simon JA, Lin A, McKenna GJ, Baden HP, Halperin AJ, Ponten J. A role for sunlight in skin cancer: UV-induced p53 mutations in squamous cell carcinoma. Proc Natl Acad Sci 1991; 88:10124–8.
17. Pierceall WE, Mukhopadhyay Y, Goldberg LH, Ananthaswamy HN. Mutations in the p53 tumor suppressor gene in human cutaneous squamous cell carcinomas. Mol Carcinog 1991; 4:445–9.
18. Yuspa SH, Morgan DL. Mouse skin cells resistant to terminal differentiation associated with initiation of carcinogenesis. Nature 1981; 293:72–4.

19. Boutwell RK. Model systems for defining initiation, promotion, and progression of skin neoplasms. In: Slaga TJ, Klein-Szanto AJP, Boutwell RK, Stevenson DE, Spitzer HL, D'Motto B, eds. Skin carcinogenesis. Mechanisms and human relevance. New York: Alan R. Liss, Inc., 1989:3–15.
20. Slaga TJ, Fischer SM, Nelson K, Gleason GL. Studies on the mechanism of skin tumor promotion: evidence for several stages in promotion. Proc Natl Acad Sci USA 1980; 77:3659–63.
21. Slaga TJ, Fischer SM, Weeks CE, Nelson K, Mamrack M, Klein-Szanto AJP. Specificity and mechanism(s) of promoter inhibitors in multistage promotion. In: Hecker E, Kunz W, Fusenig NE, Marks F, Thielmann HW, eds. Carcinogenesis 7, cocarcinogenesis and biological effects of tumor promoters. New York: Raven Press, 1982:19–34.
22. Epstein JH. Ultraviolet carcinogenesis. In: Giese AC, ed. Photophysiology. New York: Academic Press, 1970:235–73.
23. Black HS, Mathews-Roth MM. Protective role of butylated hydroxytoluene and certain carotenoids in photocarcinogenesis. Photochem Photobiol 1991; 53:707–16.
24. Proctor PH, Reynolds ES. Free radicals and disease in man. Physiol Chem Phys Med NMR 1984; 16:175–95.
25. Demopoulos HB, Pietronigro DD, Flamm ES, Seligman ML. The possible role of free radical reactions in carcinogenesis. J Environ Path Toxicol 1980; 3:273–303.
26. Freeman BA, Crapo JD. Biology of disease: free radicals and tissue injury. Lab Invest 1982; 47:412–26.
27. Jöbsis FF, LaManna JC. Kinetic aspects of intracellular redox reactions: *in vivo* effects during and after hypoxia and ischemia. In: Robin ED, ed. Extrapulmonary manifestations of respiratory disease. New York: Marcel Dekker, 1978:63–106.
28. Turrens JF, Freeman BA, Crapo JD. Hyperoxia increases H_2O_2 release by lung mitochondria and microsomes. Arch Biochem Biophys 1982; 217:411–21.
29. Rosen H, Klebanoff S. Formation of singlet oxygen by the myeloperoxidase-mediated antimicrobial system. J Biol Chem 1977; 252:4803–10.
30. Spikes JD. Photosensitization. In: Smith KC, ed. The science of photobiology. New York: Plenum Press, 1977:87–112.
31. Girotti A. Mechanisms of lipid peroxidation. J Free Radic Biol Med 1985; 1:87–95.
32. Halliwell B. Oxidants and human disease: some new concepts. FASEB J 1987; 1:358–64.
33. Hemler M, Cook H, Lands W. Prostaglandin biosynthesis can be triggered by lipid peroxides. Arch Biochem Biophys 1979; 193:340–5.
34. Black HS. Role of reactive oxygen species in inflammatory process. In: Lowe NJ, Hensby CN, eds. Nonsteroidal anti-inflammatory drugs. Basel: S. Karger, 1989:1–20.
35. Meffert H, Diezel W, Sonnichsen N. Stable lipid peroxidation products in human skin: detection, ultraviolet light-induced increase, pathogenic importance. Experientia 1976; 32:1397–8.

36. Jamieson D. Oxygen toxicity and reactive oxygen metabolites in mammals. Free Radic Biol Med 1989; 7:87–108.
37. Alper T, Howard-Flanders P. Role of oxygen in modifying the radiosensitivity of *E. coli B*. Nature 1956; 178:978–9.
38. Gerschman R, Gilbert DL, Mye SW, Dwyer P, Fenn WO. Oxygen poisoning and X-irradiation: A mechanism in common. Science 1954; 119:623–6.
39. Cerutti PA. Prooxidant states and tumor promotion. Science 1985; 227:375–81.
40. Allen RG, Balin AK. Oxidative influence on development and differentiation: an overview of a free radical theory of development. Free Radic Biol Med 1989; 6:631–61.
41. Reddy JK, Rao MS. Peroxisome proliferation-related oxidative stress and hepatocarcinogenesis. In: Cerutti PA, Nygaard OF, Simic MG, eds. Anticarcinogenesis and radiation protection. New York: Plenum Press, 1987:85–91.
42. Emerit I. Clastogenic factors, a link between chronic inflammation and carcinogenesis. In: Cerutti PA, Nygaard OF, Simic MG, ed. Anticarcinogenesis and radiation protection. New York: Plenum Press, 1987:59–62.
43. Mason RP, Kalyanaraman B, Tainer BE, Eling TE. A carbon-centered free radical intermediate in the prostaglandin synthetase oxidation of arachidonic acid: spin trapping and oxygen uptake studies. J Biol Chem 1980; 255:5019–22.
44. Lands W, Brynes M. The influence of ambient peroxides on the conversion of 5,8,11,14,17-eicosapentaenoic acid to prostaglandins. Prog Lipid Res 1981; 20: 287–90.
45. Kuehl FA, Dougherty HW, Ham EA, Egan RW, Miller DK. Effects of oxygen radicals on eicosanoid production. In: Russo-Marie F, Mencia-Huerta JM, Chignard M, eds. Advances in inflammation research 10. New York: Raven Press, 1986:179–82.
46. Cadenas E, Sies H. Singlet oxygen formation detected by low-level chemiluminescence during enzymatic reduction of prostaglandin G_2 to H_2. Hoppe-Seyler's Z Physiol Chem 1983; 364:519–28.
47. Sun Yi. Free radicals, antioxidant enzymes, and carcinogenesis. Free Radic Biol Med 1990; 8:583–99.
48. McCord JM, Fridovich I. Superoxide dismutase. An enzymatic function for erythrocuprein (hemocuprein). J Biol Chem 1969; 244:6049–55.
49. Schallreuter KU, Wood JM. Free radical reduction in the human epidermis. Free Radic Biol Med 1989; 6:519–32.
50. Fuchs J, Mehlhorn RJ, Packer L. Free radical reduction mechanisms in mouse epidermis skin homogenates. J Invest Dermatol 1989; 93:633–40.
51. Slater TF. Free radical mechanisms in tissue injury. London: Pion Ltd., 1972.
52. Udenfriend S, Clark CT, Axelrod J, Brodie BB. Ascorbic acid in aromatic hydroxylation. I. A model system for aromatic hydroxylation. J Biol Chem 1954; 208:731–9.
53. McCay PB, Lai EK, Brueggemann G, Powell SR. A biological antioxidant function for vitamin E: electron shuttling for a membrane-bound "free radical reductase." In: Galli C, Fedeli E, eds. Fat production and consumption. Technologies and nutritional implications. New York: Plenum Press, 1987:145–56.

54. Trush MA, Kensler TW. An overview of the relationship between oxidative stress and chemical carcinogenesis. Free Radic Biol Med 1991; 10:201–9.
55. Ames BN. Endogenous DNA damage as related to cancer and aging. Mutat Res 1989; 214:41–6.
56. Pryor WA. The involvement of free radicals in chemical carcinogenesis. In: Cerutti PA, Nygaard OF, Simic MG, eds. Anticarcinogenesis and radiation protection. New York: Plenum Press, 1987:1–9.
57. Imlay JA, Linn S. DNA damage and oxygen radical toxicity. Science 1988; 240: 1302–9.
58. Ito A, Watanabe H, Naito M, Naito Y. Induction of duodenal tumors in mice by oral administration of hydrogen peroxide. Gann 1981; 72:174–5.
59. Troll W, Wiesner R. The role of oxygen radicals as a possible mechanism of tumor promotion. Ann Rev Pharmacol Toxicol 1985; 25:509–28.
60. Repine JE, White JG, Clawson CC, Holmes BM. The influence of phorbol myristate acetate on oxygen consumption by polymorphonuclear leukocytes. J Clin Lab Med 1974; 83:911–20.
61. Kensler TW, Egner PA, Moore KG, Taffe BG, Twerdok LE, Thrush MA. Role of inflammatory cells in the metabolic activation of polycyclic aromatic hydrocarbons in mouse skin. Toxicol Appl Pharmacol 1987; 90:337–46.
62. Fischer SM, Cameron GS, Baldwin JK, Jasheway DW, Patrick K. Reactive oxygen in the tumor promotion stage of skin carcinogenesis. Lipids 1988; 23:592–7.
63. Perchellet EM, Abney NL, Perchellet JP. Stimulation of hydroperoxide generation in mouse skins treated with tumor-promoting or carcinogenic agents *in vivo* and *in vitro*. Cancer Lett 1988; 42:169–77.
64. Solanki V, Rana RS, Slaga TJ. Diminution of mouse epidermal superoxide dismutase and catalase activities by tumor promoters. Carcinogenesis 1981; 2: 1141–6.
65. Perchellet JP, Perchellet EM, Orten DK, Schneider BA. Inhibition of the effects of 12-O-tetradecanoylphorbol-13-acetate on mouse epidermal glutathione peroxidase and ornithine decarboxylase activities by glutathione level-raising agents and selenium-containing compounds. Cancer Lett 1985; 26:283–93.
66. Perchellet JP, Perchellet EM, Orten DK, Schneider BA. Decreased ratio of reduced/oxidized glutathione in mouse epidermal cells treated with tumor promoters. Carcinogenesis 1986; 7:503–6.
67. Taffe BG, Takahashi N, Kensler TW, Mason RP. Generation of free radicals from organic hydroperoxide tumor promoters by mouse epidermal cells. Pharmacologist 1986; 28:175.
68. Zimmerman R, Cerutti PA. Active oxygen acts as a promoter of transformation in mouse embryo C3H/10T 1/2 /C18 fibroblasts. Proc Natl Acad Sci USA 1984; 81:2085–7.
69. Weitzman SA, Weitberg AB, Clark EP, Stossel TP. Phagocytes as carcinogens: malignant transformation produced by human neutrophils. Science 1985; 227: 1231–3.
70. Wattenberg LW. Inhibition of chemical carcinogenesis by antioxidants. In: Slaga TJ, ed. Carcinogenesis. Vol. 5. Modifiers of chemical carcinogenesis. New York: Raven Press, 1980:85–98.

71. Sporn MB. Retinoids and cancer prevention. In: Slaga TJ, ed. Carcinogenesis. Vol. 5. Modifiers of chemical carcinogenesis. New York: Raven Press, 1980: 99–109.
72. Egner PA, Kensler TW. Effects of a biomimetic superoxide dismutase on complete and multistage carcinogenesis in mouse skin. Carcinogenesis 1985; 6:1167–72.
73. Black HS. Potential involvement of free radical reactions in ultraviolet light-mediated cutaneous damage. Photochem Photobiol 1987; 46:213–21.
74. Smith KC, Hannawalt PC. Molecular photobiology. New York: Academic Press, Inc., 1969:203–5.
75. Norins AL. Free radical formation in the skin following exposure to ultraviolet light. J Invest Dermatol 1962; 39:445–8.
76. Pathak MA, Stratton K. Free radicals in human skin before and after exposure to light. Arch Biochem Biophys 1968; 123:468–76.
77. Tyrrell RM. UVA (320-380 nm) radiation as an oxidative stress. In: Sies H, ed. Oxidative stress: oxidants and antioxidants. London: Academic Press, Ltd., 1991:559–83.
78. Peak MJ, Peak JG, Carnes BA. Induction of direct and indirect single-strand breaks in human cell DNA by far- and near-ultraviolet radiations: action spectrum and mechanisms. Photochem Photobiol 1987; 45:381–7.
79. Peak MJ, Johnson JS, Tuveson RW, Peak JG. Inactivation by monochromatic near-UV radiation of an *Escherichia coli hemA8* mutant grown with and without δ-aminolevulinic acid: the role of DNA *vs* membrane damage. Photochem Photobiol 1987; 45:473–8.
80. Peak MJ, Peak JG. Hydroxyl radical quenching agents protect against DNA breakage caused by both 365-nm UVA and gamma radiation. Photochem Photobiol 1990; 51:649–52.
81. Carraro C, Pathak MA. Studies on the nature of in vitro and in vivo photosensitization reactions by psoralens and porphyrins. J Invest Dermatol 1988; 90: 267–75.
82. Roza L, van der Schans GP, Lohman PHM. The induction and repair of DNA damage and its influence on cell death in primary human fibroblasts exposed to UV-A or UV-C irradiation. Mutat Res 1985; 146:89–98.
83. Kelfkens G, de Gruijl FR, van der Leun JC. Tumorigenesis by short-wave ultraviolet-A: papillomas versus squamous cell carcinomas. Carcinogenesis 1991; 12: 1377–82.
84. Fuchs J, Packer L. Photooxidative stress in the skin. In: Sies H, ed. Oxidative stress: oxidants and antioxidants. London: Academic Press, Ltd., 1991:559–83.
85. Connor MJ, Wheeler LA. Depletion of cutaneous glutathione by ultraviolet radiation. Photochem Photobiol 1987; 46:239–45.
86. Fuchs J, Huflejt ME, Rothfuss LM, Wilson DS, Carcamo G, Packer L. Impairment of enzymatic and nonenzymatic antioxidants in skin by UVB treatment. J Invest Dermatol 1989; 93:769–73.

87. Fuchs J, Huflejt M, Rothfuss LM, Wilson DS, Carcamo G, Packer L. Acute effects of near ultraviolet and visible light on the cutaneous antioxidant defense system. Photochem Photobiol 1989; 50:739–44.
88. Pence BC, Naylor MF. Effects of single-dose ultraviolet radiation on skin superoxide dismutase, catalase, and xanthine oxidase in hairless mice. J Invest Dermatol 1990; 95:213–16.
89. Punnoneu K, Jansen CT, Puntala A, Ahotupa M. Effects of in vitro UVA irradiation and PUVA treatment on membrane fatty acids and activities of antioxidant enzymes in human keratinocytes. J Invest Dermatol 1991; 96:255–9.
90. Miyachi Y, Imamura S, Niwa Y. Decreased skin superoxide dismutase activity by a single exposure of ultraviolet radiation is reduced by liposomal superoxide dismutase pretreatment. J Invest Dermatol 1987; 89:111–12.
91. Hanada K, Gange RW, Connor MJ. Effect of glutathione depletion on sunburn cell formation in the hairless mouse. J Invest Dermatol 1991; 96:838–40.
92. Miyachi Y, Horio T, Imamura S. Sunburn cell formation is prevented by scavenging oxygen intermediates. Clin Exp Dermatol 1983; 8:305–10.
93. Bissett DL, Chatterjee R, Hannon DP. Photoprotective effect of superoxide-scavenging antioxidants against ultraviolet radiation-induced chronic skin damage in the hairless mouse. Photodermatol Photoimmunol Photomed 1990; 7:56–62.
94. Mathews MM. Protective effect of β-carotene against lethal photosensitization by haematoporphyrin. Nature 1964; 203:1092.
95. Foote CS, Denny RW. Chemistry of singlet oxygen. VII. Quenching by beta-carotene. J Am Chem Soc 1968; 90:6233–5.
96. Mathews-Roth MM. Carotenoids quench evolution of excited species in epidermis exposed to UV-B (290-320) light. Photochem Photobiol 1986; 43:91–3.
97. Krinsky NI. Mechanisms of inactivation of oxygen species by carotenoids. In: Cerutti PA, Nygaard OF, Simic MG, eds. Anticarcinogenesis and radiation protection. New York: Plenum Press, 1987:41–6.
98. Epstein JH. Effects of beta-carotene on ultraviolet induced cancer formation in the hairless mouse skin. Photochem Photobiol 1977; 25:211–13.
99. Mathews-Roth MM. Antitumor activity of β-carotene, canthaxanthin, and phytoene. Oncology 1982; 39:33–7.
100. Mathews-Roth MM, Krinsky NI. Carotenoid dose level and protection against UV-B induced skin tumors. Photochem Photobiol 1985; 42:35–8.
101. Mathews-Roth MM, Pathak M. Phytoene as a protective agent against sunburn (>280 nm) radiation in guinea pigs. Photochem Photobiol 1975; 21: 261–3.
102. Sayre RM, Black HS. Beta carotene does not act as an optical filter in skin. J Photochem Photobiol 1992; 12:83–90.
103. Black HS, Douglas DR. A model system for the evaluation of the role of cholesterol α-oxide in ultraviolet light-induced carcinogenesis. Cancer Res 1972; 32:2630–2.

104. Black HS, Douglas DR. Formation of a carcinogen of natural origin in the etiology of ultraviolet light-induced carcinogenesis. Cancer Res 1973; 33:2094–6.
105. Ogura R, Sugiyama M, Sakanashi T, Hidaka T. Role of oxygen in lipid peroxide of skin exposed to ultraviolet light. In: Hayaishi O, Imamura S, Miyachi Y, eds. The biological role of reactive oxygen species in skin. New York: Elsevier, 1987: 37–41.
106. Lo WB, Black HS. Inhibition of carcinogen formation in skin irradiated with ultraviolet light. Nature 1973; 246:489–91.
107. Black HS, Chan JT. Suppression of ultraviolet light induced tumor formation by dietary antioxidants. J Invest Dermatol 1975; 65:412–14.
108. Pauling L, Willoughby R, Reynolds R, Blaisdell BE, Lawson S. Incidence of squamous cell carcinoma in hairless mice irradiated with ultraviolet light in relation to intake of ascorbic acid (vitamin C) and D,L-α-tocopheryl acetate (vitamin E). Int J Nutr Res 1982; 23:53–82.
109. De Rios G, Chan JT, Black HS, Rudolph AH, Knox JM. Systemic protection by antioxidants against UVL-induced erythema. J Invest Dermatol 1978; 70:123–25.
110. Black HS, Chan JT, Brown GE. Effects of dietary constituents on ultraviolet light-mediated carcinogenesis. Cancer Res 1978; 38:1384–7.
111. Peterson AO, McCann V, Black HS. Dietary modification of UV-induced epidermal ornithine decarboxylase. J Invest Dermatol 1980; 75:408–10.
112. Dunham WB, Zuckerkandl E, Reynolds R, Willoughby R, Marcuson R, Barth R, Pauling L. Effects of intake of L-ascorbic acid on the incidence of dermal neoplasms induced in mice by ultraviolet light. Proc Natl Acad Sci USA 1982; 79:7532–6.
113. Koone MD, Black HS. A mode of action for butylated hydroxytoluene-mediated photoprotection. J Invest Dermatol 1986; 87:343–7.
114. Black HS, Kleinhans CM, Hudson HT, Sayre RM, Agin PP. Studies on the mechanism of protection by butylated hydroxytoluene to UV-radiation damage. Photobiochem Photobiophys 1980; 1:119–23.
115. Harman D. Role of free radicals in mutation, cancer, aging and maintenance of life. Radiat Res 1962; 16:753–63.
116. Harman D. The free radical theory of aging. In: Pryor WA, ed. Free radicals in biology V. New York: Academic Press, 1982:255–75.
117. Black HS, Lenger WA, Gerguis J, Thornby JI. Relation of antioxidants and level of dietary lipid to epidermal lipid peroxidation and ultraviolet carcinogenesis. Cancer Res 1985; 45:6254–;9.
118. Mathews-Roth MM, Krinsky NI. Effect of dietary fat level in UV-B induced skin tumors, and anti-tumor action of β-carotene. Photochem Photobiol 1984; 40:671–3.
119. Orengo IF, Black HS, Kettler AH, Wolf Jr JE. Influence of dietary menhaden oil upon carcinogenesis and various cutaneous responses to ultraviolet radiation. Photochem Photobiol 1989; 49:71–7.
120. Henderson CD, Black HS, Wolf Jr JE. Influence of omega-3 and omega-6 fatty acid sources on prostaglandin levels in mice. Lipids 1989; 24:502–5.

121. Fischer MA, Black HS. Modification of membrane composition, eicosanoid metabolism, and immunoresponsiveness by dietary omega-3 and omega-6 fatty acid sources, modulators of ultraviolet-carcinogenesis. Photochem Photobiol 1991; 54:381–7.
122. Urbach F. Anatomy and pathophysiology of skin tumor capillaries. In: Urbach F, ed. International Conference on the Biology of Cutaneous Cancer. Bethesda, MD: National Cancer Institute Monograph No. 10, 1962:539–59.
123. Warburg O. On the origin of cancer cells. Science 1956; 123:309–14.

14

Changes of Antioxidant Enzymes and Effects of Active Oxygens on Aged Human Skin Fibroblasts

Yasuko Shindo*
Naganoken-Koseiren Shinonoi General Hospital, Shinonoi, Japan

Junichi Akiyama and Yoshio Takase
Shinshu University School of Medicine, Asahi, Matsumoto, Japan

I. INTRODUCTION

The free radical theory (1) is one of the important hypotheses proposed for mechanisms of aging. Many investigators (2–4) have examined antioxidant enzymes in various animal organs in order to prove the theory. However, they have not succeeded in demonstrating that these enzyme activities necessarily decrease with aging. In human fibroblasts, antioxidant enzymes such as catalase, glutathione peroxidase, and superoxide dismutase were examined in cells of different population-doubling levels (5,6), but cells obtained from persons of different ages have never been compared. Although cells from different population-doubling levels have been taken as a model of aging, changes in enzymatic activities with this method may have reflected the culture process and not aging itself. Therefore, we measured three antioxidant enzymes of human skin fibroblasts derived from persons of various ages using cultured cells of the early passage.

More recently, it has been considered important to compare the responses of young and old tissues under oxidative stress, such as induction of antioxidant enzymes (7), in order to prove the above theory. Therefore, we also investigated the activities of antioxidant enzymes and synthesis of DNA, protein, and collagen under oxidative stress to young and old cells.

**Present affiliation*: Shinshu University School of Medicine, Asahi, Matsumoto, Japan.

II. MATERIALS AND METHODS

A. Cultured Skin Fibroblasts

Normal nonexposed skin from young and old people (0–88 years old) were subjected to explant culture. We used first- or third-passage cells in all experiments. The cells were cultured at 37°C in Eagle's minimum essential medium (MEM) (NISSUI NO-1) supplemented with 10% fetal calf serum (FCS) (GIBCO) in a humidified atmosphere with 5% CO_2. The medium was changed twice a week. The cells were usually cultured in large flasks (12 × 6 cm, Nunc) and subcultured once every 3 weeks.

B. Analysis of Enzyme Activities

The cultured cells in large flasks (12 × 6 cm) were washed with phosphate-buffered saline (PBS). 0.3 Milliliters of 50 mM potassium phosphate, pH 7.5, containing 0.5% Triton X-100 was added to the flask, and after 30 minutes at 5°C, the lysed cells were harvested. After centrifugation at 10,000 × *g* for 5 minutes, the supernatent was removed and used for analysis. The activities of the following enzymes were assayed: catalase (8), superoxide dismutase (SOD) (9), glutathione peroxidase (GPO) (10), and glutathione reductase (11). One enzyme unit is equivalent to 1 μmol of product formation or 1 μmol of substrate disappearance per minute under the defined condition, except for superoxide dismutase. The latter was assayed by inhibition of xanthine plus xanthine oxidase–mediated reduction of epinephrine. Amounts of protein were determined by a modified method (12) of Lowry et al. (13), using bovine serum albumin as a standard. Reduced glutathione (GSH) and oxidized glutathione (GSSG) (14) were measured in confluent cultured cells in 6-cm dishes.

C. Rate of Catalase Synthesis

Cells from a 3-year-old female and an 86-year-old female were used (15). The medium in 6-cm dishes was replaced by a methionine-free medium containing 10% dialyzed FCS and 5 μCi/ml of [^{35}S]methionine, and the cells were cultured for 24 hours. They were either extracted immediately or chased for 72 hours in Eagle's MEM with 10% FCS. After washing with PBS, the cells from one dish were collected in 0.4 ml of 10 mM Tris-Cl, pH 7.4, containing 0.1% sodium dodecyl sulfate (SDS), 2 mM EDTA, and 0.1% Triton X-100. This fraction was treated with anti-(catalase) immunoglobulin-G, and the immunoprecipitate was subjected to SDS-polyacrylamide gel electrophoresis and fluorography. The fluorographic band was scanned by a densitometer.

D. H_2O_2 Stress

H_2O_2 (0–0.1 mM), which was a dose that did not kill the cells, was added to 6-cm dishes with 1 ml Eagle's MEM. After 4 hours, the media were removed and the dishes were used for analysis of activities of antioxidant enzymes, DNA synthesis, protein synthesis, and collagen synthesis. Cytotoxicity was checked by both trypan blue staining of the cells and the leakage of lactate dehydrogenase from the cells into the medium.

E. O_2^- Stress (Xanthine-Xanthine Oxidase)

Xanthine (0.5 mM) and xanthine oxidase (0–0.1 μl, or 2.5 mU), the dose of which did not kill the cells, were added into 6-cm dishes with 1 ml Eagle's MEM. After 1 hour, the media were removed and the cells were used for analysis of activities of antioxidant enzymes, DNA synthesis, protein synthesis, and collagen synthesis.

F. UVA Radiation

The medium in 6-cm dishes was replaced by Eagle's MEM containing 25 mM Hepes, pH 7.2. Each dish was irradiated with UVA at 4.8 J/cm^2 (BLB lamp, Toshiba, peak wave length 365 nm) using a UV-35 filter (Toshiba) to cut off radiation below 320 nm. In the case of catalase activity, cells in one dish were harvested immediately, then the media in three other dishes were replaced with the usual medium, and the cells were cultured further and harvested after 24, 48, and 72 hours. The harvested cells were centrifuged at 10,000 × *g* for 5 minutes, and the supernatant was used for assay of catalase activity.

After UV irradiation treatment to another dish with α-tocopherol (100 μM), catalase activity in these cells was measured. To examine the effect of irradiation on catalase, purified catalase (3 units/ml) in Eagle's MEM was irradiated with UVA.

For analysis of activities of other antioxidant enzymes and collagen synthesis after irradiation, the media were removed and the cells from these dishes were assayed for activity.

G. DNA Synthesis

One μCi of [^{3}H]thymidine was added to cultured fibroblasts in a 6-cm dish. The cells were harvested after 24 hours. Incorporation of the label into the acid-insoluble fraction was measured with a liquid scintillation spectrometer.

H. Protein Synthesis

0.5 μCi of [^{14}C]leucine was added to the cultured fibroblasts in a 6-cm dish. The cells were harvested after 24 hours. Incorporation of the label into the acid-insoluble fraction was measured with a liquid scintillation spectrometer.

I. Collagen Synthesis

In 6 cm dishes, 50 μg/ml of ascorbic acid, 2 μCi/ml of [^{3}H]proline and 5% dialyzed FCS were added to confluent cells in Eagle's MEM. After 24 hours the medium was removed. Acid-insoluble collagen in the medium was measured to estimate collagen synthesis by the methods of Peterkofsky (16) and Hata (17).

[^{3}H]Thymidine (20 Ci/mmol), [^{35}S]methionine (600 Ci/mmol), [^{14}C]leucine (300 mCi/mmol), and [^{3}H]proline (30 Ci/mmol) were purchased from Amersham, Japan. The purified catalase and the antibody to catalase from human blood cells were provided by T. Hashimoto (Shinshu University).

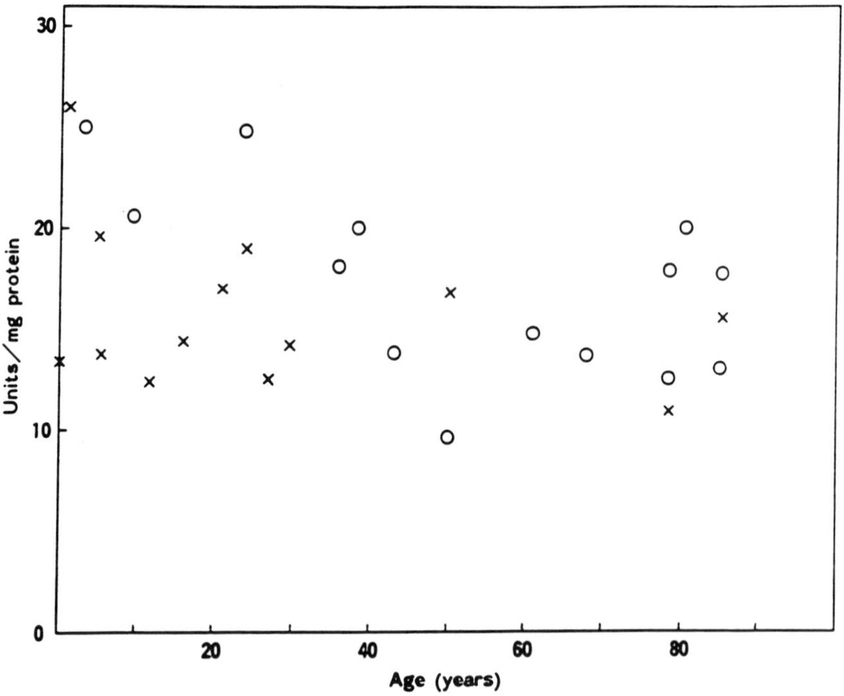

Figure 1 Superoxide dismutase activity. Superoxide dismutase activity was assayed in normal skin fibroblasts from people of various ages (X, male; O, female).

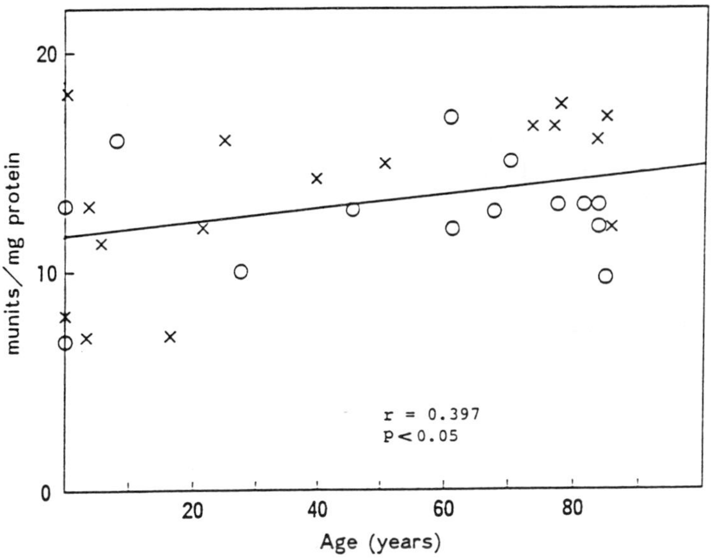

Figure 2 Glutathione peroxidase activity. Glutathione peroxidase activity was assayed in normal skin fibroblasts from people of various ages (X, male; O, female).

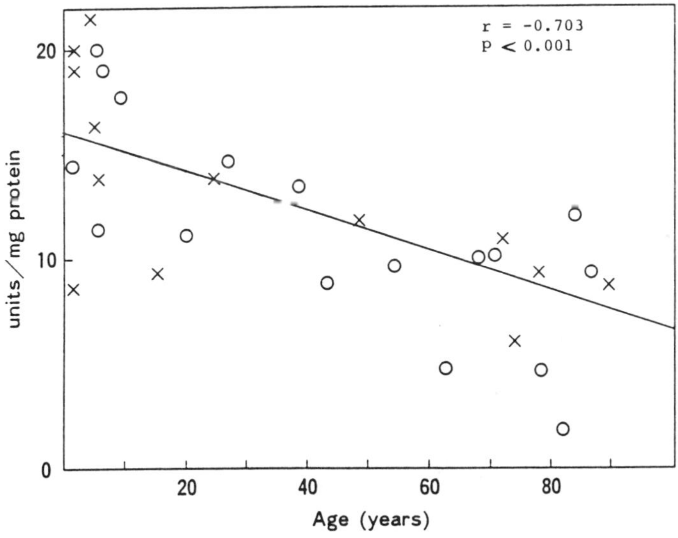

Figure 3 Catalase activity. Catalase activity was assayed in normal skin fibroblasts from people of various ages (X, male; O, female).

III. RESULTS

A. Antioxidant Enzymes

Three antioxidant enzymes were measured in fibroblasts derived from 30 persons of various age. Total SOD activity showed no change with aging (Fig. 1), GPO activity slightly increased (Fig. 2), and catalase activity decreased (Fig. 3). The catalase activity in cells from young subjects (young cells, under 20 years of age) was about twice as high as that in the cells from old subjects (old cells, over 70 years).

B. Rate of Catalase Synthesis

After 24 hours of pulse labeling, the amount of incorporation of [^{35}S]methionine into the enzyme was 45% in the old cells, compared with 100% in the

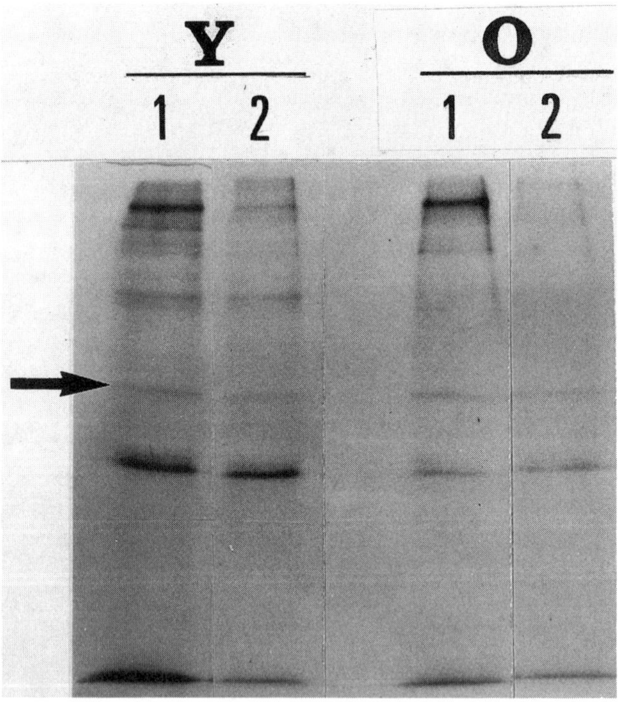

Figure 4 Rate of catalase synthesis. Y, young; O, old. 1, 24-h pulse; 2, 72-h chase after 24-h pulse. Duration of fluorography was 30 days. Quantification was conducted by densitometric measurement of the band on x-ray film. Arrow indicates protein of catalase. Radioactivity of total protein used was 1,700,000 cpm for all groups.

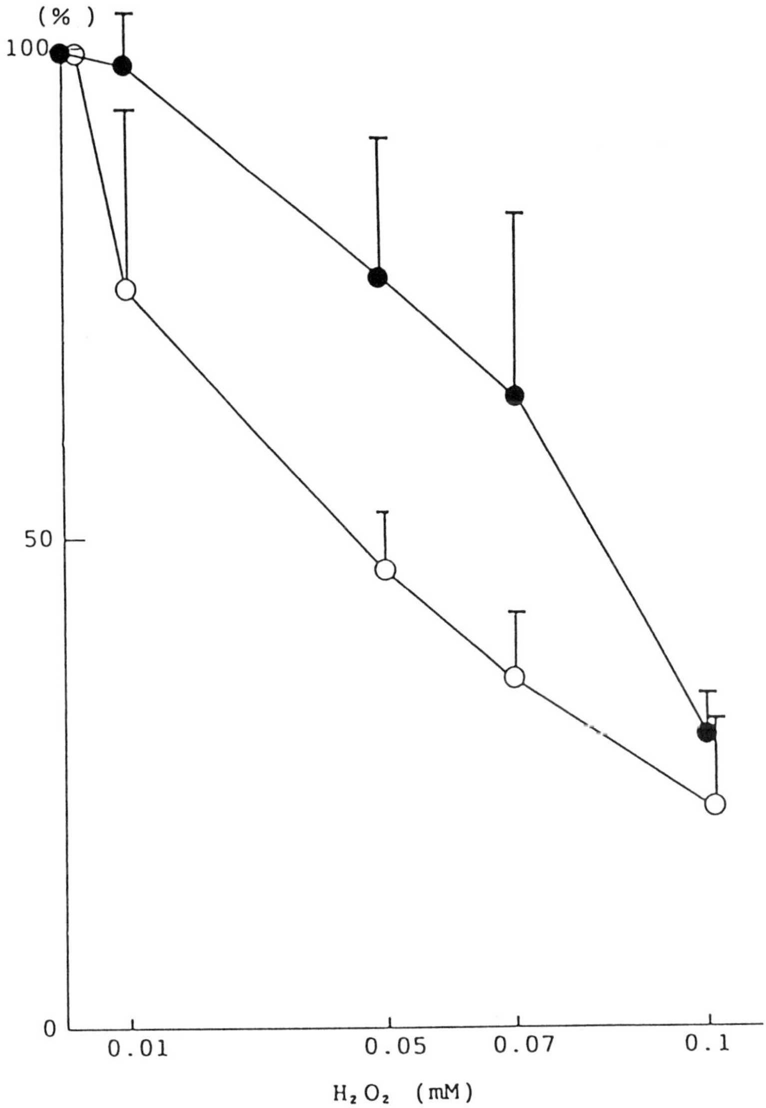

Figure 5 DNA synthesis with H_2O_2. H_2O_2 was added into a 6-cm dish with 1 ml Eagle's MEM. After 4 hours, the media was exchanged for usual Eagle's MEM. Furthermore, 1 μCi [^{3}H]thymidine was added into the dishes. After 24 hours, the cells were harvested and counted the incorporation of the label into the acid-insoluble fraction. ●, Young group (4-, 7-, and 8-year-old males and 6-week-old female). ○, Old group (71- and 83-year-old males, and 70- and 84-year-old females). Circles and bars represent the means and SDs. $p < 0.02$ vs. young and old cells on 0.05 mM H_2O_2. $p < 0.05$ vs. young and old cells on 0.07 mM H_2O_2.

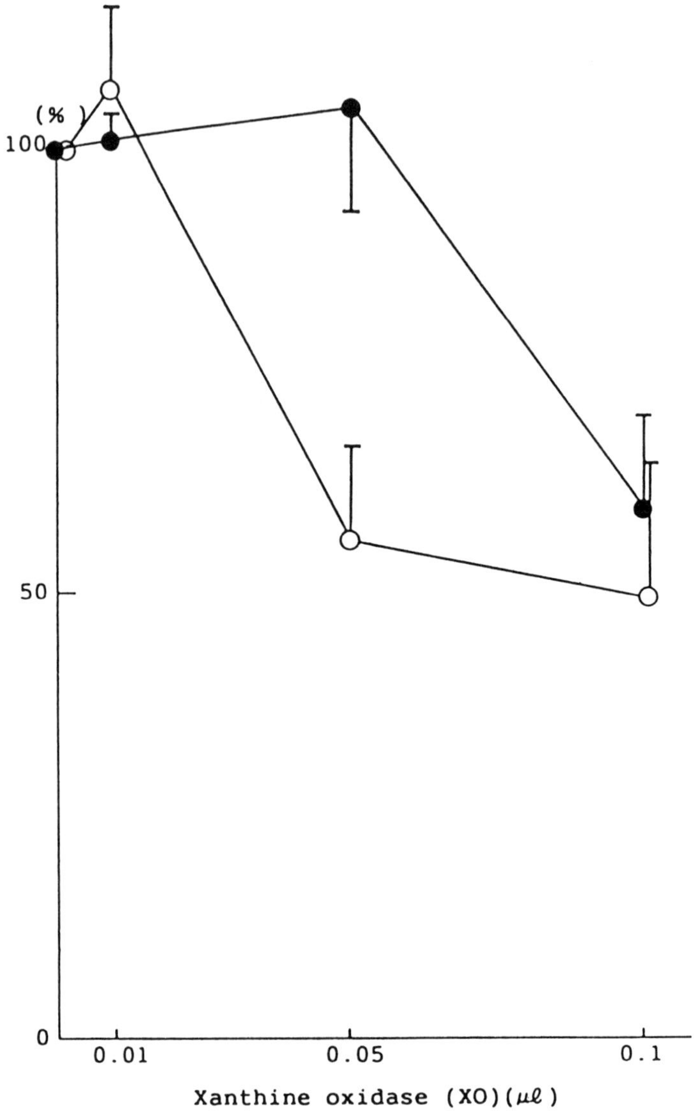

Figure 6 DNA synthesis with O_2^- (xanthine-xanthine oxidase). Xanthine (0.5 mM) and xanthine oxidase were added into a 6-cm dish with 1 ml Eagle's MEM. After 1 hour, the media was exchanged for usual Eagle's MEM. Furthermore, 1 μCi [^{3}H]thymidine was added into the dishes. After 24 hours, the cells were harvested and counted the incorporation of the label into the acid-insoluble fraction. ●, Young group (8-year-old male, and 6- and 70-week- and 2-year-old females). ○, Old group (71- and 89-year-old males, and 70- and 84-year-old females). Circles and bars represent the means and SDs. $p < 0.01$ vs. young and old cells on 0.05 μl of xanthine oxidase.

young cells (Fig. 4). In the 24-hour pulse 72-hour chase experiment, the residual radioactivity in the enzyme was 63% in the young cells and 18% in the old cells. The ratio of degradation of catalase in the young and old cells was 63:40 (40 = 18/0.45). The rate of synthesis was 100/45 (2.2), and that of degradation was 63/40 (1.5). Though the rates of both synthesis and degradation decreased in the old cells, these results imply that the lower catalase level in the old cells is mainly due to a lower rate of catalase synthesis. We performed a dilution experiment (nonlabeled catalase was added sufficiently) and confirmed that the band was catalase.

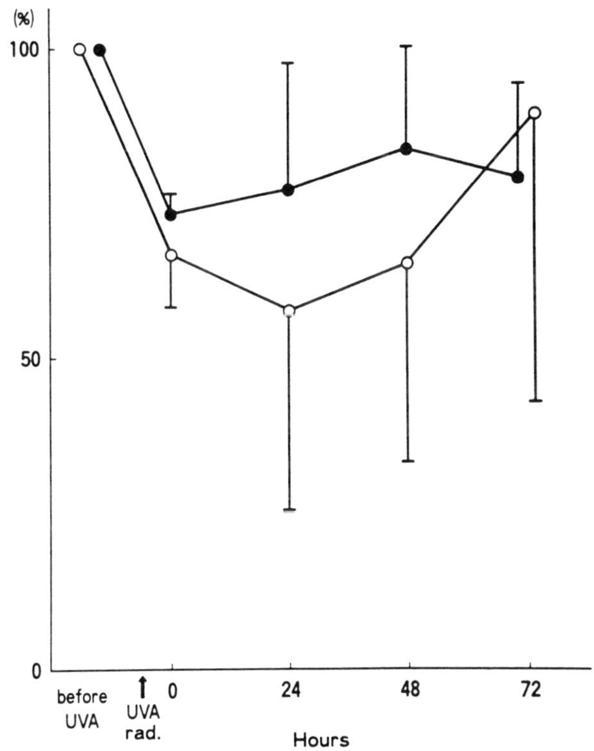

Figure 7 Catalase activity after UVA radiation. Each dish with 25 mM Hepes, pH 7.2, in Eagle's MEM was radiated by UVA (4.8 J/cm²). Catalase activity was assayed after radiation and after further incubation. ●, Young group (4-year- and 6-week-old males and 5- and 4-year-, and 6-week-old females). ○, Old group (83-, 77-, and 73-year-old males, and 83- and 81-year-old females). Circles and bars represent the means and SDs.

C. H_2O_2 and O_2^- Stress

Activities of antioxidant enzymes, protein synthesis, and collagen synthesis of fibroblasts under oxidative stress did not change. Only DNA synthesis was decreased in H_2O_2 (Fig. 5) and O_2^- (Fig. 6) stress. The rate in the old cells, in particular, was decreased more than that in the young cells.

D. UVA Radiation

Catalase activity immediately after irradiation decreased to about 70% of that in untreated material and returned to the previous levels after 24–72 hours in both young and old cells (Fig. 7). These decreased activities were statistically significant ($p < 0.001$).

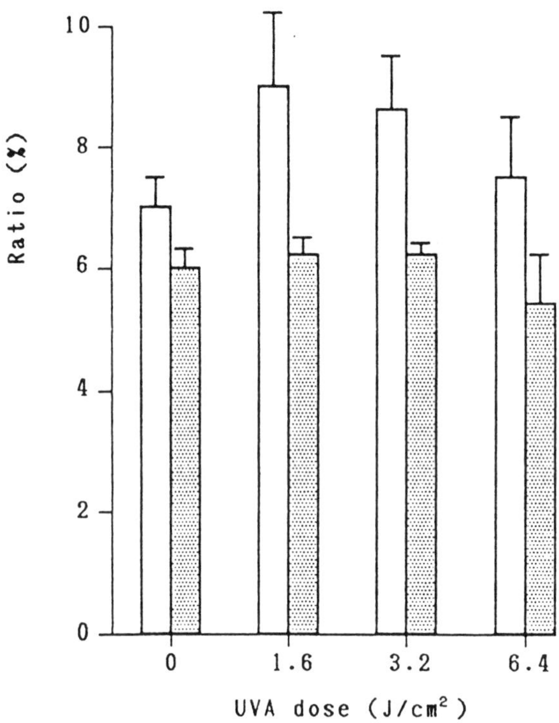

Figure 8 Collagen synthesis after UVA radiation. Each dish with 25 mM Hepes, pH 7.2, in Eagle's MEM was radiated by UVA. After radiation, medium was exchanged for Eagle's MEM and 50 μg/ml ascorbic acid and 2 μCi/ml [^{3}H]proline were added into the dishes. And acid-insoluble collagen was counted to estimate collagen synthesis. □, Young (4-week-old male). ▩, Old (88-year-old male). Bars represent SDs of the triplicate data.

Only the activity of purified catalase in the same medium after UVA irradiation was unchanged. Similarly, the catalase activity of fibroblasts with 100 μM α-tocopherol did not change after UVA irradiation. The activities of other enzymes (SOD, GPO, glucose 6-phosphate dehydrogenase, 6-phosphogluconate dehydrogenase, lactate dehydrogenase, citrate synthase, fumarase, and glucosephosphate isomerase) did not decrease after UVA irradiation. The activities of these antioxidant enzymes were not changed after UVB irradiation (28.8 mJ/cm^2, Toshiba FL 20SE, peak wave length 315 nm).

After UVA irradiation (1.6, 3.3, and 6.6 J/cm^2), the rate of collagen synthesis of the fibroblasts increased. The rate in young cells, in particular, increased more than in old cells (Fig. 8).

IV. DISCUSSION

The free radical theory (1), one of the most popular theories on aging, proposes that active oxygens and free radicals react with cell components such as cell membranes or proteins, impairing their functions. It is generally accepted that free radicals promote cutaneous aging, because the exposed skin of farmers or fishermen is very rough and looks much older than the true age of their skin. To detect free radicals directly on the skin or other tissues is impossible. Therefore, researchers have examined lipid peroxide or activities of antioxidant enzymes to indirectly prove the theory.

In 1980 Tolmasoff et al. (18) assayed SOD, which is an antioxidant enzyme, in various tissues in more than 10 species of animals with different life spans. They showed that a long life span paralleled high SOD activity. Since then, SOD has attracted the attention of many researchers. Antioxidant enzymes (SOD, catalase, and GPO) and other antioxidant agents have been examined in various animal organs. Unexpectedly, however, the activity of antioxidant enzymes in various tissues of mice and rats did not always decrease with aging (2–4). At the cellular level, many researchers have used fibroblasts of various population-doubling levels, and they merely found that SOD and catalase activities were not changed through the cell doublings (5) and that GPO decreased only in the early stages of cell culture (6). Taken on the whole, researchers were not able to demonstrate that activities of antioxidant enzymes decreased with aging.

Aging may be considered to be a process of decreasing capacity to adapt to external and internal stresses. It would then become more important to compare responses of young and old tissues after exposure to active oxygens than to simply analyze the activity of antioxidant enzymes themselves. In the short-term experiment (e.g., 24 h), Yoshioka et al. (19) reported that SOD activity of the skin of old guinea pigs decreased after hypoxanthine-xanthine oxidase injection, whereas that of young ones did not change. In the long-term experiment, Ueda et al. (20) reported that skin exposed to UV in young people showed induction of SOD, which did not change when examined in old people. So we examined not only the activities of antioxidant enzymes of human fibroblasts, but also the difference in responses between young and old cells under oxidative stress.

The doubling level of fibroblasts may not be an adequate model of aging (so-called "in vitro aging"), because any change through population doubling could be caused by the culture process itself. Schneider and Mitsui (21) also reported that the cells derived from young and old people were better as a model of aging than "in vitro aging." We have used human skin fibroblasts derived from young and old persons in early (1–3) passage numbers. We also analyzed enzyme activities in passage level 4–5 fibroblasts, but the data were almost identical to those at passage level 1–3, showing that our data reflect aging of the original fibroblasts and not the passage level. Furthermore, we have used only fibroblasts, not epidermal cells, from several reasons:

1. In the preliminary experiments, epidermal cells of Werner syndrome or old persons showed little difference in biological or morphological findings when compared with young cells.
2. Toyoshima et al. (22) reported that there was no relationship between the DNA synthetic activity and aging of epidermis in the guinea pig.
3. In the histopathological findings of the skin, changes of the dermis are much bigger than that of the epidermis with aging.

In our investigations SOD did not change, and GPO activity was slightly enhanced with aging. Preliminary data also showed that glutathione reductase (young 5.9 ± 1.1 U/mg, old 6.9 ± 2.3 U/mg, $n = 5$) and glutathione (GSH + GSSG) (young 13.4 ± 3.6 nmol/mg, old 10.4 ± 2.5 nmol/mg, $n = 5$) did not change either. Activities of six important enzymes of glycolysis (glucose-6-phosphate dehydrogenase, phosphogluconate dehydrogenase, glucose phosphate isomerase, lactate dehydrogenase, fumarase, citrate synthese) were measured in 20 human skin fibroblasts derived from persons of various age (23). There was no change with aging in each enzyme. On the other hand, only catalase activity decreased markedly with aging.

Ultraviolet (UV) light was used in the present experiments, since it should be the most important factor in skin aging. We used UVA, which reaches the dermis, at the highest dose (4.8 J/cm^2) that would not kill the cells. After irradiation, levels of none of the enzymes but catalase changed. The activity of purified catalase in the same medium, however, was not lost even after UVA radiation, and catalase activity was also preserved with 100 μM α-tocopherol in the medium. On the other hand, the activities of three antioxidant enzymes did not change after irradiation with UVB (2.8 mJ/cm^2, the largest dose that does not kill the cells).

Until now, catalase has been thought to be not so important for human life, because patients with acatalasemia survive without serious problems. The Km value of catalase for H_2O_2 is as high as 1.1 M (24). At the same time catalase is located only in peroxisomes and is unable to metabolize H_2O_2 produced anywhere else. On the other hand, since the Km value of GPO for H_2O_2 is as low as 1–3 μM in the presence of 0.1 M GSH and 25–60 μM with 5 mM GSH (25), GPO has been generally taken as the main metabolizing agent for H_2O_2 in physical concentrations in the cytoplasm. However, while the GPO activity depends on the concentration of GSH and NADPH, catalase metabolizes H_2O_2 in the absence of any other coenzymes. Furthermore, catalase reacts with H_2O_2, which is produced in high concentrations in peroxisomes, thus suppressing the leakage of H_2O_2 into cytoplasm. Although GPO was reported to be indispensable to the normal life of erythrocytes (26), its role has not been clarified in the cells of other organs: whether GPO activity is also predominant or whether GPO and catalase function together, compensating each other.

Furthermore, DNA synthesis under H_2O_2 or O_2^- stress in old cells was lower than that in young cells. These data seem to be related with lower activity of catalase in old cells, because the O_2^- generated from xanthine-xanthine oxidase was supposed to have been metabolized to H_2O_2 immediately in the medium. After UVA irradiation, collagen synthesis in young cells was higher than that in old cells. The enhancement of collagen synthesis after UVA will induce protection against UVA. Defense mechanisms against UVA in old cells are thereby supposed to be declining.

From our data, we consider catalase to be important in aging. Catalase activity decreases with aging, and it is inactivated by active oxygens produced by UVA. It can be concluded, therefore, that defense mechanisms against active oxygens are lowered with aging and that decreased activity of catalase would be responsible for such impaired defense. Although these findings support the free radical theory of aging, the true role of catalase and the importance of active oxygens in vivo during aging remain to be further elucidated.

REFERENCES

1. Harman D. A theory based on free radical and radiation chemistry. J Gerontol 1956; 11:298–300.
2. Richter V, Rotzsch W. Turnover of rat liver and rat kidney catalase as a function of age. Mech Age Der 1972; 1:427–30.
3. Nohl H, Hegner D. Responses of mitochondrial superoxide dismutase, catalase and glutathione peroxidase activities to aging. Mech Age Dev 1979; 11:145–51.
4. Lammi-Keefe CJ, Swan PB, Hagaety PVJ. Cooper-zinc and manganese superoxide dismutase activities in cardiac and skeletal muscles during aging in male rats. Gerontol 1984; 30:153–8.

5. Hayflick L. Recent advances in the cell biology of aging. Mech Age Dev 1980; 14:59–79.
6. Mbemba F, Houbion A, Raes M, Remacle J. Subcellular localization and modification with aging of glutathione, glutathione peroxidase and glutathione reductase activities in human fibroblasts. Biochim Biophys Acta 1985; 838:211–20.
7. Vanella A, Geremia E, D'urso G, Tiriolo P, Di Silvestro I, Grimaldi R, Pinturo R. Superoxide dismutase activities in aging rat brain. Gerontol 1982; 28:108–13.
8. Aebi H. Catalase. In: Bergmeyer HU, ed. Methods of enzymatic analysis. New York: Academic Press, 1974:673–84.
9. McCord JM, Fridovich I. Superoxide dismutase. J Biol Chem 1969; 244:6049–55.
10. Little C. Steroid hydroperoxides as substrates for glutathione peroxidase. Biochim Biophys Acta 1972; 284:375–81.
11. Bergmeyer HU, Gawehn K, Grassl M. Glutathione reductase. In: Bergmeyer HU, ed. Methods of enzymatic analysis. New York: Academic Press, 1974: 465–6.
12. Markwell MAK, Haas SM, Bieber LL, Tolbert NE. A modification of Lowry procedure to simplify protein determination in membrane and lipoprotein samples. Anal Biochem 1978; 87:206–10.
13. Lowry OH, Rosebrough NJ, Farr AL, Randall RJ. Protein measurement with the Folin phenol reagent. J Biol Chem 1951; 193:265–75.
14. Tietze F. Enzymatic methods for quantitative determination of nanogram amounts of total and oxidized glutathione: applications to mammalian blood and other tissues. Anal Biochem 1969; 27:502–22.
15. Mori M, Miura S, Tatibana M, Cohen PP. Cell-free synthesis and processing of a putative precursor for mitochondrial carbamyl phosphate synthetase I of rat liver. Proc Natl Acad Sci USA 1979; 76:5051–75.
16. Peterkofsky B, Diegelmann R. Use of a mixture of protease-free collagenase for the specific assay of radioactive collagen in the presence of other proteins. Biochemistry 1971; 10:988–94.
17. Hata R, Ninomiya Y, Nagai Y, Tsukada Y. Biosynthesis of interstitial types of collagen by albumin producing rat liver parenchymal cell (hepatocyte) clones in culture. Biochemistry 1980; 19:169–76.
18. Tolmasoff JM, Ono T, Cutler RG. Superoxide dismutase correlation with lifespan and specific metabolic rate in primate species. Proc Natl Acad Sci USA 1980; 77:2777–81.
19. Yoshioka A, Miyachi Y, Imamura S. Mechanisms of reactive oxygen species-induced skin erythema and superoxide dismutase activities in guinea pigs. J Dermatol 1987; 14:569–75.
20. Ueda H, Sugiura K. Lipid peroxide and superoxide dismutase in human skin. In: Kligman AM, Takase Y, eds. Cutaneous aging. Tokyo: University of Tokyo Press, 1988:449–57.
21. Schneider EL, Mitsui Y. The relationship between in vitro cellular aging and in vivo human age. Prot Natl Acad Sci USA 1976; 73:3584–8.
22. Toyoshima H, Hori M, Yoshida H. Analysis of DNA synthetic activity of aged epidermis. Jpn Dermatol 1991; 101:697–700.

23. Shindo Y, Akiyama J, Yamazaki Y, Saito K. Takase Y. Changes in enzyme activities in skin fibroblasts derived from persons of various ages. Exp Geront 1991; 26:29–35.
24. Nicholls R, Schonbaum GR. Catalase. In: Boyer PB, Lardy H, Myrback K, eds. The enzyme. Vol. 8. New York: Academic Press, 1963:147–225.
25. Foohé L, Braud I. Kinetics of glutathione peroxidase. Biochim Biophys Acta 1969; 191:541–9.
26. Cohen G, Hochstein P. Glutathione peroxidase: the primary agent for the elimination of hydrogen peroxide in erythrocytes. Biochemistry 1963; 2:1420–8.

15

Electron Donor and Acceptor Properties of Melanin Pigments in the Skin

Krzysztof Reszka
National Institute of Environmental Health Sciences, Research Triangle Park, North Carolina

Kowichi Jimbow
University of Alberta, Edmonton, Alberta, Canada

I. INTRODUCTION

Alterations of the skin in response to exposure to environmental factors, e.g., photoaging and the development of skin diseases (cancers), are often linked to peroxidative reactions that produce imbalanced redox status beyond the protective capacities of detoxifying enzymes in the skin (oxidative stresses). The main species involved in these oxidative stresses are free radicals including various radical forms of active oxygen, e.g., superoxide, hydroxyl, peroxyl, and alkoxyl radicals (1–3). Other species involved are nonradical, but still potentially harmful, e.g., hydrogen peroxide, organic hydroperoxide, and singlet oxygen (4–7).

In recent years melanin pigments have received significant attention for their ability to produce and react with radical species (8–13). Melanins appear to be multifunctional polymers predestined to play certain physiological roles in the skin. They are omnipresent in the plant and animal kingdom. In animals they are primarily responsible for the coloration of hair and skin. In addition, they are present in nonexposed sites such as the inner ear, the central nervous system, and the eye (14). Two basic types of melanin pigment are present in animals: brown-black eumelanin and yellow-red pheomelanin.

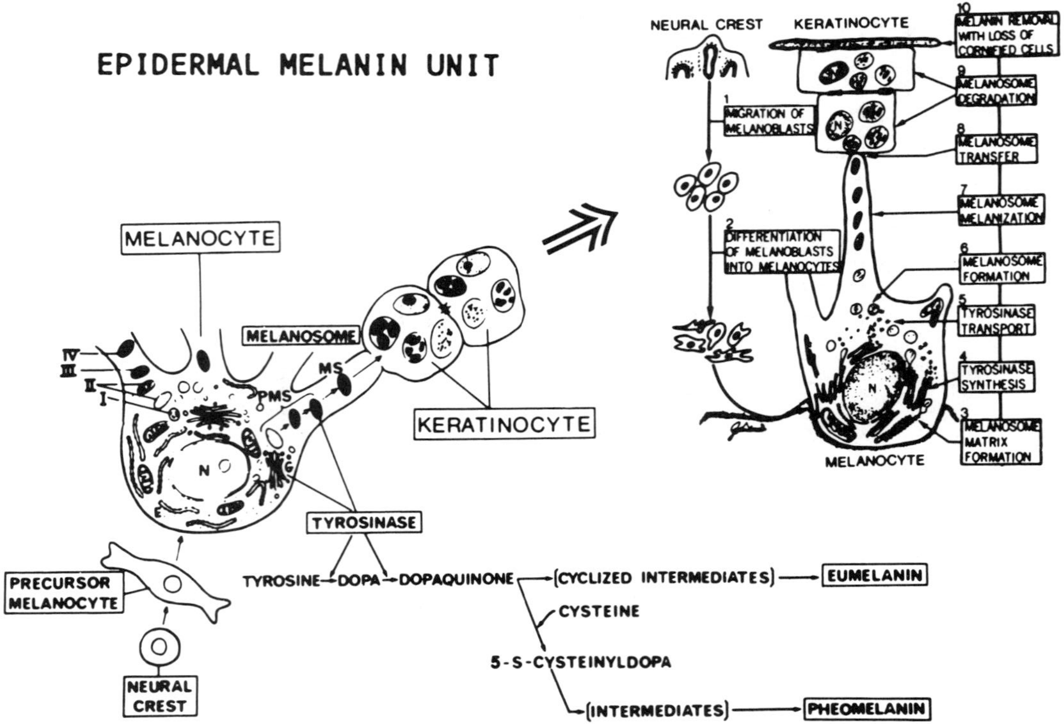

Figure 1 The synthesis of melanin and melanosomes and their interaction in the epidermal melanin unit.

Melanin synthesis occurs in specialized organelles called melanosomes. Melanosomes are the product of an exocrine unicellular gland, the melanocyte, that rests on the basal lamina of the epidermis and projects the dendrites into epidermal cells, keratinocytes. The melanocyte transfers its products, melanosomes, into the malpighian cells (keratinocytes), and melanosomes are thereby distributed throughout the epidermis by the outward movement of keratinocytes (Fig. 1). In the human epidermis, each melanocyte is associated with about 36 viable keratinocytes that transport and, in some cases, degrade melanin pigments received from the melanocyte. Together the melanocyte and associated keratinocytes form an epidermal melanin unit (15).

Melanin pigmentation of human skin will be conveniently divided into two components. Constitutive skin color is the amount of cutaneous melanin pigmentation generated according to the cellular genetic programs, without any direct effect by radiations of solar origin. It is the level of pigmentation acquired in those parts of the body that are habitually shielded from light. Facultative (inducible) skin color, or "tan," includes the short-lived immediate tanning and delayed tanning reactions, both of which are elicited by direct exposure of the skin to ultraviolet light (UVL). Since hyperpigmentation of the skin tends to decrease towards the constitutive level when exposure to UVL is discontinued, facultative color change is reversible. The hyperpigmentation induced by endocrine changes during pregnancy and Addison's disease is another type of facultative color change. Alterations in endocrine functions can significantly affect the response of human skin to UVL. Facultative color changes in humans, then, arise from the complex relationship between light, hormones, and genetic potential of the epidermal melanin unit, the basic multicellular "organ" of melanin metabolism (Figs. 2 and 3). This chapter will discuss the electron-donating and electron-accepting properties of melanin pigments during reduction and oxidation reactions related to the generation of potentially cytotoxic species during melanin pigmentation in the skin.

II. ELECTRON DONORS AND ACCEPTORS IN MELANIN BIOSYNTHESIS

A. Tyrosinase-Catalyzed Oxidation of Melanogenic Substances

The biosynthetic pathway of melanin pigments is catalyzed by the enzyme tyrosinase (Fig. 3) (14). Tyrosinase requires oxygen for its enzymic activity, and it catalyzes two-electron oxidation processes (16). The primary function of tyrosinase is to convert phenols into *ortho*-dihydroxybenzenes (DHB) by

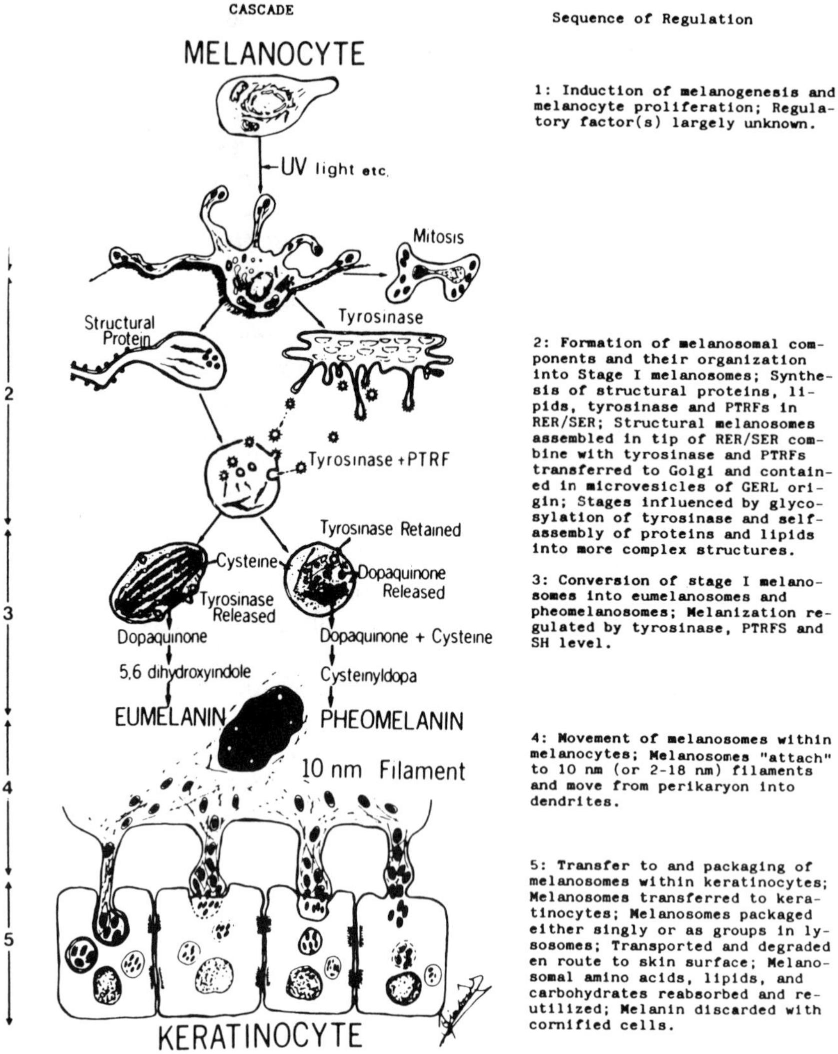

Figure 2 Cascade of melanin and melanosome synthesis in the epidermal melanin unit.

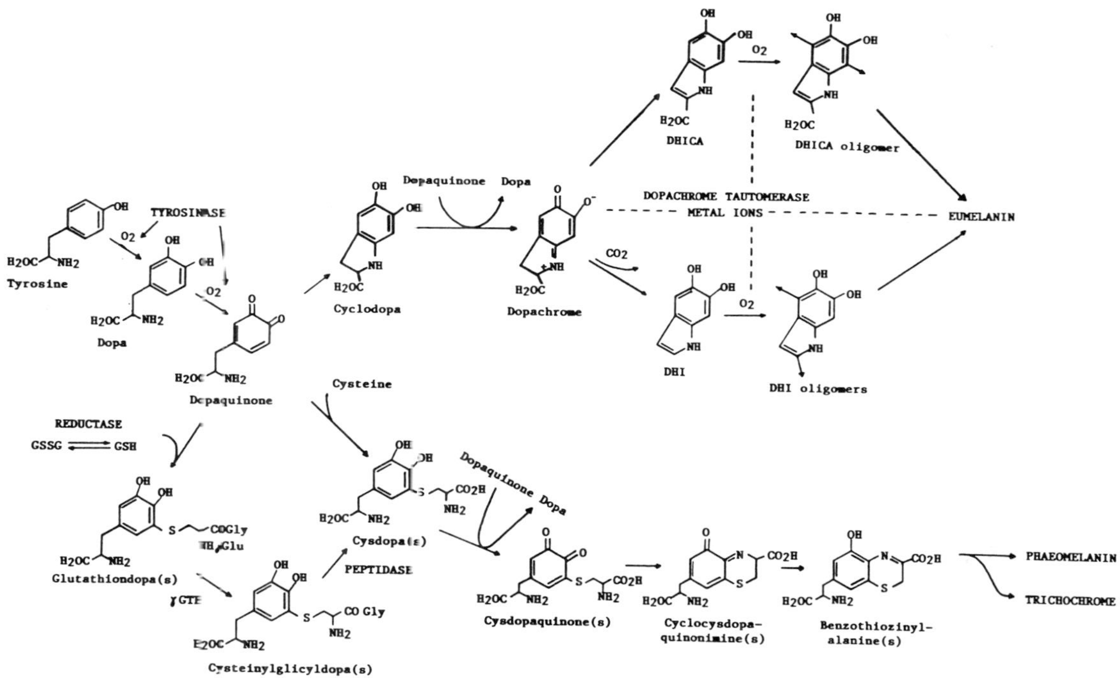

Figure 3 Chemical and biochemical reactions of melanin synthesis, i.e., eumelanin, pheomelanin, and trichochrome from tyrosine.

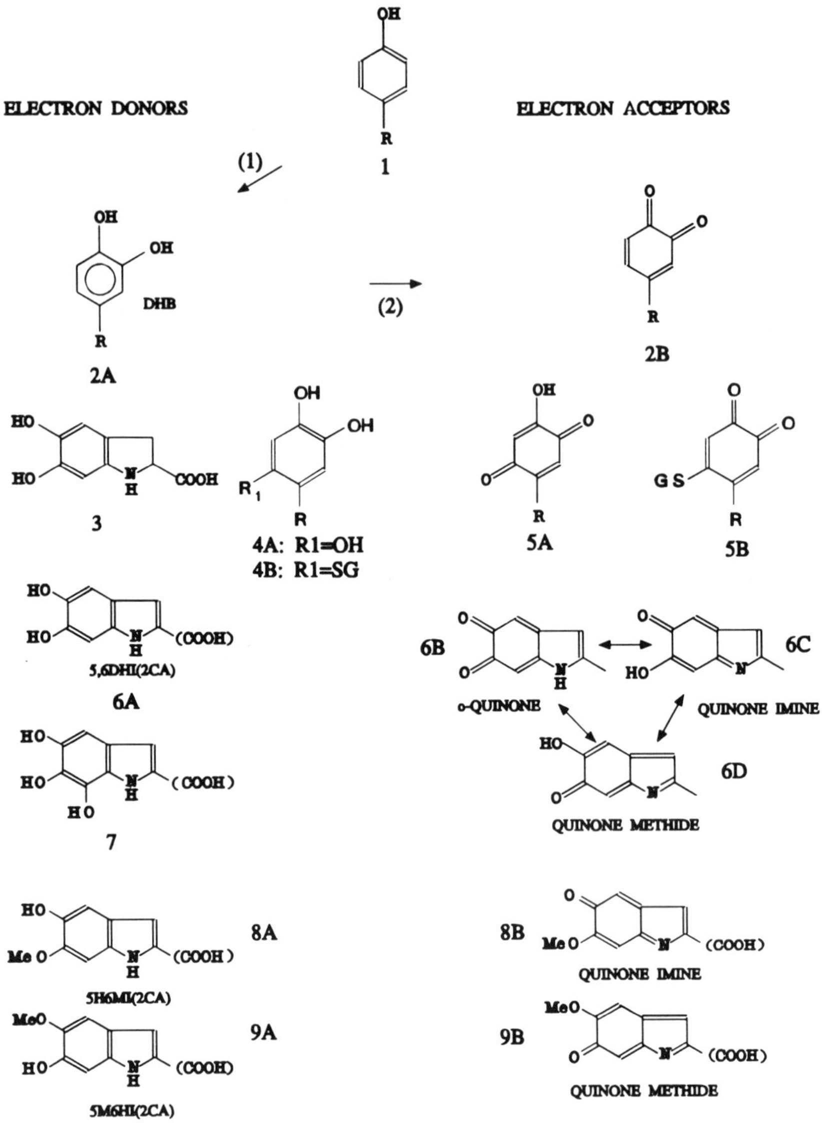
ELECTRON DONORS
ELECTRON ACCEPTORS
OH
R
(1)
1
OH
OH
DHB
R
2A
(2)
O
O
R
2B
HO
HO
N
H
COOH
3
OH
OH
R1
R
4A: R1=OH
4B: R1=SG
OH
O
O
R
5A
O
O
GS
R
5B
HO
HO
N
H
(COOH)
5,6DHI(2CA)
6A
6B
O
O
N
H
o-QUINONE
O
HO
N
6C
QUINONE IMINE
HO
O
N
6D
QUINONE METHIDE
HO
HO
N
H
(COOH)
HO
7
HO
Me O
N
H
(COOH)
8A
5H6MI(2CA)
8B
O
Me O
N
(COOH)
QUINONE IMINE
MeO
HO
N
H
(COOH)
9A
5M6HI(2CA)
9B
MeO
O
N
(COOH)
QUINONE METHIDE

Scheme 1

hydroxylation (Scheme 1-2A) (hydroxylase activity) and then to oxidize DHB to *ortho*-quinones (Scheme 1-2B) (oxidase activity). Eumelanin synthesis is initiated by tyrosinase-mediated hydroxylation of tyrosine (1: $R = CH_2CHNH_2COOH$) to DOPA (catechol derivative) (Scheme 1-2A) and oxidation of the latter to dopaquinone (Scheme 1-2B), which cyclizes to form leukodopachrome (Scheme 1-3). Leukodopachrome is then converted to 5,6-dihydroxyindole-2-carboxylic acid (5,6DHI2CA) (Scheme 1-6A) and its derivatives, which form the polymers of eumelanin. Pheomelanin is produced by the same metabolic pathway as is eumelanin, but in the presence of cysteine or reduced glutathione (GSH), which adds to dopaquinone (Scheme 1-2B) to produce cysteinyldopas, 5-S-cysteinyldopa (analog of 4B in Scheme 1) being a major isomer. In subsequent steps, 5-S-cysteinyldopa undergoes tyrosinase-catalyzed oxidation to 5-S-cysteinylquinone (Scheme 1-5B), which cyclizes to give rise to pheomelanin subunit, 4-hydroxy-benzothiazine (see Fig. 3).

B. Tyrosinase-Mediated Generation of Free Radicals

Melanins are composed of several types of electron donors and acceptors. Electron donors are primarily phenols and polyhydroxybenzene or indole derivatives. Electron acceptors are mostly quinones and quinone amines (see Scheme 1).

The interaction between electron donors and acceptors may result in electron transfer and production of free radical products. The likelihood of such an event is determined by one-electron redox potentials of both reactants. Values of redox potentials of some melanogenic compounds are summarized in Table 1. The values of the one-electron oxidation potentials of di- and trihydroxybenzene derivatives are much lower than those of phenol or tyrosine. Thus tyrosinase triggers the formation of better electron donors than phenol itself. Using electron paramagnetic resonance (EPR) spectroscopy, Mason et al. (17) demonstrated that tyrosinase catalyzes the formation of free radicals from catechol. It was found that catechol is oxidized directly to *o*-quinone by tyrosinase. This implies that the enzyme itself does not produce free radicals directly but promotes their formation indirectly through the dismutation of catechol and *o*-quinone:

$$QH_2 + Q \rightleftarrows 2Q^{\cdot -} + 2H^+ \qquad (1)$$

Semiquinones produced by one-electron oxidation of catechol (amine)s or one-electron reduction of the respective quinones are generally referred to as *primary radicals.* In many chemical, photochemical, and enzymic oxidation processes EPR spectra of catechol (amine)s change with time, indicating

Table 1 One-Electron Redox Potential (mV vs. NHE) of Some Redox Couples at pH 7.4

Redox couple	E_7	Ref.
$Q^{\cdot -}/QH_2$		
Catechol	530	37
p-Hydroquinone	459	37
DOPA	460	113
6-OH-DOPA	−110 (pH 13.5)	36
$Q/Q^{\cdot -}$		
o-Benzoquinone	210	102
p-Benzoquinone	99	102
$ArO^{\cdot}/ArOH$		
Phenol	950	113
Tyrosine	940	38
4-Hydroxyanisole (4-HA)	600	37
5-Hydroxyindole (5HI)	216 (pH 13.5)	37
5-Hydroxytryptophan (5HT)	208 (pH 13.5)	
$O_2/O_2^{\cdot -}$	−155	102

that the primary radicals are transformed into secondary free radicals. Mechanisms by which these secondary semiquinones are formed are complex and may vary depending on the types of the phenol/catechol substrate, pH of the incubation medium, and the presence of nucleophiles (e.g., cysteine or glutathione). The key intermediate in such a transformation is quinone (Scheme 1-2B), the ultimate metabolite of tyrosinase. Three possible mechanisms for this transformation have been identified. The first is the cyclization of primary quinones (Scheme 1-2B) similar to that observed during oxidation of DOPA to give rise to leukodopachrome *3* (Scheme 2, path A) or during oxidation of dopamine (18,19). The second involves the formation of a trihydroxy derivative (Scheme 1-4A) through nucleophilic addition of water to quinones. This process is particularly pertinent to quinones which do not cyclize, e.g., those derived from 4-methylcatechol (Scheme 2, path B) or *N*-acetyltyrosine (20,21). The third one is the alkylation of primary quinones by strong nucleophile, e.g., cysteine or glutathione, to give conjugate 4B (Scheme 2, path C). Electrochemical studies revealed that the rate of this addition process is three orders of magnitude faster than that of the intramolecular cyclization (18). Oxidation of these secondary products gives rise to *secondary semiquinone radicals*.

Recent EPR studies demonstrated the formation of free radicals from phenols such as 4-hydroxyanisole (4-HA) and phenolic thioether amines

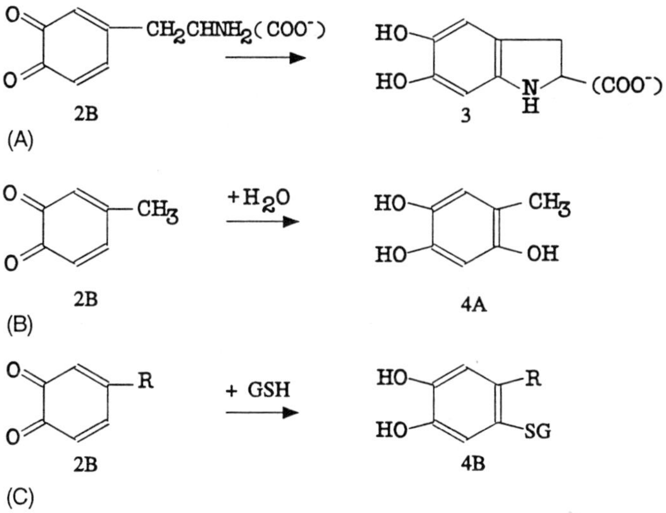

Scheme 2

(PTEAs), e.g., 4-S-cysteaminylphenol (CAP), 4-S-cysteinylphenol (CP), N-Ac-4-S-CAP (Scheme 3), and from catechol (amine)s during their oxidation by tyrosinase. 4-HA has been extensively studied as this phenol is quite efficient in producing depigmentation of the skin. Although the primary cytotoxic species produced by tyrosinase was the 4-HA–derived *ortho*-quinone (22), the phenol-derived free radicals might also be involved (10,22). In the presence of mushroom tyrosinase, 4-HA gave rise to both primary and secondary semiquinones (23,24). The two free radicals were, however, short-lived

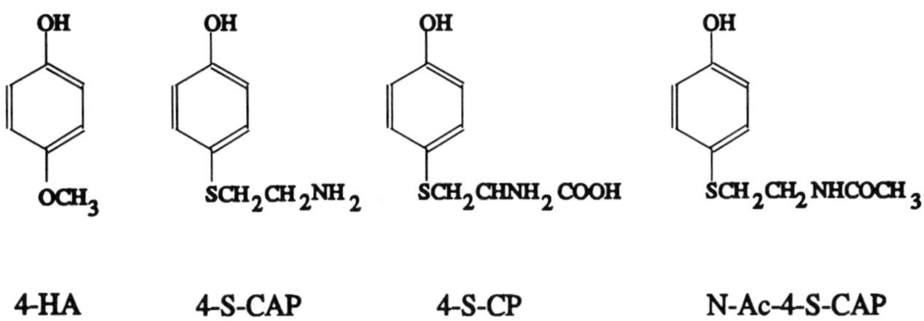

Scheme 3

and could be observed only in a flow system. The primary free radical appeared to be formed by disproportionation of 1,2-dihydroxyanisole and *o*-quinone (Scheme 1-2A and 2B, respectively, R = OCH_3) in a reaction analogous to that shown in Eq. 1. The secondary radical could not be unambiguously identified, but it might be formed either by oxidation of a trihydroxy derivative of 4-HA (Scheme 1-4A, R = OCH_3) to hydroxy-*p*-semiquinone or by oxidation of a dimer formed through oxidative coupling of the respective quinones (23,24). No evidence for the formation of superoxide ion was found (24). A similar conclusion was reached by a pulse radiolysis study, indicating that, although 3,4-dihydroxyanisole produced a semiquinone radical, the latter species was not oxidized by molecular oxygen to superoxide radical and quinone (25). The finding was in agreement with the known higher electron affinity of single benzoquinones compared to that of molecular oxygen (26) (see Table 1). Results of these experiments suggested that cytotoxicity and depigmenting potency of 4-HA are not associated with the formation of the semiquinone radicals and their redox cycling.

We found that free radicals from phenol treated with tyrosinase can be detected using the EPR-metal ion spin stabilization approach. Figure 4A shows the EPR spectrum of the primary *o*-semiquinone/Zn^{2+} complex superimposed on that of a secondary radical. Plausible events leading to the radical's formation are depicted in Figure 4B.

Tyrosinase-mediated activation of cyclizing and noncyclizing catecholamines to produce free radicals has been studied, using flow system and EPR spectroscopy (27). Kinetic measurements of the formation and decay of primary radicals from DOPA, dopamine, *N*-acetyldopamine, and 4-methylcatechol showed that semiquinones formed from noncyclizing quinones decayed more rapidly than those from cyclizing quinones.

Using a static system, we found that the formation of secondary radical from catechol treated with tyrosinase is a rapid process because the EPR spectrum recorded approximately 1 minute from the start of the reaction contained both the primary and secondary radicals (Fig. 5A). Figure 5B shows the EPR spectrum of a semiquinone radical from DOPA treated with mushroom tyrosinase. This spectrum was more stable, which is in agreement with earlier reports (27).

Phenolic thioether amines (PTEAs) (see Scheme 3) have been shown to be selectively melanocytotoxic (28–31). The mechanism of this action was associated with the formation of cytotoxic metabolites, primarily electron-deficient *ortho*-quinones through tyrosinase oxidation (28) or toxic aldehydes by action of monoamine oxidase (32). In our recent EPR experiments PTEAs became activated and generated semiquinone free radicals after exposure to mushroom tyrosinase (33). Spectra of these radicals were observed in a static system even in the absence of stabilizing metal ions. However,

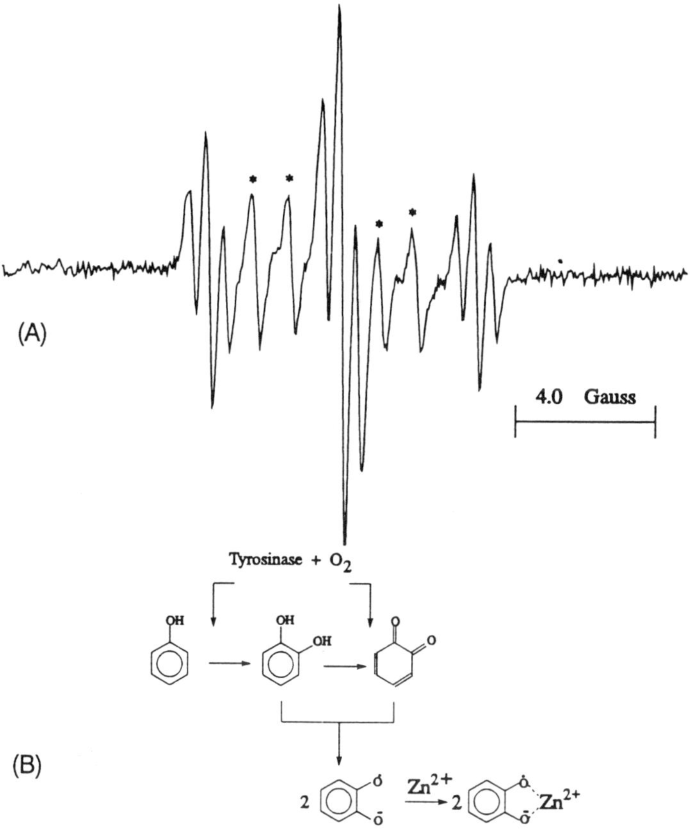

Figure 4 (A) EPR spectrum of semiquinone free radicals generated from phenol (10 mM) by mushroom tyrosinase (770 U/mL) in the presence of zinc acetate (90 mM) in acetate buffer pH 4.5. Sample was saturated with oxygen for 2 minutes before enzyme addition. Primary $SQ^{\cdot}/Zn^{2+}$ complex: 0.48 G (2H), 3.89 G (2H). Secondary radical (marked with asterisks): 0.89 G (H), 3.68 G (H). Zinc ions were added to increase stability of the radicals. (B) Scheme of the enzymic transformation of PhOH into catechol and into primary semiquinone/Zn^{2+} complex.

when Mg^{2+} or Zn^{2+} ions were present, these EPR spectra became much stronger, indicating that the radicals had *ortho*-semiquinone symmetry (19,34).

The crucial role of electron donor and acceptor interactions during melanin pigmentation has been demonstrated by studying the mechanism of inhibitory action of *p*-hydroquinone (H_2Q) on melanin formation from DOPA (35). It was found that, in a system composed of tyrosine, tyrosinase, O_2,

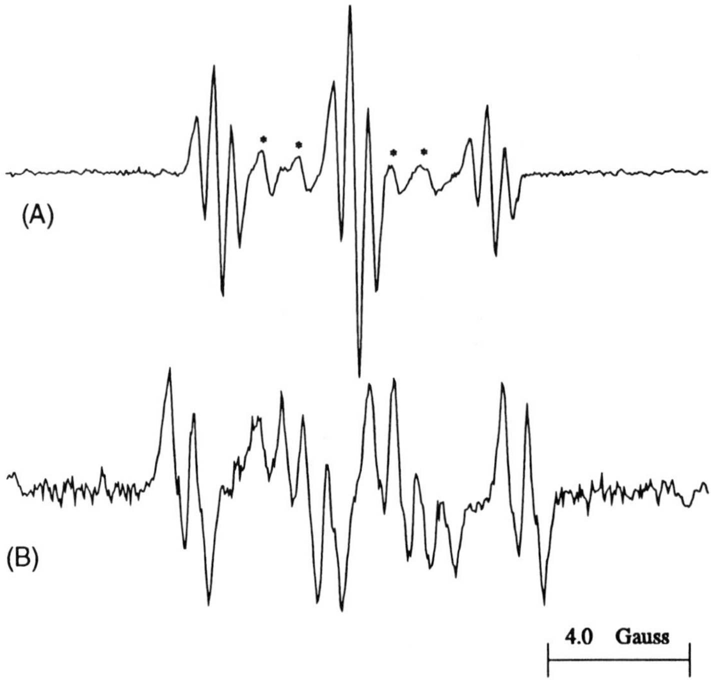

Figure 5 EPR spectra of semiquinone/Zn^{2+} from catechol (A) hyperfine couplings 0.48 G(2H), 3.89 G(2H) (primary radical) and 0.89 G(H), 3.68 G(H) (secondary radical indicated by asterisks) and from DOPA (B) hyperfine couplings of 0.24 G, 3.79 G, 0.7 G, 2.59 G, 3.14 G (all hydrogens). Samples were prepared in aerated acetate buffer pH 4.5 and contained mushroom tyrosinase (155 U/mL), zinc acetate (40 mM), and catechol (2 mM) (A) or DOPA (1.8 mM) (B).

and H_2Q, dopaquinone was constantly reduced back to DOPA by H_2Q, and H_2Q was simultaneously converted into 1,2,4-trihydroxyquinone in a concerted action of the enzyme and DOPA. Therefore, as long as H_2Q concentration was high, DOPA could not be consumed and melanin synthesis was inhibited.

III. OXIDATIVE ACTIVATION OF MELANOGENIC TYROSINASE SUBSTRATES

Melanogenic phenols, catechol (amine)s, and hydroxyindoles that are the substrates of tyrosinase can be oxidized to free radicals by nontyrosinase-mediated reaction.

A. Radiolytic Oxidation

Phenoxyl radicals ($PhO^{\cdot}$) are formed during radiolytic oxidation of phenol or tyrosine (36–38). They are very strong oxidizing agents, as indicated by their redox potential (see Table 1). They may, therefore, oxidize many biological electron donors. Because $PhO^{\cdot}$ is a one-electron oxidant, the semiquinone radical from hydroquinone is an obligatory intermediate. EPR studies have shown that phenoxyl radicals can oxidize catechol to 1,2-benzosemiquinone radical (19). Phenoxyl radicals oxidize NADH (39) and ascorbate, and they react with superoxide radical and microsomal electron transport system (36,40). Phenoxyl radicals dimerize through the formation of C–C and C–O bonds. These dimers are better electron donors than the starting phenols (41,42).

5,6-Dihydroxyindole-2-carboxylic acid (5,6DHI2CA) and its decarboxylated analog 5,6DHI (Scheme 1-6A) are produced during melanin pigmentation (see Fig. 3). By the action of the enzyme O-methyl transferase (or dopachrome tautomerase), these dihydroxyindoles may be transformed into mono- and dimethoxylated derivatives, 5H6MI(2CA), 5M6HI(2CA), and 5,6DMI(2CA), respectively (Scheme 1-8A and 8B). Oxidation of these indoles may lead to free radical formation. 5,6DHI, 5,6DHI2CA, and 5,6-dihydroxyindole-1-methyl (5,6DHI1Me) gave rise to semiquinone and semiquinone imine cation radicals when oxidized by radiolytically produced azidyl radical, $N_3^{\cdot}$ (43,44). Free radicals derived from carboxylated DHI were more stable than radicals from decarboxylated DHI. The increased stability may be the consequence of the presence of the ionized carboxylic group in the molecule. 5,6-Indole quinone, quinone imine, and quinone methide (Scheme 1-6B, 6C, and 6D, respectively) were found to be the secondary products of 5,6DHI oxidation (43). In Scheme 1, 6B was formed by dismutation of the initially produced semiquinones and then rearranged to 6C and 6D. Decay of quinones was followed by the formation of trihydroxyindole derivative THI (Scheme 1-7). Using selectively methoxylated hydroxyindoles, the intermediate responsible for the formation of THI (Scheme 1) was identified to be quinone methide (45) (Scheme 1-6D).

Thus, based on radiolytic reactions, melanin pigmentation process may be viewed in such a way that the melanin synthesis is initiated by the formation of semiquinone radical from a dihydroxyindole followed by its transformation to quinone indole, its tautomers, and THI product. THI reacts rapidly with 5,6-indolequinone, quinone imine, or quinone methide to give rise to dimers or oligomer products. In contrast to the mechanism of melanin formation suggested by photochemical reactions (46), this model of melanin synthesis does not attribute any essential role to phenoxyl radicals (43).

B. Electrochemical and Chemical Oxidation

The oxidation of catechol(amine)s to quinones, followed by the formation of cyclized products and the nucleophilic attack of quinones, can be studied electrochemically (18,47). The oxidation potentials of catecholamines have been polarographically determined, and the rates of their auto-oxidation, the ability to form cyclized products, and their cytotoxicity were compared (48). It was inferred that a readily oxidized trihydroxy benzene derivative, e.g., 6-OH dopamine, exerts its cytotoxicity through the production of free radical species (semiquinone, superoxide radical, hydroxyl radical), while the cytotoxicity of other catecholamines such as N-acetyldopamine, which possesses a higher oxidation potential, may be related largely to its reaction with the nucleophilic group within the cell. Di- and trihydroxylated derivatives of 5-hydroxytryptophane were formed through activation by electro-oxidation at an acidic pH (49). It has been shown that the electrochemically produced chlorpromazine radical cation is reduced by catecholamines (50).

Catechol(amine)s can be oxidized by superoxide radical and then produce semiquinone radicals and hydrogen peroxide (21,51–55). Trihydroxybenzenes such as 6-OH-DOPA are particularly reactive because they have low oxidation potentials (see Table 1) and readily undergo oxidation to form 5-hydroxy-*p*-semiquinone and superoxide radicals. This process requires a metal catalyst (55,56).

Auto-oxidation of 5-hydroxyindole, serotonin, and 5-hydroxytryptophan at an alkaline pH (1 N NaOH) was studied using EPR spectroscopy (57). The observed EPR spectra were consistent with those of semiquinone imine radicals with partial localization of the unpaired electron on the indole nitrogen. The presence of the 5-hydroxyl group was necessary for this radical formation because tryptophan and tryptamine, which contained an indole ring but no 5-OH group, melatonin, and 5-methoxyserotonin gave no detectable EPR spectra. In the EPR experiments the oxidation of 4-methyl catechol by periodate in the presence of nucleophiles (amino acids, peptides, and proteins) was shown to produce ring-substituted *o*-semiquinones (58).

C. Photochemical and Photosensitized Oxidations of Melanin Intermediates

The formation of semiquinone radicals during UVL radiation of catechol (amine) has been confirmed by a number of recent EPR experiments (19, 34). The radicals were produced by photoionization or by photohomolysis of the phenolic OH groups (59,60). From the EPR spin trapping experiments the proportion of photoionization to photohomolysis of DOPA was estimated to be 2.6 (59). A more recent study, using laser-flash photolysis (265 nm) of DOPA, indicated that these two processes occur with approximately equal yields (61).

UV photolysis of 5,6DHI (Scheme 1-6A) generated the EPR spectrum of *ortho*-semiquinone radical (34). Photolysis of 5,6DHI1Me in methanol produced a dark-brown polymeric material (46) and various oxidation products. It was postulated that this photolysis process is initiated by UVL-induced generation of phenoxyl radicals, which either form oligomers by oxidative coupling or react with oxygen to give rise to hydroperoxides.

EPR and the spin trapping studies have shown that, upon exposure to UVL radiation (>300 nm), 5,6-dihydroxyindole (5,6DHI) primarily photoionized while 5,6-dihydroxyindole 2-carboxylic acid (5,6DHI2CA) (Scheme 1-6A) and 5-methoxy-6-hydroxyindole-2-carboxylic acid (5M6HI2CA) (Scheme 1-9A) underwent photohomolysis (62). Our recent EPR study, however, provided direct evidence for the formation of the indoloxyl free radical from monomethoxy-monohydroxyindoles (Scheme 1-8A) during UVL photolysis (>300 nm) (63).

Intermediates (or metabolites) of pheomelanin pigment are more photolabile than eumelanic compounds. 5-S-Cysteinyldopa (5SCD), a major pheomelanin intermediate, was found to be about one order of magnitude more photolabile than DOPA during 254 nm photolysis (60). UVL irradiation of 5SCD at 280 nm and 290 nm resulted primarily in the homolysis of phenolic OH groups to release hydrogen atom and in the cleavage of the carbon-sulfur bond to produce alanyl, $CH_2CHNH_3^+COO^-$, and aryl thiyl, Ar-S$^\cdot$, radicals (Scheme 4). Formation of the alanyl radical has been confirmed using 2-methyl-2-nitrosopropane as a spin trap, but the thiyl radical was not detected presumably due to its high reactivity or instability of its adducts (60). Photoionization was found to be a dominating event during photolysis (260 nm) of 5SCD, 2,5-S,S-dicysteinyldopa (2,5SCD), 4-hydroxybenzothiazole, and pheomelanin itself (64). Upon laser-flash photolysis (267 nm), 5SCD and

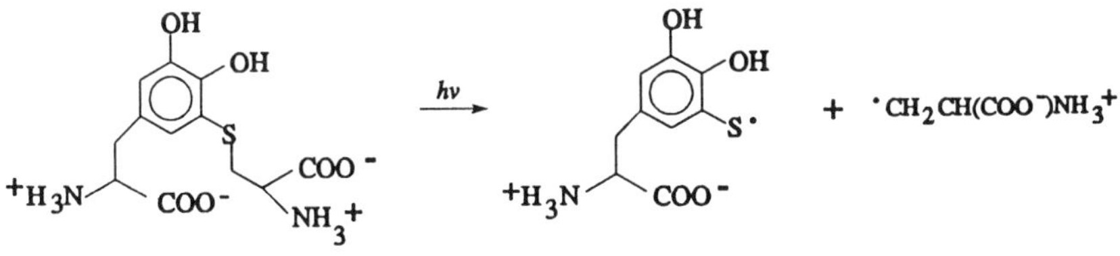

5-SCD Ar-S·

Scheme 4

2,5SCD produced less hydrated electrons as compared to those of DOPA photolysis and yielded a transient radical species which was not semiquinone (61). It is likely that the latter species was formed by the photohomolysis of carbon-sulfur bond in 5SCD (60). Thus, the difference between the eumelanin and pheomelanin intermediates (metabolites) is primarily based upon the fact that the UVL photolysis of DOPA generates semiquinone radicals, whereas 5-S-CD yields alanyl and aryl thiyl radicals.

Melanin intermediates quench the excited triplet state of photosensitizers. This process was shown for lumiflavin and eosine (65), rose bengal (66), hematoporphyrin (67,68), tris(2,2′-bipyridyl) ruthenium (II) complex (69), and anthrapyrazoles (70,71). Using EPR spectroscopy it was possible to observe the formation of free radicals from sensitizing compounds and/or from catechols, indicating that the quenching occurs via an electron transfer from catechol(amine) to a sensitizer in the triplet state.

IV. MELANIN AS AN ELECTRON DONOR AND ELECTRON ACCEPTOR

A. Free Radicals in Melanin Pigmentation

Melanin pigmentation produces persistent free radicals. This was evidenced by EPR measurements (8,72). The melanin free radicals are generated by the electron exchange between electron donors and acceptors produced in situ during melanin synthesis. Eumelanin and pheomelanin possess distinct EPR spectroscopic characteristics (9,73). The major difference between them is that dopa-eumelanin radicals are *ortho-O,O*-semiquinones derived mostly from dihydroxyindoles, while pheomelanin radicals (e.g., 5SCD) are *ortho-N,O*-semiquinones derived from hydroxy-1,4-benzothiazine (HBZT) subunits (Scheme 5).

The EPR spectrum of eumelanin is a single, structureless and slightly asymmetric line with linewidth of 4–10 G and g ca. 2.004. In contrast, the EPR spectrum of cysteinyldopa pheomelanin possesses three lines of hyperfine structure due to the interaction of the unpaired electron with the ^{14}N nucleus present in HBZT. Its overall linewidth is ca. 30 G with a g value of ca. 2.005 (73). It has been proposed that the majority of melanin free radicals exists in equilibrium with nonradical groups (9,74). These equilibria are depicted in Scheme 5 for eumelanin (A) and pheomelanin (B) subunits.

The functional groups localized at or near the polymer surface are susceptible to environmental changes. Any change in the environment may lead to some shifts in these equilibria and consequently may change the melanin EPR signal intensity corresponding to a new free radical concentration. The increase of the melanin EPR signal upon alkalization or upon addition

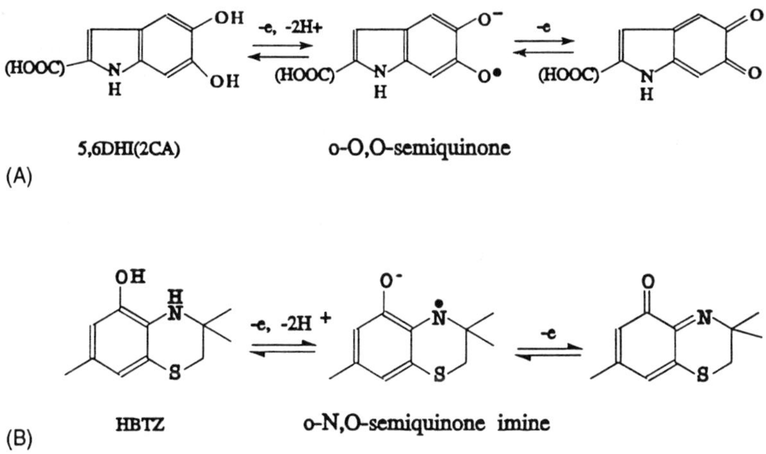

Scheme 5

of diamagnetic di- and trivalent metal ions are associated with the formation of new equilibria with a higher proportion of semiquinone radicals. Illumination and temperature changes also increase the free radical content in melanins (8). Figure 6 shows EPR spectra of DOPA melanin in the dark (A) and during exposure to UVL ($>$300 nm) (B). Generally, the formation

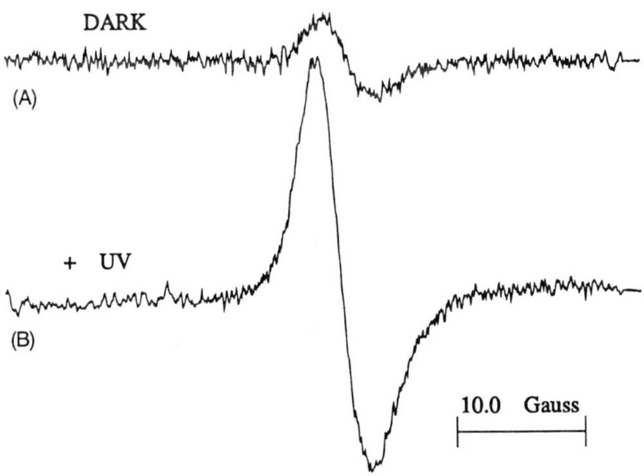

Figure 6 EPR spectra of synthetic DOPA melanin in phosphate buffer pH 7.0 in dark (A) and during exposure to UV light ($>$300 nm) (B).

of new radicals from melanin pigments is a reversible process, although prolonged incubation of melanins at an alkaline pH and their exposure to an intense light may lead to an irreversible degradation of the polymer and loss of its natural EPR signal.

B. Electron Transfer Processes

Because of the simultaneous presence of both electron-donating and electron-accepting groups in melanin pigments, the polymer is capable of exchanging electrons with various electron donors and acceptors, hence, transferring electrons from one compound to another. In an early electrochemical study (75), it was found that synthetic DOPA melanin in an aerated aqueous solution at a physiological pH exists largely in the oxidized form. It was calculated that the number of exchanged electrons per indole unit is 0.02 for the oxidation process and 0.48 for the reduction process. This estimate was qualitatively in agreement with the fact that an untreated melanin pigment is more prone to be reduced than oxidized.

Free radicals can also be produced from melanin pigments by one-electron reduction of quinone subunits and one-electron oxidation of hydroquinone subunits. Simultaneous electrochemical and EPR measurements have shown that melanin pigments can exchange electrons with mercury and platinum electrodes (76). Electro-reduction and electro-oxidation of natural and synthetic melanin pigments in aqueous solutions strongly affected the intensity of their EPR signals. Distinct phases of the increase and decrease of melanin EPR signal intensity were observed during melanin reduction. These changes were interpreted as one-electron reduction of melanin-quinone groups (Q) to give rise to semiquinone free radicals ($Q^{\bullet -}$), which were then reduced to EPR-silent hydroquinones (QH_2) by transfer of a second electron at more negative potentials. Electro-oxidation of such a reduced pigment produced first an increase and then a decrease in the intensity of EPR signal which corresponded to stepwise oxidation of QH_2 to $Q^{\bullet -}$ and then to Q. These observations indicated that melanin pigments accept and donate electrons in two discrete one-electron steps in agreement with the quinonoid nature of melanin polymers. Changes in the EPR signal intensity were much larger during electro-reduction than electro-oxidation, which might be related to a higher proportion of oxidized groups in the polymers.

1. Chemical Reduction and Oxidation

Melanin pigments can accept electrons from various reducing agents including ascorbic acid, sodium borohydride (77), NADH, and NADPH (78,79), and donate electrons to ferricyanide, dichlorophenolindophenol, and oxidized forms of cytochrome *c* and oxygen (80,81). Melanin pigments catalyze the oxidation of NADH coupled with reduction of ferricyanide (80). The rate of NADH oxidation, when all three components were present, was higher than the sum of rates of reduction of ferricyanide by NADH alone and oxidation of NADH by melanin pigments. It was also observed that the same melanin sample after being "discharged" during reduction of ferricyanide could be reactivated by NADH, which enabled it to continue the reduction process. This procedure could be repeated several times. It was therefore postulated that melanin pigment can function as an electron carrier, which transfers electrons from a donor (NADH) to an acceptor (ferricyanide). This particular behavior is common to melanin pigments derived from various substrates (DOPA, dopamine, adrenaline, adrenochrome, hydroquinone) and obtained by both enzymic oxidation and auto-oxidation processes (81).

Interaction of melanin pigments with electron donors or electron acceptors may generate new melanin radicals, presumably through one-electron reduction of quinone or one-electron oxidation of hydroquinone subunits in the polymers. Reduction of melanins by ascorbic acid was reported either to have no effect on the melanin EPR signal (82) or to increase its intensity (77). This discrepancy might result from different experimental conditions and/or from the fact that the electron-acceptor capacity of melanins can be significantly modified during their preparation.

DOPA melanin was shown to generate new free radicals when reduced by NADH (79). This process was significantly enhanced when the DOPA melanin plus NADH system was irradiated with UVL. The formation of new melanin radicals was related to the enhanced oxygen consumption, superoxide, and hydrogen peroxide production during exposure of melanin to UVL (83). Addition of Cu^{2+} ions to the melanin and NADH sample generated more free radicals in melanin as compared to those of a metal ion–free sample (79). In a recent EPR study employing redox-active metal ions, melanin generated hydroxyl radicals at the expense of hydrogen peroxide and melanin hydroquinone groups (84). It is therefore likely that the catalytic effect of Cu^{2+} ions on NADH oxidation and radical formation might derive from the intermediacy of $OH^{\cdot}$ radicals, which oxidized NADH to a strongly reducing $NAD^{\cdot}$ radical. This radical could reduce quinone groups of melanin to semiquinone radicals, or, alternatively, $NAD^{\cdot}$ could react with oxygen (85) to produce superoxide radicals, which then react with melanin (86,87).

2. Reaction with Phenols

Another group of compounds known to interact with melanin are phenols. The melanin-phenol interaction may have two components. The first is the binding of a phenol to melanin surface in a way similar to that of melanin binding to other aromatics (88). The second is the electron-exchange process with melanin, which then functions as an electron acceptor. It is likely that the binding of an electron donor to an electron acceptor may physically facilitate an electron transfer process. This type of interaction was reported earlier for melanin-chlorpromazine (CPZ) interaction (89). The melanin EPR signal was decreased upon CPZ binding, suggesting the reduction of polymer semiquinone radicals by CPZ (89).

Many phenols and catechols possess melanocytotoxic and depigmenting properties (10,90). It has been shown that *p-tert*-butylcatechol (tBC) binds to and reduces DOPA melanin, which acts as a catalyst, transferring electrons from tBC to ferricyanide (91). The melanocytotoxic action of catechols may thus be attributed not only to accumulation of tyrosinase-mediated cytotoxic products, but also to a decreased ability of the pigment to function as an electron carrier after binding to catechols. Therefore, homologs of phenols and catechols with different depigmenting potencies were tested for their ability to reduce ferricyanide in the presence of DOPA melanin (92). The tested compounds included catechol (CAT), 4-*t*-butylcatechol (tBC), 4-methylcatechol (MCAT), DOPA, 3,4-dihydroxyphenylacetic acid (DHPA), hydroquinone (H_2Q), 4-hydroxyanisole (4-HA), 4-*t*-butylphenol (*t*BP), and 2,6-*di-t*-butyl-4-methylphenol (BHT). The catalytic action of DOPA melanin was characterized quantitatively by comparing the rate of ferricyanide reduction in the presence and absence of DOPA melanin (EF factor) and also by measuring the binding constant of the phenols to melanin. This resulted in three major groups of agents. The first group included those agents with the most efficient depigmenting potency, i.e., *t*BC, MCAT, and 4HA, and they had relatively high binding constants of 2800, 2200, and 850 M^{-1}, respectively, and EF > 1. The second group of compounds included agents with the least depigmenting potency, i.e., DOPA and DHPA, which bound to melanin but then retarded ferricyanide reduction (EF < 1). The third was *t*BP, which revealed depigmentation potency but did not reduce ferricyanide. From this study a direct relationship between the magnitude of EF and the depigmenting potency was obtained. In addition, some consequences for the in vivo situation were envisaged. It was postulated first that the binding of a depigmenting phenol to melanin may physically facilitate the interaction of the agent with tyrosinase and second that melanin can modify the phenol-tyrosinase-O_2 reaction.

The possible participation of melanin pigment itself in the interaction between tyrosinase and the depigmenting agent was studied using 4HA as a model compound (93). The oxygenation of 4HA by tyrosinase was markedly accelerated in the presence of sepia and synthetic (from Sigma Chemicals). Therefore, it was postulated that melanin pigments might have participated directly in the control of pigmentation in vivo. The catalytic effect of melanin depended on its redox state, and the impact of the latter factor was different in the phenol hydroxylation and dehydrogenation reactions. Compared to the untreated pigment, ferricyanide-oxidized melanin was less efficient for the catalysis of hydroxylation but more efficient for the dehydrogenation.

3. *Reaction with Stable Free Radicals*

Melanin pigments eliminate the EPR signal of the stable radical, α,α-diphenyl-β-picrylhydrazyl (DPPH) (94). Because there was no change in the EPR signal of melanin itself, melanin radicals did not appear to be involved in DPPH reduction. A stable nitroxide radical, such as 4-trimethylammonium-2,2,6,6-tetramethylpiperidine-1-oxyl (TTPO), was shown to be reduced to the EPR-silent hydroxylamine by melanin (95) (Eq. 2). The process was reversible because the EPR signal was recovered when an excess of the one-electron oxidizing agent, ferricyanide, was added to the system of melanin plus TTPO.

$$>N-O + Mel_{red} \rightleftarrows >N-OH + Mel_{ox} \tag{2}$$

Nitroxides bearing a positive charge were reduced more easily than those bearing a negative charge. This reflected a negative charge of carboxylate groups at a neutral pH on the surface of melanin, which inhibited the interaction with any anionic species but facilitated interaction with cationic species.

Melanin pigments differ in their abilities to reduce nitroxides. DOPA melanin prepared by enzymic oxidation reduced TTPO at a faster rate and to a larger extent than that prepared by auto-oxidation. Natural melanin (from bovine eyes) was less reactive than synthetic DOPA melanin obtained by auto-oxidation. Synthetic pheomelanin appeared to be even less reactive than the natural melanin. In either case, however, the rate of the reduction was proportional to the melanin concentration, i.e., to the number of redox active groups in the polymer.

The redox status of melanin affects its ability to reduce the nitroxide (95). Reduced DOPA melanin completely eliminated the nitroxide EPR signal and at a rate much faster than that of the control, untreated polymer.

The effect of illumination on the melanin-nitroxide interaction was investigated in the presence and absence of oxygen in the system. In the absence of air, the irradiation of melanin with UVL (365 nm) promoted the elimination of the EPR signal of TTPO converting the nitroxide into hydroxylamine (Eq. 2). In the presence of air, the UVL irradiation promoted the recovery of that portion of the nitroxide that was reduced to hydroxylamine in the dark. The process was reversible because in the dark the signal returned to the initial level (dark equilibrium). Since the initial rate of hydroxylamine oxidation was the same, both in the presence and in the absence of SOD or catalase, it was concluded that this process does not involve species derived from oxygen but rather from melanin. These observations were consistent with the generation of both reducing and oxidizing species during the exposure of melanin to UVL.

4. *Reaction with Superoxide Radical*

One of the mechanisms by which melanin inhibits peroxidative processes is through scavenging the pro-oxidant superoxide radical. Goodchild and coworkers (96) demonstrated that melanin decreases the concentration of superoxide radical generated by the xanthene plus xanthene oxidase system (X-XO) and that the efficacy of this process is decreased upon completion of the pigment with chlorpromazine. EPR study showed that the low-temperature signal of $O_2^{\cdot -}$ (150 K) generated by X-XO or KO_2 is abolished by the addition of frog liver melanin (97). Because amphibian liver was devoid of SOD, it was postulated that the melanin pigment may play a role in the action of this detoxifying enzyme. Although all other tests consistently confirmed the ability of melanin to scavenge $O_2^{\cdot -}$, it should be remembered that the superoxide-generating system X-XO is not suitable for studying this reaction because melanin pigment inhibits XO activity (87).

Lyophilized melanoma tissue scavenged $O_2^{\cdot -}$ produced by DMSO-solubilized KO_2 while nonpigmented muscle tissue had no effect on the EPR signal of the superoxide radical (77 K) (98). $O_2^{\cdot -}$ was scavenged only by DMSO-soluble fraction (low molecular mass) of melanin, but not by DMSO-insoluble melanin (high molecular mass). Differences in the reactivity of the two fractions of melanin were shown by measuring their rates of reduction for the stable nitroxide radical TEMPOL. The ability to reduce the nitroxide EPR signal was strong in the supernatant of homogenized melanoma tissue (99% reduction within 5 minutes compared to 12% reduction by muscle tissue). Reduction was slower in nontreated melanoma cells and was much slower in melanoma tissue after removal of DMSO-soluble melanin. These findings may be relevant to the in vivo situation. Because of the permeability of melanoma cell membrane, cells may be surrounded by the low molecular weight soluble intermediates of melanin. It was therefore suggested that this soluble form of melanin pigment can play an important role in the protection of melanoma cells against macrophages, granulocytes, and other effector cells. The same mechanism may be responsible for the high resistance of melanoma cells against radiotherapy and chemotherapy, which produce reactive oxygen agents.

Because melanin pigments contain both reducing H_2Q and oxidizing Q groups, it was postulated that $O_2^{\cdot -}$ can interact with the pigment along two possible pathways:

$$H_2Q + O_2^{\cdot -} \rightleftarrows Q^{\cdot -} + H_2O_2 \qquad (3)$$

$$Q + O_2^{\cdot -} \rightleftarrows Q^{\cdot -} + O_2 \qquad (4)$$

In both cases new free radicals ($Q^{\cdot -}$) from melanin pigments were generated together with either hydrogen peroxide or oxygen (Eqs. 3 and 4). However, the occurrence of reaction 3 in aqueous solutions is questionable (99). The formation of new melanin radicals has been confirmed by EPR, which showed an increase of the melanin signal upon addition of $O_2^{\cdot-}$ (in the form of DMSO-solubilized KO_2). The newly formed free radicals decayed rapidly with bimolecular rate constants of 1.2–1.6 $\times$ 10^6 $M^{-1}s^{-1}$ depending on type of melanin pigment employed (87). The yield of radical formation was pH dependent and increased by a factor of 2.4 along with the increase of pH from 7.5 to 9.7. It was suggested that this effect may be due to the increased lifetime of $O_2^{\cdot-}$ at a more alkaline pH, which assures a more effective interaction between the pigment molecule and $O_2^{\cdot-}$. Diamagnetic metal ions further enhanced superoxide-induced formation of melanin radicals. To account for this observation, two possible mechanisms were proposed. First, it might be attributed to the known metal-ion spin stabilization of the newly formed semiquinone radicals (19,34). Second, it could be the result of complexation of $O_2^{\cdot-}$ with metal ions (100), which extended its lifetime and consequently exposed melanin pigment to a higher effective concentration of superoxide ions.

The ability of melanin pigment to react with $O_2^{\cdot-}$ has been compared to that of SOD. From the relative effectiveness of the two systems to quench the DMPO/$O_2^{\cdot-}$ EPR signal, it was found that the apparent rate constant k(mel $-$ $O_2^{\cdot-}$) is 2 $\times$ 10^5 $M^{-1}s^{-1}$. A value obtained from pulse radiolysis measurements is 5 $\times$ 10^5 $M^{-1}s^{-1}$ (12). A significantly higher value, k (DOPA melanin $-$ $O_2^{\cdot-}$) = 6 $\times$ 10^8 $M^{-1}s^{-1}$, was obtained by Lapina et al. (101) from an experiment in which DOPA melanin competed for $O_2^{\cdot-}$ with adrenalin. It was assumed that k(adrenalin $-$ $O_2^{\cdot-}$) is 4 $\times$ 10^4 $M^{-1}s^{-1}$. The superoxide radical was produced using X-XO and electrochemical reduction of molecular oxygen on a mercury electrode.

The ability of the superoxide radical to react with melanin may be related to the ability of this species to react with melanin precursors—catechol(amine)s (51–54). It was reported that $O_2^{\cdot-}$ oxidizes adrenalin to *ortho*-semiquinone radical and hydrogen peroxide (Eq. 3) with bimolecular rate constants of 5.6 $\times$ 10^4 $M^{-1}s^{-1}$ at pH 7.8 and 7.0 $\times$ 10^3 $M^{-1}s^{-1}$ at pH 9.5 (53). Thus, some $O_2^{\cdot-}$ can be scavenged by melanin pigment through the intermediacy of the pigment hydroquinone groups (Eq. 3).

Simple quinone, such as *para*-benzoquinone, possesses a much higher electron affinity than does molecular oxygen (25,36). For example one-electron redox potentials of 1,2-benzoquinone, 1,4-benzoquinone, and methyl-1,4-benzoquinone were reported to be +210, +99, and +23 mV, respectively (vs. NHE) while that of O_2/$O_2^{\cdot-}$ couple was $-$155 mV (102). Therefore for such quinones equilibrium described by reaction 4 was shifted markedly to the right. Because most melanin subunits exist in the oxidized form, the scavenging of the superoxide radicals via reaction with quinone groups (Eq. 4) is very likely. It was estimated that at a physiological condition of pH 7.0, approximately 30% of the superoxide produced during photo-oxidation of DOPA melanin was removed via reduction of $O_2^{\cdot -}$ to H_2O_2 (Eq. 3), and ca. 70% was removed via oxidation pathway (Eq. 4).

Superoxide and hydrogen peroxide were produced both during dark auto-oxidation of DOPA melanin (84) and during photo-oxidation of DOPA melanin (103,104). We found that UVL (>300 nm) photolysis of DOPA melanin generates readily a EPR spectrum of DMPO-O_2H (Fig. 7) under acidic conditions (e.g., pH 4.5). The acidic pH facilitated the formation of this adduct because the O_2H radical (pK_a = 4.7) adds to DMPO approximately 100 times faster than the superoxide radical. The formation of superoxide radical was coupled to oxygen consumption. The oxidation process was accelerated both by diamagnetic and by paramagnetic metal ions, with the latter ions being more effective, especially than Mn^{2+} ion. Dependence of

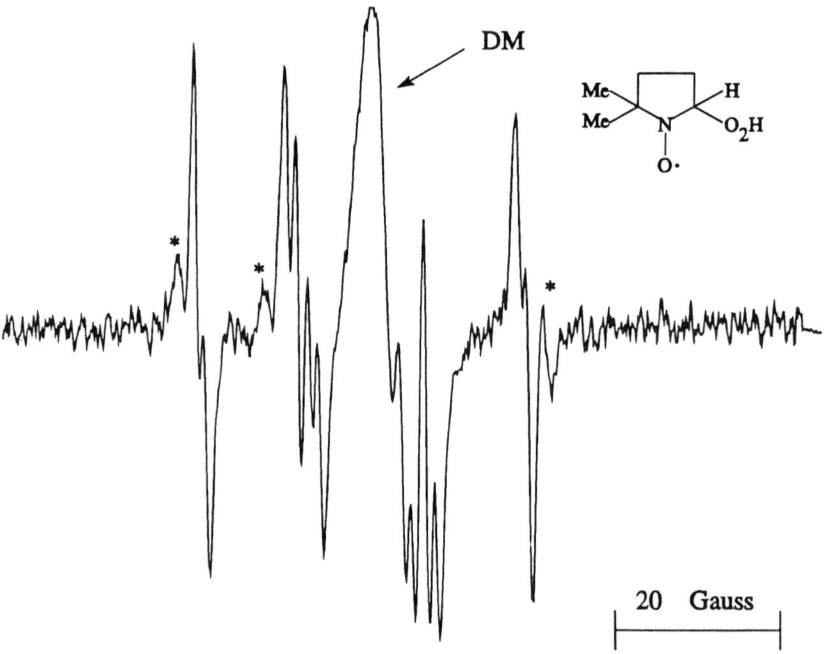

Figure 7 EPR spectrum of DMPO/O_2H adduct produced by UV (>300 nm) irradiation of synthetic DOPA melanin in aerated acetate buffer pH 4.5 [DMPO] = 80 mM. The central broad signal comes from DOPA melanin (DM), and lines from DMPO/OH adduct are marked with asterisks.

the amplitude of melanin EPR signal in the system of melanin plus Mn^{2+} on EDTA concentration indicates that the pigment competes with EDTA for the metal ions; the signal was more intense as EDTA concentration increased. Redox-active metal ions (Fe^{3+} and Cu^{2+}) decreased the rate of formation of H_2O_2, the main reason being that these ions catalyze decomposition of hydrogen peroxide via a Fenton-type cycle with the formation of hydroxyl radicals (84). Using 1,10-phenanthroline to chelate ferrous iron, it was shown that melanin is capable of reducing Fe^{3+} to Fe^{2+}. Further evidence for the occurrence of the Fenton cycle was obtained by the spin trapping method, which showed the formation of hydroxyl radicals. Possible steps leading to hydroxyl radical formation are shown in Equations 5–9 (DM = DOPA melanin):

$$Fe^{3+}\text{-EDTA} + DM \rightleftarrows Fe^{3+}\text{-DM} + EDTA \quad (5)$$

$$Fe^{3+}\text{-DM} \pm e \rightarrow Fe^{2+}\text{-DM} \quad (6)$$

$$Fe^{2+}\text{-DM} + EDTA \rightleftarrows Fe^{2+}\text{-EDTA} + DM \quad (7)$$

$$Fe^{2+}\text{-DM} + H_2O_2 \Rightarrow Fe^{3+}\text{-DM} + OH^{\cdot} + OH^{-} \quad (8)$$

$$Fe^{2+}\text{-EDTA} + H_2O_2 \Rightarrow Fe^{3+}\text{-EDTA} + OH^{\cdot} + OH^{-} \quad (9)$$

The superoxide radical mediates the formation of hydroxyl radicals during melanin photo-oxidation. In the presence of Fe^{3+}-EDTA complex, the illumination of melanin leads to hydroxyl radical formation, observed as DMPO/OH spin adducts (105). Because superoxide radicals were reported to mediate hydroxyl radical formation in the presence of ferric iron (Eqs. 10–12) (106), similar processes may occur during photo-oxidation of DOPA melanin.

$$O_2^{\cdot-} + Fe^{3+}\text{-EDTA} \Rightarrow Fe^{2+}\text{-EDTA} + O_2 \quad (10)$$

$$O_2^{\cdot-} + O_2^{\cdot-} + 2H^{+} \Rightarrow H_2O_2 + O_2 \quad (11)$$

$$Fe^{2+}\text{-EDTA} + H_2O_2 \Rightarrow Fe^{3+}\text{-EDTA} + OH^{\cdot} + OH^{-} \quad (12)$$

It was observed however, that SOD stimulated both the production of H_2O_2 and of DMPO/OH adduct. Therefore, it was suggested that Fe^{3+}-EDTA might be reduced either by melanin itself or by a melanin-derived product. Specifically it was postulated that reactions analogous to those occurring during dark auto-oxidation of melanin pigments in the presence of Fe^{3+} or Cu^{2+} could operate in this system (Eqs. 5–9) (84,107).

Photochemically generated semiquinone radicals in melanin pigment may contribute to the $OH^{\cdot}$ generation reducing Cu^{2+} or Fe^{3+} ions. Flash photolysis and pulse radiolysis studies have shown that several different types of semiquinones form radical–metal ion complexes with Cu^{2+} and Fe^{3+} ions (108). Similar reactions were observed with radical anions of quinoid anticancer drugs, which reduced ferric iron complexes of desferrioxamine, DETAPAC, EDTA, and ATP (109).

C. Melanin as Quencher of Excited States

Interaction of melanin pigment with the excited state of other molecules generates melanin free radicals. It has been observed that the illumination of rose bengal in the presence of DOPA melanin under anaerobic conditions increases free radical content of the pigment approximately 20-fold (110). The process was reversible. In the presence of air, illumination induced oxygen consumption and irreversible radical production. The rate of oxygen consumption was two orders of magnitude higher when compared to the rate of unsensitized reaction. Further stimulation of oxygen consumption was observed in D_2O, and the reaction was inhibited by azide, suggesting that singlet oxygen might be involved in this process. Comparative studies using various melanin pigments have shown that reversible radical production photosensitized by rose bengal is highest for cysteinyldopa melanin and much smaller for DOPA melanin, red hair melanin, red feather melanin, bovine eye melanin, immobilized DOPA melanin, and bovine eye melanosomes (111). The rate of the sensitized oxygen consumption was highest for cysteinyldopa melanin but decreased sharply for other melanin pigments. The value of the rate constant for the quenching of singlet oxygen by cysteinyldopa melanin was estimated to be approximately 6×10^5 $(mg/mL)^{-1}s^{-1}$, which was about five times greater than that estimated for DOPA melanin (110). From measurements of quenching of the singlet oxygen luminescence by DOPA melanin, the value for the rate constant was estimated to be $(12 \pm 6) \times 10^7$ $M^{-1}s^{-1}$ (112). This value was approximately one order of magnitude higher than that found for DOPA itself—$k(DOPA/{}^1O_2) = 3 \times 10^7$ $M^{-1}s^{-1}$ (68).

V. CONCLUSIONS

Melanin pigmentation of the skin has been generally regarded to possess a protective role against exposure to UVL. Melanin pigment may, therefore, provide a protective role against development of sunburn, solar degeneration and even skin cancer (14,15). However, evidence accumulated recently suggests that the photoprotective role of the melanin pigment may be restricted to eumelanin as well as to the visible and UVA ranges and that in the UVB and UVC ranges both eumelanin and pheomelanin components may behave as photosensitizers. UV photolysis studies using catechols, catecholamines, and hydroxylated indole derivatives (i.e., the eumelanin precursors) have demonstrated the formation of hydrated electrons, hydrogen atoms, semiquinone/semiquinone imine radicals, and indoloxyl radicals. The photolysis of pheomelanin precursors leads to the formation of hydrated electrons, hydrogen atoms, and to alanyl and aryl thiyl radicals. Thus, while eumelanin may act as a photoprotector, its precursors may be phototoxic. The photoprotection by melanin pigment can be accomplished by scavenging the reactive free radicals, by quenching excited states, and by decreasing oxygen concentration in exposed tissues. The phototoxicity is related to the generation of reactive free radicals by low molecular eumelanic and pheomelanic intermediates. It may, however, be possible to modulate the photosensitivity of the pigmentary system by affecting the redox status of endogenous melanin pigments, or by switching melanin pigmentation from eumelanogenesis to pheomelanogenesis, or vice versa, or by using sensitizers activated by red light, which is poorly absorbed by melanin.

ACKNOWLEDGMENTS

The authors wish to express their gratitude to Dr. Colin F. Chignell for his review of this chapter. This study was supported in part by grants from the National Cancer Institute of Canada, the Medical Research Council of Canada, the Alberta Cancer Board of Canada, and the Alberta Heritage Foundation for Medical Research.

REFERENCES

1. Cerutti PA. Prooxidant states and tumor promotion. Science 1985; 227:375–81.
2. Kensler TW, Taffe BG. Free radicals in tumor promotion. Pergamon Journals Ltd, Adv Free Rad Biol Med 1986; 2:347–87.
3. Black HS. Potential involvement of free radicals reactions in ultraviolet light-mediated cutaneous damage. Photochem Photobiol 1987; 40:213–21.
4. Yamaguchi T, Yamashita Y. Mutagenicity of hydroperoxides of fatty acids and some hydrocarbons. Agric Biol Chem 1980; 44:1675–8.
5. Moody CS, Hassan HM. Mutagenicity of oxygen free radicals. Proc Natl Acad Sci USA 1982; 79:2855–9.
6. Taffe BG, Takahashi N, Kensler TW, Mason RP. Generation of free radicals from organic hydroperoxide tumor promoters in isolated mouse keratinocytes. J Biol Chem 1987; 262:12143–9.
7. Guyton KZ, Bhan P, Kuppusamy P, Zweier JL, Trush MA, Kensler TW. Free radical-mediated quinone methide mediates skin tumor promotion by butylated hydroxytoluene hydroperoxide: expanded role for electrophiles and multistage carcinogenesis. Proc Natl Acad Sci USA 1991; 88:946–50.
8. Sealy RC, Felix CC, Hyde JS, Swartz HM. Structure and reactivity of melanins: influence of free radicals and metal ions. In: Pryor WA, ed. Free radicals in biology. Vol. IV. New York: Academic Press, 1980:209–59.

9. Sealy RC. Free radicals in melanin formation, structure and reactions. In: Armstrong D, Sohal RS, Cutler RGC, Slater TR, eds. Free radicals in molecular biology, aging and disease. New York: Raven Press, 1984:67–76.
10. Riley PA. Radicals and melanomas. Phil Trans R Soc London 1985; 311:679–89.
11. Ranadive NS, Menon IA. Role of reactive oxygen species and free radicals from melanins in photoinduced cutaneous inflammations. Pathol Immunopathol Res 1986; 5:118–39.
12. Sarna T, Pilas B, Land EJ, Truscott TG. Interaction of radicals from water radiolysis with melanin. Biochim Biophys Acta 1986; 883:162–7.
13. Crippa R, Horak V, Prota G, Svoronos P, Wolfram L. Chemistry of melanins. In: Bross A, ed. The alkaloids. Vol. 36. New York: Academic Press, 1989:253.
14. Jimbow K, Fitzpatrick TB, Quevedo WC Jr. Formation, chemical composition and function of melanin pigments. In: Matoltsy AG, ed. Biology of the integument. Vol. II. Springer, 1984:278–92.
15. Jimbow K, Salopek TG, Dixon WT, Searles GE, Yamada K. The epidermal melanin unit in the pathophysiology of malignant melanoma. Am J Dermatopath 1991; 13:179–88.
16. Yamazaki I. Free radicals in enzyme-substrate reactions. In: Pryor WA, ed. Free radicals in biology. Vol. III. New York: Academic Press, 1977:183–218.
17. Mason HS, Spencer E, Yamazaki I. Identification by electron spin resonance spectroscopy of the primary product of tyrosinase-catalyzed catechol oxidation. Biochem Biophys Res Commun 1961; 4:236–8.
18. Tse DCS, McCreery RL, Adams RN. Potential oxidative pathways of brain catecholamines. J Med Chem 1976; 19:37–40.
19. Kalayanaraman B, Felix CC, Sealy RC. Semiquinone anion radicals of catechol(amine)s, catechol estrogens, and their metal complexes. Environ Health Perspect 1985; 64:185–98.
20. Stone TJ, Waters WA. Aryloxy-radicals. Part IV. Electron spin resonance of some ortho-monobenzo semi-quinones and secondary radicals derived therefrom. J Chem Soc 1965:1488–94.
21. Dybing E, Nelson SD, Mitchell JR, Sasame HA, Gillette JR. Oxidation of α-methyldopa and other catechols by cytochrome P-450-generated superoxide anion: possible mechanism of methyldopa hepatitis. Molec Pharmacol 1976; 12:911–20.
22. Naish S, Cooksey CJ, Riley PA. Initial mushroom tyrosinase-catalyzed oxidation product of 4-hydroxyanisole is 4-methoxy-ortho-benzoquinone. Pigment Cell Res 1988; 1:379–81.
23. Nilges MJ, Swartz HM, Riley PA. Identification by electron spin resonance of free radicals formed during the oxidation of 4-hydroxyanisole catalyzed by tyrosinase. J Biol Chem 1984; 259:2446–51.
24. Tomasi A. Radical activation of 4-hydroxyanisole by mushroom tyrosinase. In: Riley PA, ed. Hydroxyanisole. Oxford: IRL Press Ltd, 1984:15–24.
25. Land EJ, Cooksey CJ, Riley PA. Reaction kinetics of 4-methoxy ortho benzoquinone in relation to its mechanism of cytotoxicity; a pulse radiolysis study. Biochem Pharmacol 1990; 39:1133–5.

26. Patel KB, Willson RL. Semiquinone free radicals and oxygen. J Chem Soc Faraday Trans I 1973; 69:814–25.
27. Korytowski W, Sarna T, Kalyanaraman B, Sealy RC. Tyrosinase-catalyzed oxidation of dopa and related catechol(amine)s: a kinetic electron spin resonance investigation using spin-stabilization and spin label oximetry. Biochim Biophys Acta 1987; 924:383–92.
28. Ito S, Kato T, Ishikawa K, Kasuga T, Jimbow K. Mechanism of selective toxicity of 4-S-cysteinylphenol and 4-S-cysteaminylphenol to melanocytes. Biochem Pharmacol 1987; 36:2007–11.
29. Jimbow K, Miura T, Ito S, Ishikawa K. Phenolic melanin precursors provide a rational approach to the design of antitumor agents for melanoma. Pigment Cell Res 1989; 2:34–9.
30. Inoue S, Ito S, Wakamatsu K, Jimbow K, Fujita K. Mechanism of growth inhibition of melanoma cells by 4-S-cysteaminylphenol and its analogues. Biochem Pharmacol 1990; 39:1077–83.
31. Alena F, Jimbow K, Ito S. Melanocytotoxicity and antimelanoma effects of phenolic amine compounds in mice in vitro. Cancer Res 1990; 50:3743–7.
32. Pankovich JM, Jimbow K, Ito S. 4-S-Cysteaminylphenol and its analogues as substrates for tyrosinase and monoamine oxidase. Pigment Cell Res 1990; 3:146–9.
33. Reszka K, Miyake Y, Alena F, Jimbow K. Generation of radical species from melanocytotoxic agents, phenolic thioether amines upon tyrosinase activation: an ESR study. 52nd Annual Meeting of the Society for Investigative Dermatology, Seattle, WA, May 1–4, 1991.
34. Felix CC, Sealy RC. Photolysis of melanin precursors: formation of semiquinone radicals and their complexation with diamagnetic metal ions. Photochem Photobiol 1981; 34:423–9.
35. Palumbo A, d'Ishcia M, Misuraca G, Prota G. Mechanism of inhibition of melanogenesis by hydroquinone. Biochim Biophys Acta 1991; 1073:85–90.
36. Steenken S, Neta P. Electron transfer rates and equilibria between substituted phenoxide ions and phenoxyl radicals. J Phys Chem 1979; 83:1134–7.
37. Steenken S, Neta P. One-electron redox potentials of phenols. Hydroxy- and aminoxyphenols and related compounds of biological interest. J Phys Chem 1982; 86:3661–7.
38. DeFelippis MR, Murthy CP, Faraggi M, Klapper MH. Pulse radiolysis measurement of redox potentials: the tyrosine and tryptophan radicals. Biochem 1989; 28:4847–53.
39. Forni LG, Willson RL. Thiyl and phenoxyl free radicals and NADH. Biochem J 1986; 240:897–903.
40. Kagan VE, Serbinova EA, Packer L. Generation and recycling of radicals from phenolic antioxidants. Arch Biochem Biophys 1990; 280:33–9.
41. Neta P, Steenken S. Phenoxyl radicals: formation, detection, and redox properties in aqueous solutions. In: Rodgers MAJ, Powers EL, eds. Oxygen and oxy-radicals in chemistry and biology. New York: Academic Press, 1981:83–7.
42. Simic MG, Desrosiers MF, Hunter EPL. Antioxidant properties of antioxidant dimers. In: Hayaishi O, Niki E, Kondo M, Yoshokawa T, eds. Medical, biochemical and chemical aspects of free radicals. Amsterdam: Elsevier, 1989:433–40.

43. Lambert C, Chacon JN, Chedekel MR, Land EJ, Riley PA, Thompson A, Truscott TG. A pulse radiolysis investigation of the oxidation of indolic melanin precursors: evidence for indolequinones and subsequent intermediates. Biochim Biophys Acta 1989; 993:12–20.
44. Al-Kazwini AT, O'Neill P, Adams GE, Cundall RB, Jacquet B, Lang G, Junino A. One-electron oxidation of methoxylated and hydroxylated indoles by N_3. 1. Characterization of the primary indolic radicals. J Phys Chem 1990; 94:6666–70.
45. Lambert C, Land EJ, Riley PA, Truscott TG. A pulse radiolysis investigation of the oxidation of methoxylated metabolites of indolic melanin precursors. Biochim Biophys Acta 1990; 1035:319–24.
46. d'Ischia M, Prota G. Photooxidation of 5,6-dihydroxy-1-methyl-indole. Tetrahedron 1987; 43:431–4.
47. Young TE, Babbitt BW. Electrochemical study of the oxidation of α-methylnoradrenaline and dopamine. J Org Chem 1983; 48:562–6.
48. Graham DG, Tiffany SM, Bell WR Jr, Gutknecht WF. Autooxidation versus covalent binding of quinones as the mechanism of toxicity of dopamine, 6-hydroxydopamine, and related compounds toward C 1300 neuroblastoma cells in vitro. Molec Pharmacol 1978; 14:644–53.
49. Wrona M, Lemordant D, Lin L, Blank L, Dryhurst G. Oxidation of 5-hydroxytryptamine and 5,7-dihydroxytryptamine. A new oxidation pathway and formation of a novel neurotoxin. J Med Chem 1986; 29:499–505.
50. Mayausky JS, McCreery RL. Spectroelectrochemical examination of charge transfer between chloropromazine cation radical and catecholamines. Anal Chem 1983; 55:308–12.
51. Mishra HP, Fridovich I. The role of superoxide anion in the autooxidation of epinephrine and a simple assay for superoxide dismutase. J Biol Chem 1972; 247:3170–5.
52. Cohen G, Heikkila RE. The generation of hydrogen peroxide, superoxide radical, and hydroxyl radical by 6-hydroxydopamine, dialuric acid, and related cytotoxic agents. J Biol Chem 1974; 249:2447–52.
53. Bors W, Michel C, Saran M, Lengfelder E. The involvement of oxygen radicals during the autooxidation of adrenalin. Biochim Biophys Acta 1978; 540: 162–72.
54. Bors W, Saran M, Lengfelder E, Michel C, Fuchs C, Frenzel C. Detection of oxygen radicals in biological systems. Photochem Photobiol 1978; 28:629–38.
55. Floyd RA, Wiseman B. Spin-trapping free radicals in the autooxidation of 6-hydroxydopamine. Biochim Biophys Acta 1979; 586:196–207.
56. Gee P, Davison AJ. 6-Hydroxydopamine does not reduce molecular oxygen directly, but requires a coreductant. Arch Biochem Biophys 1984; 231:164–8.
57. Perez-Reyez E, Mason RP. Characterization of the structure and reactions of free radicals from serotonin and related indoles. J Biol Chem 1981; 256:2427–32.
58. Kalyanaraman B, Premovic PI, Sealy RC. Semiquinone anion radicals from addition of aminoacids, peptides and proteins to quinones derived from oxidation of catechols and catecholamines. J Biol Chem 1987; 262:11080–7.

59. Kalyanaraman B, Felix CC, Sealy RC. Photoionization of melanin precursors: an electron spin resonance investigation using the spin trap 5,5-dimethyl-1-pyrroline-1-oxide (DMPO). Photochem Photobiol 1982; 36:5–12.
60. Koch W, Chedekel MR. Photoinitiated DNA damage by melanogenic intermediates in vitro. Photochem Photobiol 1986; 44:703–10.
61. Land EJ, Thompson A, Truscott TG, Subbarao KV, Chedekel MR. Photochemistry of melanin precursors: dopa, 5-S-cysteinyldopa and 2,5-S,S′-dicystienyldopa. Photochem Photobiol 1986; 44:697–702.
62. Koch WH, Chedekel MR. Photogeneration of free radicals from eumelanogenic intermediates and metabolites. Biochim Biophys Acta 1987; 924:458–66.
63. Reszka K, Salopek TG, Jimbow K, Chignell CF. Photolysis of pheomelanin, its precursor 5-cysteinyldopa and a eumelanin metabolite, 5-hydroxy-6-methoxy-indole-2-carboxylic acid with UV light above 300 nm: an EPR and spin trapping study. 52nd Annual Meeting of the Society for Investigative Dermatology, Baltimore, Maryland, April 29–May 2, 1992.
64. Pilas B, Felix CC, Sarna T, Kalyanaraman B. Photolysis of pheomelanin precursors: an ESR-spin trapping study. Photochem Photobiol 1986; 44:689–96.
65. Rizzuto F, Spikes JD, Coker GD. The lumiflavin-sensitized photooxidation of substituted phenylalanines and tyrosines. Photobiochem Photobiophys 1986; 10:149–62.
66. Sarna T, Zajac J, Bowman NK, Truscott TG. Photoinduced electron transfer reactions of rose bengal and selected electron donors. J Photochem Photobiol A Chem 1991; 60:295–310.
67. Felix CC, Reszka K, Sealy RC. Free radicals from photoreduction of hematoporphyrin in aqueous solution. Photochem Photobiol 1983; 37:141–7.
68. Reszka K, Sealy RC. Photooxidation of 3,4-dihydroxyphenylalanine by hematoporphyrin in aqueous solutions: an electron spin resonance study using 2,2,6,6-tetramethyl-4-piperidone-1-oxyl (Tempone). Photochem Photobiol 1984; 39: 293–9.
69. Plancherel D, von Zelewsky A. Photon-triggered complex formation: radical complexes of *o*-benzosemiquinone, DOPA, Dopamine and adrenaline formed by electron transfer reaction from excited tris (2,2′-bipyridyl) ruthenium (II). Helv Chim Acta 1992; 65:1929–40.
70. Reszka K, Kolodziejczyk P, Lown JW. Photosensitization by antitumor agents. 2. Anthrapyrazole-photosensitized oxidation of ascorbic acid and 3,4-dihydroxyphenylalanine. J Free Rad Biol Med 1986; 2:203–11.
71. Reszka K, Lown JW, Chignell CF. Photosensitization by anticancer agents. 10. Ortho-semiquinone and superoxide radicals produced during anthrapyrazole-sensitized oxidation of catechols. Photochem Photobiol 1992; 55:359–66.
72. Commoner B, Townsend J, Pake GE. Free radicals in biological materials. Nature London 1954; 174:689.
73. Sealy RC, Hyde JS, Felix CC, Menon IA, Prota G, Swartz HM, Persad S, Haberman HF. Novel free radicals in synthetic and natural pheomelanins: distinction between dopa melanins and cysteinyldopa melanins by ESR spectroscopy. Proc Natl Acad Sci USA 1982; 79:2885–9.

74. Felix CC, Hyde JS, Sarna T, Sealy RC. Interactions of melanin with metal ions. Electron spin resonance evidence for chelate complexes of metal ions with free radicals. J Am Chem Soc 1978; 100:3922-6.
75. Horak V, Gillette JR. A study of the oxidation-reduction state of synthetic 3,4-dihydroxy-dl-phenylalanine melanin. Molec Pharmacol 1971; 7:429-33.
76. Lukiewicz S, Reszka K, Matuszak Z. Simultaneous electrochemical-electron spin resonance (SEESR) studies on natural and synthetic melanins. Bioelectrochem Bioenergetics 1980; 7:153-65.
77. Chauffe L, Windle JJ, Friedman M. Electron spin resonance study of melanin treated with reducing agents. Biophys J 1975; 15:565-72.
78. Van Woert MH. Oxidation of reduced nicotinamide adenine dinucleotide by melanin. Life Sci 1967; 6:2605.
79. Crippa PR, Mazzini A, Salmelli D. Oxidation of NADH by melanin: effect of UV light and copper ions. Physiol Chem Physics 1979; 11:491-9.
80. Gan EV, Haberman HF, Menon IA. Electron transfer properties of melanin. Arch Biochem Biophys 1975; 173:666-72.
81. Gan EV, Lam KM, Haberman HF, Menon IA. Oxidizing and reducing properties of melanins. Br J Dermatol 1977; 96:25-9.
82. Blois MS, Zahlan AB, Maling JE. Electron spin resonance studies on melanin. Biophys J 1964; 4:471.
83. Felix CC, Hyde JS, Sarna T, Sealy RC. Melanin photoreactions in aerated media: electron spin resonance evidence for production of superoxide and hydrogen peroxide. Biochem Biophys Res Commun 1978; 84:335-41.
84. Hintz P, Kalyanaraman B. Metal ion-induced activation of molecular oxygen in pigmented polymers. Biochim Biophys Acta 1986; 883:41-5.
85. Willson RL. Pulse radiolysis studies of electron transfer reactions in aerobic solution. J Chem Soc Chem Commun 1970; 1005.
86. Korytowski W, Hintz P, Sealy RC, Kalyanaraman B. Mechanism of dismutation of superoxide produced during autooxidation of melanin pigments. Biochem Biophys Res Commun 1985; 131:659-65.
87. Korytowski W, Kalyanaraman B, Menon IA, Sarna T, Sealy RC. Reaction of superoxide anions, with melanins: electron spin resonance and spin trapping studies. Biochim Biophys Acta 1986; 882:145-53.
88. Blois MS, Taskovich L. The reversible binding of some aromatic and cyclic compounds to biopolymers in vitro. J Invest Dermatol 1969; 53:344.
89. Forrest IS, Gutmann F, Keyzer H. *In vitro* interaction of chlorpromazine and melanin. Agressol 1966; 7:147.
90. Jimbow K, Obata H, Pathak MA, Fitzpatrick TB. Mechanism of depigmentation by hydroquinone. J Invest Dermatol 1974; 62:436-49.
91. Menter JM, Willis I. The interaction of L-dopa melanin with p-tert-butylcatechol. J Invest Dermatol 1980; 75:257-60.
92. Menter JM, Willis I. Interaction of several mono- and dihydroxybenzene derivatives of various depigmenting potencies with L-3,4-dihydroxyphenylalanine-melanin. Arch Biochem Biophys 1986; 244:846-56.
93. M,enter JM, Townsel ME, Moore CL, Williamson GD, Soteres BJ, Fisher MS, Willis I. Melanin accelerates the tyrosinase-catalyzed oxygenation of p-hydroxyanisole (MMEH). Pigment Cell Res 1990; 3:90-7.

94. Swartz HM. Cells and Tissues, In: Swartz HM, Bolton JR, Borg DC, eds. Biological application of electron spin resonance. New York: Wiley-Interscience, 1972:155–95.
95. Sarna T, Korytowski W, Sealy RC. Nitroxides as redox probes of melanins: dark-induced and photoinduced changes in redox equilibria. Arch Biochem Biophys 1985; 239:226–33.
96. Goodchild NT, Kwock L, Lin PS. Melanin: a possible cellular superoxide scavenger. In: Rodgers MAJ, Powers EL, eds. Oxygen and oxy-radicals in chemistry and biology. New York: Academic Press, 1981:645–8.
97. Geremia E, Corsaro C. Bonomo R, Giardinelli R, Pappalarado P, Vanella A, Sichel G. Eumelanins as free radicals trap and superoxide dismutase activities in amphibia. Comp Biochem Physiol 1984; 79B:67–9.
98. Schwabe K, Lassman G, Damerau W, Naundorf H. Protection of melanoma cells against superoxide radicals by melanins. J Cancer Res Clin Oncol 1989; 115:597–600.
99. Afanas'ev IB. Superoxide ion: chemistry and biological implications. Vol. I. CRC Press 1989:51.
100. Ilan YA, Czapski G. The reaction of superoxide radical with iron complexes of EDTA studied by pulse radiolysis. Biochim Biophys Acta 1976; 498:386.
101. Lapina VA, Dontzov AE, Ostrovski MA, Emanuel NM. Interaction of anion radical of oxygen with eye melanins and ommochroms. Doklady AN SSSR 1985; 280:1463–5.
102. Swallow AJ. Physical chemistry of semiquinones. In: Trumpower BL, ed. Function of quinones in energy conserving systems. New York: Academic Press, 1982:59–72.
103. Korzhova LP, Frolova EV, Romakov YA, Kuznetsova NA. Photoinduced destruction of dopa-melanin. Biochem (USSR) 1989; 54:805–11.
104. Korytowski W, Sarna T. Bleaching of melanin pigments. J Biol Chem 1990; 265:12410–16.
105. Korytowski W, Pilas B, Sarna T, Kalyanaraman B. Photoinduced generation of hydrogen peroxide and hydroxyl radicals in melanins. Photochem Photobiol 1987; 45:185–90.
106. Richmond R, Halliwell B, Chauhan J, Darbre A. Superoxide-dependent formation of hydroxyl radicals: detection of hydroxyl radicals by the hydroxylation of aromatic compounds. Anal Biochem 1981; 118:328–35.
107. Pilas B, Sarna T, Kalyanaraman B, Swartz HM. The effect of melanin on iron associated decomposition of hydrogen peroxide. Free Rad Biol Med 1988; 4:285–93.
108. Khudyakov IV, Kuzmin VA, Emanuel NM. Decay kinetics of aryloxy and semiquinone radicals in the presence of copper ions. Int J Chem Kinetics 1978; 10:1005–18.
109. Butler J, Hoey BM, Swallow AJ. Reactions of the semiquinone free radicals of anti-tumor agents with oxygen and iron complexes. FEBS Lett 1985; 182: 95–8.
110. Sealy RC, Sarna T, Wanner EJ, Reszka K. Photosensitization of melanin: an electron spin resonance study of sensitized radical production and oxygen consumption. Photochem Photobiol 1985; 42:529–32.

111. Sarna T, Menon IA, Sealy RC. Photosensitization of melanins: comparative study. Photochem Photobiol 1985; 42:529–32.
112. Yegorov SY, Dontsov AY, Krasnovski AA Jr, Ostrovski MA. Quenching of singlet molecular oxygen by screening pigments—melanins and ommochromes. Biophysics 1988; 32:740–2.
113. Jovanovic SV, Simic MG. Redox properties of oxy and audiooxidant radicals. In: Simic MG, Taylor KA, Ward JF, von Sonntag C, eds. Oxygen radicals in biology and medicine. New York: Plenum Press, 1988:115–21.

II

CLINICAL DERMATOPATHOLOGY, DERMATOPHARMACOLOGY, AND DERMATOTOXICOLOGY

The rapid development in this fundamental area of cutaneous biology has also attracted the attention of clinical dermatologists. The high prevalence of skin diseases and their frequently unresolved pathogenesis make them an urgent priority for intensive research efforts. Although the data from experimental dermatology, numerous case reports, and some clinical studies support the concept that oxidative stress is a major determinant in clinical skin pathology, we are still beginning to understand the fascinating world of biochemistry of human skin and the meaning of oxidative skin injury "in real life." In this part skin diseases associated with oxidative injury will be described and several aspects of antioxidant skin pharmacotherapy and skin toxicology discussed.

16

Skin Diseases Associated with Oxidative Injury

Yoshiki Miyachi*
Tenri Hospital, Nara, Japan

I. SKIN AS A TARGET ORGAN OF OXIDATIVE STRESS

Skin is probably one of the target organs most susceptible to environmental oxidative stress because it is always in contact with oxygen and provides the outermost barrier against numerous physicochemical substances to produce reactive oxygen species (ROS) (1). ROS are produced by various stimuli, however, the most important sources of reactive oxidants directly related to cutaneous damage are photochemical reactions with ultraviolet (UV) light. UV as well as ionizing radiation produce ROS such as superoxide anion radicals (O_2^-), hydroxyl radicals ($OH^{\cdot}$), and singlet oxygens.

Lipid peroxidation by UV light is also mediated by radical reactions involving oxygen. Plasma membrane containing a large amount of polyunsaturated fatty acids is susceptible to UV exposure, and the fatty acids are easily oxidized to lipid peroxides. Also, since skin is a sebum-secreting organ and receives cosmetics and topical agents with possible oxidative degradation, lipid peroxides are inevitably produced with subsequent deleterious effects and damage to cellular functions.

**Present affiliation*: Gunma University School of Medicine, Maebashi, Gunma, Japan.

Invasion of the skin by pathogenic microorganisms attracts neutrophils to the site of inflammation, and these cells are known to transform molecular oxygen to superoxide anion via a membrane enzyme system and coupled oxidation. A further respiratory burst results in the formation of highly reactive oxidants. These ROS are primarily produced beneficially by neutrophils to kill ingested microorganisms as one of the self-defense mechanisms, however, excessive generation of such potent oxidants can induce insult to surrounding tissue called auto-oxidative damage with resultant inflammatory reactions. Physical burns and ischemia-reperfusion cause alterations in metabolism culminating in the biosynthesis of O_2^- through xanthine oxidase, which will be described later.

In view of these facts, the skin is intermittently threatened by various oxidative attacks from both inside and outside, making the skin a target organ of oxidative stress (2). The modulation of a redox balance by reducing oxidative stress and increasing antioxidant mechanisms seems to underlie the key issues to be solved in cutaneous biology.

In this chapter, several skin diseases associated with oxidative stress are briefly reviewed.

II. SKIN DISEASES ASSOCIATED WITH PHYSICOCHEMICAL OXIDATIVE STRESS

A. Ultraviolet Light

Acute actinic damage to the skin is expressed clinically as sunburn reaction peaking at 24 hours after sun exposure with subsequent tanning reaction. Although sunburn erythema is closely related to prostaglandins, sunburn cell formation is considered to be induced by photoreactive oxidants (3,4). Individual keratinocyte necrosis (sunburn cell) may be produced when ROS hit cellular membranes and/or DNA of keratinocytes if they are in the sensitive phase of the cell cycle. Impairment of the cutaneous antioxidant defense system has been reported (5,6), which must contribute to the further actinic tissue injury. The cumulative effect of this photo-oxidative stress may affect the cutaneous photoaging process, suggesting possible pharmacological interventions (7). At present, sunscreen agents seem to be superior to topical antioxidants (8).

Photoreactive oxidants are readily produced in the presence of photosensitizers, and photodynamic chemotherapies basically utilize this photo-oxidative toxicity, which sometimes induces adverse effects. Likewise, ROS are involved in the formation of photoallergens in contact photosensitivity (9). Impaired antioxidant enzymes in xeroderma pigmentosum fibroblasts are still controversial (10,11), however, there is a very good possibility that ROS participate in UV carcinogenesis of xeroderma pigmentosum.

B. Heat

It has been reported that experimental thermal injury to the skin results in a remarkable increase in the lipid peroxide level of the injured skin, which leaks into the blood and induces tissue injury in other organs (12). Skin SOD activity was found to be decreased immediately after experimental burns. In contrast, both serum and skin lipid peroxide levels peaked 3 hours after burns, returning to the normal level by 2 weeks. Of great interest was that topical administration of SOD was clinically effective, providing a promising approach for the future treatment of burns (13). A similar decrease in cutaneous SOD activity has been reported even in the site of hyperthermia treatment (14).

C. Chemicals

In experimental contact dermatitis to dinitrochlorobenzene (DNCB) in guinea pigs, decreased skin SOD activity was observed after induction of irritant contact dermatitis but not after elicitation of allergic contact dermatitis (15). It seems likely that attracted neutrophils at the lesional skin of irritant contact dermatitis are activated and generate ROS at sites of inflammation. Although no significant changes in skin SOD activity were noticed in a single induction of experimental allergic contact dermatitis, severe skin manifestations in patients with chronic occupational contact dermatitis can be partly explained by a defective capacity for elevating SOD activity (16). These findings suggest that oxidant tissue damages are closely related to the expression of contact dermatitis. In this connection, antioxidant agents seem to be a reasonable approach to the treatment of contact dermatitis.

III. SKIN DISEASES ASSOCIATED WITH NEUTROPHILIC OXIDATIVE STRESS

Neutrophils play an important role in cutaneous inflammation by secreting a variety of inflammatory products. Neutrophils are known to transform molecular oxygen to superoxide anion (O_2^-) via a membrane enzyme system and coupled oxidation. A further respiratory burst results in the formation of highly reactive oxidants, which are produced by stimulated neutrophils primarily in order to aid in the killing of ingested microorganisms. However, ROS generated by activated neutrophils can induce auto-oxidative tissue damage at the sites of inflammation. A membrane-bound NADPH oxidase is considered to lead to the production of superoxide resulting in chemical insult to surrounding healthy tissue.

A. Cutaneous Vasculitis

Cutaneous vasculitis is characterized by vessel wall necrosis with neutrophil infiltration, which must play an important role in tissue injury. Since dapsone, which has been successfully used for the treatment of cutaneous vasculitis, revealed a potent ROS scavenger (17), reactive oxidants produced by recruited neutrophils may provide a pathogenetic mechanism in cutaneous vasculitis. Sera from untreated patients induced increased hydroxyl radical generation (18), and thus ROS are probably involved in the formation of cutaneous lesions. Because elevated immune complex levels and perivascular deposition of immunoreactants were observed in some patients, it is speculated that circulating immune complexes in the patients are partly responsible for enhanced generation of ROS, which is again supported in the studies of erythema multiforme (19) and experimental Arthus reaction (20). Also, the mechanism of action of thalidomide in erythema nodosum leprosum, a typical Arthus-type reaction, is ascribable to its ROS-scavenging activity (21).

B. Behcet's Disease

Potentiated neutrophil functions and successful clinical treatment with colchicine were reported in Behcet's disease (22). It was found that the production of ROS was increased in patients with Behcet's disease because significantly higher ^{51}Cr release was observed when patients' neutrophils were incubated with ^{51}Cr-labeled endothelial cells (23). Of great interest is the fact that the simultaneous addition of SOD and catalase in the cell culture decreased the ^{51}Cr release to control levels. These findings suggest the importance of neutrophilic oxidative stress in the pathogenesis of Behcet's disease.

C. Acne and Rosacea

Several different mechanisms are involved in inflammatory acne, however, the most important event in the inflammatory process is the release of extracellular inflammation-producing products, especially a low molecular weight, chemotactic factor by *Propionibacterium acnes*-attracting neutrophils. Tetracyclines and erythromycin are used for treatment, and these drugs were proven to inhibit neutrophil chemotaxis and ROS generation by neutrophils (24). Thus, in inflammatory acne, neutrophilic oxidative stress seems to play a critical role.

Similarly, in papulopustular rosacea, metronidazole has recently been introduced as a new treatment of choice and has an antioxidant effect (25).

D. Psoriasis

Several pieces of evidence indicate the possible involvement of neutrophils in the pathogenesis of psoriasis. They include the existence of psoriatic chemotactic factor, activated neutrophil accumulation in the stratum corneum, and the generation of ROS (26). The recruitment of neutrophils may affect the cytokine network in psoriasis. In this connection, the antioxidant action of retinoids (27,28) seems meaningful.

IV. SKIN DISEASES ASSOCIATED WITH OTHER OXIDATIVE STRESS

A. Infections

Since ROS are important defense sources and are primarily produced by phagocytes to kill ingested microorganisms, once the ability to produce ROS is decreased, these organisms may easily invade the skin, inducing clinical manifestation. In superficial mycosis such as *Tinea rubrum* and *Candida albicans* infections, neutrophil infiltration is noticed and the opsopnized fungus enhances the chemiluminescence of neutrophils. In sporotrichosis, neutrophils play a major role in tissue injury because the densely infiltrated neutrophils are reported to produce ROS (29).

In contrast, neutrophils from patients with leprosy produce decreased levels of ROS that may result in the defective defense against M. leprae (30). Similar impairment in ROS production has been noticed in other viral infections (31,32).

B. Ischemia-Reperfusion Injury

It has been established that ROS play important roles in causing postischemic reperfusion necrosis of both island and random pattern skin flaps (33, 34). The effect of SOD on salvaging the necrosis is also reported even with topical SOD ointment (35). In the clinical field of plastic and reconstructive skin surgery, how to enhance flap survival using drugs is an urgent issue to be solved, and the concept of ischemia-reperfusion injury should be introduced.

C. Autoimmune Diseases

Sera from patients with systemic lupus erythematosus (SLE) are known to produce increased levels of ROS, and neutrophils from SLE patients induced endothelial cell injury when they were co-incubated, which was canceled by SOD and catalase (36). ROS are also suspected to be involved in the photosensitivity in SLE patients (37). Several drugs used for the treatment of autoimmune connective tissue diseases are known to have some antioxidant activities (38–40). In a certain autoimmune bullous skin disease, involvement of ROS is indicated (41), and interestingly, dapsone, an ROS scavenger, is very effective.

D. Aging

One attractive theory of aging at the biochemical level is based on the production of free radicals and lipid peroxidation leading to membrane damage that may impair cellular functions. Though human skin has evolved several defensive lines including both nonenzymatic and enzymatic antioxidant systems, if one fails to increase antioxidant activity in response to continual oxidative stress, the cumulative harmful effects may result in cutaneous aging. In cutaneous inflammation, it is speculated that the prognosis of the skin lesion may in part be governed by the degree to which the patient's antioxidant activity is up-regulated in response to generation of tissue-damaging substances such as ROS and lipid peroxides (42). SOD activity of the skin remains apparently normal in the aged, but defective in inducing SOD activity when oxidative stress occurs (43,44). It seems likely that the capacity to play a protective role in defending against overall oxidative damage to cells and augment antioxidant activity under a variety of oxidative stress condition is subject to age-related alterations. Thus, use of exogenous antioxidants to attenuate oxidant toxicity should be considered as one of the important issues of skin aging.

V. CONCLUSIONS

As briefly reviewed in this chapter, skin is one of the best target organs for oxidative attacks, which may affect virtually all aspects of cutaneous functions. Though the skin has an extensive antioxidant network, many oxidants escape this antioxidant surveillance and induce critical damage to cutaneous constituents, resulting in not only inflammation but aging and even carcinogenesis. The use of antioxidants should be considered in order to modulate the imbalance between environmental oxidant attacks and antioxidants.

REFERENCES

1. Hayaishi O, Imamura S, Miyachi Y, ed. The biological role of reactive oxygen species in skin. Tokyo: University of Tokyo Press, 1987.

2. Miyachi Y, Yoshioka A, Imamura S, Niwa Y. The cumulative effect of continual oxidative stress to the skin and cutaneous aging. In: Kligman AM, Takase Y, eds. Cutaneous aging. Tokyo: University of Tokyo Press, 1988:435–47.
3. Miyachi Y, Horio T, Imamura S. Sunburn cell formation is prevented by scavenging oxygen intermediates. Clin Exp Dermatol 1983; 8:305–10.
4. Danno K, Horio T, Takigawa M, Imamura S. Role of oxygen intermediates in UV-induced epidermal cell injury. J Invest Dermatol 1984; 83:166–8.
5. Miyachi Y, Imamura S, Niwa Y. Decreased skin superoxide dismutase activity by a single exposure of ultraviolet is reduced by liposomal superoxide dismutase pretreatment. J Invest Dermatol 1987; 89:111–12.
6. Fuchs J, Huflejt ME, Rothfuss LM, Wilson DS, Carcamo G, Packer L. Acute effects of near ultraviolet and visible light on the cutaneous antioxidant defense system. Photochem Photobiol 1989; 50:739–44.
7. Miyachi Y, Imamura S. Photo-oxidative skin damage and anti-oxidants (editorial). Photodermatol Photoimmunol Photomed 1990; 7:49–50.
8. Hamanaka H, Miyachi Y, Imamura S. Photoprotective effect of topically applied superoxide dismutase on sunburn reaction in comparison with sunscreen. J Dermatol 1990; 17:595–8.
9. Miyachi Y, Imamura S, Niwa Y, Tokura Y, Takigawa M. Mechanisms of contact photosensitivity in mice. VI. Oxygen intermediates are involved in contact photosensitization, but not in ordinary contact sensitization. J Invest Dermatol 1986; 86:26–8.
10. Crawford D, Zbinden I, Moret R, Cerutti P. Antioxidant enzymes in xeroderma pigmentosum fibroblasts. Cancer Res 1988; 48:2132–4.
11. Nishigori C, Miyachi Y, Imamura S, Takebe H. Reduced superoxide dismutase activity in xeroderma pigmentosum fibroblasts. J Invest Dermatol 1989; 93: 506–10.
12. Nishigaki I, Hagihara M, Hiramatsu M, Izawa Y, Yagi K. Effect of thermal injury on lipid peroxide levels of rat. Biochem Med 1980; 24:185–9.
13. Ono Y, Miyachi Y, Moriguchi T. Evaluation of topical SOD treatment on experimental burn. Proceedings of 5th International Congress of Oxygen Radicals, 1992:581–584.
14. Yoshioka A, Miyachi Y, Imamura S. Effects of local hyperthermia treatment on skin SOD activity in mice. J Clin Biochem Nutr 1989; 6:109–115.
15. Miyachi Y, Imamura S, Niwa Y. SOD activity in skin of guinea pigs with experimental allergic or irritant contact dermatitis. J Clin Biochem Nutr 1987; 3:113–16.
16. Miyachi Y, Uchida K, Komura J, Asada Y, Niwa Y. Auto-oxidative damage in cement dermatitis. Arch Dermatol Res 1985; 277:288–92.
17. Miyachi Y, Niwa Y. Effects of potassium iodide, colchicine and dapsone on the generation of polymorphonuclear leukocyte-derived oxygen intermediates. Br J Dermatol 1982; 107:209–14.
18. Miyachi Y, Yanase K, Imamura S, Niwa Y. Increased hydroxyl radical generation by normal polymorphonuclear leukocytes incubated in sera from patients with leukocytoclastic vasculitis. Arch Dermatol Res 1982; 274:65–71.

19. Miyachi Y, Imamura S, Niwa Y. Erythema multiforme: A possible role of increased reactive oxygen species. J Clin Lab Immunol 1986; 19:11–14.
20. Yoshioka A, Miyachi Y, Imamura S, Niwa Y. Role of reactive oxygen species in Arthus reaction. J Clin Biochem Nutr 1988; 5:215–20.
21. Miyachi Y, Ozaki M, Uchida K, Niwa Y. Effects of thalidomide on the generation of oxygen intermediates by zymosan-stimulated normal polymorphonuclear leukocytes. Arch Dermatol Res 1982; 274:363–7.
22. Miyachi Y, Taniguchi S, Ozaki M, Horio T. Colchicine in the treatment of cutaneous manifestations of Behcet's disease. Br J Dermatol 1981; 104:67–9.
23. Niwa Y, Miyake S, Sakane T, Shingu M, Yokoyama M. Auto-oxidative damage in Behcet's disease—endothelial cell damage following the elevated oxygen radicals generated by stimulated neutrophils. Clin Exp Immunol 1982; 49:247–55.
24. Miyachi Y, Yoshioka A, Imamura S, Niwa Y. The effect of antibiotics on the generation of reactive oxygen species. J Invest Dermatol 1986; 86:449–53.
25. Miyachi Y, Imamura S, Niwa Y. Anti-oxidant action of metronidazole: A possible mechanism of action in rosacea. Br J Dermatol 1986; 114:231–4.
26. Miyachi Y, Niwa Y. Effects of psoriatic sera on the generation of oxygen intermediates by normal polymorphonuclear leukocytes. Arch Dermatol Res 1983; 275:23–6.
27. Yoshioka A, Miyachi Y, Imamura S, Niwa Y. Anti-oxidant effects of retinoids on inflammatory skin diseases. Arch Dermatol Res 1986; 278:177–83.
28. Miyachi Y, Imamura S, Niwa Y. Effects of a novel synthetic vitamin A-like compound (a polyprenoic acid derivative, E-5166) on neutrophil-derived reactive oxygen species. J Clin Biochem Nutr 1988; 5:111–16.
29. Yoshioka A, Miyachi Y, Imamura S, Niwa Y. The effect of the supernatants obtained from Sporothrix schenckii and Candida albicans culture on the generation of reactive oxygen species by polymorphonuclear leukocytes. Mycopathologia 1987; 100:43–8.
30. Niwa Y, Sakane T, Miyachi Y, Ozaki M. Oxygen metabolism in phagocytes of leprotic patients; enhanced endogeneous superoxide dismutase activity and hydroxyl radical generation by clofazimine. J Clin Microbiol 1984; 20:837–42.
31. Niwa Y, Sakane T, Miyachi Y, Kanoh T. Decrease in generation of reactive oxygen species by neutrophils from patients with infectious mononucleosis: role of suppressor T lymphocytes. Blood 1984; 64:994–9.
32. Niwa Y, Sakane T, Somiya K, Miyachi Y. Decreased oxygen radical generation by neutrophils from patients with measles presumably owing to activation of suppressor T lymphocytes. J Clin Microbiol 1985; 21:318–22.
33. Suzuki S, Miyachi Y, Niwa Y, Isshiki N. Significance of reactive oxygen species in distal flap necrosis and its salvage with liposomal SOD. Br J Plast Surg 1989; 42:559–64.
34. Suzuki S, Yoshioka N, Isshiki N, Hamanaka H, Miyachi Y. Involvement of reactive oxygen species in post-ischemic flap necrosis and its prevention by antioxidants. Br J Plast Surg 1991; 44:130–4.
35. Suzuki S, Matsushita Y, Isshiki N, Hamanaka H, Miyachi Y. Salvage of distal flap necrosis by topical superoxide dismutase (SOD). Ann Plast Surg 1991; 27: 253–7.

36. Niwa Y, Sakane T, Shingu M, Miyachi Y. Role of stimulated neutrophils from patients with systemic lupus erythematosus in tissue injury, with special reference to serum factors and increased active oxygen species generated by neutrophils. Inflammation 1985; 9:163–72.
37. Emerit I, Michelson AM. Mechanism of photosensitivity in systemic lupus erythematosus patients. Proc Natl Acad Sci USA 1981; 78:2537–40.
38. Miyachi Y, Yoshioka A, Imamura S, Niwa Y. Anti-oxidant action of anti-malarials. Ann Rheum Dis 1986; 45:244–8.
39. Miyachi Y, Yoshioka A, Imamura S, Niwa Y. The effect of sulphasalazine and its metabolites on the generation of reactive oxygen species. Gut 1987; 28:190–5.
40. Miyachi Y, Yoshioka A, Imamura S, Niwa Y. Anti-oxidant effects of gold compounds. Br J Dermatol 1987; 116:39–46.
41. Niwa Y, Sakane T, Shingu M, Yanagida I, Komura J, Miyachi Y. Neutrophil-generated active oxygens in linear IgA bullous dermatosis. Arch Dermatol 1985; 121:73–8.
42. Niwa Y, Kanoh T, Sakane T, Soh H, Kawai S, Miyachi Y. The ratio of lipid peroxides to superoxide dismutase activity in the skin lesions of patients with severe skin diseases: an accurate prognostic indicator. Life Sci 1987; 40:921–7.
43. Niwa Y, Kasama T, Kawai S, Komura J, Sakane T, Kanoh T, Miyachi Y. The effect of aging on cutaneous lipid peroxide levels and superoxide dismutase activity in guinea pigs and patients with burns. Life Sci 1988; 42:351–6.
44. Yoshioka A, Miyachi Y, Imamura S. Mechanisms of reactive oxygen species-induced skin erythema and superoxide dismutase activities in guinea pigs. J Dermatol 1987; 14:569–75.

17

Biology of Aging Skin: Bioengineering Mensuration of Cutaneous Properties

Jeff Harvell and Howard I. Maibach
University of California, San Francisco,
San Francisco, California

I. INTRODUCTION

The subject of skin aging is of interest in terms of cosmetology, topical drug delivery systems, and pathological processes such as contact dermatitis. Our ability to understand the effects of aging on each of these parameters depends upon an enhanced ability to quantitate the various independent factors involved in cutaneous senescence. Such factors include changes in skin surface topography, skin thickness, epidermal regenerative capacity, transepidermal water loss, stratum corneum hydration, inflammatory capability, elastic properties, and frictional properties. The objective measurements of these factors have been enhanced by continuing developments in the field of skin bioengineering. Devices capable of objective and reproducible measurements of the various properties of normal skin function are presently available, many of them commercially so (Table 1). This chapter discusses the mensuration of these various properties with reference to the aging skin and relates these measures to our current knowledge of the effects of senescence on such aspects as percutaneous absorption, inflammation, and mechanical properties. The first three sections discuss the principles behind the most widely used cutaneous bioengineering devices. The next sections define the utility of these and other instruments in the mensuration of age-associated cutaneous parameters.

Table 1 Bioengineering Techniques/Devices for Measurements of Cutaneous Properties

Measuring technique/device	Property measured
Evaporimeter™	Transepidermal water loss
Laser doppler velocimetery	Blood flow
Capacitance	Stratum corneum hydration
Photoplethysmography	Blood volume
Ultrasound	Skin thickness
Cutometer™	Cutaneous elasticity
Friction meter	Skin friction coefficient
Sebumeter™	Surface sebum content

II. PRINCIPLES BEHIND CUTANEOUS BIOENGINEERING DEVICES

A. Stratum Corneum Hydration (Capacitance)

Normal stratum corneum serves as a reservoir for water. An equilibrium exists for the passive transport of water from the deeper layers of the dermis to the environment. In the midst of this diffusion path is the nonviable epidermis (i.e., stratum corneum), which serves as a depot for water. The water content of the stratum corneum is an important property, which influences such aspects as suppleness and pliability of the skin, as well as friction, barrier function, and percutaneous absorption (1,2).

The degree of stratum corneum hydration can be assessed by means of commercially available devices that relate the amount of water present in the stratum corneum to the ability to produce an electric current. In other words, these devices determine the *conductance* or *capacitance* of the stratum corneum. As hydration increases, resistance to the flow of current is decreased, and the measured capacitance is therefore greater.

The Skin Surface Hygrometer (IBS Co., Hamamatsu, Japan and Acaderm, Menlo Park, CA) and the Corneometere SM 820 PC (Courage and Khazaka, Cologne, Germany and Acaderm, Menlo Park, CA) are two similar devices, which contain a measuring probe that is applied directly to skin. The measuring probe works as a condensor, the capacitance of which is influenced by a change in the water content of the stratum corneum.

B. Transepidermal Water Loss (TEWL)

As mentioned previously, water diffuses passively from the dermis to the environment. The diffusion path consists of the following: dermis to viable epidermis, viable epidermis to nonviable epidermis (i.e., stratum corneum, which serves as a water depot), and finally stratum corneum to the environment. The loss of water from the stratum corneum to the environment is an equilibrium phenomenon, which can be approximated by the vapor pressure gradient that exists immediately adjacent to the surface of the skin (3). This approximation is the fundamental basis for measures of transepidermal water loss (i.e., vapor pressure gradient ≈ transepidermal water loss).

The most widely used instrument for the measurement of transepidermal water loss (TEWL), the Evaporimeter EP1 (ServoMed, Stockholm, Sweden and Courage and Khazaka, Germany), determines the vapor pressure gradient by measuring the vapor pressure at two fixed points immediately adjacent to the skin surface. The Evaporimeter probe consists of two sensors vertically stacked at a known distance from the skin surface (Fig. 1) Based on the vapor pressure values obtained by these separate sensors, the vapor pressure gradient is calculated, and the result displayed as the TEWL in $g/m^2/h$ (3,4).

TEWL measurements are dependent upon the water content of the stratum corneum, as an increase in stratum corneum hydration increases the membrane diffusion constant from the stratum corneum to the environment (2). Also, TEWL measures are reflective of the water barrier function of the stratum corneum. Main components of the water barrier to passive diffusion include protein and stratum corneum lipids, especially glycolipids, free fatty acids, ceramides, sphingolipids, and free and esterified sterols (5). Any perturbation that depletes these lipids or alters protein conformation results in an increased TEWL.

Many exogenous substances that induce irritant reactions in the skin also lead to damage to the stratum corneum membrane barrier. TEWL measurements, therefore, offer an objective means of quantifying such damage in assessments of clinical contact dermatitis (6).

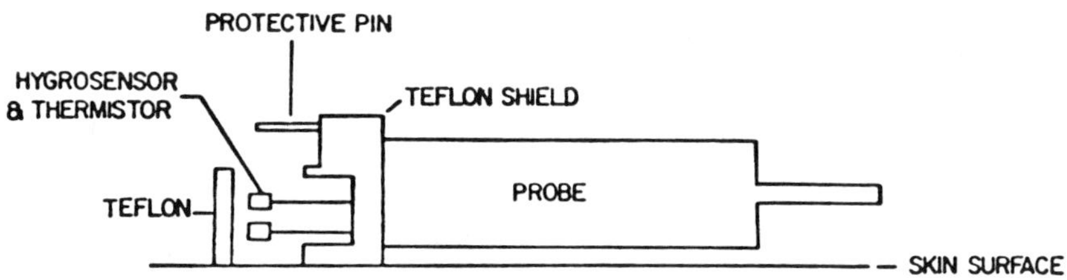

Figure 1 An evaporimeter consists of an open tube of circular cross section placed against the skin. Two relative humidity sensors are positioned within the tube at known distances above the skin surface; then, from the measured relative humidity gradient, TEWL is calculated. (From Ref. 4.)

The values obtained for this instrument depend upon a number of factors, both subject-related and environmental. Most published reports agree that TEWL measures are independent of sex and race (6). (The influence of age will be discussed separately in section VII.) TEWL measurements differ according to anatomic site and are subject to degrees of inter- and intraindividual variation. On forearm skin, the interindividual variation was estimated as 35–48% (7). In a separate study, the intraindividual coefficient of variation by sites on normal, nonirritated skin was 18% (8). Finally, since sweating represents a form of transepidermal water loss which is additive to that which occurs diffusely throughout the stratum corneum, emotional and physical factors that increase a subject's propensity to perspire may affect the measured TEWL value (6).

TEWL measures are strongly dependent upon skin temperature. The effect is direct and logarithmic, in that an increase in skin temperature causes an increased TEWL. From permeation studies, Mathias et al. (9) derived an equation for correction of TEWL to a standard skin surface temperature of 30°C. This equation is widely utilized in most experiments that use TEWL measures.

Some environmental factors that influence TEWL values include ambient temperature; however, this effect is largely due to the fact that increases in ambient air temperature cause increases in skin temperature, thus affecting the measured value as described above. Other environmental factors that may influence the derived TEWL value include air currents and ambient air humidity. Due to these varying influences, which are beyond the control of the experimenter, Pinnagoda et al. (6) recommend that measurements be performed in a room with decreased air currents and an ambient temperature regulated to about 20°C, possibly by an air conditioner. Additionally, to decrease the effect of sweating due to emotional and/or physical factors, a 15- to 30-minute rest period is recommended for subjects prior to beginning measurements. Additional recommendations and guidelines for TEWL measurements are reviewed by the Standardization Group of the European Society of Contact Dermatitis (6).

C. Laser Doppler Velocimetry and Photoplethysmography

Laser doppler velocimetry (LDV) and photoplethysmography (PPG) represent noninvasive optical means of measuring changes in blood flow, in the case of LDV, and blood volume, in the case of PPG. In measures of the effect of aging on the microcirculation, LDV has been more widely utilized than PPG. LDV has additional clinical uses in medicine in addition to its extensive research applications, especially in the field of dermatology. Such uses include objective evaluation of patch test reactions and erythematous diseases, as well as monitoring the viability of skin flaps.

The principle for LDV measurements is based on the ability to detect a frequency shift in a beam of laser light and equate this shift with flow. Specifically, a beam of laser light from a 2 or 5 mW helium-neon source is directed down an optical fiber, which rests lightly on the subject's skin. The beam has a penetration depth of approximately 1 mm, thus for cutaneous measures, only the superficial, dermal microcirculatory flow is sampled. As the beam penetrates the skin, it becomes reflected by both moving and stationary components. The stationary components reflect the beam at the same incident frequency, whereas the moving components, such as red blood cells, reflect the beam at a shifted frequency that is proportional to the number and velocity of red blood cells (i.e., flow). This is the well-known Doppler effect. A second optical fiber detects both the shifted frequency (from blood flow) and the reflected incident frequency (from stationary tissue). An internal photodetector then converts the Doppler shifted frequency to an output in millivolts (10).

Another device to measure cutaneous blood flow is known as the laser doppler flowmeter (LDF), which is simply a variation of the same basic method used in LDV. In fact, the instruments are virtually interchangeable for cutaneous blood flow measures. One study that measured the perfusion in canine island flaps showed a high correlation coefficient (0.92) between the two instruments (11).

Similar to TEWL measures, many subject-related and environmental factors should be considered when performing cutaneous blood flow measurements. Those subject-related factors include subject position, mental activity, smoking, vasoactive drugs, caffeine intake, and physical exertion. Environmental factors include room temperature and relative humidity, however, these are less important than subject-related variables (10).

In contrast, PPG measures not changes in blood flow, but rather changes in blood volume. In PPG measurements, a light-emitting diode (LED) focuses a beam of infrared radiation into the skin. The incident radiation is scattered, reflected, and absorbed as it passes through the tissue. The absorption of this radiation is mostly via red blood cell hemoglobin. That portion of the incident radiation that is not absorbed (i.e., the reflected and scattered portion) returns to a phototransistor, where it is detected. Based on this measurement, the amount of incident radiation absorbed by hemoglobin can be determined, and correlated to the volume of blood present in the sample (10).

III. SKIN SURFACE CHANGES

From a macroscopic perspective, it is obvious that as individuals age, their skin is subject to considerable changes. Such changes as wrinkling, increased laxity, and pigmentation anomalies are obvious to the scientist and nonscientist alike. In an attempt to quantify and provide objective means of assessing these changes, however, investigators have resorted to studying the changes in the skin surface microrelief.

Many investigators employ a method of obtaining skin surface replicas, or "negatives" of the skin surface, from which various parameters associated with aging skin can then be objectively discerned. The various procedures by which these replicas are generated are similar in that an emulsion (usually methacrylate or silicon rubber) is applied to the skin, allowed to dry, and then peeled away, giving a negative imprint of the complex pattern of ridges and furrows that characterizes the skin surface (12,13).

Grove et al. (12) made subjective assessments of dermatoglyphic pattern changes in the skin surface of aged individuals as compared to the young by utilizing replicas of polyvinylidine chloride emulsion. These subjective investigations indicated that the skin of the elderly lacks the orderly arrangement of geometric patterns seen in young skin. The geometric patterns that make up the two-dimensional picture of the skin surface are due to skin folds which form a pattern of intersecting lines. These lines form a two-dimensional pattern of triangles and quadrangles, as seen in Figure 2. Primary lines are defined as the deepest skin folds and are thought to be attributable to the structure of the underlying superficial dermis. Secondary lines are less deep than primary lines and are thought to represent the underlying structure of the granular layer of the epidermis (14). Grove et al. attributed the less orderly pattern of geometry seen in aged skin to the loss of secondary lines (12).

In another approach, which uses the concept of skin surface geometry changes, Voros et al. (14) studied the rather sun-protected skin of the inner surface of the forearm in individuals from 8 to 89 years of age. Replicas were obtained with a silicon rubber product and measurements made with the use of a microscope. Since the intersecting primary and secondary lines form polygons, objective information as to changes in the geometry of these polygons was gleaned by measuring the following parameters: 1) number of polygons per field, 2) average surface area of the polygons, 3) average perimeter of the polygons, and 4) average equivalent diameter of the polygons. In general, the size of the individual polygons tended to increase with age, and therefore the number of polygons per microscopic field decreased with age. Accordingly, there were increases in surface area, average perimeter, and average diameter. Finally, these authors noted that the geometry in male skin appears to change at a faster rate than female skin (14).

This same method of obtaining negative replicas, with slight modifications, has allowed others to investigate unique parameters associated with the skin surface profile (SSP). Methods utilized to study the furrowing pattern of the skin use two techniques based on obtaining negative replicas:

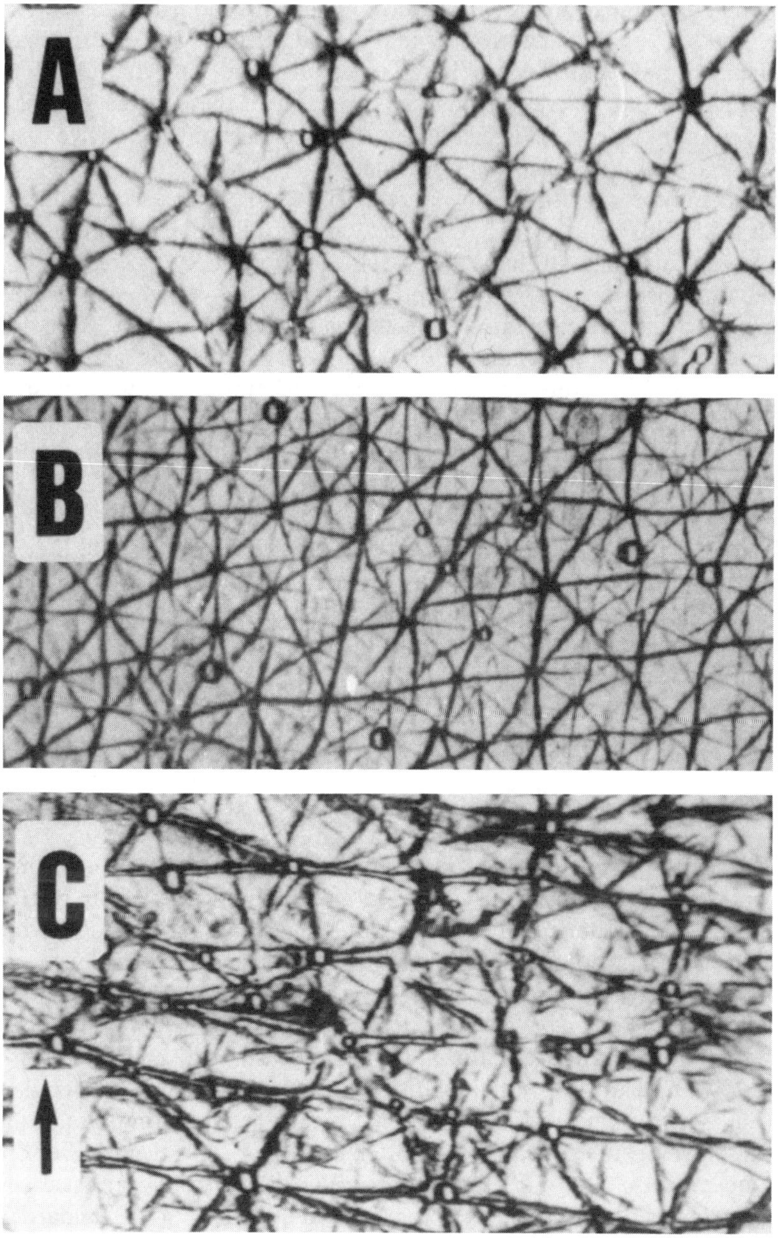

Figure 2 Skin surface pattern for the male, supraumbilical area. A, 25 years; B, 55 years; C, 80 years. Photographs from negative replicas. (From Ref. 13.)

profilometry and image analysis. The technique of profilometry was developed by Makki et al. (15) and involves making a positive cast from the negative replica. By making such a positive cast, the furrowing pattern of the skin is then converted into a series of ridges. The profile of these ridges is subsequently determined by running a stylus over the surface of the positive cast and then analyzing the parameters obtained.

In image analysis, as described by Corcuff et al. (16), replicas are lit laterally with floodlamps at a particular incident angle of lighting. The result of light falling on the ridges of the replica is shadow production, where the shadow's length is proportional to the depth of furrows (i.e., height of ridges). Parameters that can be calculated from such a procedure include 1) the direction of furrows, 2) the so-called "line density," or length of furrows per unit area, and 3) the average depth of furrows. With this technique, Corcuff et al. showed that furrow depth does not change until the age of 40 years, at which point there is a progressive increase in the depth of furrows with age (16).

While still somewhat theoretical, it is thought that the skin surface profile is reflective of the structure of the underlying superficial dermis and, therefore, that changes in this structure with age will be reflected in skin surface profile changes (13). Certainly, histological examination tells us that the superficial dermis does undergo obvious morphological change as the individual ages. Likewise, the use of skin surface studies also show that this complex pattern of ridges and furrows also changes with age; however, further study is needed to determine to what extent these two processes are corelated.

IV. SKIN THICKNESS

The measurement of skin thickness is an important parameter when one considers that there exists a great deal of variation in skin thickness from site to site in a single individual. Such site-to-site differences are of obvious mechanical importance. Additionally, site-to-site differences in the thickness of individual layers of the skin (especially the epidermis) are of importance in the study of percutaneous absorption.

Many techniques have been utilized to make measures of both total skin thickness and measures of individual layers (stratum corneum, epidermis, dermis, subcutis). Skin calipers have been a means of measuring total skin thickness; however, these measures are often subject to some degree of error due to subcutaneous fat and "pinching force" (17). Histological exam of skin biopsy material has, in the past, represented a type of "gold standard" for the measurement of skin thickness; however, this method also has inherent sources of error in the form of 1) fixation artifacts, 2) relaxation of skin tension due to excision, and 3) sectioning. Xerography has been used for determining skin thickness but is seldom used today due to expense, site limitations, and the potentially harmful effects that exposure to radiation may induce (18).

One of the methods becoming more widely used today employs pulsed ultrasound measurements of cutaneous structures. The ultrasound method involves the generation of high-frequency sound impulses, which are directed into the skin. The incident sound impulse is reflected by tissue components, the degree of reflection depending upon the acoustical nature of the tissue encountered. Thus, a series of signals (or reflections) is produced that corresponds to the acoustical nature of the different types of cutaneous tissue encountered. In A-scan, or amplitude display, ultrasound, the scanner maps the intensity of the reflections encountered along a single thin incident beam applied perpendicular to the skin surface. The result is displayed as a graph that plots the amplitude of each reflection versus the penetration depth of the beam in millimeters. B-scan, or brightness display, ultrasound is the scan most familiar to the physician, and essentially takes the data from many individual A-scans and converts the measured amplitudes into a digitized two-dimensional figure; bright pixels represent strong reflections and dark pixels represent a relative absence of reflections (18).

Ultrasound has found dermatological use in the noninvasive assessment of mass lesions (19) as well as the measurement of skin thickness in such disease processes as psoriasis (20), corticoid-induced dermal atrophy (21), scleroderma, and scleredema (18). Current dermatological ultrasound machines operate at a frequency of 15–25 MHz. The higher-frequency devices (i.e., above 20 MHz) are capable of resolving the epidermal-dermal interface, as well as discerning adnexal structures.

With this method, various investigators have made measures of skin thickness. Tan et al. (22), using the ultrasound A-scan system on the flexor aspect of the right midforearm in subjects aged 1–78 years, showed that total skin thickness (i.e., epidermis to subcutaneous fat) increases linearly with age up to 20 years and then decreases linearly with age afterwards. With a higher-frequency device, Branchet et al. (23) made skin thickness measurements of both the epidermis and dermis in the sun-protected region of the upper inner arm in males and females. The results of this study revealed that epidermal thickness decreased with increased age and that this age-associated decrease was faster in men than in women. In contrast, the dermal thickness decreased at approximately the same rate in both men and women with increased age. Such age-associated dermal thickness changes were in accordance with other studies (24–28) which, likewise, reported decreased dermal thickness with increased age. Such an age-associated decrease is hypothesized to be due to a reduction in the number of fibroblasts as well as a decrease in individual fibroblast cell biosynthetic activity. This same reduction in fibroblast activity and number is also purported to be the reason for the impairment in wound healing associated with increasing age (29).

As for the epidermal changes, Branchet et al. (23) conjectured that such a decrease in epidermal thickness may be related to the decrease in keratinocyte proliferative capacity with increased age, as reported by Grove et al. (12). Using a fluorescent marker (dansyl chloride), which stains only the horny layer, they estimated the epidermal transit time as the time required for this marker to fully disappear from the stained stratum corneum. In this investigation, it was discerned that the epidermal transit time for older subjects was approximately 8–9 days longer than for younger subjects.

Indeed, if such a mechanism is responsible for the observed decrease in epidermal thickness, then one may expect that older individuals would exhibit fewer epidermal cell layers; however, at least for the stratum corneum, this does not appear to be the case. Christophers and Kligman (30) and Grove et al. (31) could find no difference in the number of cell layers in the stratum corneum when comparing older individuals to younger individuals. Thus, the reason for the observed decrease in epidermal thickness with age is still not definitively known, and more investigation is required.

Finally, with the increased use of cutaneous ultrasound, a number of investigators have reported a new structural phenomenon, which appears to be age associated. The identification of what is known as a "sub-epidermal non-echogenic band" was first noted by de Rigal et al. in 1989 (32). This group reported that this subepidermal band was absent in young children, appeared in early adulthood, and gradually became thicker with increasing age. Schatz et al. (33) reported finding a similar subepidermal band, but emphasized that the band's age association was an observed decrease in echogenicity with increasing age. Although the subepidermal band's anatomic location roughly correlates with the papillary layer of the dermis, its structural basis and the pathophysiology underlying its changes remain largely unknown (18).

V. MICROCIRCULATION

Most of the methods utilized to study age-associated changes in the microcirculation include the use of the laser doppler velocimeter, the principles of which were described earlier. Certainly, histological observations tell us that, at least from a morphologic point of view, the microcirculation appears compromised: the number of dermal capillary loops is decreased and the dermoepidermal junction is flattened, which essentially decreases the area of absorption (34); however, the question of interest is whether or not such microscopic changes have a profound effect on the cutaneous efficiency. In other words, to what degree do the observed histological changes influence the ability of the microcirculation to deliver nutrients to the skin, remove the products of metabolism, or absorb topical medicaments that have penetrated the epidermis and dermis?

In answering such questions, one must first discern whether or not there exists a baseline difference in blood flow between the young and old. Hatanaka et al. (35) investigated this question by measuring the fingertip blood flow in fingers 2, 3, and 4 of the nondominant hand. This group found no significant differences in fingertip blood flow with regard to age, sex, time of day, food intake, or number of red blood cells. A different study, which included the impact of the autonomic nervous system changes on vessel reactivity, also studied blood flow in the digits of aged and young subjects. In this study conducted, by Oimomi et al. (36), the reaction time of the cutaneous microcirculation after such influences as Valsalva's maneuver and postural change was slower in the aged population, suggesting that older individuals have a decrease in autonomic nervous system function.

Other studies have looked at the effect of age on the reactivity of blood vessels due to a pharmacological stimulus. Guy et al. (37) used methyl nicotinate, a potent vasodilator, as the stimulus and made observations in a young and old population. After a 15-second topical application of a 100-mM aqueous solution of methylnicotinate, the vasodilatory response was followed by means of laser doppler velocimetry and photoplethysmography. Four parameters were chosen to characterize the pharmacodynamic response: 1) time of maximum response, 2) magnitude of maximum response, 3) area under the time-response curve, and 4) time required for the response to decay to 75% of the maximum value. The results of this study indicated no siginficant difference in the pharmacodynamic response between the young and old groups for three of the four parameters; only the magnitude of the peak response as determined by laser doppler velocimetry measurements was statistically different for the two populations. Therefore, it was suggested that young and old groups exhibit a remarkable similarity in the response to this potent vasodilator. In a very similar study, which included the additional parameter of "time of onset of action," again, no significant differences in the pharmacodynamic response were seen for like concentrations of methylnicotinate in the two age groups (38).

Of course, the measured parameter in this type of study, namely vasodilation, depends on two independent parameters: 1) penetration of the topically applied methylnicotinate through the epidermis and 2) reactivity of the blood vessel. It was therefore suggested that the similar response seen in these two age groups may imply one of two biological scenarios. Either penetration and reactivity are in fact the same in the two groups, or a differential penetration is compensated by an altered vascular reactivity. This issue will be discussed in greater depth in Sec. VII.

VI. INFLAMMATORY CAPACITY

The ability to generate an inflammatory response in response to an exogenous substance is, in general, decreased in the elderly. This observation derives from reports that have used various compounds applied to the skin, which produce their inflammatory effects via either primary irritant reactions or allergic (particularly Type I or Type IV) reactions. Such differences in inflammatory capacity depend on differences in percutaneous penetration in old and young skin and on differences in the microcirculatory efficiency, which serves as the route by which immunocompetent cells make their way to the site of inflammation.

In a procedure devised by Frosch and Kligman, Grove et al. (31) investigated the time taken to raise a blister on the skin of young and old subjects after the application of a 1:1 aqueous solution of ammonium hydroxide. Two parameters were visually assessed in this study: the initial response time (IRT), which was defined as the time taken after application of NH_4OH for the appearance of tiny vesicles around hair follicles, and the blister filling time (BFT), the time required for a fully tense blister to form. The results of this study indicated that the IRT was faster in older subjects, while the BFT was considerably longer in the older subjects. The explanation for these results was hypothesized to reflect both 1) a decreased barrier function in the older subjects (which would mean a faster formation for the initial vesicles) and 2) a reduced microcirculatory function in the elderly (which would mean a compromised ability to fill a blister with the circulatory exudate). Such a study is also of relevance in discussions of percutaneous absorption, which will be addressed in the next section. Overall, the study indicates that older individuals take a longer time to raise a fully formed blister after the application of 50% ammonium hydroxide.

In another study by Grove et al. (12), these investigators looked at differences in the reactivity to different excitants applied to the skin of subjects in two age groups. The excitants applied were DMSO, histamine, 48/80, ethyl-nicotinate, chloroform-methanol, and lactic acid. As depicted in Figure 3, the elderly showed a marked reduction in the inflammatory reactions to these various substances.

When considering Type IV reactions, also known as delayed hypersensitivity reactions, the elderly also exhibit a lessened response. Lejman et al. (39) investigated the reaction of young and aged individuals to the Rhus allergen, which is the culprit in poison ivy dermatitis. This study revealed that in previously sensitized individuals, the reaction to Rhus allergen took longer

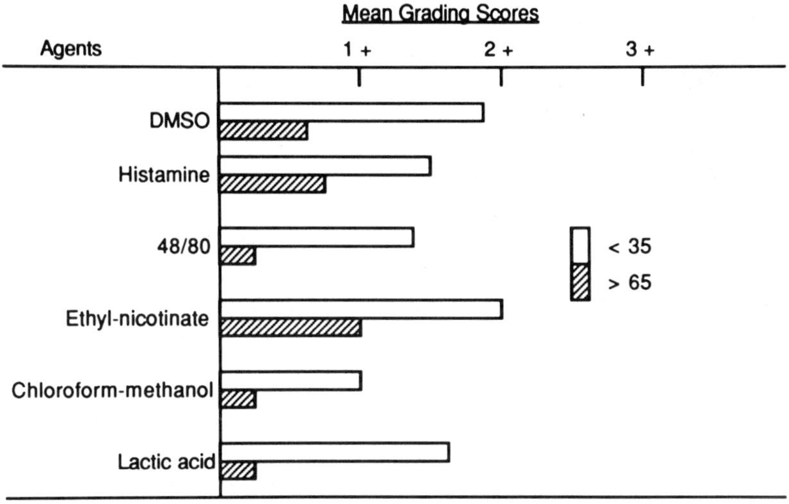

Figure 3 Integumental reactivity tests. Values represent mean grading from three independent observers. (From Ref. 12.)

to appear (Fig. 4), and was decreased in severity in the older population (Fig. 5). As can be seen from these figures, 8 of 9 young subjects reacted by 24 hours, in contrast to only one of 6 elderly subjects. Also, whereas 6 of 9 young subjects reacted with a 2+ reaction, only one of 6 aged patients reacted to this degree. These results concur with earlier studies, which showed an age-associated decline in the reactions to Rhus allergen (40–42).

Although the reactivity to common allergens appears to decrease with increasing age, at least two reports have shown that older individuals may, in fact, react to a wider variety of allergens than the young. Coenraads et al. (43), studying reactions to a panel of common allergens in the young and old, found that older persons responded to a greater number of allergens present in the panel than younger persons. This observation was in accord with a similar investigation by Kreissl and Wiedemann (44), which found a positive correlation between the number of patch test reactions and age. Although the reason for this observation has not been definitively proven, such an observation may reflect an increased propensity to become sensitized with increasing age. In other words, the aged population had more allergen "experience."

The inflammatory response to a model primary irritant, sodium lauryl sulfate, has also shown differential effects in populations of different age. Cua et al. (45) treated various body sites in young and old volunteers with a 24-hour closed patch test of 0.25% sodium lauryl sulfate (SLS). After the 24-

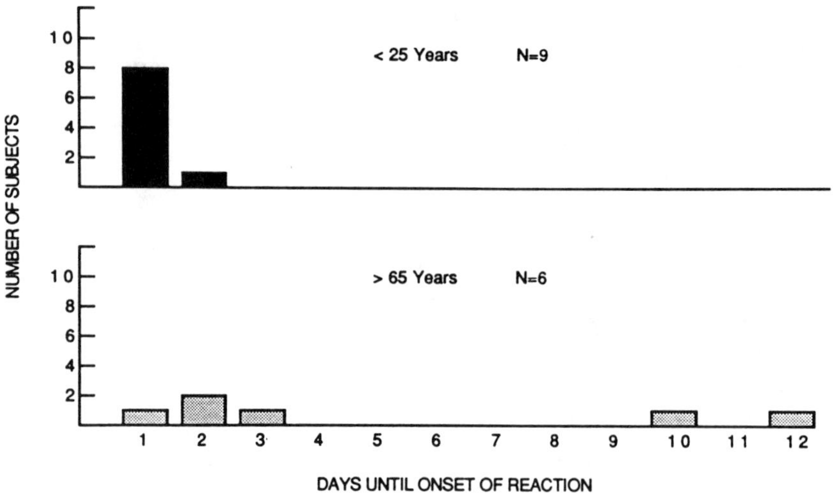

Figure 4 Age-associated differences in time required to develop at least grade 1 response after experimental exposure to Rhus oleoresin. (From Ref. 39.)

hour period, the sites were evaluated for irritant response with visual scoring and TEWL. For most body sites, the older group exhibited both a lower mean visual score as well as a lower percentage of responders.

Additionally, the TEWL, as an indication of barrier impairment due to irritation, was significantly lower in the aged group. Thus, the elderly group was significantly less susceptible to SLS irritation.

Another study using SLS looked at the differential response to this irritant in both the vulva and forearm skin in premenopausal and postmenopausal women. For the forearm site, the results were similar to those reported by Cua et al. above. Specifically, 8 of 10 premenopausal women reacted to the irritant with some degree of erythema, whereas only 6 of 10 postmenopausal women reacted. Additionally, the increases in TEWL and capacitance as indicators of barrier impairment after irritation were delayed in the older group, and the TEWL increase from baseline (postirritation) was significantly lower in this group. In contrast to the age differentiation apparent on the forearm skin, no such distinction between the pre- and postmenopausal women could be made in vulvar skin. Such site-to-site variability was hypothesized to reflect either the effect of photoaging, or possibly hormonal influences (46,47).

Similar to the observation that skin sites differ as to their ability to differentiate the aged and the young, compounds used as model irritants also

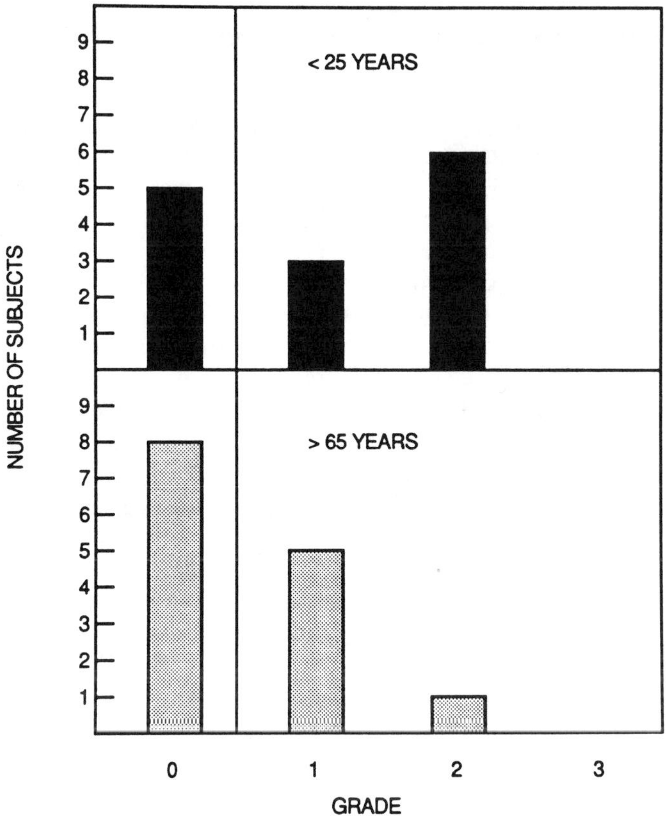

Figure 5 Age-associated differences in the intensity of the allergic response to experimental exposure to Rhus oleoresin at the peak of the response. (From Ref. 39.)

differ in their capability to distinguish age groups. For instance, in a study by Coenraads et al. (43), which tested the cutaneous response to croton oil, thymoquinone, and crotonaldehyde, an age-dependent difference in reactivity was apparent only for croton oil, and as expected, the reactivity decreased with age.

Thus the ability of a compound to exhibit differential reactivity in young and old populations depends on both the nature of the compound as well as the site tested. Such factors may have been operative in earlier studies, such as that of Nilzen and Voss-Lagerland (48), which were in opposition to current thinking and showed that the sensitivity to soaps and detergents increased with age.

VII. BASELINE TEWL, BASELINE CAPACITANCE, AND PERCUTANEOUS ABSORPTION

As alluded to earlier, the subject of percutaneous absorption (PA) is of upmost importance in considering potential age-related differences in the irritant reactivity of individuals. This is of course due to the fact that the first step in the inflammatory sequence requires that the responsible compound become cutaneously absorbed to some degree. Percutaneous absorption consists of two steps: 1) penetration through the layers of the epidermis, and 2) absorption into the blood stream. The first step is dependent on such endogenous factors as degree of stratum corneum barrier function and hydration of the stratum corneum. The degree of stratum corneum barrier function is, in turn, dependent upon such factors as the number of cell layers and the amount and type of intracellular lipids. The second step of PA is dependent upon the integrity of the microcirculation (see Sec. V).

The elucidation of these components is of importance in the development of topical drug delivery systems, which are becoming more commonplace in the medical armamentarium. The importance of age-related variance in these parameters becomes evident when one considers that three of the four FDA-approved drugs for transdermal delivery find their greatest use in those disease processes commonly associated with the older patient. These include nitroglycerin for the treatment of angina, clonidine for hypertension, and estradiol for osteoporosis prophylaxis (49).

The degree of barrier function imparted by the stratum corneum is a reflection of the number of cell layers, the type and amount of stratum corneum lipids, and the water content of the stratum corneum. The studies concerning age-related differences in the number of cell layers agree that the number of cell layers does not appear to change with age (12,30,50). This is despite the fact that, as reported earlier, the thickness of the epidermis appears to decrease with age.

As discussed earlier, the degree of barrier function is assessed by transepidermal water loss measurements; however, such measurements are dependent upon the endogenous barrier function as well as the hydration of the stratum corneum: when the endogenous barrier function is decreased, the TEWL measurement is greater, and when the hydration state, or water content of the stratum corneum is increased, the TEWL will be greater (see Secs. II. A, B). Several studies have made TEWL measurements in both the aged and the young to investigate possible age-related differences in the barrier function. Although the results of these studies are conflicting, the prevailing notion appears to be that TEWL decreases with age. It is not clear at this time whether such age-related TEWL changes reflect a relatively greater degree of endogenous barrier function with age or whether they may be due to an associated decrease in the hydration state of the stratum corneum with age. Indeed, it is generally accepted that aged skin appears drier than youthful skin and has a greater propensity to xerosis. Additionally, Potts et al. (51) have shown through a visoelastic technique that the water content of the stratum corneum is decreased in the aged. In contrast, age-related differences in the water content of the stratum corneum have not been borne out by many in vivo studies, which have utilized capacitance measures. Tagami (52), Cua et al. (45,53), Elsner et al. (47), and Wilhelm et al. (54) all report that no significant age-related differences exist in the hydration state of the stratum corneum as measured with the capacitance-type device. The discrepancies between these different observations and what is generally thought to be the case in the clinical setting—namely, that elderly skin tends toward dryness—requires further investigation.

Several studies utilizing TEWL measurements report that baseline TEWL (i.e., on nonirritated skin) is lower in aged subjects than in young subjects (47,52–57). The report by Wilhelm et al. (54) measured TEWL on 11 body regions in young and old volunteers and found that TEWL was significantly decreased in all body regions measured in the aged (Fig. 6).

As indicated, not all earlier investigations utilizing TEWL measures suggested these same observations. Several studies, including those of Grice and Bettley (58), Kligman (59), Rougier et al. (60), and Tupker et al. (61), uncovered no such age-related differences. In a study by Roskos (4), no significant difference was observed for baseline TEWL in old and young subjects on a single-site measurement (the forearm). Nevertheless, this same investigation did reveal a difference in the TEWL curve as a function of time. In this study, normal skin of old and young subjects was occluded with a plastic chamber for 24 hours. As a result of such occlusion, excess water builds up beneath the chamber. When the chamber is removed, one typically sees a marked increase in the transepidermal eater loss, which represents the evaporation of the "trapped" excess water. In following the evaporation of this water excess back to baseline over time, Roskos observed that the evaporation process was slower in aged skin (4). The same observation was made by Oriba et al. (62) in their measurements of the evaporation rate in vulvar skin of pre- and postmenopausal females, which had been previously delipidized.

The role of lipids in stratum corneum water barrier function and in stratum corneum water-holding capacity is becoming well characterized. Some studies have investigated the possible influence of the lipid molecule in determining the observed age-related changes in baseline TEWL or water desorption kinetics. Using the technique of infrared spectroscopy, Roskos (4) studied differences in lipid content and structural organization as a function of increasing age. The results indicated a consistent decrease in the amount of

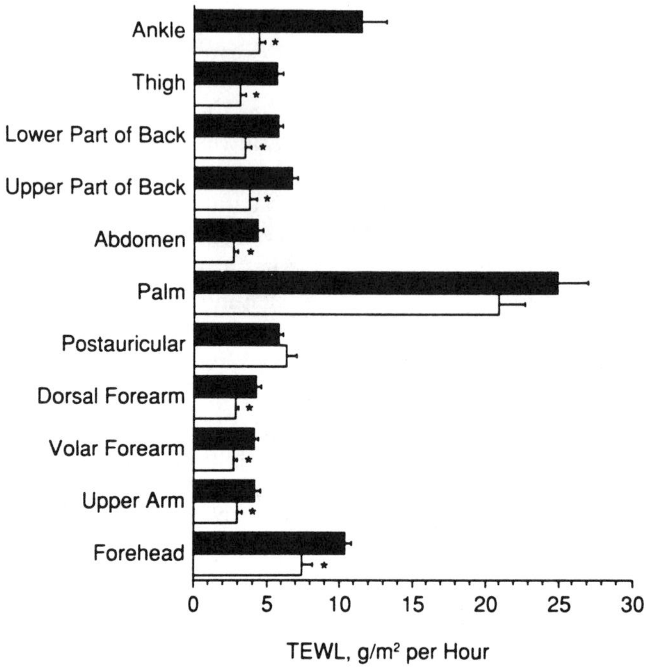

Figure 6 Age differences in transepidermal water loss (TEWL) rates at different anatomical locations. Solid bars indicate young group; open bars, aged group. Shown are means +/− 1 SEM (n = 14–15). Asterisk indicates statistically significant difference between the age groups ($P > 0.05$).

lipid contained in the stratum corneum of the older subjects as compared to the young (4).

An extension of this study then investigated the differential percutaneous absorption of six compounds of varying lipophilicity (49). The six compounds tested were testosterone, estradiol, hydrocortisone, benzoic acid, acetylsalicylic acid, and caffeine. The results indicated that for the least lipophilic compounds (hydrocortisone, benzoic acid, acetylsalicylic acid, and caffeine) there was significantly less absorption in the aged subjects as compared to the young subjects. In contrast, the lipophilic compounds exhibited essentially the same amount of absorption in both groups (Fig. 7). The observed differences in percutaneous absorption of these compounds based on lipophilicity were attributed to the fact that aged stratum corneum contains less lipid, thus creating a "hostile" environment for relatively lipophilic molecules as compared to hydrophilic compounds (49).

The topic of differences in percutaneous absorption in the aged and young is still subject to debate. Older studies on this subject conducted by Christophers and Kligman (30) reported that 1) penetration of fluorescein through isolated stratum corneum is significantly greater in aged stratum corneum than in young stratum corneum 2) clearance of intradermal injections of Na^{+22} is significantly prolonged in the elderly population (hypothesized to be due to the "compromised" microcirculation of the aged), and 3) the percutaneous absorption of testosterone was significantly greater in the young. The reduced clearance of intradermally injected Na^{+22} was confirmed by Tagami (63) but contradicted by DeSalva and Thompson (64). Furthermore, as discussed earlier, the response of the microvasculature to a vasodilatory stimulus (in the form of methylnicotinate) does not appear to differ as a function of age, nor do baseline laser doppler velocimetry measures. Thus, it is apparent that more work needs to be done in the realm of "absorption" in order to determine if, in fact, the histological appearance of the aged microcirculation (which suggests "compromise") correlates to a decreased absorptive efficiency.

The same is true of "penetration," in that more work is needed to characterize the basis for 1) differences in the ability of certain compounds to penetrate the skin and 2) differences in baseline transepidermal water loss. Such differences may reflect the role of lipids, keratin, and other components of the stratum corneum.

VIII. MECHANICAL PROPERTIES

The mechanical properties to be discussed in this section include friction and elastic deformation. Changes in elastic deformation are anatomically obvious to most, as it is a common observation that skin becomes loose and sags with increased age. Such changes are due to changes in the composition of elements in the dermis and hypodermis, which are apparent from histological examination. These elements include collagen, elastic fibers, and reticulin fibers, all of which are embedded in ground substance.

A. Elasticity

One device that measures changes in elastic properties of skin is known as the Cutometer (SEM 474, Courage and Khazaka, Cologne, Germany and Acaderm, Menlo Park, CA). The device is based on the principles of suction/elongation. Briefly, a vacuum is applied to the skin via a suction cup applied perpendicularly to the skin surface. The resulting deformation inside the suction cup chamber is detected as the amount of skin elevation in this chamber and is measured by an optical measuring unit. Four parameters are typically determined: Ue, immediate distention; Uv, delayed distention; Ur, im-

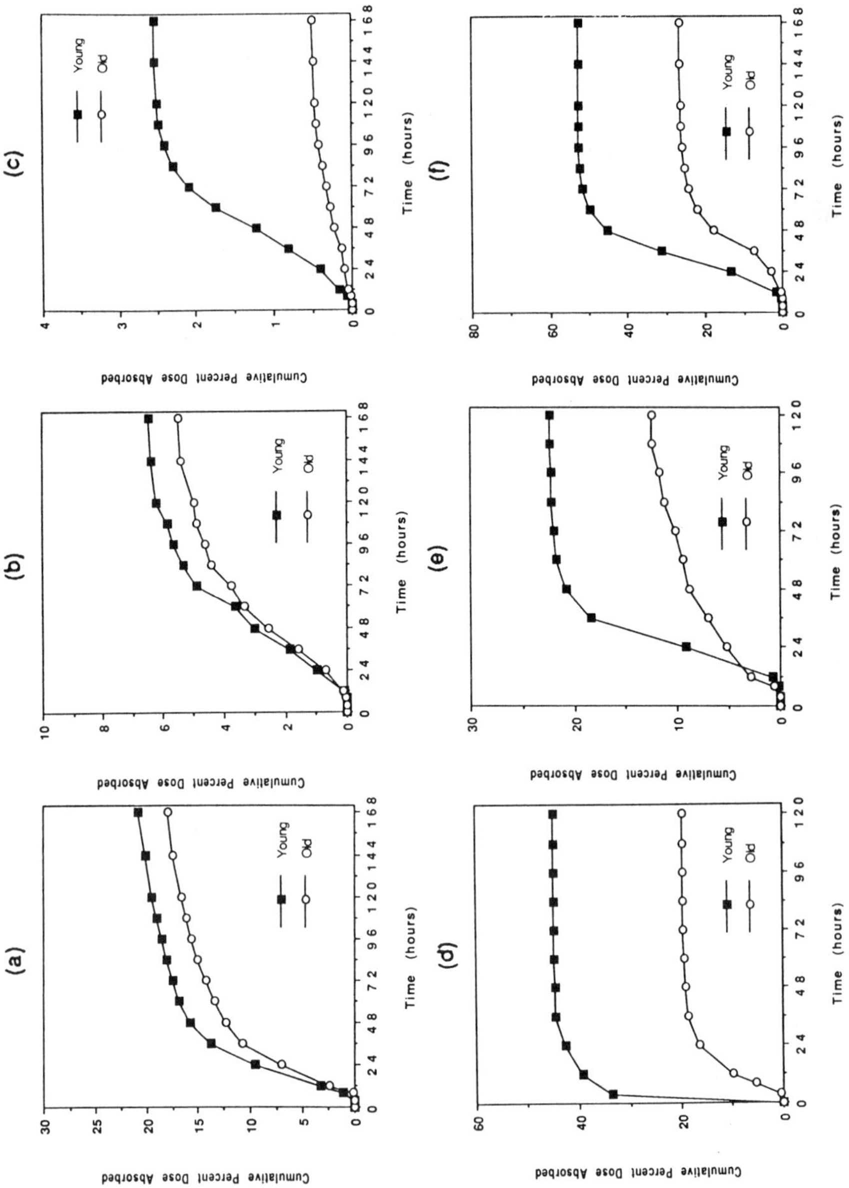
(a)
(b)
(c)
(d)
(e)
(f)
Young
Old
Time (hours)
Cumulative Percent Dose Absorbed

mediate retraction; and Uf, final distention. These parameters, graphically represented in Figure 8, are dependent upon skin thickness, and thus direct comparisons between subjects and different anatomical regions cannot be made; however, certain biologically relevant ratios, reported to be independent of skin thickness differences, do permit comparison between individuals (65). These ratios are Uv/Ue and Ur/Uf. Uv/Ue is thought to be representative of the skin's visoelastic properties, while Ur/Uf represents the ability of the skin to return to its initial position after deformation. Furthermore, the histological correlate of visoelastic properties (Uv/Ue) is thought to be represented by the ground substance of the dermis, while Ur/Uf is dependent upon the elastic fiber network (66).

In a study by Cua et al. (66), Uv/Ue increased with advanced age, while Ur/Uf decreased with age. These results agree with similar studies by Agache et al. (65), Elsner et al. (67), Leveque et al. (68), and Pierard and Lapiere (69). Furthermore, the results obtained from this study were hypothesized to be due to histological changes that accompany normal skin aging, namely, 1) alterations in the components of ground substance (glycosaminoglycans, proteoglycans, and uronic acid) (70) and 2) alterations in the elastic fiber component (71).

B. Friction Coefficient

The next mechanical property, friction, is defined as the force opposing movement between two solid objects. Skin friction is the force that allows one to grasp objects, and excessive frictional forces result in such skin pathology as erosions, abrasions, and blisters. Measurements of friction have both a static component, which is the force resisting the *initial* movement between surfaces, and a kinetic component, which is the frictional force *during* movement. In measurements of skin friction, only the kinetic component is measured. This component is measured by means of a commercially available device known as the Newcastle Friction Meter (Design Unit, Department of Mechanical Engineering, Newcastle University, Newcastle-upon-Tyne, U.K. and Acaderm, Menlo Park, CA). This is a hand-held device consisting of a spring-loaded teflon wheel that rotates at constant speed. Application of the wheel to a surface (skin) results in the generation of a frictional force between the wheel and the surface, from which a friction coefficient (μ) can be determined (53).

Figure 7 Representative curves of the cumulative percentage dose absorbed as a function of time for (a) testosterone, (b) estradiol, (c) hydrocortisone, (d) benzoic acid, (e) acetylsalicyclic acid, and (f) caffeine, following topical administration. The young (filled squares) and old (open circles) data are presented on each graph for comparison. (From Ref. 49.)

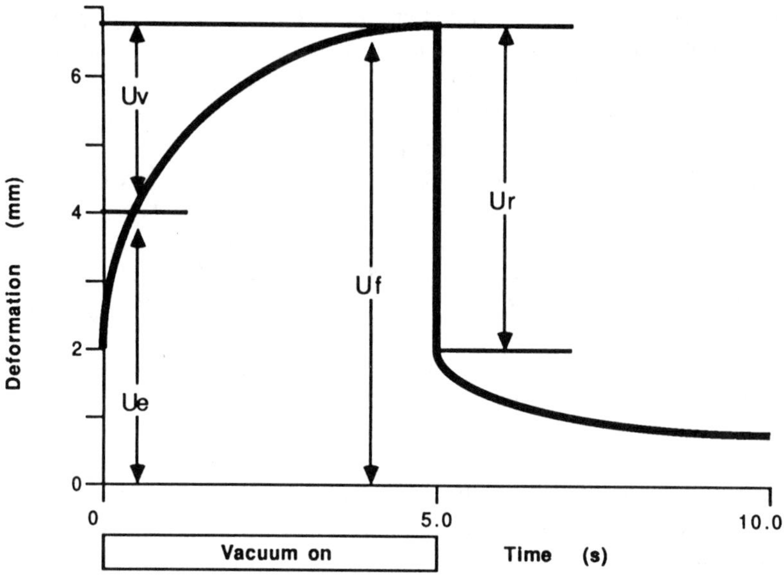

Figure 8 Skin deformation plotted as a function of time. (From Ref. 53.)

Measurements of skin friction coefficients are heavily dependent upon skin surface hydration. Generally, extremely dry or wet skin has a lower frictional coefficient than skin with an intermediate degree of hydration (72–74). This dependence was confirmed in a study by Elsner et al. (67), which showed that the frictional coefficient was higher in the more hydrated vulvar skin of females than in the forearm skin. This same study revealed that at the forearm site, the frictional coefficient was lower in postmenopausal women than in premenopausal women. This decrease in the frictional coefficient with age was attributed to an apparent decrease in stratum corneum hydration with increased age, but as has been discussed earlier, such an age-dependent decrease in water content may not actually exist.

In contrast, in an investigation by Cua et al. (53), which measured frictional coefficients and capacitance at various body regions in young and old subjects, no age-related differences in friction coefficient could be demonstrated. Furthermore, there were no age-related differences in the hydration state of the stratum corneum (53). Thus, while it is established that friction coefficients vary with the hydration state of the stratum corneum, it has not been unequivocally shown that friction coefficients vary with age.

IX. SEBUM SECRETION

Sebum production varies considerably in the first decades of life. Changes in sebum production depend upon steroid hormones, thus sebum secretion is high in the newborn due to transplacental transfer of maternal steroid hormones, and in childhood steroid hormone levels are relatively low. At puberty steroid secretion begins to rise to adult levels, and at this time one observes a concomitant increase in sebum excretion rate with the attendant disease manifestation of acne.

Changes in sebum excretion after puberty are less well characterized. It was previously thought that after puberty, sebum excretion rates remained relatively constant throughout adult life, decreasing in women after menopause and in men at some point very late in life. Downing et al. (75), however, has described a steady decline in sebum excretion rates in both men and women with increasing age. The method employed for these measurements utilized absorption of sebum onto a film of betonite clay applied to the skin surface. The absorbed lipids were then analyzed both quantitatively and qualitatively by thin layer chromatography. From these studies it was suggested that the decreased sebum secretion rate may actually be due to the age-dependent increase in the cellular turnover time, reported earlier (75).

A commercially available device known as the Sebumeter (SM 810 PC, Courage & Khazaka and Acaderm, Menlo Park, CA) is capable of measuring relative amounts of sebum present on the skin surface. The device works on the principle of photometry. Briefly, an opaque plastic strip is applied to the skin surface with a constant pressure of 6 N, and skin surface lipids absorb to the plastic strip. The absorption of lipids to this strip results in greater translucency of the strip, and the relative change in translucency is related to the amount of sebum absorbed. The units of measurement are expressed as $\mu g/cm^2$.

In studies utilizing this instrument to investigate possible age-associated changes in quantitative sebum levels, no age dependence was observed (54). Few studies of this type have been carried out; however, it is likely that the sensitivity of this device is less than those quantitative measures made by Downing et al. (75).

X. CONCLUSION

The previous discussion has focused on the mensuration of cutaneous properties that change with increasing age (i.e., from adulthood onward). Such properties include measures of transepidermal water loss, blood flow, stratum corneum hydration, skin thickness, cutaneous elasticity, skin friction, and sebum secretion.

Since the subject of cutaneous aging is still relatively new, it is difficult to make firm conclusions regarding changes in the above-numerated parameters. This is due to the fact that in many instances there is no uniformity of opinion as to the direction of observed change with age. For instance, as is evident from Table 2, there appears to be a general consensus that the thick-

Table 2 Changes in Cutaneous Parameters with Age

Parameter	Change with age	Ref.
Transepidermal water loss	Unchanged	4,58–61
	Decreased	47,52–57
Blood flow	Unchanged	35
	No change in pharmacologic responsiveness	37,38
Stratum corneum hydration	Unchanged	45,47,52–54
	Decreased	51
Skin thickness	Decreased	22–28
Sebum content	Decreased	75
	Unchanged	54
Friction coefficient	Unchanged	53
	Decreased	67

ness of both the epidermis and the dermis decreases with age. In contrast, in reviewing the topic of transepidermal water loss, there exist approximately equal numbers of investigations that report either 1) no change in TEWL with increased age or, alternatively, 2) decreased TEWL with increasing age. Furthermore, when attempting to synthesize our knowledge of these independent factors to understand something as complex as percutaneous absorption, the picture remains blurred.

Clearly, our ability to understand the processes of human cutaneous aging will depend upon 1) further investigation with the presently available bioengineering devices, 2) refinement of these devices (specifically, by defining their limitations), and 3) development of other devices for objective measurement of cutaneous properties. As the life expectancy of the human population continues to increase, it becomes essential that the scientific community devote more time to the effect of aging on various organ systems. Such research will enhance our ability to both prevent and treat skin disease, as well as diseases of other organs.

REFERENCES

1. Berardesca E, Maibach HI. Bioengineering technology for the noninvasive evaluation of the effects of clinical materials applied to skin. Clinical Materials 1987; 2:251–7.
2. Berardesca E, Maibach HI. Noninvasive bioengineering assessment of psoriasis. Int J Dermatol 1989; 28(3):157–60.
3. Nilsson GE. Measurement of water exchange through skin. Med Biol Eng Comput 1977; 15:209–18.

4. Roskos KV. The effect of skin aging on the percutaneous penetration of chemicals through human skin. Ph.D. dissertation. San Francisco: University of California, 1989.
5. Wertz PW, Downing DT. Glycolipids in mammilian epidermis: structure and function in the water barrier. Science 1982; 217:1261–2.
6. Pinnagoda J, Tupker RA, Agner T, Serup J. Guidelines for transepidermal water loss (TEWL) measurement. A report from the standardization group of the European Society of Contact Dermatitis. Contact Dermatitis 1990; 22:164–78.
7. Blichmann CW, Serup J. Reproducibility and variability of transepidermal water loss measurements. Acta Dermato-venereologica 1987; 67:206–10.
8. Agner T, Serup J. Individual and instrumental variation in irritant patch test reactions. Clin Exp Dermatol 1990; 15(1):29–33.
9. Mathias CGT, Wilson DM, Maibach HI. Transepidermal water loss as a function of skin surface temperature. J Invest Dermatol 1981; 77:219–20.
10. Bircher AJ, Maibach HI. Laser doppler velocimetry and photoplethysmography. In: Kligman A, Takase Y, eds. Cutaneous Aging. Tokyo: University of Tokyo Press, 1988:521–37.
11. Fischer JC, Parker PM, and Shaw WW. Microsurgery 1983; 4:164–70.
12. Grove GL, Lavker RM, Hoelzle E, Kligman AM. Use of nonintrusive tests to monitor age-associated changes in human skin. J Soc Cosmet Chem 1981; 32: 15–26.
13. Agache PG, Mignot J, Makki S. Microphotography of the skin and aging. In: Kligman A, Takase Y, eds. Cutaneous Aging. Tokyo: University of Tokyo Press, 1988:475–99.
14. Voros E, Robert C, Robert AM. Age-related changes of the human skin surface microrelief. Gerontology 1990; 36:276–85.
15. Makki S, Barbenel JC, Agache PA. A quantitative method for the assessment of microphotography of human skin. Dermatologica 1991; 182:1–6.
16. Corcuff P, deRigal J, Leveque JL. Image analysis of the cutaneous microrelief. Bioeng Skin Newslett 1982; 4:16–31.
17. Alexander H, Miller DL. Determining skin thickness with pulsed ultrasound. J Invest Dermatol 1979; 72:17–19.
18. Perednia DA. What dermatologists should know about digital imaging. J Am Acad Dermatol 1991; 25(1):89–108.
19. Breibart EW, Rehpenning R. Möglichkeiten und Grenzen der Ultraschalldiagnostik zur in vivo Bestimmung der Invasionstiefe des malignen Melanomas. Z Hautkr 1983; 58:975–87.
20. Hermann RC, Ellis CN, Fitting DW, Ho VC, Voorhees JJ. Measurement of epidermal thickness in normal skin and psoriasis with high-frequency ultrasound. Skin Pharmacol 1988; 1:128–36.
21. Gomez EC, Berman B, Miller DL. Ultrasonic assessment of cutaneous atrophy caused by intradermal corticosteroids. J Dermatol Surg Oncol 1982; 8:1071–4.
22. Tan CY, Statham B, Marks R, and Payne PA. Skin thickness measurement by pulsed ultrasound: its reproducibility, validation and variability. Br J Dermatol 1982; 106:657–67.
23. Branchet MC, Boisnic S, Frances C, Robert AM. Skin thickness changes in normal aging skin. Gerontology 1990; 36:28–35.

24. Shuster S, Black MM, McVitie E. The influence of age and sex on skin thickness, skin collagen and density. Br J Dermatol 1975; 93:639–43.
25. Montagna W, Carlisle K. Structural changes in aging human skin. J Invest Dermatol 1979; 73:47–53.
26. Hall DA, Blackett AD, Zajac AR, Switala S, Airey CM. Changes in skinfold thickness with increasing age. Age Aging 1981; 10:19–23.
27. Hastie IR, Millard PH. The quantification of skin transparency. Br J Dermatol 1983; 108:315–20.
28. Fenske NA, Lober CW. Structural and functional changes of normal aging skin. J Am Acad Dermatol 1986; 15:571–85.
29. Goodson WH, Hunt TK. Wound healing and aging. J Invest Derm 1979; 73:88–91.
30. Cristophers E, Kligman AM. Percutaneous absorption in aged skin. In Montagna, W ed., Advances in Biology of Skin, Vol 6, Aging, Oxford: Pergamon Press, 1965:163–75.
31. Grove GL, Duncan S, Kligman AM. Effect of ageing on the blistering of human skin with ammonium hydroxide. Br J Dermatol 1982; 107:393–400.
32. de Rigal J, Eschoffier C, Querleux B. Assessment of aging of the human skin by in vivo ultrasonic imaging. J Invest Dermatol 1989; 93:621–5.
33. Schatz H, Stoudemayer T, Gabriel K, et al. Visualization of aging and photodamage in human skin by 20 MHz ultrasound. Presentation to the Eighth International Symposium on Bioengineering and the Skin, Stresa, Italy, June 13-16, 1990.
34. Ryan TJ. Cutaneous circulation. In Goldsmith LA, ed. Biochemistry and Physiology of the Skin. New York: Oxford University Press, 1983:817–77.
35. Hatanaka H, Matsumoto S, Ishikawa K, Kawasaki T, Kubota S, Takagi K, Tanke G, Yoshimura Y, Oimomi M, Baba S. Rinsho Byori 1984; 32:1025–8.
36. Oimimi M, Hatanaka H, Maeda Y, Hata F, Kitamura Y, Matsumoto S, Baba S. Autonomic nervous function determined by changes of periflux blood flow in the aged. Arch Gerontol Geratr 1986; 5:159–63.
37. Guy RH, Tur ET, Bjerke S, Maibach HI. Are there age and racial differences to methyl nicotinate-induced vasodilation in human skin? J Am Acad Dermatol 1985; 12:1001–6.
38. Roskos KV, Bircher AJ, Maibach HI, Guy RH. Pharmacodynamic measurements of methyl nicotinate percutaneous absorption: the effect of aging on microcirculation. Br J Dermatol 1990; 122:165–71.
39. Lejman E, Stoudemayer T, Grove G, Kligman AM. Age differences in poison ivy dermatitis. Contact Dermatitis 1984; 11:163–7.
40. Shelmire B. Contact dermatitis form weeds: patch testing with their oleoresins. JAMA 1939; 113:1085–93.
41. Kligman A. Poison ivy (Rhus) dermatitis. Am Med Assoc Arch Dermatol 1958; 77:149–80.
42. Smith JG Jr., Kiem JM. Allergic contact sensitivity in the aged. J Gerontol 1961; 16:118–19.
43. Coenraads PJ, Bleumink E, Nater JP. Susceptibility to primary irritants: age dependence and relation to contact allergic reactions. Contact Dermatitis 1975; 1:377–81.

44. Kreissl H, Wiedemann G. Epikutane Testungen bei hautgesunden Kindern und Jugendlichen. Dermatolog Wochenschr 1963; 148:536–41.
45. Cua AB, Wilhelm KP, Maibach HI. Cutaneous sodium lauryl sulphate irritation potential: age and regional variability. Br J Dermatol 1990; 123:607–13.
46. Elsner P, Wilhelm D, Maibach HI. Irritant effect of a model surfactant on the human vulva and forearm: age related differences. J Reproductive Med 1990; 35(11):1035–9.
47. Elsner P, Wilhelm D, Maibach HI. Sodium luryl sulfate-induced irritant contact dermatitis in vulvar and forearm skin or premenopausal and postmenopausal women. J Am Acad Dermatol 1990; 23:648–52.
48. Nilzen A, Voss-Lagerland K. Epicutaneous tests with detergents and a number of other common allergens. Dermatologica 1962; 124:42–52.
49. Roskos KV, Maibach HI, Guy RH. The effect of aging on percutaneous absorption in man. J Pharmacokin Biopharm 1989; 17(6):617–30.
50. Balin AK, Pratt LA. Physiological consequences of human skin aging. Cutis 1989; 43:431–6.
51. Potts RO, Buras EM, Chrisman DA. Changes with age in the moisture content of the human skin. J Invest Dermatol 1984; 82:92–100.
52. Tagami H. Aging and the hydration state of the skin. In Kligman A, Takase Y, eds. Cutaneous aging. Tokyo: University of Tokyo Press, 1988:99–109.
53. Cua AB, Wilhelm KP, Maibach HI. Frictional properties of human skin: relation to age, sex, and anatomical region, stratum corneum hydration and transepidermal water loss. Br J Dermatol 1990; 123:473–9.
54. Wilhelm KP, Cua AB, Maibach HI. Skin aging: effect of transepidermal water loss, stratum corneum hydration, skin surface pH, and casual sebum content. Arch Dermatol 1991; 127:1806–9.
55. Leveque JL, Corcuff P, deRigal J, Agache P. In vivo studies of the evolution of physical properties of the human skin with age. Int J Dermatol 1984; 23:322–9.
56. Baker H. Deperdition d'eau par voie transepidermique. Ann Dermatol Syphiligr 1971; 98:289–96.
57. Thune P, Nilsen T, Hansted K, Gustavsen T, Dahl HL. The water barrier function of the skin in relation to the water content of the stratum corneum, pH and skin lipids. Acta Dermatolvenereologica 1988; 68:277–83.
58. Grice KA, Bettley FR. Skin water loss and accidental hypothermia in psoriasis, icthyosis, and erythroderma. Br Med J 1967; 4:195–201.
59. Kligman AM. Perspectives and problems in cutaneous gerontology. J Invest Dermatol 1979; 73:39–46.
60. Rougier A, Lotte C, Corcuff P, Maibach HI. Relationship between skin permeability and corneocyte size according to anatomical site, age, and sex in man. J Soc Cosmet Chem 1988; 39:15–26.
61. Tupker RA, Coenraads PJ, Pinnagoda J, Nater JP. Baseline transepidermal water loss (TEWL) as a prediction of susceptibility to sodium lauryl sulfate. Contact Dermatitis 1989; 20:265–69.
62. Oriba H, Maibach HI. Vulvar transepidermal water loss (TEWL) decay curves: effect of occlusion, delipidation, and age. Acta Derm Venereol (Stockh) 1989; 69:461–5.

63. Tagami H. Functional characteristics of aged skin. 1. percutaneous absorption. Acta Derm (Kyoto) 1972; 67:19–21.
64. DeSalva SJ, Thompson G. $Na^{22}Cl$ skin clearance in humans and its relation to skin age. J Invest Dermatol 1965; 45:315–18.
65. Agache P, Monneur C, Leveque JL, deRigal J. Mechanical properties and Young's modulus of human skin in vivo. Arch Dermatol Res 1980; 269:221–32.
66. Cua AB, Wilhelm KP, Maibach HI. Elastic properties of human skin: relation to age, sex, and anatomical region. Arch Dermatol Res 1990; 282:283–8.
67. Elsner P, Wilhelm D, Maibach HI. Frictional properties of human forearm and vulvar skin: influence of age and correlation with transepidermal water loss and capacitance. Dermatologica 1990; 181:88–91.
68. Leveque JL, de Rigal J, Agache PG, Monneur C. Influence of ageing on the in vivo extensibility of human skin at a low stress. Arch Dermatol Res 1980; 269: 127–35.
69. Pierard GE, Lapiere CM. Physiopathological variations in the mechanical properties of skin. Arch Dermatol Res 1977; 260:231–9.
70. Johnston KL, Oikarinen A, Lowe NJ. Ultraviolet-induced connective tissue changes in the skin: models for actinic damage and cutaneous aging. In Maibach HI, Lowe NJ, eds. Models in dermatology, vol. 1. Basel: S. Karger, 1985:69–76.
71. Fazio MJ, Olsen DR, Uitto JJ. Skin aging: lessons from cutis laxa and elastoderma. Cutis 1989; 43:437–44.
72. Zimmerer RE, Lawson KD, Calvert CJ. The effects of wearing diapers on the skin. Pediatr Dermatol 1986; 3:95–101.
73. Wolfman LJ. Friction of skin. J Soc Cosmet Chem 1983; 34:465–76.
74. Nacht S, Close JA, Yeung D, et al. Skin friction coefficient: changes induced by skin hydration and emollient application and correlation with perceived skin feel. J Soc Cosmet Chem 1981; 32:55–6.
75. Downing DT, Stewart ME, Strauss JS. Changes in sebum secretion and sebaceous glands during aging. In Kligman A, Takase Y, eds. Cutaneous aging. Tokyo: University of Tokyo Press, 1988:135–48.

18

Mechanism of Photosensitivity Due to Porphyrins: Chemical Species Involved and Relevance to Clinical Porphyrias

Herbert F. Haberman and I. Aravind Menon
University of Toronto, Toronto, Ontario, Canada

I. CLINICAL PORPHYRIAS

A. Definition

Porphyrins are interesting and important compounds. Porphyrins were probably some of the earliest biopolymers and have been found in several fossils. They occur widely in the mammalian body as metabolites or derivatives of natural metabolites. They are essential for life on this planet and are necessary for the biosynthesis of chlorophyll, hemoglobin, cytochromes, catalase, peroxidase, and superoxide dismutase. Abnormally high amounts of porphyrins produce photosensitivity (1–7). Exogenously administered porphyrins, porphyrin precursors, or derivatives are used in photodynamic therapy of cancers (8,9).

Porphyrins are tetrapyrrole structures in which four pyrrole rings are joined together by methene bridges. They have conjugated double bonds running through the pyrrole rings and the methene bridges. The tetrapyrroles occur in reduced form—porphyrinogens—and in oxidized form—porphyrins. Both the oxidized and reduced forms occur as two isomers, namely isomer I and isomer III. The heme synthetic pathway involves the formation of pyrroles and subsequent linking of these by methene bridges to form a cyclic tetrapyrrole. The first porphyrinogen formed, uroporphyrinogen has eight carboxyl groups. Further steps in the pathway involve decarboxylation resulting in the formation of coproporphyrinogen (having four carboxyl groups) and protoporphyinogen (having two carboxyl groups). These porphyrinogens are oxidized to produce the corresponding porphyrins. As decarboxylation proceeds, the water solubility of the three major porphyrins progressively decreases and their lipid solubility increases. This is reflected in the excretion of these porphyrins, the lipid-soluble porphyrins being found mostly in feces, whereas the water-soluble ones are present in urine.

Porphyrias are acquired or inherited disorders of the heme synthetic pathway in which the tissue levels of porphyrins or porphyrin precursors are elevated. Each of the porphyrias is associated with changes in specific enzyme activities. These enzyme abnormalities cause the accumulation of one or more porphyrins or porphyrin precursors, mostly due to a partial block in the heme synthetic pathway (Table 1). However in addition, δ-aminolevulinic acid (ALA) synthetase can be induced, resulting in the overproduction of ALA and subsequently all the intermediates in the pathway. At the present time it is thought that the high levels of porphyrins or porphyrin precursors are responsible for causing the symptoms and signs of each disease. Currently the diagnosis of porphyrias is also made largely by the determination of porphyrins and porphyrin precursors in blood, urine, and feces (9,10). The usual findings are shown in Table 2.

Table 1 Decreased Enzyme Activities in Porphyrias

Diagnosis	Decreased enzyme activity
Acute intermittent porphyria (AIP)	Porphobilinogen deaminase (PBGD)
Acute coproporphyria (ACP)	Coproporphyrinogen oxidase (CPGO)
Variegate porphyria (VP)	Protoporphyrinogen oxidase (PPGO)
Porphyria cutanea tarda (PCT)	Uroporphyrinogen decarboxylase (UPGD)
Hepatoerythropoietic porphyria (HEP)	Uroporphyrinogen decarboxylase (UPGD)
Congenital erythropoietic porphyria (CEP)	Uroporphyrinogen cosynthetase (UPGCoS)
Erythropoietic protoporphyria (EPP)	Heme synthetase (HS)

Table 2 Clinical Laboratory Testing of Porphyrias

		Blood			Urine				Feces		
Type	Enzyme	UP	CP	PP	ALA	PBG	UP	CP	UP	CP	PP
AIP	PBGD				+ +	+ + + +	+	+			
ACP	CPGO				+	+ + +	+ +	+ + + +		+ + + +	
VP	PPGO				+	+ + +	+ + +	+ +			+ + + +
PCT	UPGD						+ + + +	+ + +		+	
HEP	UPGD	+		+			+ + + +	+ + +			
CEP	UPGCoS	+ + + +					+ + + +				
EPP	HS			+ + + +							+ + +
Pb	ALD,HS			+	+ + + +	+ +		+			

Pb: Lead intoxication; ALD: ALA dehydrase.

B. Classifications

There are two main target organs in porphyrias; the effects produced on these organs are responsible for most of the clinical picture. An increase in porphyrin precursors, e.g., ALA or decreased central nervous system heme, is thought to be responsible for the neurological changes, whereas cutaneous photosensitivity is due to increased porphyrins in the skin.

There are a number of classifications of the porphyrias, none of which seems to be rationally comprehensive because of overlapping symptoms, manifestations, and affected tissues. It should be noted that all porphyrias, with the exception of acute intermittent porphyria (AIP) and lead intoxication, are associated with photosensitivity.

From a clinically practical point of view we like to divide the porphyrias into two types.

Type I: the nervous system is a major target (the inducible acute porphyrias)
- Ia: inducible acute porphyria without cutaneous photosensitivity, i.e., AIP
- Ib: inducible acute porphyria with neurological and/or cutaneous manifestations, i.e., acute coproporphyria or hereditary coproporphyria (ACP) and variegate porphyria (VP)

Type II: non-neurological, primarily cutaneous porphyrias
- IIa: principally hepatic, in which uroporphyrinogen decarboxylase is decreased, namely, acquired porphyria cutanea tarda (PCT), mainly due to estrogens or ethanol, hereditary PCT, hepatoerythropoietic porphyria (HEP), the porphyrias due to toxic chemicals, PCT associated with hepatoma
- IIb: principally erythropoietic, namely, congenital erythropoietic porphyria (CEP), erythropoietic protoporphyria (EPP)

The neurological porphyrias are characterized by acute attacks with systemic signs and symptoms, i.e., abdominal pain, peripheral and autonomic neuropathies, psychiatric symptoms, etc. These attacks can be induced by fasting, infection, hormones, or drugs, probably by their effects on the rate-limiting enzyme ALA synthetase. These diseases can result in severe morbidity and even mortality. Since ACP and VP can have cutaneous findings similar to PCT, it is important to differentiate between these diseases. It should also be noted that AIP is the only inherited porphyria with no association with photosensitivity.

C. Photosensitivity in Porphyrias

The most common porphyrias with photosensitivity are PCT and EPP. The cutaneous signs and symptoms in these two conditions are markedly different.

In PCT, the common findings are friability of the skin, bullae on light-exposed areas, scarring, hypertrichosis, hyperpigmentation, a sclerodermoid picture, and milia. The patient is often unaware of the relationship to sun exposure. With EPP, however, the patient is well aware of the relationship to light, because soon after exposure to sun the patient notices pruritus, burning, and/or pain. This goes on to erythema, swelling, or edema. Scarring does occur, but is usually much less pronounced than in PCT. Frequently there is pitting or thickening on the nose, forehead and cheeks or dorsum of hands. There can be raggedies around the lips. In addition to PCT, the same cutaneous findings can also be found in VP, ACP, CEP, and HEP. They are usually quite severe in CEP and HEP.

The cutaneous manifestations, the tissue levels of porphyrins, and their urinary as well as fecal excretion patterns differ in the various porphyrias. For instance, EPP is characterized by the accumulation of abnormally high levels of protoporphyrin (PP) in red blood cells and often in the plasma and feces, whereas high levels of uroporphyrin (UP) and coproporphyrin (CP) in plasma and urine are characteristic of PCT (10,11). The reasons for the differences in the clinical picture in the different photosensitizing porphyrias are not clearly understood at the molecular level. These differences may result from the different localization of the various porphyrins. The differences in the accumulation of these porphyrins within the cells and in the extracellular fluids can be expected to be composite functions of their solubility in the aqueous and lipid phases as well as subcellular localization (12, 13). Evidence supporting this postulate has been reported from the photobiological investigations on leukocytes (14), red blood cells (15), fibroblasts (16), and mitochondria (17).

II. ROLE OF REACTIVE SPECIES IN PHOTOSENSITIZATION

Photosensitization produces the reactive oxygen species, singlet oxygen ($^1O_2$), superoxide (O_2^-), hydrogen peroxide (H_2O_2), and hydroxyl radical ($\cdot OH$). Since $^1O_2$ is short-lived, its effects are restricted to the cellular location where it is formed (18,19). On the other hand, both O_2^- and H_2O_2 can survive for relatively longer periods and can exert the oxidative damage at relatively distal locations. Photosensitization has been reported to produce a variety of biological effects, e.g., lipid peroxidation, DNA cross-linking, protein cross-linking, and formation of prostaglandins (20–29). When more than one reactive oxygen species are produced, the resultant biological effects are probably due to the combined effects of all these species. Under these conditions the kinetics of the overall effect may be rather complex (30–34).

III. REACTIVE OXYGEN SPECIES FORMED BY PORPHYRIN PHOTOSENSITIZATION

Free prophyrins are potent photosensitizers (19). The action spectra of porphyrins consist of a major peak at approximately 405 nm (Soret band) and minor peaks between 530 and 630 nm. The exact positions of the peaks vary from one porphyrin to another. The available evidence shows that the action spectrum for photosensitization parallels the absorption spectrum. The primary reactive species formed during porphyrin photosensitization seems to be singlet oxygen. However, there is evidence to show the formation of H_2O_2 and $\cdot OH$; the formation of these species is especially enhanced in the presence of reducing agents, such as sulfhydryl compounds and ascorbic acid (35,36).

A. Singlet Oxygen

There is abundant evidence to show that singlet oxygen is formed during the irradiation of the porphyrins, presumably by the transfer of the excess energy of the porphyrins in the excited states to oxygen. Several reports have indicated that singlet oxygen may be involved in the photosensitization by the porphyrins (19,28,37).

B. Hydrogen Peroxide

A comparison of the formation of H_2O_2 during the irradiation of PP, CP, and UP showed that UP produced the largest amount, PP was the least effective and CP was intermediate between UP and PP. It is obvious that the H_2O_2 formation is parallel to the water solubility of the porphyrins and inversely related to their solubility in lipid (38). The H_2O_2 formation is also inversely related to the photolytic efficacy of these porphyrins, where the potency decreased in the following order: PP $\rightarrow$ CP $\rightarrow$ UP (39).

The formation of H_2O_2 was inhibited by sodium azide, a strong quencher of singlet oxygen (38). Two other singlet oxygen quenchers, namely, 1,4-diazabicyclo[2.2.2]-octane (DABCO) and dimethyl furan (DMF), were found to interfere with the assay for H_2O_2; hence these could not be employed in these experiments. The inhibition of H_2O_2 formation by sodium azide indicates that the formation of H_2O_2 during irradiation of the porphyrins was mediated by singlet oxygen. However, this conclusion is at present tentative and has to be confirmed using other singlet oxygen quenchers.

C. Hydroxyl Radical

There is no direct evidence for the formation of hydroxyl radical during porphyrin photosensitization. Since Fe would catalyze the production of

·OH from H_2O_2 and the ·OH thus formed may have a significant role in the uroporphyrin photosensitivity, the role of Fe in porphyrin photosensitivity has been investigated. This was examined by studying the effects of the Fe-chelating compound deferoxamine in an in vitro system. These results showed that deferoxamine inhibited the uroporphyrin photosensitivity, but not the photosensitivity due to protoporphyrin. These findings indicate indirectly that ·OH may be involved in UP photosensitization, but not in PP photosensitization (40).

IV. REACTIVE SPECIES INVOLVED IN PORPHYRIN PHOTOSENSITIZATION

A. In Vitro Cytotoxicity

There is a gradation in the in vitro and in vivo photosensitizing effects of the three porphyrins. PP is the most potent, UP is the least effective, and CP is intermediate between PP and UP (12,39).

The use of scavengers of superoxide and H_2O_2, namely, superoxide dismutase (SOD) and catalase, shows that the lysis due to PP is not mediated by these species. On the other hand, the photolysis due to CP and UP was shown to be mediated at least partially by H_2O_2, formed from superoxide. It is interesting to recall that small amounts of superoxide and hydroxyl radical are formed during the photosensitization of hematoporphyrin (41,42).

It is obvious that the above gradation in the photosensitizing efficacy is parallel to the lipid solubility of the porphyrins and inversely related to their solubility in water. These findings are in agreement with some earlier reports (17,38). UP is a weak photosensitizer both in vitro and in vivo (12,39). However, the clinical findings in PCT, erythrohepatic porphyria, and congenital erythropoietic porphyria (conditions with increased levels of UP) are characterized by very dramatic clinical findings. The photosensitizing effects of PP on several cell lines in vitro seem to be induced by the porphyrin localized within the cell membrane (43,44). The accumulation of the porphyrins within the plasma membrane may be expected to follow the pattern of their relative lipid solubility. Since it appears that the photoinduced damage is due to the singlet oxygen within the cells (17,45,46), it is understandable that the amounts of singlet oxygen that may be formed from the different porphyrins within the cells would decrease in the order PP → CP → UP. The cell lysis under the experimental conditions follows the same order. Therefore, these results indicate that the primary reactive species responsible for the cytotoxicity is singlet oxygen produced within the cells.

B. Release of Mediators and Activation of Complement During Porphyrin Photosensitization

Irradiation of rat peritoneal mast cells has been reported to induce the release of histamine. At low concentrations of PP, this proceeds by a cytotropic mechanism without causing complete lysis of the cells. However, at high concentrations of PP, the histamine release is accompanied by cell lysis and therefore proceeds by a cytotoxic mechanism (47). Similarly, PP and ultraviolet light (UVA) induces histamine release from rat peritoneal mast cells, mouse bone marrow mast cells, and human cutaneous mast cells in a dose- and temperature-dependent manner. The mast cell activation was associated with loss of membrane integrity and inhibited by the hydrogen peroxide scavenger catalase. Histamine release was independent of extracellular calcium in the rodent mast cells, but was markedly reduced in the absence of calcium in human cells. These findings indicate that PP and UVA induce mast cell mediator release by a process that may involve hydrogen peroxide formation. There appear to be differences in response to PP and UVA between rodent and human mast cells (48).

Porphyrin photosensitization has been reported to induce activation of the complement system, release of mediators from mast cells, dermal accumulation of neutrophils, and activation of factor XII–dependent pathways and has profound effects on the eicosanoid metabolism of mast cells, endothelial cells, macrophages, and epidermal and dermal homogenates. While it is clear that the pathogenesis of cutaneous lesions in porphyria cutanea tarda and erythropoietic protoporphyria involves interactions among inflammatory mediators and various cells in skin, much still needs to be done to further our understanding of their pathophysiology (49).

V. RELATIONSHIP BETWEEN REACTIVE OXYGEN SPECIES FORMED DURING PORPHYRIN PHOTOSENSITIZATION AND CLINICAL PORPHYRIAS

The in vitro findings have been used to explain the differences in the mechanisms of photosensitivity in EPP and PCT. The mediation of singlet oxygen in the photosensitivity due to PP is parallel to the alleviation of the clinical symptoms in EPP by β-carotene, an effective singlet oxygen quencher (18,50–52). On the other hand, β-carotene does not protect against the photosensitivity in PCT. This is understandable if the photosensitivity due to UP is mainly mediated by superoxide, H_2O_2, and possibly by $\cdot OH$. However, the differences in the solubility of PP and UP may also have a role in the effectiveness of β-carotene. It would be interesting to see whether patients having PCT would respond to scavengers of these reactive oxygen species.

The clinical picture of porphyria cutanea tarda differs considerably from that of erythropoietic protoporphyria. The above differences in the mechanism of photosensitivity due to UP and PP are probably responsible for some of the differences in the clinical photosensitivity of these two porphyrias.

The classical therapy for porphyria cutanea tarda is phlebotomy, thought to be due to iron removal. A potent Fe chelator, deferoxamine (DM), has also been reported to be effective (53–55). There are frequently increased levels of ferritin in the serum and probably in the skin of patients with porphyria cutanea tarda.

It has been reported that DM inhibits the cell lysis produced by UP but not that by PP. The lack of inhibition by DM on the PP-induced cell lysis shows that the inhibitory effect of DM on the UP-induced lysis is not due to a nonspecific effect such as absorption of the radiation, interference with singlet oxygen, or the photoinduced formation of PP triplet (39). These findings indicate that DM may not affect the formation of a similar triplet

Table 3 Photosensitivity and Clinical Picture in Cutaneous Porphyrias: A Simplified Classification into Two Types

Category	Type I	Type II
Porphyria	EPP	PCT, VP, ACP, CEP, HEP
Porphyrin responsible	Protoporphyrin	Uroporphyrin
Solubility	Lipid	Water
Reactive species	Singlet oxygen	Singlet oxygen, H_2O_2, hydroxyl radical
Role of Fe	No evidence	Indirect evidence for mediation by Fe
Main target	Vascular endothelium	Subepidermal
Patient's awareness of etiology of sun	Very high	Much less aware
Clinical picture	Less destructive, some scarring, thickening, raggedies, pitting	More destructive with more scarring, can have scleradermoid picture
Signs and symptoms	Early onset of pain, burning after exposure going on to swelling and erythema	Friability of skin bullae, milia, hyperpigmentation, hypertrichosis

from UP or the possible singlet oxygen–mediated component of cell lysis in the UP photosensitization. These results indirectly point to the possibility that UP photosensitized cell lysis in vitro is at least partially mediated by ·OH. If the results from this in vitro model could be extrapolated to in vivo conditions, such as clinical porphyrias, it may be concluded that the cutaneous photosensitivity due to UP, as in the case of porphyria cutanea tarda, may also be at least partially mediated by ·OH.

There have been numerous studies on the role of Fe in porphyria cutanea tarda. The disease is frequently associated with increased levels of Fe in plasma and tissues, e.g., liver. It has been shown that Fe inhibits liver uroporphyrinogen decarboxylase (56,57). It has also been suggested that Fe could oxidize uroporphyrinogen to uroporphyrin, thereby increasing the uroporphyrin levels in the blood and skin (57). However, the complete role(s) of Fe in porphyria cutanea tarda is not clear. Recent results indicate that Fe, in addition to its effects on the liver, may play a role in the UP photosensitization in the skin by accelerating the formation of ·OH, which may be the major reactive species responsible for the photosensitization in porphyria cutanea tarda (39).

Table 3 shows a simplified classification of the cutaneous porphyrias into two types. This highlights the differences in clinical picture and some of the factors in the pathogenesis of these diseases.

VI. CONCLUSIONS

In general, cutaneous photosensitization would be the result of a series of photochemical and biochemical events. Initially the photosensitization of a particular compound would produce the primary reactive species (e.g., singlet oxygen). Subsequently this would lead to the formation of secondary species (e.g., superoxide, H_2O_2, hydroxyl radical). These chemical agents would initiate a series of biological reactions, e.g., protein cross-linking, lipid peroxidation, membrane damage, disorganization of intracellular structure, and destruction of subcellular organelles. These effects could lead to the release of biological mediators such as cytokines, prostaglandins, and histamine. The overall biological effects may be due to the direct action on the target cells produced by the above-mentioned chemical and/or biological agents.

ACKNOWLEDGMENTS

We would like to thank Hansa Menon for assistance in the preparation of the manuscript. We acknowledge the grants from the Medical Research Council of Canada, Ontario Ministry of Health, and Philip Smith Foundation. H. F. Haberman is an Associate of the Ontario Cancer Treatment and Research Foundation.

REFERENCES

1. Moore M, McColl K, Fitzsimons E, Goldberg S. The porphyrias. Blood Rev 1990; 4:88–96.
2. Elder G. The cutaneous porphyrias. Semin Dermatol 1990; 9:63–9.
3. Poh-Fitzpatrick MB. Porphyrin-sensitized cutaneous photosensitivity—pathogenesis and treatment. In: Disler PB, Moore MR, eds. Clinics in Dermatology—Porphyria. Vol. 3. Philadelphia: J.B. Lippincott; 1985:41–82.
4. Poh-Fitzpatrick M. The porphyrias. Dermatol Clin 1987; 5:55–61.
5. Pimstone N. Porphyria cutanea tarda. Semin Liver Dis 1982; 2:132–42.
6. Doss M, Frank M. Hepatobiliary implications and complications in protoporphyria, a 20-year study. Clin Biochem 1989; 22:223–9.
7. Moore MR, McColl KEL, Rimington C, Goldberg A. Disorders of porphyrin metabolism. New York: Plenum Publishing Company, 1987.
8. Dougherty TJ. Photosensitizers: therapy and detection of malignant tumors. Photochem Photobiol 1987; 45:879–89.
9. Doiron DR, Keller GS. Porphyrin photodynamic therapy: principles and clinical applications. In: Honigsmann H, Stengl G, eds. Therapeutic photomedicine. Basel: Karger; 1986:85–93.
10. Eales L. Clinical chemistry of the porphyrias. VI. Dolphin D. The porphyrins. New York: Academic Press; 1979:663–804.
11. Hindmarsh JT. The porphyrias. Clin Biochem 1986; 32:1255–63.
12. Sandberg S, Romslo I, Hovding G, Bjorndal T. Porphyrin-induced photodamage as related to the subcellular localizing of the porphyrins. Acta Dermatovener Suppl 1982; 100:75–80.
13. Brun A, Sandberg S. Mechanisms of photosensitivity in porphyric patients with special emphasis on erythropoietic protoporphyria. J Photochem Photobiol B Biol 1991; 10:285–302.
14. Sandberg S, Glette J, Hoppen G, Solberg CO, Romslo I. Porphyrin-induced photodamage to isolated human neutrophils. Photochem Photobiol 1981; 34:471–5.
15. Sandberg S, Romslo I. Porphyrin-sensitized photodynamic damage of isolated rat liver mitochondria. Biochim Biophys Acta 1980; 593:187–95.
16. Wakulchik SD, Schiltz IR, Bickers DR. Photolysis of protoporphyrin-treated human fibroblasts in vitro: studies on the mechanism. J Lab Clin Med 1980; 96:158–67.
17. Sandberg S, Romslo I. Porphyrin-induced photodamage at the cellular and the subcellular level as related to the solubility of the porphyrin. Clin Chim Acta 1981; 109:193–201.
18. Krinsky NI. Biology and photobiology of singlet oxygen. Bors W, Saran M, Tait D. Oxygen radicals in chemistry and biology. Berlin: Walter de Gruyter, 1984:455–64.
19. Valenzeno DP. Photomodification of biological membranes with emphasis on singlet oxygen mechanisms. Photochem Photobiol 1987; 46:147–60.
20. Aust SD, Svingen BA. The role of iron in enzymatic lipid peroxidation. In: Pyror WA, ed. Free radicals in biology. Vol. 5. New York: Academic Press, 1982:1–28.

21. Girotti AW. Photodynamic lipid peroxidation in biological systems. Photochem Photobiol 1990; 51:497–509.
22. Lands WEM, Kulmacz RJ, Marshall PJ. Lipid peroxide actions in the regulation of prostaglandin biosynthesis. In: Pryor WA, ed. Free radicals in biology. Vol. 6. New York: Academic Press, 1984:39–61.
23. Mead JF. Free radical mechanisms of lipid damage and consequences for cellular membranes. In: Pryor WA, ed. Free radicals in biology. Vol. 1. New York: Academic Press, 1976:51–68.
24. Tappel AL. Measurement of and protection from in vivo lipid peroxidation. In: Pryor WA, ed. Free radicals in biology. Vol. 4. New York: Academic Press, 1976:49–94.
25. Tyrell RM. A common pathway for protection of bacteria against solar UVA and oxidizing agent H_2O_2. Mutation Res 1985; 145:129–36.
26. Vliegenthart JFG, Veldink GA. Lipoxygenase. In: Pryor WA, ed. Free radicals in biology. Vol. 5. New York: Academic Press, 1982:29–64.
27. Wang SH. Photochemistry and photobiology of nucleic acids. Vols. 1 and 2. New York: Academic Press, 1976.
28. Girotti AW. Mechanisms of photosensitization. Photochem Photobiol 1983; 38:745–51.
29. Girotti AW. Photosensitized cross-linking of erythrocyte membrane protein. Biochim Biophys Acta 1980; 602:45–56.
30. Menon IA, Shirwadkar S, Ranadive NS. Nature of the oxygen species generated by xanthine oxidase involved in secretory histamine release from mast cells. Biochem Cell Biol 1989; 67:397–403.
31. Halliwell B, Gutteridge JMC. Free radicals in biology and medicine. 2nd ed. New York: Oxford University Press, 1989.
32. Schraufstatter IU, Hyslop PA, Ninshaw DB, Spragg RG, Sklar LA, Cochrane CG. Hydrogen peroxide induced injury of cells and its prevention by inhibitors of polyADP-ribose polymerase. Proc Natl Acad Sci USA 1986; 83: 4908–12.
33. Freeman BA, Turrens JF, Mirza Z, Crapo JD, Young SL. Modulation of oxidant lung injury by using liposome entrapped superoxide dismutase and catalase. Fed Proc 1985; 44:2591–5.
34. Korthuis RJ, Granger DN, Townsley MI, Taylor AE. The role of oxygen-derived free radicals in ischemia-induced increases in canine skeletal muscle vascular permeability. Circ Res 1985; 57:599–609.
35. Girotti AW, Thomas JP, Jordan JE. Lipid photooxidation in erythrocyte ghosts: sensitization of the membranes toward ascorbate- and superoxide-induced peroxidation and lysis. Arch Biochem Biophys 1985; 236:238–51.
36. Menon IA, Basu PK, Persad SD, Das A, Wiltshire JD. Reactive oxygen species in the photosensitization of retinal pigment epithelial cells by rose bengal. J Toxicol Cutan Ocular Toxicol 1992; 8:35–42.
37. Spikes JD. Photosensitization in mammalian cells. Parrish JA, Kripke ML, Morison WL. Photoimmunology. New York: Plenum Medical Book Co., 1983: 23–49.

38. Menon IA, Becker MAC, Persad SD, Haberman HF. Quantitative determination of hydrogen peroxide formed during UV-visible irradiation of protoporphyrin, coprophyrin and uroporphyrin. Clin Chim Acta 1989; 186:375–81.
39. Menon IA, Persad S, Haberman HF. A comparison of the phototoxicity of protoporphyrin, coproporphyrin and uroporphyrin using a cellular system in vitro. Clin Biochem 1989; 22:197–200.
40. Menon IA, Becker MAC, Haberman HF. Role of iron in the photosensitization by uroporphyrin. Clin Chim Acta 1991; 202:237–42.
41. Peterson DA, McKelvery S, Edmondson PR. A hypothesis for the molecular mechanism of tumor killing by porphyrins and light. Med Hypothesis 1982; 7:201–5.
42. Hariharan PV, Courtney J, Eleczko S. Production of hydroxyl radicals in cell systems exposed to haematoporphyrin and red light. Internatl J Radiat Biol 1980; 37:691–4.
43. Malik Z, Dialdetti M. Cytotoxic effect of hemin and protoporphyrin on chronic lymphocytic leukemia lymphocytes. Exp Hematol 1980; 8:867–79.
44. Malik Z, Dialdetti M. Destruction of erythroleukemia, myelocytic leukemia and burkitt lymphoma cells by photoactivated protoporphyrin. Int J Cancer 1980; 26:495–500.
45. Moan J, Pettersen EO, Christensen T. The mechanism of photodynamic inactivation of human cells in vitro in the presence of hematoporphyrin. Br J Cancer 1979; 39:398–407.
46. Spikes JD. Photosensitization. Smith KC. The science of photobiology. New York: Plenum Press, 1989:79–110.
47. Ranadive NS, Menon IA, Haberman HF. Photolysis of mast cells in the presence of protoporphyrin. Acta Derm Venereol Stockh 1979; 59:493–7.
48. Glover R, Bailey C, Barrett K, Wasserman S, Gigli I. Histamine release from rodent and human mast cells induced by protoporphyrin and ultraviolet light: studies of the mechanism of mast-cell activation in erythropoietic protoporphyria. Br J Dermatol 1990; 122:501–12.
49. Lim H. Mechanisms of phototoxicity in porphyria cutanea tarda and erythropoietic protoporphyria. Immunol Scr 1989; 46:671–85.
50. Mathews-Roth MM. Photobiology of carotenoid protection. In: Regan JD, Parrish JA, eds. The science of photomedicine. New York: Plenum Press, 1982: 409–10.
51. Burton GW, Ingold KU. β-Carotene: an unusual type of lipid antioxidant. Science 1984; 224:569–73.
52. Sayre RM, Black HS. Beta-carotene does not act as an optical filter in skin. J Photochem Photobiol B: Biol 1992; 12:83–90.
53. Rocchi E, Gibertini P, Cassanelli M, Pietrangelo A, Borghi A, Pantaleoni M. Iron removal therapy in porphyria cutanea tarda: plebotomy versus slow subcutaneous desferrioxamine infusion. Br J Dermatol 1986; 114:621–9.
54. Gibertini P. Rocchi E, Cassanelli M, Pietrangelo A, Ventura E. Advances in the treatment of porphyria cutanea tarda. Effectiveness of slow subcutaneous desferrioxamine infusion. Liver 1984; 4:280–4.

55. Marchesi L, Di PC, Cainelli T. A comparative trial of desferrioxamine and hydrochloroquine for treatment of porphyria cutanea tarda in alcoholic patients. Photodermatol 1984; 1:286–92.
56. Mukerji S, Pimstone N, Burns M. Dual mechanism of inhibition of rat liver uroporphyrinogen decarboxylase activity by ferrous iron: its potential role in the genesis of porphyria cutanea tarda. Gastroenterology 1990; 87:1248–54.
57. Mukerji S, Pimstone N. Free radical mechanism of oxidation of uroporphyrinogen in the presence of ferrous iron. Arch Biochem Biophys 1990; 281:177–84.

19

Determination of Blood Cell Chemiluminescence: A Method to Investigate Selected Skin Diseases

Klaus Hohlmaier
Schwaerzberg Clinic for Allergies, Asthma, and Psoriasis, Bad Rappenau, Germany

I. BLOOD CELLS AS A SOURCE OF REACTIVE OXYGEN SPECIES

The main source of reactive oxygen species (ROS) in human blood are polymorphnuclear cells (PMNs) and monocytes (MCs). Among the PMNs, neutrophil granulocytes are the most efficient producers of ROS; much less is known about eosinophilic granulocyte ROS production (1). The potency of ROS synthesis is also present in lymphocytes, as transformation by the Epstein-Barr virus enables them to yield ROS in response to the tumor-promoting phorbol ester TPA (2). ROS production is also subject to variation during differentiation, for example, in maturating monocytes in vitro (3) and during terminal differentiation to macrophages in vivo (4).

II. REACTIVE OXYGEN MOLECULES PRODUCED IN BLOOD CELLS

The cellular reaction cascade of ROS is a complex and diverse pattern of synchronous reaction products: superoxide radicals, hydrogen superoxide, hypochlorous acid, hydroxyl radicals, and singlet oxygen. The synthesis of these oxygen species is initiated by the production of superoxide anion radicals

$$2\ O_2 + NADPH \rightarrow 2\ {}^{\bullet}O_2^- + NADP^+ + H^+ \quad (1)$$

$$2\ {}^{\bullet}O_2^- + 2\ H^+ \rightarrow H_2O_2 + O_2 \quad (2)$$

$$H_2O_2 + Cl^- + H^+ \rightarrow HOCl + H_2O \quad (3)$$

$${}^{\bullet}O_2^- + HOCl \rightarrow {}^{\bullet}OH + Cl^- + O_2 \quad (4)$$

$${}^{\bullet}O_2^- + H_2O_2 \rightarrow OH^- + {}^{\bullet}OH + O_2 \quad (5)$$

Figure 1 Biological pathways for the generation of reaction oxygen species. Equation 1 describes the action of NADPH-oxidase; spontaneously, hydrogen peroxide is formed, too (Eq. 2). Hydrogen peroxide can be converted to hypochlorous acid by myeloperoxidase (e.g., in neutrophils) as described by Eq. 3. Superoxide can be transformed to the highly aggressive hydroxyl ion according to Eq. 4 or via the Haber-weiss reaction (Eq. 5).

($\cdot O_2^-$) by NADPH-oxidase. This enzyme could be localized in the plasma membrane of human neutrophils (5).

Superoxide radicals can give rise to hydrogen superoxide (Fig. 1). In turn, the more toxic and microbicidal hypochlorous acid can be generated in the presence of myeloperoxidase, which is highly active in neutrophils; a further reaction sequence leads to the production of singlet oxygen, if hydrogen superoxide is present. The latter pathway is less important in monocytes and macrophages because they contain low levels of myeloperoxidase (6). In these cells, as well as in neutrophils, bacteriotoxic hydroxyl ions are produced via the Haber-Weiss reaction (Fig. 1).

III. IDENTIFICATION OF REACTIVE OXYGEN SPECIES IN CELLS BY LUMINESCENT PROBES

Two different "enhancers" are commonly used to monitor cellular ROS by luminescence: luminol and lucigenin (Fig. 2). Although singlet oxygen is itself able to produce chemiluminescence, the signal is too weak for precise determination. Luminol, probably oxidized by hypochloric acid, changes the free energy produced into luminescence, which can be determined by appropriate apparatus to give a measure for hypochloride present (if luminol is used in excess).

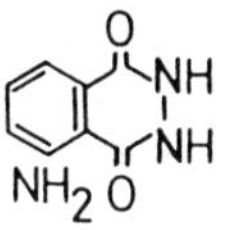

(a)

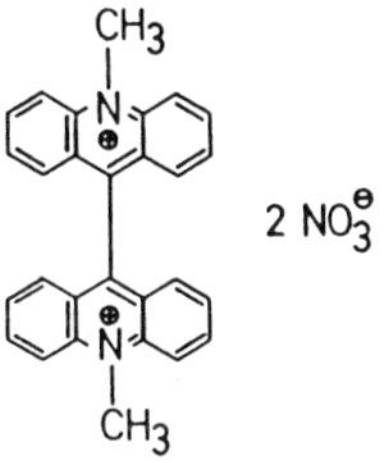

(b)

Figure 2 Chemical constitution of (a) luminol and (b) lucigenin.

The acridinium salt lucigenine produces luminescence independent from myeloperoxidase. It could be shown that it is able to react specifically with superoxide radicals yielding the luminescent intermediate *N*-methyl-acridone (7).

Substrate specifity for superoxide can be concluded, for example, by inhibition of lucigenine luminescence after addition of superoxide dismutase (8,19) or by the lack of lucigenine luminescence production in cell free systems synthesizing hypochloride only (9).

IV. EXPERIMENTAL SYSTEMS AND INTERPRETATION OF RESULTS OF LUMINESCENCE DETERMINATIONS

Chemiluminescence can be determined in different preparations. Most commonly used are either the whole blood system or isolated cells (neutrophils, eosinophils, monocytes, macrophages).

The native blood system is relatively stable, allows quick determinations, and includes the interaction of blood cells during incubation. On the other hand, hemoglobin has considerable quenching properties, and sedimentation of cells in the sample may be a source of trouble. Isolated cells have limited viability, e.g., neutrophils (in spite of the use of appropriate media and storage on ice) may decline in luminescence production as soon as 2 hours after isolation.

Clumping and aggregation phenomena during the isolation procedure may occasionally cause problems, especially in the presence of activating factors (immune complexes, acute phase proteins) in the patient's plasma. For most uses, heparin is taken as an anticoagulant with concentrations of 20–100 I.U./mL. Higher concentrations suppress chemiluminescence, and at 1000 I.U./mL inhibition may reach as much as 40% (19).

To obtain reproducible and meaningful results it is most important to avoid problems arising from the patient or volunteer. Food ingestion, physical exercise, time of day, smoking, and (subclinical) infectious diseases all influence chemiluminescence; it is necessary, therefore, to control these parameters as much as possible. Even then, day-to-day variations of up to 20% when comparing absolute counts per minute may be observed in the same patient. Therefore, the use of appropriate controls, e.g., unstimulated values and pool plasma, is essential and facilitates the determination of stimulation/inhibition ratios relative to controls.

Commonly used particulate stimulators are latex beads, zymosan (opsonized with patient and control serum or not opsonized), and also different less standardized preparations from bacteria like lipopolysaccharides and fungi. Often used well-defined stimulators are the bacterial tetrapeptide FMLP, concanavalin A, LTB 4, the tumor promotor TPA, and recently also cytokines (10). As controls for the substrate specifity of luminol-dependent chemiluminescence the myeloperoxidase inhibitor azide, for lucigenine-dependent chemiluminescence superoxide dismutase, can be used.

Making use of these precautions, measurement of cellular chemiluminescence is a reproducible and very sensitive method.

V. OTHER EXPERIMENTAL SYSTEMS FOR THE DETERMINATION OF REACTIVE OXYGEN SPECIES

Other often used methods to determine reactive oxygen species are 1) the cytochrome *c* assay (11) and 2) the nitroblue tetrazolium method (12). With the nitroblue tetrazolium method, it is possible to determine ROS (by nitroblue tetrazolium reduction to formazan) and phagocytosis (intracellular accumulation of tetrazolium dye) at the same time. The procedure is relatively simple to perform, but whole blood has to be used, which makes it impossible to discern between neutrophil, monocytes, and eosinophil ROS production. The cytochrome *c* assay is able to discriminate between extracellular superoxide (which is quantitated) and intracellularly generated superoxide (which is not determined in this assay).

VI. APPLICATIONS OF CELLULAR CHEMILUMINESCENCE IN SIGNAL TRANSDUCTION RESEARCH AND CLINICAL DERMATOLOGY

A. Signal Transduction Research

It is an important question whether the extent of superoxide production with its many important biological effects may be regulated by GTP-dependent regulatory proteins. Some recent publications investigate or discuss this issue (13–15). Interestingly, with GM-CSF as neutrophil activator, an affinity shift of the receptor similar to other receptors regulated by G-proteins could be observed. Activators of G-proteins like fluoride or the persistent activating GTP analog GTPyS are able to produce remarkable amounts of superoxide. Furthermore, toxin activators or inactivators of G-proteins like pertussis or cholera toxin have antagonistic effects on ROS synthesis in whole blood (15).

Of course, these observations are all indirect and at most infer a role for G-proteins in the regulation of superoxide production, for example, via phospholipase C (Fig. 3). If true, new perspectives would emerge in our understanding of diseases with impaired or altered neutrophil function (discussed below).

B. Clinical Dermatology

As already mentioned, whole cell chemiluminescence is a laboratory method of high sensitivity. Comparing the chemiluminescence of groups of patients with probable dysfunction of the neutrophil or eosinophil granulocyte or of the monocyte is, regrettably, sometimes hampered by the wide overlap of results because of the influence of the aforementioned individual factors and dangers of error.

As for the neutrophil disorders, it now appears established that in psoriasis an activated basal level of ROS generation occurs (16–18), caused both by heat-labile factors in the psoriatic sera and by an inherent defect of the neutrophil. There is some evidence that this "preactivation" effect correlates with disease activity (16); it is not known if it is also present in patients with arthropathic psoriasis and no skin symptoms. In these cases it could become a diagnostic help to associate unclear causes of arthralgia to a psoriatic background.

On the other hand, chemoattractant-stimulated lucigenine luminescence is suppressed (possibly down-regulated) compared to normal healthy ones; our investigations so far show that both effects (suppression of stimulated

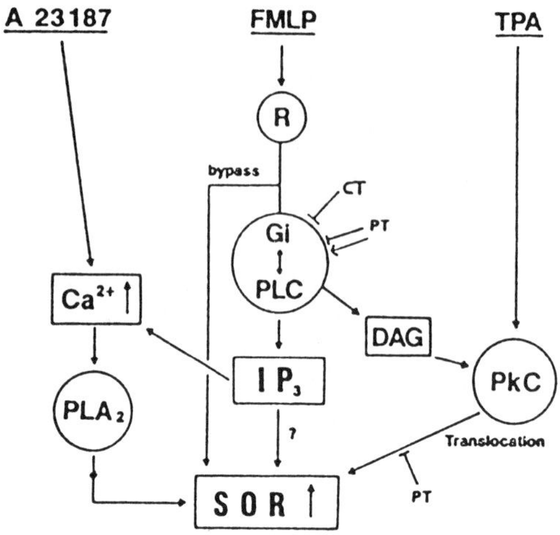

Figure 3 Hypothetical, schematic presentation of signal transduction pathways contributing to superoxide generation in human neutrophil granulocytes. A 23187 = calcium ionophore; FMLP = formyl-methionyl-leucyl-phenylalamine; TPA = tetradecanoylphorbolacetate; R = FMLP receptor; Gi = "inhibiting" signal transduction G protein; PLC = phospholipase C; PLA_2 = phospholipase A_2; DAG = diacylglycerol; PkC = proteinkinase C; IP_3 = inositol-1,3,5-triphosphate; SOR = superoxide radicals; PT = pertussis toxin; CT = cholera toxin; |—— = inhibitory; ← = stimulatory.

and elevation of basal luminescence) are normalized by pretreatment with cholera toxin, pointing to a possible defect of G-proteins in psoriatic neutrophils (K. Hohlmaier, unpublished observations). Interestingly, the antipsoriatic compound dithranol has pro-oxidative properties activating NADPH-oxidase. After addition to neutrophils in vitro, a transient peak of extracellular ROS appears, but the production of intracellular radicals is inhibited (18). Probably, an autoinactivation of the enzyme is induced by the pro-oxidative drug, giving some "molecular analogy" of the clinicians' rule that "psoriasis must burn in the fire of dithranol."

Although difficult to assess, in erisypelas with documented streptococcus infection, luminol-enhanced chemiluminescence is increased during the fever period (19). This may be interpreted as an IL-1 effect, as this endogenous pyrogen is present during fever episodes. Preliminary investigations by us show that during severe and persistent infection or sepsis the chemiluminescence pattern may well be different from the reaction in controlled, transient bacterial infection.

Other well-known defects of oxygen radical production in the neutrophil like myeloperoxidase deficiency, glucose-6-phosphate deficiency, and NADP(H)-oxidase deficiency result in recurrent candidosis or, in the case of low NADP(H) levels, in "cold abscesses" or "chronic granulomatosis" rarely seen by the dermatologist.

Abnormalities of the ROS-generating system in the eosinophil may be important in atopic diseases and the late-phase reaction of type I allergies. It has been reported that lucigenine-dependent chemiluminescence in eosinophils of atopic individuals are "supersensitive" to stimuli like FMLP or PAF (1).

Aberrations of monocyte superoxide generation are another field for pathophysiologic investigations in skin diseases. In two cases of dermatitis ulcerosa well documented by clinical and laboratory data, I observed accelerated and excessively enhanced luminol-dependent chemiluminescence (20). After therapy with dapsone and glucocorticoids, luminescence rapidly returned to normal levels and the skin lesions healed.

Contrary to other reports (21), granulocyte chemiluminescence was not changed compared to normals. One may conclude that dermatitis ulcerosa, often occuring in Morbus Crohn or after immunosuppressive therapy, has a more complex cellular pathology than expected.

Taken together, determination of cellular luminescence will contribute to pathophysiological investigations in skin diseases as long as neutrophil and eosinophil granulocytes or monocytes are involved such that an underlying systemic factor (e.g., cytokines, complement, immune complexes) will impair their function. Other applications are the investigation of the role of the aforementioned cell types in the pathophysiology of asthma and in the immune defense against neoplasms. It may also turn out to be a convenient screening system to investigate effects of antioxidative drugs in vitro.

REFERENCES

1. Koendermann L, Bruijnzeel PLB. Increased sensitivity of the chemoattractant-induced chemiluminescence in eosinophils isolated from atopic individuals. Immunology 1989; 67:534–6.
2. Kapp A, Wolff-Vorbeck G, Peter HH. Chemiluminescence response of human B-cell lines. Free Radic Res Comms 1987; 2:337–41.
3. Schmidt J, Thielkes F. Chemiluminescence pattern of rat alveolar macrophages during maturation in vitro. In: Stanley P, Kricka L, eds. Bioluminescence and chemiluminescence: Current status. Chichester: Wiley and Sons, 1991:321–4.
4. Williams AJ, Cole PJ. Human bronchoalveolar lavage cells and luminol-dependent chemiluminescence. J Clin Pathology 1981; 34:167–71.
5. Dewald B, Baggiolini M. Subcellular localization of the superoxide-forming enzyme in human neutrophils. J Clin Invest 1979; 63:21–9.

6. Sagone AL, King GW, Metz EN. A comparison of the metabolic response to phagocytosis in human granulocytes and monocytes. J Clin Invest 1976; 57: 1352–8.
7. Hosker HSR, Kelly C, Corris PA. Assessment of phagocytic function using chemiluminescence. Blood Rev 1989; 3:88–93.
8. Williams AJ, Cole PJ. Investigation of alveolar macrophage function using lucigenin-dependent chemiluminescence. Thorax 1981; 36:866–99.
9. Aniansson H, Stendahl O, Dahlgren C. Comparison between luminol- and lucigenin-dependent chemiluminescence of polymorphonuclear leucocytes. Acta Path Microbiol Immunol Scand (C) 1984; 92:357–61.
10. Kapp A. Aktivierte Granulozyten als Produzenten reaktiver Sauerstoff-Spezies—ihre Bedeutung für die Entzündungsreaktion. Hautarzt 1990; 41:196–203.
11. Barbior BM, Kippnes RS, Curnutte JT. Biological defence mechanism—the production by leucocytes of superoxide, a potent bactericidal agent. J Clin Invest 1973; 52:741.
12. Baehner RL, Nathan DG. Quantitation of nitroblue tetrazolium test in chronic granulomatous disease. N Engl J Med 1968; 278:971.
13. McColl SR, Beauseigle D, Gilbert C, Naccache PH. Priming of the human neutrophil respiratory burst by granulocyte-macrophage colony stimulating factor and tumor necrosis factor-a involves regulation at a post-cell surface receptor level. J Immunol 1990; 145:3047–53.
14. Coffey RG. Mechanism of GM-CSF stimulation of neutrophils. Immunol Res 1989; 8:236–48.
15. Hohlmaier K, Gürenci J, Holzmann H. Superoxide radical production in human blood is stimulated by pertussis and inhibited by cholera toxin: evidence for G-protein functional activity related to superoxide generation. In: Stanley PE, Kricka LJ, eds. Bioluminescence and chemiluminescence: Current status. Wiley & Sons, 1991:449–54.
16. DeSole P. Recent developments in clinical applications of cellular luminescence. In: Stanley PE, Kircka LJ, eds. Bioluminescence and chemiluminescence: Current status. Wiley and Sons, 1991:245–52.
17. Bloomfield FJ, Young MM. Enhanced chemiluminescence production by phagocytosing neutrophils in psoriasis. Inflammation 1988; 12(2):153–9.
18. Anderson R, Lukey PT, Dippenaar U, Eftychis HA, Finlay GH, Wooten MW, Nel AE. Dithranol mediates pro-oxidative inhibition of polymorphonuclear leucocyte migration and lymphocyte proliferation. Br J Derm 1987; 117:405–18.
19. Pauksens K, Sjölin J, Venge P. Chemiluminescence of polymorphonuclear leucocytes and whole blood during acute bacterial infection. Scand J Infect Dis 1989; 21:277–84.
20. Hohlmaier K, Marsch WC, Holzkopf H. Hyperaktive Monozyten bei Dermatitis ulcerosa: Chemilumineszenz-Untersuchungen an peripheren Monozyten. Zentralbl Haut- u Geschl Krankh 1989; 156:764.
21. Fowler JB, Callen JP. Pyoderma gangraenosum. Dermatol Clin 1983; 1:615–23.

20

Reactive Oxygen Species in the Pathophysiology of Ischemia/ Reperfusion Injury in Skin

Michael F. Angel and Paul N. Manson
The Johns Hopkins School of Medicine, Baltimore, Maryland

I. INTRODUCTION

There has been an explosion of information on the importance of free radicals in biology over the past decade. There are several excellent global reviews of the field both in this volume and elsewhere (1–8). The intent of this chapter, however, is to discuss recent observations of the importance of free radical-mediated injury to skin in several ischemia-related conditions.

Ischemia can broadly be defined as blood flow inadequate to meet aerobic metabolic needs (9). Several parameters are important in assessing ischemia's effect on a tissue: composition of tissue, duration of ischemic episode, nature of onset of ischemia, degree of ischemia (complete vs. incomplete), and primary vs. secondary ischemia.

Composition of the tissue is very important. Tissue such as brain (10) is quite sensitive to ischemia; skin (11) is very resistant, while muscle (12) is intermediate in sensitivity. Duration of ischemia is also very important and is related to the tissue composition.

The total disruption of blood flow with complete cessation of oxygen delivery and waste excretion is complete (total) ischemia. As blood flow is reduced by a graded obstruction of either arterial inflow or venous outflow, a point is reached at which blood supply is insufficient to meet the aerobic energy means of the involved tissue. One can continue to decrease blood flow to the point of complete vascular obstruction. Partial (incomplete) ischemia is, therefore, a spectrum of states beginning initially after the point of insufficient blood flow and progressing to the point of total occlusion (4,10).

Primary ischemia is defined as the first ischemic episode in a tissue. Primary ischemia occurs following a traumatic disruption of blood flow such as during free tissue transfer. If a second episode occurs, such as following vascular thrombosis, this process is termed secondary ischemia (13). Skin flaps are less tolerant of secondary ischemia than primary ischemia (13,14).

II. FREE RADICALS IN ISCHEMIA FLAPS

A. Definition of Skin Flaps

A skin flap is defined as a tissue transferred from one site to another, maintaining an intact blood supply (15). This is in contrast to a skin graft, which has no blood supply; it subsequently develops one 3–5 days after transfer by neovascularization.

The vascular structures in a flap's pedicle are the source of a flap's blood supply. If the vascular pedicle stays intact, the flap is defined as a pedicle flap. If the pedicle is divided, the flap must be transferred by microvascular anastomoses. This is defined as a free tissue transfer.

Pedicle skin flaps can be broadly divided into two categories: random pattern and axial pattern (16).

1. Random Pattern Flaps

Random pattern flaps receive their blood supply via the dermal and subdermal plexuses. Clinically these flaps are either rotated or advanced into defects. Because of their anatomy, there is a limitation in the length of a flap that can survive, and these flaps cannot be distantly transferred. Clinically, these flaps have wide use, especially in the face or head and neck region. Ways to increase flap length survival will be discussed in subsequent sections.

2. Axial Pattern Flaps

Axial pattern flaps, in contrast to random ones, receive their blood supply from a vessel that runs throughout the flap. Because of this anatomical relationship, an axial flap has better blood supply than a random flap and, therefore, can survive to longer lengths.

The blood supply to skin was first documented by Manchot (17) in the last century. However, it took investigators years to come to the realization that skin flaps can be nourished by direct cutaneous, musculocutaneous (18, 19), fasciocutaneous (20), or septocutaneous vessels. It is beyond the scope of this chapter to more closely discuss this area, and the interested reader is referred to these articles. Suffice it to say, the introduction of these flaps has greatly enlarged the armamentarium of the reconstructive surgeon. Flaps can be transferred locally or can be transferred distantly as a free tissue transfer using microvascular techniques.

B. Axial Flap Models

1. The No-Reflow Phenomenon

One major experimental model involves the axial flap. The flap based on inferior epigastric vessels in the rat and rabbit and the flank flap in the pig are examples of this model. Blood flow is interrupted for various amounts of time by vascular clamping. Once flow is resumed, these flaps ultimately survive. A duration of ischemia is finally reached, however, where, despite restored flow, the flaps ultimately die. This has been described as the "no-reflow phenomenon." It was first noted by Wilm-Kretschmer and Majno in skin (17). The phenomenon has subsequently been found to exist in a wide variety of tissues (pulmonary, brain, cardiac, and renal).

May and coworkers (21) in 1978 first described this condition in experimental skin flaps. There have been several theories to explain causes of the no-reflow phenomenon and subsequent necrosis in experimental flaps. Intravascular sludging (22) occurs along with cellular swelling (23). Fibrinolysis (24) is decreased and the vasculature becomes vasospastic (25).

2. Free Radicals in Primary Total Ischemia

Recently, the role of pathogenic free radicals in this process has begun to be elucidated. Manson and coworkers (26) first noted the importance of free radicals in ischemic flap surgery. They used the epigastric flap in the rat. Primary total ischemia was induced by placement of microvascular clamps on the entire pedicle. Prior to clamp release, superoxide dismutase (a free radical scavenger) was administered and resulted in increased flap survival. The hypothesis was that during total ischemia, precursors of free radicals accumulate as products of anaerobic metabolism. During reperfusion, in this case after clamp release, superoxide radicals (O_2^-) were produced which interconverted to various free radical species. Similar results were seen with total primary ischemia in a free flap model treated with SOD (27).

3. Free Radicals in Primary Venous Ischemia

Harashina and coworkers (29) first noted that primary venous obstruction was more damaging to experimental flaps than complete pedicle interruption. This has been confirmed by others (14). There are several possible reasons for this observation. During venous obstruction some degree of arterial inflow continues. Vascular congestion and tissue edema lead to a diffusion barrier for oxygen. Second, Su and coworkers (30) speculated that venous congestion altered cellular metabolism, clotting mechanisms, and the regional microcirculation. These factors favor tissue damage, which, if left unchecked, may lead to necrosis. Third, continued arterial inflow permits low levels of glucose and oxygen in the flap. This can lead to accumulation of lactate and free radicals, which are damaging to the flap.

Work first presented in the neurosurgical literature showed the increased deleterious effect of incomplete ischemia on neural tissue (4,10). Angel and coworkers (31) hypothesized that primary venous obstruction is a partially ischemic state in which dangerous free radicals are produced both during the venous obstruction and during the reperfusion. Biochemical analyses in experimental flaps were consistent with this hypothesis (32).

4. *Sources of Free Radicals in Ischemic Tissue*

During ischemia, there are various potential sources of free radicalk production. Figure 1 shows a simplified scheme of purine catabolism. Nucleic acids found in the diet, senescent endogenous nucleic acids and high energy metabolites (ATP), share a common degradation pathway. The xanthine

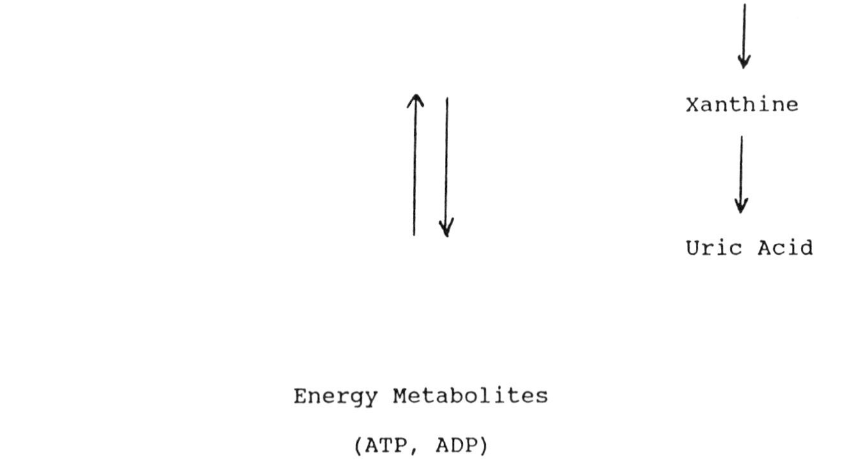

Figure 1 Purine catabolism.

oxido-reductase system catalyzes a two-step conversion: hypo-xanthine → xanthine → uric acid. Xanthine oxidine is an enzyme widely distributed throughout mammalian systems (33). It is found in high concentrations in the liver and small bowel. In the resting state, it is in a dehydrogenous form. During ischemia it becomes an oxide capable of producing superoxide free radicals that interconvert to more damaging forms, i.e., the hydroxyl free radical (2,6). Im and coworkers demonstrated that ischemic rodent skin flaps had increased levels of xanthine oxidase (34). They further found that administration of allopurinol, a specific inhibitor of xanthine oxidase, significantly increased flap survival in a primary venous ischemia model (34,35).

The hydroxyl free radical is a very reactive and damaging compound. It can be generated by one of two reactions. Beauchamp and Fridovich (36) suggested the Haber-Weiss reaction:

$$O^{2-} + H_2O_2 \rightarrow OH + OH^- + O_2$$

In this reaction the superoxide free radical (O_2^-) combines with hydrogen peroxide (H_2O_2), resulting in the production of the damaging hydroxyl free radical (OH^-). Some question the importance of this reaction because it is very slow (3). Therefore, it is thought that O_2^- and H_2O_2 interact to produce OH^- through a metal catalyst (Mn) in the Fenton reaction (3,37):

$$Mn^n + O_2^- \rightarrow Mn^{n-1} + O_2$$
$$Mn^{n-1} + H_2O_2 \rightarrow Mn^nOH + OH^-$$

The overall reaction is $O_2^- + H_2O_2 \rightarrow O_2 + OH + OH^-$, emphasizing the role of metals in catalyzing the formation of OH^-.

Iron has been found to be important in this reaction (38,39). In the presence of superoxide free radical, Fe^{+3} is converted to Fe^{+2}, which is available for the above-noted reactions.

Deferrioxamine is a specific free radical scavenger (40) and specific chelator of iron (41). In an axial model of ischemia, it has been found to favorably affect the flap survival (42,43).

The NADPH-dependent oxidase system present in neutrophils is a highly efficient system of superoxide generation (5,6). This enzyme is normally dormant. When activated, however, it catalyzes the sudden reduction of oxygen to hydrogen peroxide and superoxide free radical. This is the "respiratory burst." It is thought that in a variety of ischemic tissue that neutrophils, by production of free radicals, can amplify tissue damage that has already begun (6).

Through molecular genetics, several monoclonal antibodies have been developed that can block neutrophil activity in ischemic tissue. Antibodies (60.3, 904) have been raised against the CD11/CD18 complex, a neutrophil membrane glycoprotein complex mediating adherence to endothelial cells

(44,45). GMP140 is an unrelated glycoprotein complex which also mediates neutrophil-endothelial cell adhesion (46). Administration of these factors in primary ischemia models has resulted in improved flap survival (47).

5. *Free Radicals and Thromboxane in Ischemic Flaps*

Several potential sources of free radical production exist in cell membranes. Arachidonic acid metabolism and free radical mechanisms are closely related. The metabolism of arachidonic acid by cyclooxygenase and by lipoxygenase to produce leukotrienes involves the formation of intermediate peroxy compounds and the hydroxyl radical as by-products. On the other hand, free radicals can initiate the arachidonic acid cascade (48,50). This is not to say that all reactions involved by arachidonic acid metabolism are mediated by free radicals. However, when the action of prostaglandins and leukotrienes are studied, free radical influence should be considered.

In an axial flap, Douglas and coworkers (51) found that ibuprofen, a cyclooxygenase inhibitor, significantly increased ischemic tolerance. They concluded that prostaglandin synthesis in general and thromboxane in particular, are important contributors to flap necrosis. Others have confirmed this hypothesis (34,52).

Thromboxane is a potent vasoconstrictor and platelet aggregator. It also plays an important role in such conditions as burns (55). Some have shown (56) that upon flap elevation, norepinephrine and serotonin are released, which, besides having effects of their own, result in the release of thromboxane.

Blood thromboxane levels are elevated in ischemic limbs (57). This finding has been confirmed in ischemic flaps (52,58). Axial pattern flap survival has been enhanced by the administration of UK 38485, a specific inhibitor of thromboxane synthetase (54,59). It redirects arachidonic acid and metabolism toward prostacyclin, thereby reducing platelet aggregation and causing vasodilation.

Knight and coworkers (60) studied the interaction of thromboxane and free radicals in ischemic axial flaps. Axial flaps subjected to 4 hours of either secondary arterial or secondary venous ischemia received either placebo, superoxide dismutase, or UK 38485. After reperfusion, flaps had biochemical analysis for xanthine oxidase, malonyldialdehyde (an indicator of lipid peroxidation), superoxide dismutase, and thromboxane B. Treatment with UK 38485 predictably reduced thromboxane B levels and had no effect on the other parameters. Treatment with SOD not only affected free radical parameters, but also *reduced* thromboxane B levels. From this study it was concluded that in ischemic skin flaps, free radical mechanisms are the primary lesion. Damaged vasculature can therefore increase thromboxane and further damage the tissue. This result is consistent with other investigators (8,49,50).

When a skin flap ultimately necroses, this may be due to primary damage to the parenchymal cells (keratinocytes in skin) or to the microvasculature.

It is known that skin is very resistent to ischemia. In fact, skin can be incubated at body temperature for several days and still be successfully grafted (61). On the other hand, skin flaps clinically and experimentally can only tolerate 8–12 hours of warm ischemia (22).

The emerging hypothesis is that endothelial cell failure is the primary lesion in ischemic skin flaps. This failure is broadly suggested by two observations. In some areas the endothelium loses its integrity and edema results because of hyperpermeability. Free radicals are a likely agent for this process (5). The endothelium, besides becoming highly permeable, also becomes occluded. Free radicals can directly cause edema and vascular congestion by inducing cellular swelling (23) and vasoconstriction (25). In addition, by indirect production of lipoperoxides and prostaglandins and their derivatives (e.g., thromboxane) can occlude flow. Studies by Marzella and coworkers (62) and Feng and coworkers (63) also emphasize the importance of microvascular failure in causing ultimate flap necrosis.

6. Free Radicals and Secondary Ischemia

The concept of secondary ischemia is important clinically. Kerrigan and coworkers (13) first noted that skin flaps are less tolerant of secondary ischemia insult than primary insult in a porcine model. Angel and coworkers (14) observed this phenomenon in rodents. They found that secondary venous ischemia is quite deleterious to skin flap survival. Sixty percent of 5 hour primary ischemic skin flaps ultimately survived, while 0% of those secondarily ischemic did.

The entire pathophysiology of secondary ischemia is not totally understood at this time. Free radicals and thromboxane have been found, however, to be important agents (60).

C. Axial Random Flap Models

Another skin flap model that has been used in an axial pattern flap with random extension. The commonly used examples of this model are the abdominal flap of Finseth and Adelberg (64) and the dorsal rat flap, first described by McFarlane and coworkers (65). Investigators have tried using these models to reduce the amount of necrosis with varying success. These models simulate the clinical problem of an ischemic tip of a flap noted at the time of surgery postoperatively.

Serial histological examinations of the distal part of this flap have revealed that, as early as 6 hours postoperatively, there is moderate extravasation of red blood cells (RBCs) with mild edema. This progressed to frank necrosis with marked RBC extravasation at 24 hours postoperatively. Skin grafts studied at the time showed much less tissue damage (66).

A biochemical analysis found further differences between the distal part of the dorsal rat flap and a skin graft. The distal flap at 24 hours had significantly elevated levels of MDA, indicative of lipid peroxidation (67).

These data (histological and biochemical) were consistent with the hypothesis that the distal part of the dorsal rat flap is in a partially ischemic state in which damaging free radicals are produced. The administration of free radical scavengers allopurinol (67) and deferoxamine (68) in this model has resulted in increased flap survival.

In recent related work, it has been noted that increased levels of glutathione reductase, which can scavenge free radicals, are associated with increased flap survival (69).

D. Free Radicals and Hematomas

Skin flap necrosis due to an underlying hematoma has been clinically noted on many occasions. Explanations for this have included "internal pressure" (70) and infection (71). Mulliken and coworkers (72) found that a high molecular weight compound obtained from hemolyzed red blood cells, when placed under the skin flaps, caused subsequent flap necrosis.

Angel and coworkers (73) investigated this phenomenon in inbred Fischer rats. They found that when blood was placed under 4 × 2 cm dorsal rat flaps, flap necrosis was 63% compared to 8% for flaps with underlying plasma. When deferoxamine was mixed with the blood, flap necrosis dramatically decreased to 12%; when iron, however, was mixed with plasma, necrosis increased to 45%. They hypothesized that iron was catalyzing the Fenton reaction, resulting in increased production of the damaging OH species. Biochemical analysis was consistent. They further found that the systemic administration of deferoxamine also significantly increased flap survival (73).

III. FREE RADICALS AND FROSTBITE

Frostbite is a tissue injury resulting from freezing and thawing. Although different from ischemic injury in several respects, it has many similarities. There are several mechanisms to explain this injury: 1) direct injury to the cells by freezing, 2) postthaw vasoconstriction (74) in the microvasculature, and 3) postthaw damage (75,76) to the microvasculature. This injury is characterized by damage to endothelial cells and basement membrane, marked vasodilation, and tissue edema with an inflammatory response. After complete freezing is terminated, it is initially followed by complete vascular inflow. Circulatory flow there decreases in various areas with patchy thrombosis developing, leading to confluent areas of tissue necrosis.

Manson and coworkers (78) investigated the role of free radical-mediated reperfusion injury in frostbite. Rabbit ears were frozen at −21°C for 30, 60, 90, or 120 seconds and then rewarmed to room temperature (22°C). Animals were randomly and blindly treated with saline, SOD, deferoxamine, or allopurinol. Deferoxamine and SOD significantly increased survival compared to saline controls. Allopurinol, however, did not affect survival.

Marzella and coworkers (79) studied ultrastructural changes caused by frostbite. By electron-microscopic analysis, they concluded that non-free radical-induced endothelial damage is the initial lesion in frostbite. They hypothesized that subsequent damage was amplified by free radical mechanisms.

IV. CLINICAL STUDIES

To date there have been no controlled clinical studies to prove the efficacy of free radical scavengers in skin flap surgery. Researchers at the University of Michigan, however, have obtained some supportive evidence. Friedl and coworkers (80) found that xanthine oxidase was significantly increased in the effluent from arms that had undergone 60–120 minutes of tourniquet-induced ischemia. Wilkens and coworkers (81) found that xanthine oxidase was elevated in free tissue transfer. They also noted that the rise in xanthine oxidase was markedly reduced by cooling the flap. They found similar results in human jejunal free tissue transfers (82).

Late ulceration secondary to venous insufficiency is an important clinical problem. Angel and coworkers (83) hypothesized that these limbs are in a partially ischemic state, which contributes to the formation of damaging free radical species. A pilot study (84) is highly suggestive. In that study, there was evidence of increased lipoperoxidation and salutary effect of vitamin E in clinical course. This area, however, will require further study.

REFERENCES

1. Bannister WH, Bannister JV. Biological and clinical aspects of superoxide and superoxide dismutase. Amsterdam: Worth Holland. Elsevier, 1981.
2. McCord JM. Oxygen-derived free radicals in post ischemic tissue injury. N Engl J Med 1989; 312:159–63.
3. DelMaestro RF, Thaw HH, Bjork J, Planker M, Arfors KE. Free radicals as mediators of tissue injury. Acta Physiol Scand 1980; (Supp) 492:43–57.
4. Demopoulos HB, Flamm ES, Pietronigro DD, Seligman ML. Acta Physiol Scand 1980; (Supp) 492:91–119.
5. Reilly DM, Shiller HJ, Bulkey GB. Pharmacologic approach to tissue injury mediated by free radicals in other reactive oxygen metabolites. Am J Surg 1991; 161: 488–503.

6. Bulkey GB. The role of oxygen free radicals in human disease processes. Surgery 1983; 94:407–10.
7. Umeda F, Kato KI, Ibayashi H, Kojima M, Shibata Y, Yamato T. Inhibitory effect of vitamin E on lipoperoxide formation in the rat adrenal gland. Tohuku J Ex Med 1982; 137:369–77.
8. Halliwell B. Superoxide dismutase, catalase, and glutathione peroxidase: solutions to the problems of living with O_2. New Physiologist 1974; 73:1075.
9. Walter JB, Israel MS, eds. General pathology. 5th ed. New York: Churchill, 1979.
10. Siesjo BK, Nordstrom CH, Rehncrona S. Metabolic aspects of cerebral hypoxia-ischemia. Adv Exp Med Biol 1976; 72:261.
11. Willms-Kretschmer K, Majno G. Ischemia of the skin: electron microscopic study of vascular injury. Am J Path 1969; 54:327–54. 1969.
12. Eriksson E, Anderson VA, Replogle RL. Effects of prolonged ischemia on muscle microcirculation in the rat. Surg Forum 1974; 25:254–5.
13. Kerrigan CL, Zelt RG, Daniel RK. Secondary critical ischemia time of experimental skin flaps. Plast Reconstru Surg 1984; 74:522–4.
14. Angel MF, Mellow CG, Knight KR, O'Brien B McC. Secondary ischemia time in rodents: contrasting complete pedicle interruption with venous obstruction. Plast Reconstr Surg 1990; 85:789–93.
15. Hodges PL. Principles of flaps. Selected readings in plastic surgery 1986; 4(3).
16. MacGregor IA, Morgan G. Axial and random pattern flaps. Br J Plast Surg 1973; 26:202–13.
17. Manchot D. The Cutaneous arteries of the human body. Ristic J, Morain WD, trans. New York: Springer-Verlag, 1983.
18. Mathes SJ, Nahai F. Clinical applications for muscles and musculocutaneous flaps. St. Louis: C.V. Mosby, 1982.
19. McCraw JB, Arnold PG. McCraw and Arnold's atlas of muscle and musculocutaneous flaps. Norfolk: Hampton Press Publishing Company, Inc., 1986.
20. Tolhurst DE, Harsekee B, Zeeman RJ. The development of the fasciocutaneous flap and its clinical applications. Plast Reconstru Surg 1983; 71:597–605.
21. May JW, Chart LA, O'Brien B McC, Hurley JV. The no reflow phenomenon in experimental free flaps. Plast Reconstr Surg 1978; 61:256–64.
22. Strock PE, Majno G. Microvascular changes in acutely ischemic rat muscle. Surg Gynecol Obstet 1969; 129:1213–24.
23. Leaf A. Cell swelling: a facet of ischemic tissue injury. Circulation 1973; 48: 455–8.
24. Jacobs GR, Reinisch JR, Puckett CL. Microvascular fibrinolysis after ischemia: its relation to vascular potency and tissue survival. Plast Reconstr Surg 1981; 68:737–41.
25. Richards R, Seaber AV, Urbaniak JR. Chemically induced vasospasms: effect of ischemia, vessel occlusion and adrenergenic blockage. Plast Reconstr Surg 1985; 75:238–44.
26. Manson PN, Anthenelli RM, Im MJ, Bulkey GB, Hoopes JE. The role of oxygen free radicals in ischemic tissue injury in island skin flaps. Ann Surg 1983; 198:87–90.

27. Manson PN, Narayan KK, Im MJ, Bulkey GB, Hoopes JE. Improved survival in free skin flap transfers in rats. Surgery 1986; 99:211–14.
28. Sagi A, Ferder M, Levens D, Strauch B. Improvement of island flaps after prolonged ischemia by perfusion with superoxide dismutase. Plast Reconstr Surg 1977; 77:639–42.
29. Harashina T, Sawada Y, Watanabe S. The relationship between venous occlusion time of island flaps and flap survival. Plast Reconstr Surg 1977; 60:92–5.
30. Su CT, Im MJ, Hoopes JE. Tissue glucose and lactate following vascular occlusion in island skin flap. Plast Reconstr Surg 1982; 70:202–5.
31. Angel MF, Swartz NM, Narayanan K, Ramasastry SS, Kuhns DB, Basford RF, Futrell JW. The essential role of free radicals in ischemic tissue. Plast Surg Forum, Kansas City, MO, 1985.
32. Angel MF, Mellow CG, Knight KR, Morgan RF, O'Brien B McC. Biochemical analysis of acutely ischemic skin flaps Curr Surg 1990; 47:272–3.
33. Khaidi VAS, Chaglassian TL. The species distribution of xanthine oxidase. Biochem J 1965; 97:318–20.
34. Im MJ, Shen WH, Pak CJ, Manson PN, Bulkey GB, Hoopes JE. Effect of allopurinol on the survival of hyperemic island skin flaps. Plast Reconstr Surg 1984; 73:276–8.
35. Im MJ, Manson PN, Bulkey GB, Hoopes JE. Effects of superoxide dismutase and allopurinol on the survival of acute island skin flaps. Ann Surg 1985; 201: 357–9.
36. Beauchamp C, Fridovich I. A mechanism of the production of ethylene from methional. The generation of hydroxyl radical by xanthine oxidase. J Biol Chem 1970; 245:4641–6.
37. Freeman BA, Crapo JD. Biology of disease: free radicals and tissue injury. Lab Invest 1982; 47:412.
38. Hunter FE, Gebicki JM, Hoffsten PE, Weinstein J, Scott A. Swelling and lysis of rat liver mitochondria induced by ferrous ions. J Biol Chem 1963; 238:828–35.
39. Halliwell B. Superoxide-dependent formation of hydroxyl radicals in the presence of iron chelates. FEBS Lett 1978; 92:321–6.
40. Butler J, Halliwell B. Reaction of iron-EDTA chelates with superoxide radical. Arch Biochem Biophys 1982; 218:174–8.
41. Keberle H. The biochemistry of desferrioxamine and its relation to iron metabolism. Ann NY Acad Sci 1964; 119:753–68.
42. Green CJ, Dhami L, Prasad S, Healing G, Shury C. The effect of desferrioxamine on lipid peroxidation and survival of ischaemic island skin flaps in rats. Br J Plast Surg 1989; 42:565–9.
43. Angel MF, Mellow CG, Knight KR, O'Brien B McC. The effect of deferoxamine on tolerance to secondary ischemia caused by venous obstruction. Br J Plast Surg 1989; 42:422–4.
44. Arfors KE, Lundberg C, Lindbom L, Lundberg K, Beatty PG, Harlan JM. A monoclonal antibody to the membrane glycoprotein complex CD18 inhibits polymorphonuclear leukocyte accumulation and plasma leakage in vivo. Blood 1987; 69:338–40.

45. Simpson PJ, Todd RF, Fantone JC, Mickelson JK, Griffith JD, Lucchesi BR. Reduction of experimental canine reperfusion injury by a monoclonal antibody (anti-MO1, anti-CD116) that inhibits leukocytes adhesion. J Clin Invest 1988; 81:624–29.
46. Gamble JR, Skinner MP, Berndt MC, Vadas MA. Prevention of activated neutrophil adhesion to endothelium by soluble adhesion protein, GMP 140. Science 1990; 249:414–17.
47. Vedder NB, Winn RK, Rice CL, Chi EY, Arfors KE, Harlan JM. Inhibition of leukocyte aherence by anti-CD18 monoclonal antibody attenuates reperfusion injury in the rabbit ear. Proc Natl Acad Sci USA 1990; 87:2643–6.
48. Pryor WA, Stanky JP. A suggested mechanism for the production of malonylaldehyde during antioxidation of polyunsaturated fatty acids: nonenzymatic production of prostaglandin endoperoxides during autoxidation. J Organ Chem 1975; 40:3615–17.
49. Gullis RJ, Rowe CE. The stimulation of the phospholipase A_2 acylation system of synaptic membranes by cyclic nucleotides. Biochem J 1975; 148:567–81.
50. Deby C, Deby-Dupont G. Natural factors modulating the intervention of activated oxygen in the biosynthesis of prostanoids. Bull Eur Physiol Pathol Respir 1981; (Supp) 17:129–39.
51. Douglas B, Weinberg H, Song Y, Silverman DG. Beneficial effects of ibuprofen on experimental microvascular free flaps: pharmacologic alteration of the no-reflow phenomenon. Plast Reconstr Surg 1987; 79:366–71.
52. Edstrom LE, Balkovich M, Stotman GJ. Effect of ischemic skin flap elevation on tissue and plasma A_2 and prostacyclin production: modification by thromboxane synthetase inhibition. Ann Plast Surg 1988; 20:106–11.
53. Zachary LS, Heggers JB, Robson MC, Murphy RC. Combined prostacyclin and thromboxane synthetase inhibitor UK 38485 on flap survival. Ann Plast Surg 1986; 17:112–15.
54. Mellow CG, Knight KR, Angel MF, O'Brien B McC. Effect of a thromboxane synthetase inhibitor UK 38485 on the tolerance of skin flaps to primary ischemia. Aust NZ J Surg 1990; 60:293–7.
55. Heggers JB, Loy GL, Robson ML, Del Beccaro EJ. Histologic demonstration of prostaglandins and thromboxanes in burned tissues. J Surg Res 19809; 28: 110–17.
56. Pang CY, Forrest CR. Discussion of: Beneficial effects of ibuprofen on experimental microsurgical free flaps: pharmacologic alteration of the no-reflow phenomenon. Plast Reconstr Surg 1987; 79:372–4.
57. Furr SF, Seaber AV, Urbaniak JR. Role of thromboxane in post-ischemic microcirculatory blood flow. Surg Forum 1986; 37:533–5.
58. Angel MF, Knight KR, Mellow CG, Morrison WA, O'Brien B McC. The effect of prior elevation of skin flaps and ischemia on blood thromboxane levels. Ann Plast Surg 1989; 22:501–4.
59. Mellow CG, Knight KR, Angel MF, O'Brien B McC. The effect of thromboxane synthetase inhibition on tolerance of skin flaps to secondary ischemia caused by venous obstruction. Plast Reconstr Surg 1990; 86:329–34.

60. Knight KR, Angel MF, Lepore DA, Abbey DA, Arnold LI, Gray KA, Mellow CG, O'Brien B McC. Secondary ischaemia in rabbit skin flaps: the roles played by thromboxane and free radicals. Clin Sci 1991; 80:235–40.
61. Creech BJ, DeVito RV, Eade GG. Viability of split-skin grafts from pigs following incubation on autologous blood or serum for various periods. Plast Reconstr Surg 1973; 51:572–4.
62. Marzella L, Jesudass RR, Manson PN, Myers RAM, Bulkey GB. Functional and structural evaluation of the vasculature of skin flaps after ischemia and reperfusion. Plast Reconstr Surg 1988; 81:742.
63. Feng LT, Berger BE, Lysz TU, Shaw UW. Vasoactive prostaglandins in the impending no-reflow state: evidence in a primary disturbance in microvascular tone. Plast Reconstr Surg 1988; 81:755–64.
64. Finseth F, Adelberg MC. Prevention of skin flap necrosis by a course of treatment with vasodialating drug. Plast Reconstr Surg 1978; 61:738–43.
65. McFarlane RM, DeYoung G, Henry RH. The design of a pedicle flap to study necrosis and its prevention. Plast Reconstr Surg 1965; 35:177–82.
66. Angel MF, Ramasastry SS, Narayanan K, Zerbe A, Swartz WM, Futrell JW. Studies on the nature of the flap/bed interaction in rodents. Part II: Morphological and fluorescein kinetic data. Ann Plast Surg 1986; 77:434–40.
67. Angel MF, Ramasastry SS, Swartz WM, Narayanan K, Kuhns DB, Basford RE, Futrell JW. The critical relationship between free radicals and degrees of ischemia: evidence for tissue for tissue intolerance of marginal perfusion. Plast Reconstr Surg 1988; 81:233–9.
68. Angel MF, Ramasastry SS, Swartz WM, Narayanan K, Basford RE, Futrell JW. Augmentation of skin flap survival with allopurinol. Ann Plast Surg 1987; 18: 494–8.
69. Angel MF, Narayanan K, Swartz WM, Ramasastry SS, Kuhns DB, Basford RE, Futrell JW. Deferoxamine increases skin flap survival: additional evidence of free radical involvement in ischaemic flap surgery. Brit J Plast SUrg 1986; 39: 469–72.
70. Rees R, Hinshaw D, Cashmer B, Im MJ, Smith B, Smith DJ. Experimental evidence that oxidative stress from delay upregulates antioxidant defenses in ischemic skin. Surg Form 1991; 62:648–50.
71. Grabb WC. Introduction to clinical aspects of skin flaps. In: Grabb WC, Meyers MB, eds. Skin flaps. Boston: Little, Brown, 1975: 141.
72. Krizek TJ, Davis JG. The role of the red cell in subcutaneous infection. J Trauma 1965; 5:85–94.
73. Angel MF, Narayanan K, Swartz WM, Ramasastry SS, Basford RE, Kuhns DB, Futrell JW. The etiologic role of free radicals in hematoma-induced flap necrosis. Plast Reconstr Surg 1986; 77:795–801.
74. Haddad J, Angel MF, Abramson M. Reduction of hematoma-induced flap necrosis by deferoxamine. Surg Forum 1986; 37:564–5.
75. Rakower SR, Shahgoli S, Song L. Doppler ultrasound and digital plethysonography to determine the need for sympathetic blockage after frostbite. J Trauma 1978; 18:713–18.

76. Bourne MH, Piepkorn MW, Clayton F, Leonard LG. Analysis of microvascular changes in frostbite injury. J Surg Res 1986; 40:26–35.
77. Reite PB. Functional realities of small vessels in tissue injured by freezing and thawing. Acta Physiol Scand 1965; 63:111–20.
78. Manson PN, Jesudass RR, Marzella L, Bulkey GB, Im MJ, Narayan KK. Evidence for an early free radical-mediated reperfusion injury in frostbite. Free Rad Bio Med 1991; 10:7–11.
79. Marzella L, Jesudass RR, Manson PN, Myers RAM, Bulkey GB. Morphologic characterization of acute injury to vascular endothelium of skin after frostbite. Plast Reconstr Surg 1989; 83:67–75.
80. Friedl HP, Smith DJ, Thomson PD, Louis DS, Till GO, Ward PA. Histamine and xanthine oxidase activity in a human model of ischemia/reperfusion injury. Surg Forum 1989; 40:587–9.
81. Wilkens EG, Punch JD, Rees RS, Cashmer B, Till GO, Smith DJ. Evidence for xanthine oxidase activity in human free flaps after reperfusion. Surg Forum 1990; 41:672–3.
82. Wilkens EG, Smith D, Cashmer B, Rees RR, Smith DJ. Xanthine oxidase activity and oxy-radicals in human free jejunal flaps after head and neck operations. Surg Forum 1991; 42:562–6.
83. Angel MF, Ramasastry SS, Swartz WM, Basford RE, Futrell JW. The causes of skin ulcers associated with venous insufficiency: a unifying hypothesis. Plast Reconstr Surg 1987; 79:289–97.
84. Ramasastry SS, Angel MF, Narayanan K, Basford RE, Futrell JW. Biochemical evidence of lipoperoxidation in venous stasis ulcers: beneficial role of Vitamin E as antioxidant Ann NY Acad Sci 1989; 570:506–8.

21

Dermatopharmacology of Oxidative Stress

M. B. Roberfroid and Pedro Buc Calderon
Université Catholique de Louvain, Brussels, Belgium

I. INTRODUCTION

Radical reactions are important in a wide range of unrelated biological systems and critical for the normal operation of a wide spectrum of biological processes. Moreover, among the very many ways to chemically injure and kill cells, an important class of reactions is that producing radical intermediates. Because of the ubiquity of molecular oxygen in aerobic organisms and its ability to accept electrons, oxygen-derived radicals are often mediators and/or products of normal, pathological, or toxic reactions.

A radical is, by definition, any chemical species capable of independent existence that has an odd number of electrons in an outer atomic or molecular orbital (1–3). Radical reactions are generally chain reactions.

Initiation results from homolysis, photolysis, or radiolysis of molecules or from redox reactions (2). The most likely unimolecular bond homolysis that can take place in living cells is the so-called "molecule-assisted homolysis" (MAH) during which bond making accompanies bond breaking so that the activation energy of the overall process is lowered (4). Light, in particular ultraviolet light, photolyses chemical bonds (5). The areas of skin exposed to light and air are particular targets for such an initiation of radical reactions. Moreover, the absorption of light by photosensitizers may, in the presence of O_2, cause photo-oxidation leading to detrimental effects (6,7).

Ionizing radiation produces cation radicals that react either with H_2O to produce $H^{\cdot}$, $HO^{\cdot}$, H_2O_2 or with unsaturated lipids to generate autoxidation products (8–10). Both one-electron oxidation and one-electron reduction of normal cell components or xenobiotics produce radicals, but, in all cases, their rapid reoxidation by molecular O_2 produces the superoxide anion O_2^-, which thus appears as the most common intracellular product of one-electron redox processes (11,12). Redox reactions in which a complexed or chelated transition metal (e.g., Cu, Fe, Mn, Va) plays a role either as binding site or in the catalytic site are of particular importance for intracellular radical production (13).

Once they have been produced, radicals usually participate in a process called propagation during which their number remains constant. Four major mechanisms have been identified for the propagation of radicals: 1) atom (or group) transfer, 2) electron transfer, 3) addition, and 4) b-scission (2). Within the context of this chapter, hydrogen atom abstraction, electron transfer, and addition of a radical to an olefin or an aromatic ring deserve special attention.

The radical chain reactions terminate either when the radical disappears in a redox reaction or when two radicals cross-link. Radical scavengers are special endogenous or exogenous molecules that bind radicals to form an adduct that is prone to a termination process.

As indicated above, exposure of most radicals to molecular O_2 in aerobic living organisms yields O_2^-. However, neither O_2^- nor H_2O_2 (its dismutation product) appears to have sufficient reactivity to account for the variety of toxic consequences associated with the generation of O_2^-. The hydroxyl radical $HO^{\cdot}$ and the singlet oxygen $^1O_2$ are likely to be the major toxic intermediates (14) that can react with unsaturated fatty acyl chains, nucleic acids, or amino acid residues of polypeptide chains (11,15,16).

Due to the biological diversity of chemicals present in various structural arrays with aqueous and hydrophobic areas, control of radical reactions is a complex task. However, since the basic metabolism of living organisms generates various "physiological" radicals, innate, rigid controls over radical mechanisms exist which are operative in normal biological situations. Furthermore, since key molecular structures such as cellular membranes, DNA, and proteins contain molecular targets that are highly susceptible to radical attack, there are several factors that can protect such susceptible biomolecules from the uncontrolled radical chain reactions. The various controls over radical attacks that exist in mammalian cells are controls imposed by membrane structure and by specific molecular structure, controls exerted by metal chelators and antioxidants, and enzyme inactivation of oxygen-derived free radicals. But pathological as well as toxicological situations exist in which either radical production overwhelms the capacity of radical chain reaction's termination or cellular defenses against radical initiated processes are reduced. Such situations create what has been called an *oxidative stress* (17).

II. PHARMACOLOGY OF OXIDATIVE STRESS

It is now firmly established that radicals mediate both pathological and toxicological processes. A pharmacology of oxidative stress can thus be developed with the objective of developing therapeutic strategies to restore efficient control over radical-mediated processes whenever they overwhelm the natural endogenous control capacity. Such strategies (Fig. 1) make use of (18,19):

Structurally unrelated natural—glutathione (GSH), flavonoids—or synthetic antioxidants—butylhydroxytoluene (BHT), butylhydroxyanisole (BHA), *N*-acyldehydroalanines (AD)
Enzymes and chemical analogs of enzymes—superoxide dismutases (SOD), catalase (CAT), GSH peroxidase, Ebselen®, copper diisopropyl salicylate (CUDIPS)
Antioxidant vitamins—A, E (αTH), C
Scavengers of O_2^--derived radicals—mannitol, glucose, dimethyl sulfoxide (DMSO)
Metal ion chelators—Desferal®
Precursors of intracellular antioxidants—aminothiols (WR2721), *N*-acetylcysteine (NAC), selenium

From a mechanistic point of view, such strategies are classified as follows (Fig. 2):

1. Enzymatic or enzymaticlike redox reactions, the substrate of which is a radical (or a reactive oxygen species) that is transformed to a less reactive, nonradical product. Such enzymes (SOD, CAT, GSH peroxidase) are endogenous constituents of a large variety of cells. They often act in concert.
2. Nonenzymatic antioxidants that participate in a stoichiometric redox modification of a radical. During such a reaction, the antioxidant is transformed so as to produce a new radical that is more persistent and/or less reactive. This radical product is then the substrate of subsequent redox reactions that inactivate it. The nonenzymatic antioxidant participates in the radical chain reactions, but it leads to termination.

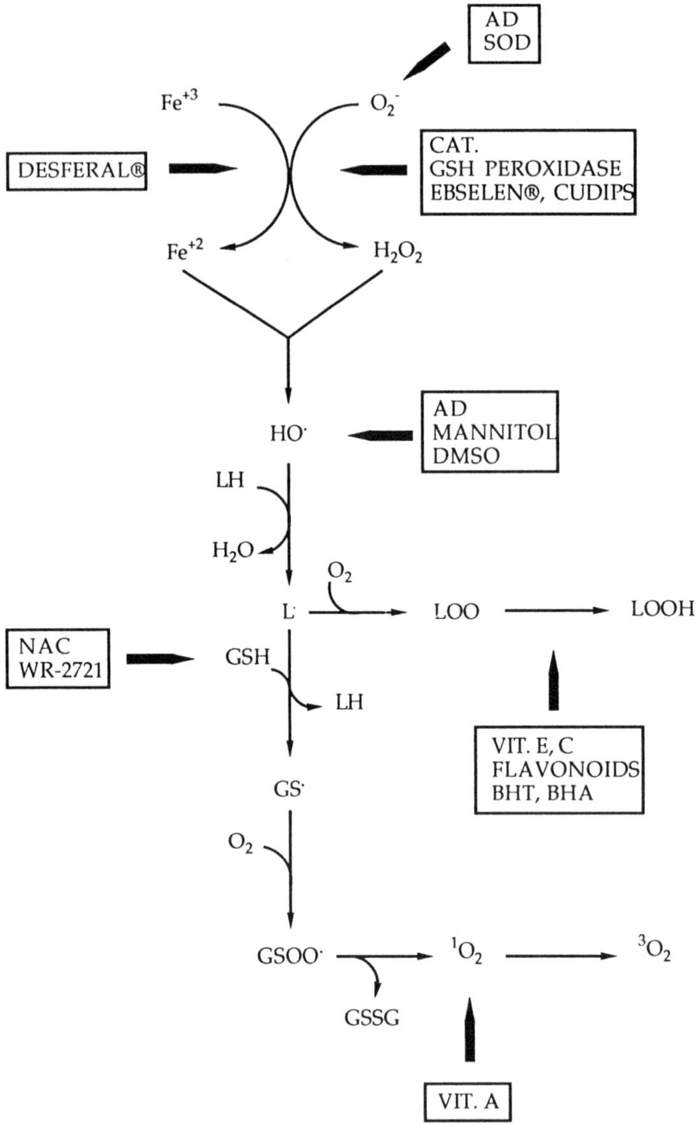

Figure 1 Schematic representations of strategies available for a pharmacology of oxidative stress. LH = unsaturated lipid$^-$, LOOH = hydroperoxide of LH.

1. Enzymatic redox reaction

$$2O_2^{\cdot-} + 2H^+ \longrightarrow H_2O_2 + O_2$$

$$H_2O_2 \longrightarrow H_2O + 1/2\,O_2$$

$$\begin{matrix} H_2O_2 \\ ROOH \end{matrix} + 2GSH \longrightarrow \begin{matrix} 2H_2O \\ ROH \end{matrix} + GSSG$$

2. Non enzymatic redox reaction

$$2\,ROO^{\cdot} + 2\,\alpha TH \longrightarrow 2\,ROOH + 2\,\alpha T$$

$$2\,\alpha T^{\cdot} + 2\,\text{vit C} \longrightarrow 2\,\text{vit C}^{\cdot} + 2\,\alpha TH$$

$$2\,\text{vit C}^{\cdot} + NADH + H^+ \longrightarrow 2\,\text{vit C} + NAD^+$$

3. Radical scavenging

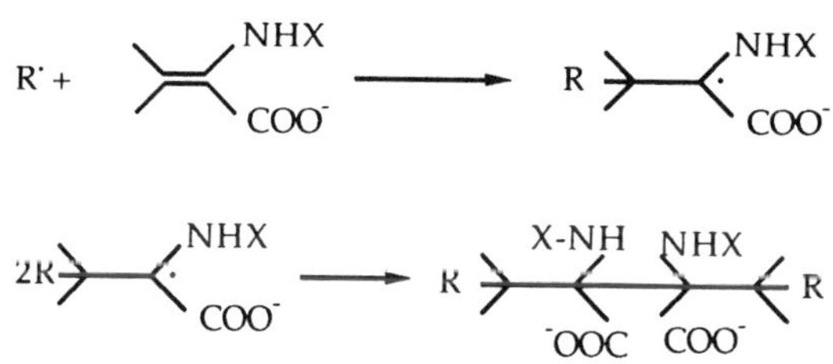

4. Precursors of intracellular antioxidants

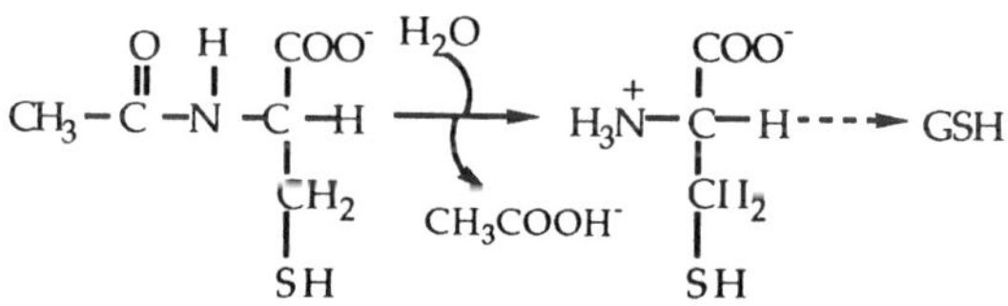

$$Se \longrightarrow \text{catalytic site of GSH peroxidase}$$

5. Metal ion chelation

Desferal® to chelates iron ions

6. Energy quenching

$${}^1O_2 + Q \longrightarrow {}^3O_2 + Q^* \longrightarrow Q + \text{heat}$$

Figure 2 Mechanistic bases for strategies applicable to control oxidative stress.

Both enzymatic and nonenzymatic "redox mediators" of termination of radical chain reactions are continuously recycled. They play a catalytic role in the inactivation of radicals.

3. Scavengers of radicals that react with radicals so as to create a new chemical bond. This new chemical entity generally keeps the radical nature of the substrate, but the new radical is either a better substrate for further redox inactivation (see Refs. 1, 2) or it dimerizes spontaneously to form a new molecular entity (20).
4. Precursors of intracellular antioxidants that help reinforce processes 1 and 2.
5. Metal ion chelators, which are used to avoid metal ion–dependent redox reactions that mainly produce O_2-derived radicals.
6. Vitamin A, retinoids, and other synthetic compounds that inactivate $^1O_2$ by quenching or energy transfer (21).

If potentially therapeutic strategies are available to control oxidative stress, the next question is "Which are the diseases for which such strategies are applicable or have already been applied?" To answer that question, one may raise a second: "Which are the disease states in which radicals are involved either as a cause or as a consequence of the disease activity?" Finally one has to ask if, in these disease states, "control of oxidative stress or any radical-mediated processes is pharmacologically feasible." Cataract and blindness, Alzheimer's and Parkinson's diseases, atherosclerosis, autoimmune diseases, cancer, and ischemia-reperfusion are some disease states in which radicals are likely to play a role. From the point of view of the pharmacology of such diseases, it is only recently that oxidative stress has been considered a fruitful basis for drug design. Up until now mostly empirical approaches have been used, which have resulted in the recommended use of SOD as a therapeutic tool to treat several pathologies related to inflammation (22). The testing of selenoorganic compounds like Ebselen® as potential anti-inflammatory drugs (23) has resulted in the a posteriori recognition that many nonsteroidal anti-inflammatory drugs are indeed radical scavengers, a property that may largely explain their pharmacological activity (24). It was recently discovered that Probucol® , an active hypocholesterolemic drug, is also a radical scavenger that inhibits the oxidation of low density lipoproteins in arterial walls, a process likely to be involved in atherosclerosis (25).

The pharmacology of oxidative stress is still at a very early stage of its development, and very few drugs are direct products of that discipline. The major reason for this is that even though we know reasonably well the biochemistry of radicals, we still lack the chemical knowledge that would allow us to design a priori radical scavengers of pharmacological interest. Moreover, the development of such drugs often poses pharmacokinetic problems because it is difficult to selectively direct the drug to the site of oxidative stress without interfering with the normal cellular metabolism for which radicals are necessary intermediates.

Because skin is easily reached for topical and well-controlled applications, it is a preferential target for such therapies. Moreover, the development of H. G. Viehe and colleagues (26) of the captodative concept has opened new avenues in the a priori synthesis of radical scavengers, the *N*-acyldehydroalanines (AD), which have interesting pharmacological properties. We would like therefore to use the rest of this chapter to describe in some detail the properties of such molecules and to discuss their potential applications in what could be called a dermatopharmacology of oxidative stress.

III. *N*-ACYLDEHYDROALANINES: A PRACTICAL EXAMPLE

A. The Captodative Effect

It has been proposed that substitution by both an electron-donating and an electron-withdrawing group (captodative substitution) stabilizes a carbon-centered radical (26). Olefins having such a captodative substituted geminal carbon atom have been shown to react with free radicals to give stabilized free radical adducts, which often disappear by dimerization or by reaction with another radical. For that reason these olefins have been called radicophiles (27).

B. *N*-Acyldehydroalanines, a Group of Pharmacologically Active Radicophiles

N-Acyldehydroalanines (AD) form an original class of captodative substituted olefins, which are identified as AD compounds (Fig. 3). In addition

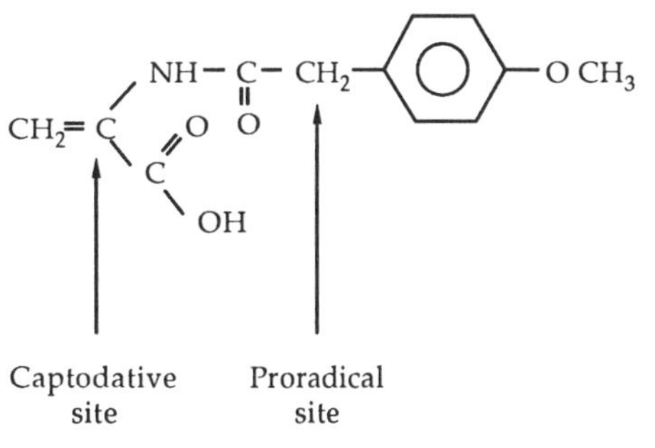

Figure 3 Chemical structure of N-acyldehydroalanine (AD).

to the radicophilic site (on the captodative substituted olefinic carbon), they have a methylene group, which has been called the "proradical site." In fact, the cleavage of the C-H bond yields a carbon-centered radical, which is stabilized by captodative effect. Since radicals react mainly by hydrogen abstraction, electron transfer, and addition to double bonds (see above), AD compounds are potential radical scavengers that offer three different sites. In reaction I in Figure 4, the addition of a radical (e.g., $HO^{\cdot}$) to the carbon-carbon double bond forms a captodative stabilized radical adduct, which dimerizes or reacts with another radical; in reaction II, the attack of $HO^{\cdot}$ results in hydrogen abstraction from the methylene group, producing another captodative stabilized free radical adduct. Moreover, the aromatic ring acts as a trap for radicals, mostly $HO^{\cdot}$ (reaction III), producing hydroxylated aromatic derivatives. Whatever the type of reaction, the propagation of the radical-mediated chain of reactions will be terminated. Considering the problem in biological fluids of trapping all kinds of free radicals with different polarities and structures, AD compounds present some interesting possibilities. Indeed, they have been shown in vitro to react with and to scavenge both O_2^- and $HO^{\cdot}$, inhibiting in this way the in vitro and in vivo deleterious effects mediated by these radicals.

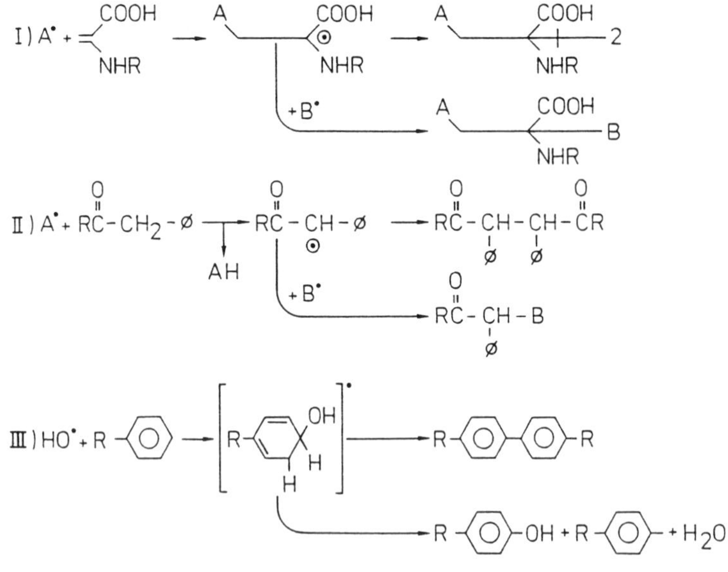

Figure 4 Schematic representation of the reactions between a radical and AD.

1. In Vitro Studies: N-*Acyldehydroalanines as Scavengers of Oxygen Radicals*

a. ESR Studies The reaction of AD compounds with $HO^\cdot$ results in the generation of a carbon-centered adduct that reacts with the spin trap DMPO, dimethyl-pyrolidine 1-on (28). Additional ESR studies to determine the structure of this carbon-centered radical were performed by analyzing the signal of the reaction product with methylnitrosopropane (MNP). Among the radical adducts that could form at each of the sites described in Figure 3, only one structure, that resulting from $HO^\cdot$ attack at the unsubstituted methylene group of the double bond, has an ESR spectrum consistent with the experimental data (29). It must be stressed, however, that observation of the captodative adduct does not totally preclude the occurrence of other radical adducts that may be shorter lived or formed more slowly and hence not detectable under the conditions used.

b. Kinetic Studies and Reactivity Towards Reactive Oxygen Species The oxidation of xanthine by xanthine oxidase and the autoxidation of epinephrine were the model systems used to generate O_2^-, while the Fenton reaction and water radiolysis were used for $HO^\cdot$ production. The effect of AD compounds was compared to that of known free radical scavengers such as superoxide dismutase, mannitol, glucose, uric acid, etc. as a means for evaluating the reactivity of AD compounds towards O_2^- and $HO^\cdot$. The *para*-methoxyphenylacetyl dehydroalanine derivative, indexed as AD-5, was used as a model compound. It inhibits the reduction of both ferricytochrome c and nitroblue tetrazolium (NBT) by O_2^-. It also inhibits the oxidation of epinephrine to adrenochrome. Moreover, the oxidative degradation of 2-deoxyribose and of linolenic acid, two chemical processes mediated by $HO^\cdot$, were also inhibited by AD-5. The rate constant of AD-5 scavenging O_2^- was calculated to be $0.5–0.7 \times 10^3\ M^{-1}s^{-1}$, while the rate constant for the reactions of AD-5 with $HO^\cdot$ was calculated to be $2.8–4.5 \times 10^9\ M^{-1}s^{-1}$ (28,30).

c. Inhibition of Free Radical–Mediated Reactions Since AD compounds react with and scavenge oxyradicals, they were tested for their capacity to prevent oxyradical-mediated biochemical effects. Microsomal lipid peroxidation was used as a typical end point of radical chain reaction. Nitrofurantoin, *tert*-butylhydroperoxide, ionizing radiations, adriamycin, iron/NADPH, and iron ascorbate were used as model systems to stimulate radical production and subsequent lipid peroxidation in rat liver microsomes. In all these experiments, even though involved radical species were different (O_2^-, $HO^\cdot$, alkoxyl radical), AD compounds inhibited lipid peroxidation but with various efficacies: $HO^\cdot$ and O_2^--mediated reactions were more efficiently inhibited than $RO^\cdot$ mediated processes (28,31–33).

2. *In Vivo Studies:* N-*Acyldehydroalanine Derivatives as Inhibitors of Radical-mediated Toxicological Processes*

In vivo, ionizing radiations induce a rapid or a delayed mortality depending on both the frequency and the level of dose administered. Single doses of x-rays varying from 6 to 8 Gy induce bone marrow toxicity with the animals dying 1 week after irradiations. AD protects mice against damage resulting from total body x-irradiation, increasing the LD_{50} (30) from 6.1 to 7.3 Gy with a dose reduction factor (DRF) of 1.2 (34).

Ionizing radiations when applied as fractionated doses (e.g., 4 × 175 rad at weekly intervals) induce the neoplastic transformation of T lymphocytes within the thymus of C57Bl/Ka mice. Genomic modification leading to mutation and cancer represents one of the most cited examples of delayed toxicity of ionizing radiations. The involvement of free radicals in such carcinogenic processes correlates well with the protective effects of free radical scavengers, as seen by the inhibition in the development of radiation-induced sarcomas by the aminothiol WR-2721 and the significant decrease in the lifetime incidence of x-ray–induced leukemia obtained by intravenous SOD. AD-20, when applied before exposure to x-rays, not only delays the appearance of but also reduces the number of thymic lymphomas (35).

Adriamycin is an anticancer drug widely used to treat several types of malignancies. Nevertheless, its clinical usefulness is limited by a selective organ toxicity and a risk of cardiomyopathy and congestive heart failure. The impairment by adriamycin of the cardiac mitochondrial respiratory chain is mainly responsible for its cardiotoxicity, and evidence from several experimental studies suggests that it may be the consequence of the production of reactive oxygen species (36). AD-20 (the *ortho*-methyoxyphenylacetyl derivative) inhibits, both in vitro and in vivo, the damaging effect of adriamycin on lipid membranes, as measured by the inhibition on lipid peroxidation of either rat liver microsomal suspensions or mouse heart mitochondrial membranes. Moreover, we have reported that changes in the mice heart mitochondrial membrane fluidity caused by adriamycin administration are not observed when AD-20 is administered 15 minutes before adriamycin (33). Such changes in fluidity results from an increase in the rigidity of the membrane, indicating that peroxidation of the membrane phospholipids may play a major role. Toxicity of adriamycin in vivo was further evaluated either by recording the lethality induced by a single or a multiple doses protocol of i.p. administration or by measuring parameters linked to specific tissue/organ necrosis, such as serum transaminases activities, heart mitochondrial enzymic activities, and total peripheral leukocyte count. Administration of AD-20 enhances the survival time of mice, whatever the i.p. protocol of administration used. In addition, cardiac, hepatic, and bone marrow toxicities induced by adriamycin are prevented by pretreatment of mice with AD-20 (33,37).

AD compounds prevent the pro-oxidant effects of choline deficiency, a sequence of early alterations observed in the liver during initiation of a carcinogenic process by feeding rats with a choline-devoid diet (38).

IV. DERMATOPHARMACOLOGY OF *N*-ACYLDEHYDROALANINES

As discussed above, the pharmacology of oxidative stress requires a valuable chemical concept to design a priori radical scavengers or effective antioxidants as well as a good pharmacokinetic approach so as to target the drug to the critical site of the deleterious effect. AD compounds are organic structures, the synthesis of which has been made possible by the captodative concept. They are effective oxyradical scavengers that inhibit, both in vitro and in vivo, mainly O_2^-- and $OH^{\cdot}$-mediated, potentially deleterious effects. Moreover, they have in vivo activities. But, once again due to pharmacokinetic limitations, the doses that must be used in experimental models are much too high to be applicable to clinical situations.

The last part of this chapter will focus on skin applications, which demonstrate that whenever a topical application is possible, AD compounds might prove to be very effective drugs to inhibit radical-mediated, potentially pathological processes. Indeed, in a recent publication, Vo et al. (39) have hypothesized that, because of its radical-scavenging activity and of the involvement of reactive O_2 species in tumor promotion, topical application of AD would inhibit, in mouse skin, TPA-induced cancer promotion as well as a cellular event known to be associated with such a pathological process, namely, skin hyperplasia. Moreover, these authors have tested the effect of AD on TPA-induced chemiluminescence and stimulation of PGE_2 formation in cultured murine epidermal cells (Table 1). Single or repeated topical applications (200 or 400 nmol) of both AD-5 and AD-19 (*para*-methoxylphenyl- and *para*-thiomethylphenyl-*N*-acyldehydroalanine, respectively) simultaneously with TPA to mouse Sencar skin significantly reduces TPA-induced hyperplasia up to 70% depending on the protocol. Even though it does not inhibit TPA-induced papilloma formation, AD-19 at a dose of 80 nmol/application reduces significantly (± by 50%) the percentage of TPA-treated Sencar mice bearing carcinomas at week 40 after DMBA application. Such data confirm our own observations on NMRI mice (data not shown).

In cultured murine epidermal cells, the same authors have demonstrated that both AD-5 and AD-19 completely inhibit TPA-induced PGE_2 synthesis (a molecular mediator of hyperplasia), approximately 50% inhibition occurring at 0.1 mM ($p < 0.001$). In the same experimental system, the *N*-acyldehydroalanines also inhibit, in a dose-dependent manner, TPA-induced chemiluminescence response, which demonstrates the induction of O_2^- production.

Table 1 Relative Effect of AD-19 on TPA-Induced Changes in Sencar Mouse Skin and Murine Epidermal Cells

In vivo			
TPA-induced hyperplasia (at 24 h)			
AD-19	1 application (200 nmol)— −50%		
	4 applications (200 nmol)— −70%		
TPA-induced papillomas yield (at week 16)			
AD-19	80 nmol/15 min after each TPA application: no differences		
TPA-induced carcinfomas (incidence at week 40)			
AD-19	80 nmol/15 min after each TPA application = −48%		
In vitro (% of TPA response)			
PGE_2 synthesis			
AD-19		0.1 mM	39%
		1 mM	14%
Chemiluminescence			
AD-19		0.1 mM	80%
		1 mM	20%

Source: Ref. 39.

The major conclusion of these authors is that AD compounds might be potent inhibitors of the progression of benign tumors to carcinomas. They hypothesize that AD could act by inhibiting the oxyradical-mediated second mutation involved in the process of malignant conversion.

V. CONCLUSIONS AND PERSPECTIVES

Oxidative stress is largely recognized as a causal event in various pathologies. The pharmacology of oxidative stress could be a stimulating and creative new discipline to design and develop new drugs. The data reported here support the validity of the concept of stabilization of radicals by captodative substitution to design and synthesize radical scavengers that have high and selective activity in trapping O_2^- and $OH^\cdot$ both in vitro and in vivo.

Topical application of the original *N*-acyldehydroalanine derivatives on skin allows one to bypass the pharmacokinetic problems after systemic dosing. Such application has been successful in inhibiting cellular and molecular events known to accompany cell hyperplasia induced by the phorbol ester TPA. Moreover, application of an AD compound has led to a 50% reduction of the conversion of papillomas to carcinomas.

N-Substituted dehydroalanines are thus new and original molecules (40) with the potential to control deleterious oxidative stress in skin. Being highly lipophilic, these compounds could find useful applications in topical drug delivery "systems."

The captodative concept has proven useful in synthesizing, a priori, pharmacologically active radical scavengers. It is thus worth considering for further development in the field of pharmacology of oxidative stress either for the synthesis of new drugs or for the structural modification of existing compounds.

REFERENCES

1. Bolton JR. In: Schwartz HM, Bolton JR, Borg DC, eds. Biological applications of electron spin resonance. New York: Wiley Interscience, 1972:11–61.
2. Pryor WA. Free radicals. New York: McGraw Hill, 1966.
3. Waters WA. The chemistry of free radicals. 2nd ed. New York: Oxford University Press, 1984.
4. Pryor WA. Free radical in pathology. Chem Eng News 1971; 49:41–5.
5. Yamazaki I. One-electron and two-electron transfer mechanisms in enzymic oxidation-reduction reactions. Adv Biophys 1971; 2:33–76.
6. Blum HF. Photodynamic action and diseases caused by light. New York: Hafner, 1964.
7. Gallo U, Santamaria L, eds. Research progress in organic biological and medicinal chemistry. Vol. III. New York: Elsevier, 1972.
8. Ausloos P, ed. Fundamental processes in radiation chemistry. New York: Wiley Interscience, 1968.
9. Bacq ZM, Alexander P. Fundamentals of radiobiology. 2nd ed. Oxford: Pergamon, 1966.
10. Altman KI, Gerber GB, Okada S. Radiation biochemistry. Vols. I and II. New York: Academic Press, 1970.
11. Freeman BA, Capro JD. Biology of disease: free radical and tissue injury. Lab Invest 1982; 47:412–26.
12. Mason RP, Chignell CF. Free radicals in pharmacology and toxicology. Pharmacol Res 1981; 33:189–211.
13. Halliwell B, Gutteridge JMC. Oxygen toxicity, oxygen radicals, transition metals and disease. Biochem J 1984; 219:1–14.
14. Slater TF. Free radical mechanisms in tissue injury. Biochem J 1984; 222:1–15.
15. Kuwabara M, Zhi-Yi Z, Yoshii G. ESR of spin-trapped radicals in aqueous solutions of pyrimidine nucleosides and nucleotides. Reactions of the hydroxyl radical. Int J Radiat Biol 1982; 41:241–59.
16. Trelstad RL, Lawley KR, Holmes LB. Nonenzymatic hydroxylations of proline and lysine by reduced oxygen derivatives. Nature 1981; 289:310–12.
17. Sies H. Oxidative stress. London: Academic Press, 1985.
18. Halliwell B, Gutteridge JMC. Free radicals in biology and medicine. 2nd ed. Oxford: Clarendon Press, 1989.
19. Obervely LW ed. Superoxide dismutase. Vol. III. Boca Raton, FL: CRC Press, 1985.
20. Wefers H, Sies H. Oxidation of glutathione by the superoxide radical to the disulphide and sulfonate yielding oxygen. Eur J Biochem 1983; 137:29–36.

21. Rabek JF. In: Frimer AA, ed. Singlet oxygen. Vol. IV. Boca Raton, FL: CRC Press, 1985.
22. McCord JM. Free radicals and inflammation. Protection of synovial fluid by superoxide dismutase. Science 1974; 185:529–31.
23. Parnham MJ, Graf E. Seleno-organic compounds and the therapy of hydroperoxide linked pathological conditions. Biochem Pharmacol 1987; 36:3095–102.
24. Roberfroid MB, Viehe HG, Remacle J. Free radicals in drug research. Adv Drug Res 1987; 16:1–84.
25. Carew TE, Schwenke DC, Steinberg D. Antiatherogenic effect of probucol unrelated to its hypocholesterolemic effect: evidence that antioxidants in vivo can selectively inhibit low density lipoprotein degradation in macrophage-rich fatty streaks and slow the progression of atherosclerosis in the Watanabe heritable hyperlipidemic rabbit. Proc Natl Acad Sci USA 1987; 84:7725–9.
26. Viehe HG, Janousek I, Merenyi R, Stella L. The capto-dative effect. Acc Chem Res 1985; 18:148–54.
27. Viehe HG, Merenyi R, Stella L, Janousek I. Capto-dative substituent effects in synthesis with radicals and radicophiles. Angew Chem (Int Ed) 1979; 18:917–32.
28. Buc Calderon P, Sipe HJ, Flitter W, Mason RP, Roberfroid M. N-Acyl dehydroalanines scavenge oxygen radicals and inhibit in vitro free radical mediated processes. Chem Biol Interact 1990; 73:78–88.
29. Sipe HS, Buc Calderon P, Roberfroid M, Mason RP. Free Rad Res Commun (submitted).
30. Buc Calderon P, Roberfroid M. Inhibition of O_2- and OH-mediated processes by a new class of free radical scavengers: the N-acyl dehydroalanines. Free Rad Res Commun 1988; 5(3):159–68.
31. Buc Calderon P, Roberfroid M. Inhibition of rat liver microsomal lipid peroxidation by tert-butyldehydroalanines: an in vitro comparative study. Arch Biochem Biophys 1989; 273:339–46.
32. Buc Calderon P, Roberfroid M. AD-5, a dehydroalanine derivative, decreases the amount of reactive oxygen species formed during nitrofurantoin microsomal metabolism. Life Sci 1990; 46:207–15.
33. Praet M, Buc Calderon P, Pollakis G, Roberfroid M, Ruysschaert JM. A new class of free radical scavengers reducing adriamycin mitochondrial toxicity. Biochem Pharmacol 1988; 37:4617–22.
34. Buc Calderon P, Roberfroid M. Increase in the survival time of mice exposed to ionizing radiation by a new class of free radical scavengers. Experientia 1990; 46:708–10.
35. Buc Calderon P, Defresne MP, Barvais C, Roberfroid M. N-Acyldehydroalanines protect from radiation toxicity and inhibit radiation carcinogenesis in mice. Carcinogenesis 1989; 10:1641–4.
36. Doroshow JH, Locker GY, Myers CE. Enzymatic defenses of the mouse heart against reactive oxygen metabolism. Alteration produced by adriamycin. J Clin Invest 1980; 65:128–35.

37. Buc Calderon P, Praet M, Ruyscchaert JM, Roberfroid M. Increasing therapeutic effect and reducing toxicity of doxorubicin by N-acyldehydroalanines. Eur J Canc Clin Oncol 1989; 25:679–85.
38. Ghoshal AK, Ruschmore TH, Buc Calderon P, Roberfroid M, Farber E. Prevention by free radical scavenger AD-5 of prooxidant effects of choline deficiency. Free Rad Biol Med 1990; 8:3–7.
39. Vo TK, Fischer S, Slaga TM. Effects of N-acyl dehydroalanines on phorbolesther-elicited tumor development and other events in mouse skin. Cancer Lett 1991; 60:25–32.
40. Viehe HG, Hervens F, Roberfroid M. Acetylated enamides and pharmaceutical compositions containing them. Patent WO 84/02523.

22

Environmental Stress and Reactive Oxidants in Skin Toxicology

Jürgen Fuchs and Rainer Milbradt
Johann Wolfgang Goethe University, Frankfurt, Germany

Nanna Schürer
Heinrich Heine Universität Düsseldorf, Düsseldorf, Germany

I. INTRODUCTION

From an evolutionary viewpoint the primeval function of skin is the biophysical separation of the "milieu interieur" from the "milieu exterieur." Skin is a biological interface, with the external environment and functions as the first line of defense against noxious stimuli and xenobiotics. As such it is the first organ to come into contact with harmful stresses and serves as a barrier against microbial, physical, and chemical attack. Furthermore, skin is continuously exposed to internal and external bioaggressive stimuli. Internal stress factors may be autoimmunological cascades, autoaggressive inflammatory reactions, and various metabolic disturbances. External stress factors include electromagnetic radiation, and various toxic environmental chemicals may impair skin functions. Endogenous and exogenous pathological factors can generate reactive oxidants, such as free radicals and reactive species, as mediators. Many different forms of environmental stress conditions can impair skin functions. Harmful environmental extremes include those brought about by oxygen depletion or excess, by extremes of temperature, by water stress, by nutrient depletion or excess, and by exposure to environmental pollutants. The involvement of free radicals and reactive oxygen species in environmental stress conditions such as hyperoxia (Frank and Massaro 1980, Nohl et al. 1981, Housset et al. 1983), hypoxia (McCord 1985,

1986, Zweier et al. 1987), heat schock (Loven 1988), and metabolism of environmental pollutants and toxic chemicals (Mason 1982, Mason and Chignell 1982, Cavalier and Rogan 1985) is well documented. Skin is a potential target organ of oxidative injury, because it is continuously exposed to visible and ultraviolet irradiation and high oxygen concentrations, and it contains a variety of oxidizable structures critical for maintenance of cellular homeostasis. It is one of the largest body organs and serves as a major portal of entry for many airborne environmental pollutants, some of which are free radical-generating agents.

Oxidative stress is a condition of imbalanced pro-oxidant/antioxidant equilibrium in favor of the former and subsequent molecular damage (Sies 1991). During the course of skin evolution, diverse primary (prevention) and secondary (interception and repair) antioxidant mechanisms developed. A battery of protective systems serves to protect humans from oxidative stress, including enzymes as well as water- and lipid-soluble antioxidants. A second line of defense consists of molecular turnover and repair systems. If all these defenses fail, cell renewal is the ultimate survival mechanism. Skin appears to be endowed with all of these protective systems. If these antioxidant defense mechanisms fail, or if an increased flux of reactive oxidants from endogenous and exogenous sources exceeds the antioxidant capacity, oxidative injury will result. Reactive oxidants appear to be involved in the pathology of various environmental skin diseases, although their direct detection is difficult or impossible due to an extremely short half-life. The objective of this chapter is to discuss aspects of the free radical pathology of selected environmental skin diseases.

II. DERMATOTOXICOLOGY

Adverse cutaneous effects of environmental agents, in particular chemicals, can in some cases be attributed to pro-oxidative properties of the compound. Many chemically reactive species can wreak havoc with living cells, but free radicals and reactive oxygen species are especially damaging because of their unique chemistry, e.g., their ability to participate in chain reactions, which, once initiated, have the potential to spread rapidly among cellular membranes. Free radical-mediated reactions have been discussed as one of the basic mechanisms of drug toxicity (Roberfroid et al. 1987), and various xenobiotics are known to induce lipid peroxidation in vitro and in vivo (Plaa 1976, Kappus and Sies 1981). Redox cycling is a common mechanism of chemical-induced cytotoxicity. Various endogenous and exogenous compounds exert toxic effects via oxygen reduction. These compounds are first enzymatically reduced before they activate oxygen by electron transfer, resulting in formation of reactive oxygen species and an oxidized drug. Various enzymes, e.g., cytochrome P-450 reductase, unspecifically maintain the redox cycle by reducing the oxidized drug. As a consequence, cellular reducing equivalents such as NADPH and NADH are diminished and steady-state concentration of reactive oxygen species is elevated. Drugs capable of redox cycling are azo-, nitro-, and nitroxide compounds, thiols/disulfides, metal chelates, quinones/hydroquinones, and bipyridyls.

The role of free radicals and reactive oxygen species in dermatotoxicology has been neglected and deserves more attention. The study of cutaneous toxicology has important health implications (Marzulli and Maibach 1983). Occupational dermatoses include irritant and allergic contact dermatitis, skin cancer, chemical leukoderma, and chemical-induced comedogenesis as well as connective tissue disorders (Mathias 1988). The most common reactions to xenobiotics absorbed through the skin are inflammatory reactions to allergens, irritants, and photosensitizers. It seems feasible to suggest that drug-induced peroxidation of cellular components may contribute significantly to toxic reactions of certain substances in the skin. Skin can be a target organ for toxic effects of systemic agents (Shupack 1976) as well as for topically applied substances. Cutaneous metabolism via cytochrome P-450, peroxidases, and flavoprotein dehydrogenases can lead to bioactivation of xenobiotics to pro-oxidant drugs and induce lipid peroxidation.

III. IRRITANT CONTACT DERMATITIS

Mediators of inflammation comprise a heterogeneous class of molecules, such as cytokines, lymphokines, prostanoid and lipid metabolites, kinines, complement, and coagulation factor–derived products. In addition, reactive oxidants are important mediators of inflammation (Del Maestro et al. 1980, Goerz and Merk 1983, Merk 1987, Black 1989, Kapp 1990). Noxious physicochemical stimuli, e.g., ultraviolet light, and pro-oxidant chemicals can directly produce reactive oxidants. Intradermal injections of xanthine oxidase/hypoxanthine–generating superoxide anion radicals (Ohmori 1978, Yoshioka et al. 1987) or glucose oxidase producing hydrogen peroxide (Schalwijk et al. 1986, Trenam et al. 1991, 1992) cause skin inflammation and tissue infiltration with inflammatory cells. Reactive oxidants may also be of endogenous origin, e.g., generated by inflammatory cells or increased cellular respiration. Reactive oxidants cause tissue injury by reaction with and modification of all basic molecular structures, e.g., lipids, proteins, carbohydrates, and nucleic acids. Lipid peroxidation is the most prominent reaction, because skin lipids are readily peroxidized and serve important physiological functions.

A. Skin Lipids

From a functional point of view epidermal lipids are essential for the cutaneous barrier to water loss (Elias and Menon 1991) and modulate desquamation and cohesion of the stratum corneum (Williams 1991), yet they are also a major energy source for the viable epidermal cell, represent a precursor for prostaglandin and leukotriene synthesis (Ruzicka 1988), contribute to the epidermal antimicrobial activity, and participate in metabolism of xenobiotics. Abnormalities in epidermal lipid content and/or metabolism are associated with increased transepidermal water loss as observed in essential fatty acid deficiency (Burr and Burr 1929) and/or disturbances of cornification with the clinical picture of ichthyosis (Williams and Elias 1987).

Skin lipids can be artifically divided into lipids of the epidermis, epidermal appendages (hair, nail), sebaceous glands, and subcutaneous tissue. Human skin surface lipids emanate from two sources: the keratinizing epidermis and the sebaceous gland (Downing et al. 1987). The composition of lipids changes markedly during the apical migration through successive epidermal layers (Kooyman 1932, Yardley 1983). In the basal cell layer of the epidermis, phospholipids account for 50% by weight of the total lipid. In contrast, about 40% of the lipid remaining in the cornified layer consists of sphingolipids, with the remainder accounted for by other neutral lipids (Lampe et al. 1983). These sphingolipids consist chiefly of glucosyl ceramides and ceramides. Of these, glucosyl ceramide A and ceramide A consist of a sphingosine base with an amide-linked long chain (C35), nonhydroxy and α-hydroxy acid with an additional ω-esterified nonhydroxy acid, primarily the essential fatty acid linoleic acid (C18:2) (Wertz and Downing 1984). This dienoic, unsaturated fatty acid has been suggested to function as a molecular rivet between adjacent acyl ceramides and acylglucosly ceramides; all other ceramides are highly saturated (Abraham et al. 1988). Sebaceous gland fatty acids consist of saturated and unsaturated, branched and unbranched structures with chain lengths ranging from 6 to 28 carbons (Nikolaides and Apon 1977). Free fatty acids may comprise up to 20% of the sebum lipid mixture in humans, and about 50% of the sebum fatty acids are unsaturated, providing substantial substrate for lipid peroxidation processes. In humans, surface wax esters and squalenes originate uniquely in the sebaceous glands (Greene et al. 1970). In contrast to most mammalian skin, large amounts of squalene, a sterol compound with six unsaturated double bonds, accumulate in human skin. Squalene represents about 10–15% of total skin surface lipids. Since 25% of the total skin surface lipids are unsaturated, physiological presence of lipid peroxidation products at the epidermal surface should be suspected. Indeed, in healthy human skin lipid peroxidation products are detected (Sugiura et al. 1985, 1986a, b).

B. Lipid Peroxidation

Lipid peroxidation is one of the molecular consequences of free radical reactions and oxidative stress in skin. Polyunsaturated fatty acids are readily oxidized in vitro. Lipid autoxidation occurs slowly in vivo and is a strictly controlled process. However, an uncontrolled and rapid peroxidation of lipids may occur in vivo if a high flux of lipid alkyl radicals formed, e.g., by hydrogen abstraction from the lipid by a reactive oxygen radical, overcomes the endogenous defense systems. Highly reactive products of lipid peroxidation, such as lipid alkyl radicals (R$^{\cdot}$), peroxyl radicals (R-O-O·) and alkoxyl radicals (R-O·), may damage all basic biomolecules in their ultimate vicinity. Due to their relative stability, lipid hydroperoxides (R-O-O-H) and their degradation products, e.g., malondialdehyde, 2-alkenals, and 4-hydroxy-alkenals, can damage cells and tissues at more distant sites not directly exposed to lipid peroxidation. Lipid peroxidation may be induced by physiological and pathophysiological reactions. Nonenzymatic peroxidation of lipids can be triggered by heat, ultraviolet and ionizing irradiation, redox active metal ions, redox cycling drugs, charge transfer complexes, and free radical initiators. Enzymatic peroxidation may involve the neutrophil plasma membrane-associated NADPH dehydrogenase, microsomal NADPH-dependent cytochrome P-450 reductase, and cyclooxygenase/lipoxygenase.

Products of lipid peroxidation are present in normal, healthy human skin and may play important physiological roles, such as in regulation of prostaglandin synthesis, chemoattraction of leukocytes, and skin antimicrobial activity. Skin content of lipid peroxidation products is significantly influenced by nutritional factors: lipid peroxidation products in rat skin increase when a vitamin E–deficient diet is fed and decrease when the diet is deficient in essential fatty acids (Glavind and Christensen 1967). The epidermal content of lipid peroxidation products is also influenced by the cell turnover rate—a high mitotic activity is thought to correlate with low levels of lipid peroxidation (Logani et al. 1982). Lipid peroxidation has been suggested to play a significant role in basic pathological processes in humans such as aging, carcinogenesis, inflammation, and xenobiotic toxicity (Yagi 1982).

C. Skin Irritation by Lipid Peroxidation Products

Systemic toxicity of various lipid peroxidation product has been documented (Horgan et al. 1957, Holman and Greenberg 1960, Andrews 1960), e.g., in mice after intraperitoneal administration, LD_{50} values varying from 4 to 1000 μmol peroxide/mouse are reported (Horgan et al. 1957). Organic hydroperoxides and lipid peroxidation products also exert toxic effects in mammalian skin (Tanaka 1979, Ogura 1981, Ohsaawa et al. 1984). The high eczematogenic activity of a certain turpentine is attributed to its high hydroperoxide content of d3-carene (Hellerström et al. 1955). Peroxidized squalene causes strong irritation of stripped human skin and growth inhibition of human keratinocytes in culture (Yamamoto et al. 1982). It is thought that epidermal damage due to epicutaneous application of lipid peroxidation products is mediated by propagation of lipid peroxidation in situ (Hayakawa 1971, Yagi 1987). In mice, subcutaneous administration of an aqueous extract of ultraviolet-irradiated linoleic acid induces necrosis of the epidermis and dermis followed by inflammatory response and disruption of collagen and elastic fibers (Waravdekar et al. 1965).

D. Anticancer and Alkylating Agents

The charge transfer mechanism for generation of reactive oxygen species is suggested as a unifying concept for the mechanism of action of anticancer agents, their toxicity (Kovacic et al. 1986), as well as for carcinogenesis by alkylating agents (Kovacic and Crawford 1986). If an electron donor interacts with an electron acceptor, a complex may form in which an electronic charge passes from the donor to the acceptor. Such a complex is called a charge-transfer complex. The "oxyradical charge transfer theory" of pharmacodynamic/toxic actions of pro-oxidant drugs entails conversion of a target biomolecule by the drug to a conducting salt, which can catalytically generate reactive oxygen species via electron transfer, and lipid peroxidation reactions may subsequently occur (Ames et al. 1986). Reactive oxygen species can be produced by charge transfer reaction from agents such as quinones, metal complexes, peroxides, and iminium-type ions. Particularly in the case of iminium ions, the positive charge enhances electron abstraction from donor molecules (Ames et al. 1985).

Many reports about skin toxicity of drugs that may induce lipid peroxidation in vivo are related to anticancer agents. Skin necrosis following inadvertant extravasation of redox-cycling antineoplastic drugs is a serious complication in cancer chemotherapy—the pathogenesis is based on a direct toxic effect and not on sensitization. Soft tissue injuries by the anticancer drug adriamycin (doxorubicin) (Svingen et al. 1981, Nobbs and Barr 1983, Daugherty and Khurana 1985), known to cause lipid peroxidation in vitro, are inhibited in vivo by antioxidants. Adriamycin is reduced by several flavin-containing reductases to adriamycin semiquinone, which subsequently reduces oxygen to reactive oxygen species.

E. Chemical Warfare Agents

Animals and plants use toxic chemicals for defense and to capture prey. Humans have used chemicals in warfare for centuries. Although the use of chemical weapons is outlawed by the Geneva protocol of 1925, they have been used on several occasions since. Chemical weapons can be characterized as: 1) lung irritants (chlorine, phosgene), 2) blood gases (hydrogen cyanide, cyanogene chloride), 3) vesicants (nitrogen and sulfur mustards, lewisite) 4) nerve agents (e.g., organophosphates), 5) vomiting gases (e.g., diphenylchlorarsine), 6) tear gases (e.g., chloroacetophenone, 2-chlorobenzalmalononitril), and 7) incapacitating agents (e.g., chinuklidinyl-benzilate). Only vesicants and tear gases usually cause dermatological problems. Vesicants (skin irritants) are alkylating agents (e.g., sulfur mustard, nitrogen mustard) or organic arsenicals (e.g., lewisite). An historically interesting description, characterization, and overview of chemical weapons was given by Gillert (1944). The chemistry of nitrogen and sulfur mustards and the biological effects on skin were recently described in detail (Krüger 1991a, b). Most of the toxic effects of sulfur and nitrogen mustards are related to DNA alkylation; molecular damage may, however, also occur via free radical–initiated processes. The "oxyradical charge transfer mechanism" (Ames 1986) may lead to increased concentrations of reactive oxygen species and contribute to the pathogenesis of irritant contact dermatitis in humans caused by alkylating agents like 2-bis-chloroethyl sulfide (Requena et al. 1988). Lipid peroxidation is observed in experimental nitrogen and sulfur mustard posioning (Elsayed et al. 1989a, b, 1991). Mustard gas causes lipid peroxidation and depletion of ATP in skin (Somani and Babu 1989). It was suggested that antioxidants might be of therapeutic benefit in patients suffering from chemical burns caused by these kind of agents. The antioxidant tocopherol (Vojvodic et al. 1985) and thiol compounds (Wilems 1989) provide some protection against sulfur and nitrogen mustard poisoning. *N*-Acetylcysteine was used in therapy of skin mustard gas poisoning in victims of the Iran-Iraq war. It was, however, suggested that there are no specific and very effective antidotes for sulfur and mustard gas because the chemical reaction with biological tissue is so unspecific and rapid as to be irreversible for practical purposes (Somani and Babu 1989, Carnes and Watson 1989).

IV. ALLERGIC CONTACT DERMATITIS

Poison ivy and poison oak (*Toxicodendron* spp.) dermatitis are the most common causes of allergic contact dermatitis to plants in humans. The induction of allergic contact dermatitis to poison ivy/oak is mediated by alkyl catechols (urushiols). The molecular mechanism of conversion of alkyl catechols (prohaptene) to protein-reactive species (haptene) involves oxidative activation and formation of a protein-reactive quinone (Byck and Dawson 1968, Dupuis 1979). Recently, however, it was suggested that free radicals and not reactive quinones are the haptenic species derived from alkyl catechols (Schmidt et al. 1990). Furthermore, formation of haptenic free radical species and redox cycling processes were implicated in the molecular mechanism of contact allergy to dinitrofluorobenzene and related halonitrobenzenes, alantolactones, and isoalantolactones (Schmidt and Chung 1990). Schmidt suggested that free rapical species are presumably the ultimate skin haptenes in the case of dinitrohalobenzenes (with the exception of the fluoro derivative). Mouse skin exposed to dinitrohalobenzenes undergoes a classical oxidative stress response by decreasing glutathione and elevating glutathione disulfide (Schmidt and Chung 1990). Recently a possible role for superoxide anion production in the pathogenesis of nickel and chromate contact dermatitis was suggested (Sharkey et al. 1991).

A. Histamine Release by Reactive Oxygen Species

Increased buffer oxygen tension (Damas et al. 1981) induces the release of histamine from isolated peritoneal mast cells. Enhanced basal release of histamine from isolated perfused rat lungs exposed to 100% oxygen is suggested to be the result of free radical-induced peroxidation of the lung mast cell membranes (Januszkiewicz and Faiman 1984, Januszkiewicz et al. 1986). Chemically generated reactive oxygen species, presumably hydrogen peroxide, induce the release of histamine from isolated peritoneal mast cells (Ohmori et al. 1979). It is postulated that reaction of hydrogen peroxide with some cellular components may produce some changes leading to histamine release, however, it remains to be elucidated whether hydrogen peroxide directly functions as a releasing stimulus. Hydrogen peroxide-induced release is not due to unspecific lysis of the cell. It requires the presence of calcium ions and is enhanced by deuterium oxide (Ohmori et al. 1980), indicating involvement of singlet oxygen. The lifetime of singlet oxygen is increased severalfold in deuterium oxide (Ito 1978). Irradiation of mast cells in the presence of pheomelanin, i.e., photochemical generation of reactive oxidants, also releases large amounts of histamine (Ranadive et al. 1986).

B. Histamine Release by Lipid Peroxidation Products

A role of lipid peroxidation products in histamine-induced skin reactions has been suggested. Histamine release of mast cells is triggered by oxidation products of linoleic (Mannaioni and Masini 1988) and arachidonic acid. In bronchial secretion of young bronchial asthma patients (5–14 years old), increased levels of hydroperoxides and malondialdehyde can be detected in the acute phase of the disease, whereas products of lipid peroxidation are decreased in the remission phase (Krylov et al. 1984). In contrast to oxidation products of arachidonic acid, histamine release induced by linoleic acid oxidation products is cytotoxic in nature. Reactive oxidants could be mediators in nonimmunological triggering (pseudoallergic reaction) of histamine release from mast cells and basophils.

C. Drug Intolerance and Allergies

A role of free radicals in the pathogenesis of intolerance reactions to aspirin has been discussed (Moneret-Vautrin 1985, Moneret-Vautrin 1986), and it was pointed out that high oral doses of the antioxidant ascorbate possess antiallergic activity in humans (Cathcart 1986). Indeed, the antioxidant vitamins E and C ameliorate pseudoallergic reactions, e.g., aspirin-induced asthma and Quincke's edema, in some patients (Moneret-Vautrin 1985, Moneret-Vautrin et al. 1986). A pathophysiological role of reactive oxygen species in airway hyperreactivity was recently discussed (Doelman 1991).

Current epidemiological data from Europe and Japan indicate an increase in allergies, such as extrinsic asthma, allergic rhinitis, atopic eczema, and allergic contact dermatitis. An excellent article on the increased incidence of allergies and its correlation with environmental pollution is by Kunz et al. (1991). Japanese investigators revealed a relationship between the prevalence of mountain cedar pollinosis and exposure to particulate air pollutants (diesel exhaust) (Ishizaki et al. 1987). A similar correlation was established between industrial air pollution and allergies of the respiratory tract and polinosis, sensitization to house dust mite. Air pollutants directly increase production of IgE and may also bind to pollen and enhance their allergenicity. Reactive oxygen species (ozone and peroxides), free radicals (nitrous oxide), and free radical precursors (sulfur dioxide, diesel soot particles) are among the hazardous air pollutants (Hippeli and Elstner 1991). It seems feasible to suggest that these species may readily convert nonallergenic compounds into reactive haptenes, and thus contribute to the increased incidence of allergies.

V. PHOTOSENSITIZATION

Photosensitization is a cutaneous reaction to electromagnetic irradiation. The reaction is distinct from normal photodermatitis by an amplification of the physiological response due to endogenous or exogenous chemical substances. Light or substance alone evokes no abnormal response, but the two in combination result in an additive or overadditive skin reaction. Several photosensitized reactions are oxygen dependent, indicating involvement of reactive oxygen species.

A. Coal Tar Products

Tar preparations, in particular coal tar, have been used in dermatological therapy for decades, and it is widely accepted that coal tar causes phototoxic reactions. The photosensitizing components of coal tar in humans are anthracene, pyrene, benzo[*a*]anthracene, fluoranthrene, and acridine. The metastable triplet state of the polyaromatic hydrocarbons is probably involved in the sensitizing reaction. It has been suggested that singlet oxygen participates significantly in the phototoxicity of tar products (Kochevar et al. 1982, Joshi and Pathak 1983).

B. Organic Dyes

Toward the end of the nineteenth century it was discovered that the microbial toxicity of acridine dyes depends on the intensity of the sunlight in the laboratory (Raab 1900). Raab also reported that eosin can cause a photosensitizing effect and noted that oxygen was involved in the photosensitivity reaction. Eosin was probably the first photosensitizing drug administered orally in humans. A French neurologist noted in 1900 that epilepsy patients treated with eosin responded with a dermatitis in light-exposed body areas. Eosin derivatives are suggested to cause photosensitization via formation of singlet oxygen (Pooler 1989). The mechanism of eosin-mediated generation of reactive oxygen species may involve triplet eosin and/or resonance stabilized eosin radicals. Other triphenylmethane dyes such as gentian violet, fuchsin, and brilliant green may also cause photosensitivity and phototoxic reactions via formation of triphenylmethane and triphenylmethylperoxyl radicals or the triplet state. Superoxide anion and hydroxyl radicals are the molecular agents of the phototoxicity mediated by certain acridine dyes (acriflavin, ethacridinlactate) (Martin and Logsdon 1987) and by the food and drug additive tartrazine (Merville et al. 1984).

C. Drugs

A variety of drugs may cause photosensitivity reactions in skin, and different mechanisms are involved in the pathogenesis of photosensitivity reactions. Drug-induced photosensitivity involves phototoxic and photoallergic reactions. Phototoxic reactions occur most frequently and are nonimmunologic. Photoallergic reactions are defined as acquired, delayed hypersensitivity dermatitis and are clinically rather uncommon. They involve photochemical modification of the haptene molecule to a reactive species that readily reacts with a macromolecule to yield an antigen. Phototoxic reactions usually occur with the first exposure to a photosensitizing substance and cause an acute, sunburnlike dermatitis. Photoallergy develops after an incubation period of about 10 days and is characterized by an urticarial, papular, or excematoidlike morphology. Formation of free radicals and reactive oxygen species may be involved in both phototoxicity and photoallergy. Reactive oxidants are involved in the pathophysiology of photosensitization induced by the xenobiotics listed in the following table:

Table 1 Reactive Oxidant Involvement in the Pathophysiology of Photosensitization Induced by Xenobiotics

Substrate	Type of photosensitization	Ref.
Penflutizide	phototoxic	Hayakawa et al. 1987
Sulfanilamide	phototoxic	Matsuo et al. 1986
	photoallergic	Chignell et al. 1980
		Chignell et al. 1981
Chlorpromazine	phototoxic	Motten and Chignell 1983
	photoallergic	Chignell et al. 1985
		Motten et al. 1985
		Schoonderwoerd et al. 1989
		Clare et al. 1947
Piroxicam	phototoxic	Buettner et al. 1986
	photoallergic	Kochevar et al. 1986
Tetracycline	phototoxic	Western et al. 1987
		Li et al. 1987 a,b
Anthracene	phototoxic	Glette and Sandberg 1986
Benzo[a]pyrene		Joshi and Pathak 1984
Pyrene		Kochevar et al. 1982
Fluoranthrene		
Acridine		
Terachlorosalicylanilide	photoallergic	Jenkins 1984
		Chignell 1989

VI. CHEMICAL-INDUCED SKIN SCLEROSIS

Sclerosis is characterized by excessive deposition of connective tissue in an mesenchymal organ, and fibrosis is defined by pronounced accumulation of fibroblasts in connective tissue. Frequently, however, the terms sclerosis and fibrosis are used synonymously.

A. Organ Fibrosis

Drug-induced organ fibrosis is described for such chemically heterogeneous compounds such as oxygen (retrolental fibroblasia), phenytoin (gingival hypertrophy), methysergide (retroperitoneal fibrosis), nitrofurantoin, amiodarone, bleomycin (lung and skin fibrosis), and practolol (sclerosing peritonitis). The herbicide paraquat serves as a model substance for producing fibrosis of the lung.

Acute paraquat toxicity results in severe alveolitis, whereas chronic toxicity is characterized by proliferative lung fibrosis. Specific toxicity to the lung by paraquat is thought to be caused by preferential uptake of the drug into alveolar epithelial cells (Smith 1985). It is not known whether paraquat-induced lung fibrosis is related to superoxide radical production by paraquat, by production of reactive oxidants by the subsequently invading inflammatory cells, or is just part of a repair process. It was, however, suggested that the mechanism of paraquat-mediated increase in lung collagen synthesis is mediated by superoxide radicals because paraquat-induced increase in collagen production in organ cultures is obviated by superoxide dismutase treatment (Hussain and Bhatnagar 1979).

The bipyridyl cation paraquat is rapidly reduced by flavin-containing oxidoreductases to the paraquat radical, which autooxidizes and yields the superoxide anion radical. By reaction of the paraquat radical with oxygen, oxidized paraquat is regenerated, thereby sustaining a redox cycling process.

The common pathophysiological mechanism of bleomycin, paraquat, nitrofurantoin, and ultraviolet/ionizing irradiation–mediated organ fibrosis could be formation of reactive oxygen species. A similar mechanism may be suggested for chemical-induced skin sclerosis. This hypothesis is strengthened by the findings that collagen synthesis is stimulated by elevated steady-state concentration of reactive oxygen species, e.g., in hyperoxic peridontal tissue culture of mice (Yen et al. 1979), hyperoxic rat lung organ culture (Bhatnagar et al. 1978), and in rat lung by ozone (Last and Greenberg 1980). The experimental finding that lipid peroxidation stimulates collagen synthesis in human dermal fibroblasts suggests an explanation for the fibrosis occurring in oxidant-induced tissue injury (Geesin et al. 1989). Although the exact molecular mechanisms of collagen gene expression and collagen synthesis are not yet known in detail, reactive oxidants are among the several mediators that affect these processes. It is important to emphasize, however, that fibroblast function and collagen synthesis are also regulated by various other factors (Krieg and Aumailley 1990) and that fibroblasts may directly be affected by drugs causing organ fibrosis. The etiopathology of various forms of fibrotic skin disease is largely unknown, and cytokines such as transforming growth factor and interleukin-1 are thought to play a significant role in the development of some of the fibrotic skin diseases (Uito and Jimenez 1990, Krieg and Aumailley 1990). Antioxidants and free radical scavengers were proposed to be of therapeutic value in treatment of fibrotic processes (Borel 1983). Tocopherol and superoxide dismutase are used clinically with varying success rates in treatment of various fibrotic skin diseases such as induratio penis plastica, keloids, and hypertrophic scars. Allopurinol, a xanthine oxidase inhibitor, was used with success in treatment of patients with Dupuytren's contracture (Murrell et al. 1990b).

In conclusion, it remains to be established whether reactive oxidants are an epiphenomenon or a significant determinant in the pathophysiology of chemical-induced skin sclerosis.

B. Skin Sclerosis

Fibrotic skin diseases comprise a large group of dermatological disorders. Sclerosis of the skin occurs in the clinically heterogeneous group of connective tissue diseases. Furthermore, sclerosis is frequently seen in a variety of dermatopathological conditions distinct from autoimmune diseases: localized scleroderma, lichen sclerosus, hypertrophic scars, keloids, dermatosclerosis due to venous insufficiency, phenylketonuria, porphyria cutanea tarda, carcinoid syndrome, and juvenile-onset diabetes mellitus.

1. Chemical-Induced Scleroderma

Several years ago it was pointed out that reversible, sclerodermalike skin lesions can be induced in humans by a variety of chemicals (Korting and Holzmann 1967). A large clinical data base on drug- and environmental agent-induced skin sclerosis has since been amassed.

Drugs responsible for skin sclerosis include the following:

Drug	Ref.
Bleomycin	Finch et al. 1980
	Bork and Korting 1983
Carbidopa, L-5-hydroxytryptophan	Sternberg et al. 1980
8-Methoxypsoralen/UVA therapy	Thürlimann and Harms 1982
Cocaine	Trozak and Gould 1984
Pentazocine	Palestine et al. 1980
	Rodnan and Medsger 1981
Vitamin K_1	Texier et al. 1975

Environmental agents known to induce skin sclerosis include:

Agent	Ref.
Epoxides, epoxy hardener (bis-4-amino-3-cyclohexylmethane)	Yamakage et al. 1980
Denaturated rapeseed oil ("toxic oil" syndrome)	Phelps and Fleischmajer 1988
Organic solvents	Haustein et al. 1985
	Haustein and Ziegler 1986
Aromatic hydrocarbons (benzene, toluene, xylene)	Walder 1965, 1981
Petroleum-derived crude solvents (naphtha, trichlorethylene)	
Alicyclic and aliphatic hydrocarbons, chlorinated hydrocarbons (naphtha, *n*-hexane, hexachloroethane)	Czirjak et al. 1987
	Yamakage and Ishikawa 1982

Agent	Ref.
Perchlorethylene	Sparrow 1977
Trichlorethylene	Reinl 1957
	Saihan et al. 1978
	Schirren 1971
	Flindt-Hansen and Isager 1987
Trichlorethane	Flindt-Hansen and Isager 1987
Pesticides (malathion, parathion, chlordane, heptochlor)	Rowell 1986
Vinyl chloride	Maricq 1985
	Veltman 1980

2. *Genetic Predisposition*

Progressive systemic sclerosis serves as a model disease for studying the pathology of sclerosis. Induction of progressive systemic sclerosis requires a certain genetic predisposition in susceptible patients (Sollberg and Holzmann 1987). Genetic susceptibility is presumably an additional significant determinant in development of skin sclerosis triggered by some chemicals, e.g., halogenated hydrocarbons, because not everyone exposed to these chemicals will acquire sclerosis. Presently it is established that vinyl chloride and toxic oil–induced scleroderma are associated with genetic susceptibility. Other chemicals, e.g., bleomycin, cause skin sclerosis in rats via an obligate toxic mechanism (Mountz et al. 1983).

3. *Chemical-Induced Progressive Systemic Sclerosis*

It is clearly established that quartz can cause progressive systemic sclerosis in some susceptible patients (Haustein and Ziegler 1986). Petroleum-derived crude solvents are also reported to trigger progressive systemic scleroderma (Czirjak et al. 1987). Various aliphatic hydrocarbons (*n*-hexane, naphthane) cause generalized morphea, sclerodactyly, Raynaud's phenomenon, esophageal dysfunction, and lung fibrosis (Yamakage and Ishikawa 1982). Trichlorethylene and trichlorethane are also reported to cause PSS-like symptoms according to the American Rheumatology Association (Flindt-Hansen and Isager 1987).

4. *Quartz*

In 1914 Byrom Bramwell, a Scottish physician, reported nine patients with diffuse scleroderma, five working as stone masons. Subsequently, increasing evidence linked the prevalence of progressive systemic sclerosis with silicosis, and it is suggested that silicosis may be a predisposing factor in the pathogenesis of this disease (Erasmus 1957, Rodnan et al. 1967, Ziegler et al. 1982, Haustein and Ziegler 1986, Ziegler et al. 1986). Silica-induced skin sclerosis fulfills the criteria of the American Rheumatism Association to establish a diagnosis of progressive systemic sclerosis. Progressive systemic sclerosis induced by silica is distinct from drug-induced sclerodermalike lesions because it does not resolve when exposition to the causative factors is stopped.

5. Reactive Oxygen Species and Quartz

Quartz particles activate macrophages, which release interleukin-1, and subsequently fibroblast collagen production is stimulated (Hepplestone and Styles 1967). Interleukin-1 also affects T-helper lymphocytes, which then release interleukin-2, finally stimulating B lymphocytes. Reactive oxygen species may be involved in this type of cell activation (Meier et al. 1989). Quartz-related free radicals may participate in the biochemical mechanisms involved in acute silicosis and presumably in quartz-induced PSS. Fresh quartz dust generates hydroxyl and possibly superoxide anion radicals in a cell-free aqueous medium (Dalal et al. 1989). In addition, phagocytosis-derived reactive oxidants may contribute to the pathophysiological process of crystal-induced inflammation (McCarty 1979). Silicates can activate the respiratory burst of leukocytes (Montsko et al. 1984). Phagocytosis of quartz particles by macrophages, subsequent macrophage cell death, and release of quartz particles and continuation of this process (Löblich 1961) can lead to a rise in steady-state levels of reactive oxygen species. Silicates may also stimulate fibroblast metabolism via immunological mechanisms (Ziegler et al. 1986). In silica-induced skin sclerosis, macrophage-derived lymphokines and monokines are suggested to activate fibroblast collagen and glycosaminoglycan metabolism (Haustein et al. 1990).

6. Silicon-Induced Skin Sclerosis

Silicon-induced skin sclerosis is described in patients carrying silicone implants (Kumagai et al. 1979, Kondo et al. 1985, Spiera 1988). In some patients with silicon-induced skin sclerosis, the disease closely resembles initial cases of progressive systemic sclerosis. The compound may act as an immunological adjuvans and thereby trigger autoimmune diseases (Haustein and Ziegler 1986). It has been demonstrated that dimethyl polysiloxane is released from the silicone implant into the tissue and macrophages. Although cyrstalline silica was not detected, it can be speculated that the pathophysiology of silicone-induced skin sclerosis closely resembles the pathophysiology of quartz-induced skin sclerosis. The fact that paraffin and other fully saturated alkanes (Yamakage and Ishikawa 1982) can induce skin sclerosis argues against a common mechanism of xenobiotic-induced skin sclerosis, e.g., free radical–induced sclerosis. However, it cannot be ruled out that paraffin or alkanes contain trace amounts of unsaturated, peroxidizable compounds (alkenes) and lipid peroxidation products, which trigger free radical reactions in susceptible patients. Industrial hydrocarbons are frequently contaminated with chlorinated hydrocarbons (Walder 1983).

C. Toxic Oil Syndrome

In 1981 reports from Spain described rapeseed oil contaminated by aniline derivatives used for preparing food. The early stage of toxic oil syndrome was characterized by vasculitis in practically every organ clinically affected (lung, GI tract, central nervous system). In the late stage, sclerodermalike skin lesions, Raynaud's phenomenon, sicca syndrome, dysphagia, myositis, and arthritis were observed. Mainly women with HLA DR3 and DR4 developed sclerodermalike skin lesion about 5–6 months after the intoxication, indicating again a genetic predisposition (Haustein and Ziegler 1986). Free radicals and lipid peroxidation caused by oil denaturants were suspected to participate in the pathogenesis of endothelial damage and skin fibrosis observed in the toxic oil syndrome (Tabuenca 1981, Fournier et al. 1982, Rush et al. 1984, Spurzem and Lockey 1984, Gilsanz et al. 1984, Haustein and Ziegler 1986). The aniline derivatives probably reacted with fatty acids and formed reactive oxidants, damaging endothelials cells and capillaries (Rush et al. 1984).

D. Halogenated Aliphatic Hydrocarbons

Halogenated aliphatic hydrocarbons are ubiquitous environmental pollutants. Toxic metabolites of chlorinated aliphatic hydrocarbons, e.g., trichlorethanol and/or metabolites of the dehalogenation process, were thought to be involved in the mechanism of halogenated hydrocarbon skin sclerosis (Walder 1983).

Toxicity of halogenated aliphatic hydrocarbons is mediated by reactive metbolites formed during dehalogenation. The substances are subject to single electron reduction by the mixed function oxidase system (reductive dehalogenation) to generate reactive alkyl and alkylperoxyl radicals. The greater the degree of chlorination, the greater the ease of conversion to a reactive derivative and the greater its toxicity. Carbon tetrachloride is readily converted by microsomal cytochrome P-450 to the trichloromethyl radical, which in the presence of oxygen readily forms the trichloromethane-peroxyl radical, which readily initiates lipid peroxidation in biological systems. Reductive metabolism of halogenated hydrocarbons can lead to formation of lipid peroxidation products in cells and tissues. Free radical derivatives of halogenated aliphatic hydrocarbons are strongly alkylative and readily deplete tissue glutathione stores, causing a pro-oxidant cellular environment.

It is important to point out that the metabolic pathways of saturated and unsaturated halogenated aliphatic hydrocarbons are different. Generally, biotransformation of saturated halogenated hydrocarbons results in the formation of reactive free radicals. Chloroform, methylene chloride, and methyl chloride are less toxic than carbon tetrachloride, and each has a different metabolic degradation pathway. Metabolites of chloroform are trichloromethanol and phosgene; of methylene chloride, dichloromethanol and formyl chloride; and of methyl chloride, chlormethanol, formaldehyde, and formic acid. Free radicals presumably do not play a major role in the mechanism of toxicity. Chlorinated ethane derivatives metabolically form free radicals, and the tendency to free radical formation increases with the degree of halogenation, reaching a maximum at tetrachloroethane. In contrast to chlorinated alkanes, chlorinated alkenes are metabolized to highly reactive (nucleophilic) epoxides, which are further converted to acyl chlorides or aldehydes. Tetrachlorethylene is the most inert compound of this class, trichloroethylene the most reactive, and 1,1-dichloroethylene (vinylidene chloride) has intermediate reactivity.

E. Vinyl Chloride Disease

Vinyl chloride disease is characterized by pulmonary dysfunction, hepatic portal fibrosis, skin sclerosis, Raynaud's phenomenon, endothelial damage, and immunological disturbances. After the first reports from the Soviet Union in 1949, the disease was described in detail as occupational acro-osteolysis (Cordier et al. 1966, Markowitz et al. 1972). Industrial workers who clean autoclaves in which vinyl chloride is polymerized have a high incidence of skin sclerosis associated with other symptoms. Vinyl chloride monomer is bioactivated by the mixed function oxidase to form chloroethylene oxide and acyclic chloroethyl dioxide (Ward et al. 1976, Infante 1981). These reactive metabolites have been suggested to initiate lipid peroxidation and form covalent adducts, leading to pathological sequelae. Involvement of vinyl chloride-derived epoxides, peroxides, and free radicals has been implicated in the pathogenesis of vinyl chloride-caused skin sclerosis (Haustein and Ziegler 1986). Vinyl chloride disease may be a representative model of free radical-mediated inflammatory/immune disease resulting from exposure to halocarbons (Levine and Reinhardt 1983). Vinyl chloride-induced skin sclerosis is reversible after exposure has stopped (Veltman 1980) and is associated with genetic susceptibility of the patient, as shown by an increase in HLA DR5 and B8 and DR3.

VII. CHEMICAL LEUKODERMA

Skin depigmentation (leukoderma) can be caused by various chemical agents, including azelaic acid (Nazzaro-Porro 1987), the glutathione reductase inhibitor carmustine (BCNU) (Hilger et al. 1974), bleomycin (Bork 1985) [bleomycin may also cause hyperpigmentation (Perrot and Ortonne 1978, Lowitz 1975)], certain mercaptoamines (Bleehen et al. 1968), Cer(IV) oxide (Rapaport 1982), alkyl and aryl phenols, and mercurials.

A. Mercurials

Topical application of mercurials was used for centuries for skin lightening. Because of nephrotoxic side effects, this kind of depigmenting therapy is obsolete. The mechanism of mercurial-induced skin depigmentation was not known. Lipid peroxidation is observed in mercury (II) chloride toxicity (Yonaha et al. 1982) and could be related to some effect of mercury (II) chloride on thiols (Gstrauenthaler et al. 1983). Depletion of cellular glutathione and other antioxidant thiols may then render the cells more susceptible to oxidative attack. Furthermore, mercury (II) chloride inhibits cellular antioxidant systems, such as superoxide dismutase, catalase, and glutathione peroxidase (Gstrauenthaler et al. 1983). The pro-oxidant state may then cause damage to the melanocytes, whose antioxidant capacity seems to be less effective than the antioxidant defense of other skin cell types (Muglia et al. 1986, Yohn et al. 1991).

B. Alkyl and Aryl Phenols

The depigmenting effect of phenolic compounds was first noted in factory workers who wore rubber gloves containing hydroquinone monobenzylester as an antioxidant (Oliver et al. 1939, 1940). Phenolic reagents probably interfere with melanin biosynthesis by free radical–type reactions. The following phenol derivatives have been reported to cause skin depigmentation:

Phenol derivative	Ref.
p-Cresol	Shelley 1974
p-Hydroquinone	
p-Hydroquinone-monobenzylester	Spencer 1961
	Snell 1964
	Bentley-Phillips and Bayles 1975
p-t-Butyl-catechol	Gellin et al. 1970
p-Isopropyl-catechol	Bleehen et al. 1968
p-Hydroxyanisol, *p*-hydroxydiphenyl	Riley 1969a,b
1,3 di-*t*-Butyl-4-methylphenol	Vollum 1971
p-t-Amylphenol, *p,t*-butylphenol	Kahn 1970

Although other phenolic compounds, such as adrenaline and noradrenalin, are rapidly oxidized by tyrosinase in vitro, they do not cause skin depigmentation. Structure activity analysis of depigmenting hydroquinones revealed that hydroxylation in the *para* position relative to a nonpolar side chain was an optimal requirement for depigmenting activity. Depigmenting compounds with an ether link in the side chain were more effective than those with hydrocarbon chains. 4-Isopropylcatechol was found to be the most potent depigmenting agent by screening 33 catechol derivatives, quinones, mercaptoamines, and thiols (Bleehen 1968). The depigmenting effect of *m*-substituted catechols is weaker than that of *p*-substituted catechols (Kahn 1970).

C. Melanocytotoxicity of Phenolic Compounds

Several mechanisms of melanocytotoxicity of phenolic compounds, including formation of phenolic free radicals, have been discussed (Bolognia and Pawelek 1988). Tyrosinase is the critical enzyme that ultimately regulates melanogenesis (Hearing and Jimenez 1987), and the pigment cell specificity of the depigmenting chemicals can be explained by the selective uptake and metabolism of these compounds by tyrosinase. Their reactive oxidation products, e.g., semiquinone radicals, may induce cytotoxicity via induction of lipid peroxidation and redox cycling. The stability and reactivity of the semiquinone radical determines the extent of the depigmenting effect.

D. *p*-Hydroxyanisol

The cytotoxic effect of *p*-hydroxyanisol is selective towards melanocytes in mixed cultures containing keratinocytes and pigmented cells. This indicates that the melanocytotoxic effect is exerted by tyrosinase oxidation products (Riley 1969b). It was suggested that resonance stabilized semiquinone type radicals of *p*-hydroxyanisol (anisyl-semiquinone), which are formed inside the melanocytes via tyrosinase, are melanocytotoxic and thus cause selective destruction of melanocytes (Riley 1969b, Jimbow et al. 1974). Recently, however, the semiquinone radical of 4-hydroxyanisol was found to be unreactive with oxygen, indicating that superoxide anion radicals are not involved in the cytotoxic mechanism. The anisyl-3,4-quinone, which readily reacts with nucleophils, or less likely the anisyl-semiquinone radical, is considered the cytotoxic species (Riley et al. 1989).

E. Antioxidants and *p*-Hydroxyanisol Melanocytotoxicity

Treatment of guinea pig skin with *p*-hydroxyanisol results in the formation of a stable free radical species in the epidermis (Riley 1970). The lipophilic antioxidant alpha-tocopherol partially inhibits depigmentation of guinea pig skin induced by *p*-hydroxyanisol, whereas the hydrophilic antioxidant ascorbate is not protective (Riley 1969a). In a melanocyte cell culture assay, ascorbate protects melanocytes from *p*-hydroxyanisol cytotoxicity, whereas alpha-tocopherol succinate does not (Riley et al. 1975). In vitro studies using guinea pig melanocytes revealed that the melanocytotoxic action of *p*-hydroxyanisol was prevented by antioxidants such as alpha-tocopherol, ubiquinone-6, cysteine, and ascorbate. These results clearly indicate that antioxidants can prevent the melanocytotoxic effects of *p*-hydroxyanisol. This favors the hypothesis of alkyl-, aryl-catechol "free radical"-induced melanocytotoxicity.

F. Hydroquinone

As a corollary, the depigmenting effect of hydroquinone is believed to result from selective action on melanocyte metabolism and involves toxic tyrosinase-mediated oxidation products of hydroquinone (Penney et al. 1984). In the presence of a peroxidase and hydrogen peroxide, *p*-semiquinone radicals are immediately formed from hydroquinone, which persist for several hours, whereas during nonenzymatic autooxidation only transient formation of small quantities of semiquinone radicals is observed (Yamazaki et al. 1960). Flavin containing cytochrome b_5 and cytochrome *c* oxidoreductase can catalyze one-electron reduction of benzoquinone to the semiquinone radical (Iyanagi and Yamazaki 1970). The semiquinone radical may also be formed by nonezymatic coproportionation of hydroquinone and quinone in an electron transfer reaction.

VIII. CHEMICAL-INDUCED COMEDOGENESIS

Chloracne was first described by Von Bettmann in 1897 (Von Bettmann 1901). Two years later Herxheimer described acne among workers manufacturing hydrochloric acid and blamed chlorine gas for causing acne; indeed, chlorinated aromatic hydrocarbons were responsible. Until the 1940s most cases of chloracne were caused by exposure to chlorinated hydrocarbons. More recently, chlorinated herbicides and dioxins (agent orange = a 50:50 mixture of 2,4-dichlorophenoxyacetic acid and 2,3,7,8-tetrachlorodibenzo-*p*-dioxin) are common causes. One of the key events of comedo formation is increased proliferation and retention of horny cells in the infrainfundibulum of the sebaceous glands. Decreased dehiscence of horny cells may be related to changes in the intercellular cement (lipids).

Comedogenesis is defined as the process of abnormal differentiation of the follicular epithelium resulting in the formation of microcomedones. The microcomedones may evolve into noninflammatory comedos or inflammatory papules, nodules, or pustules. Comedo formation occurs in the course of various skin diseases, such as acne vulgaris, Fox-Fordyce disease, and drug eruptions. Comedos can also be caused by occupational hazards. Many kinds of topical substances, such as sebum, tar, oils, and chlorinated hydrocarbons, are known to produce acneiform eruptions in human skin. Physical agents that may cause acneigenic reactions include ultraviolet and ionizing irradiation.

A. Sebum

All of the major components of sebum are comedogenic in the rabbit ear model (Kligman 1968, Kligman and Katz 1968). Glycerides are only weakly comedogenic. Comedogenicity of free fatty acids decreases with chain length from C12 to C20. Unsaturated fatty acids are more potent comedogenes than saturated fatty acids, and the hydroperoxide of oleic acid is more comedogenic than oleic acid itself. Squalene and squalene peroxides are the most potent comedogenic compounds in humans (Mills et al. 1978, Motoyoshi 1983).

It was suggested that oxidation of squalene may induce comedogenesis and cause a microaerophilic condition in the pilosebaceous unit via reducing oxygen tension in the follicle. This may favor growth of anaerobic Propionibacteria, which synthesize porphyrins. Porphyrins catalyze oxidation of squalene, thereby self-perpetuating comedogenesis (Saint-Leger et al. 1986). The total absence of highly peroxidizable squalenes from monkey skin surface lipids can be related to the absence of acne in such animals (De Luca et al. 1989).

Peroxidized fatty acids, which cause epidermal hyperplasia in mice (Klein-Szanto and Slaga 1982), were suggested to be the species responsible for follicular hyperkeratosis and hyperplasia. However, another hypothesis of the etiopathology of comedogenesis suggests that high rates of sebum production result in low levels of linoleate in the sebum, subjecting the follicular epithelium to essential fatty acid deficiency and the characteristic follicular hyperkeratosis that results in comedo formation. The proportion of linoleic acid is markedly decreased in acne comedones (Wertz et al. 1985, Perisho et al. 1988). Deficiency of essential fatty acids containing linoleic acid in humans induces a scaly dermatosis, extensive transepidermal water loss, and epidermal hyperproliferation (Ziboh and Chapkin 1987). Suppression of sebum production by drugs elevates sebum linoleate concentration and relieves follicular hyperkeratosis (Downing et al. 1987).

It was recently reported that linoleic acid inhibits phagocytocis and generation of reactive oxidants in human neutrophils without affecting chemotaxis. The authors suggested that decreased levels of linoleate in acne comedones contribute to the inflammatory process by a failure to inhibit neutrophil chemotaxis and reactive oxidant generation (Akamatsu et al. 1990).

B. Chlorinated Hydrocarbons

The role of chemicals in inducing comedogenesis is clearly established in the case of coal tar and halogenated hydrocarbons. These compounds induce comedos in the rabbit ear model and also produce acne in humans. Chlorinated aromatic hydrocarbons frequently produce chloracne in humans (Kimmig and Schulz 1957). 2,3,7,8-Tetrachlorodibenzo-*p*-dioxin (TCDD) is comedogenic by topical contact or oral systemic administration in animals. TCDD has also been demonstrated to be carcinogenic in rodents and monkeys. The mechanism of comedo formation, however, remains obscure. In 1976, after an accident at a chemical plant in Seveso, Italy, chloracne was the most frequently cutaneous manifestation in TCDD toxicity in humans (Goldmann 1973, Caputo et al. 1988), occurring 30–60 days after exposure.

Although TCDD may affect epidermal differentiation and growth by influencing a specific receptor in epidermal basal cells, a role of increased lipid peroxidation in TCDD toxicity has been discussed (Albro et al. 1978, Sweeney et al. 1979). Furthermore, it has been revealed that TCDD induces formation of reactive oxygen species and inhibits oxidative phosphorylation in mitochondria (Nohl et al. 1989). TCDD probably induces formation of pathophysiological redox cycling at the level of mitochondrial cytochrome *b* (Nohl 1989). Lipid peroxidation induced by TCDD may contribute to comedo formation in skin. In hairless mice, vitamin A deficiency increases the sensitivity of skin to TCDD-induced toxicity (Puhvel et al. 1990). Reactive oxidants may play a role in acute and chronic TCDD toxicity (Stohs 1990), although more complex mechanisms than oxidative imbalance may contribute to TCDD comedogenicity.

IX. CARCINOGENESIS

The molecular mechanisms of carcinogenesis have been extensively studied in skin as a model system. In 1761, John Hill speculated about a relationship between mucocutaneous carcinoma of the nose and the use of snuff tobacco. A correlation between scrotal skin cancer in humans and exposure to chimney soot was established by Sir Percivall Poot in 1775. The first experimental cancer in mouse skin was produced by topical application of coal tar condensate in 1915 (Yamagiwa and Ichikawa 1915), and subsequently croton oil was found to be a very effective agent to augment the skin carcinogenicity of benzpyrene (Berenblum 1941). The terms *initiation* and *promotion* were first applied to a stepwise model for eliciting tumors of the skin of rabbit ears (Rous and Kidd 1941). A multistep carcinogenesis model has been developed for many epithelial tissues including liver, lung, stomach, pancreas, and colon and is best studied in epidermis. In mouse skin three distinct stages of carcinogenesis can be defined: initiation, promotion, and progression.

A. Initiation

Initiation is a phenomenon of gene alteration. Gene alterations can be caused directly by elecromagnetic irradiation, carcinogenic chemicals, and oncogenic viruses. Skin tumor initiation probably occurs in dark basal keratinocytes, which are keratinocyte stem cells that play an important role in the first stage of skin tumor promotion. The dark basal keratinocytes are present in large numbers in animal skin during embryogenesis, in moderate numbers in newborns, in low numbers in young adults, and in very low numbers in older adults (Klein-Szanto 1981). The initiating potential of mouse skin decreases with the age of the mouse to a point at which it is very difficult to initiate mice greater than one year of age when the number of dark basal keratinocytes are very rare.

B. Promotion

Promotion is a phenomenon of gene activation. Tumor-promoting agents such as phorbol esters, the active ingredients of croton oil, and nonpromoting agents, such as calcium ionophors, as well as wounding, increase the number of dark basal keratinocytes. Promotion itself is divided into two steps: conversion and propagation (Slaga et al. 1980).

In stage I of tumor promotion, dark basal keratinocytes increase by stimulation of existing dark cells to divide and by conversion of basal cells into dark cells. In stage II of tumor promotion, the state of increased numbers of dark basal cells is maintained by continuation of dark cell proliferation, and cellular metabolism (nucleic acid and protein synthesis) is modulated and activated. As a consequence a benign tumor, a papilloma, is formed.

C. Progression

The final stage of carcinogenesis is the progression of the transformed lesion into a malignant neoplasm (squamous cell carcinoma) involving additional genetic changes, which may arise either spontaneously or mediated by certain carcinogens.

Complete carcinogens possess both initiating and promoting properties; incomplete carcinogens have only initiating activity. The promotion step is more readily reversible than the other two steps and has its greatest role in the development and maintenance of the benign state of neoplastic development.

The molecular mechanisms of tumor promotion are not fully resolved, however, it is evident that reactive oxidants are biologically significant components in this scenario (Cerutti 1985, Kensler and Taffe 1986, Fischer et al. 1988). Oxidant stress can modify regulation of genes controlling tumor promotion and may modulate the expression of genes related to cell growth and differentiation by inducing alterations in DNA structure or by epigenetic mechanisms. Indirect evidence stems from experiments demonstrating that 1) free radicals and reactive oxygen species are initiating agents, 2) reactive oxidant-generating systems mimic the effects of tumor promoters, 3) xenobiotics that produce free radicals and reactive oxygen species are potent tumor promoters, 4) tumor promoters produce reactive oxygen species in various cells, 5) tumor promoters modulate cellular antioxidants, and 6) antioxidants inhibit tumor promotion and initiation.

1. *Reactive oxidents are tumor initiators.* It is firmly established that most ultimate carcinogenic metabolites are electrophilic and specifically bind to DNA. The metabolic activation of xenobiotics to a proximate carcinogen may involve either one- or two-electron redox reactions catalyzed by the mixed function oxidase (MFO) system (cytochrome P-450, NADPH cytochrome P-450 reductase). The MFO system also produces reactive oxidants directly, which can participate in nonenzymatic xenobiotic activation. One-electron redox reactions of procarcinogens yielding free radical intermediates may play a significant role in activation of carcinogens to DNA-binding and -damaging compounds (Trush and Kensler 1991).

2. *Reactive oxidant generating systems mimic the effects of tumor promoters.* Reactive oxygen species such as superoxide anion radical and hydrogen peroxide can act in a promotional manner that mimics the effects of the mouse skin tumor promoter 12-O-*tetra*-decanoylphorbol-13-acetate (TPA) in mouse fibroblasts (Zimmerman and Cerutti 1984). Chromosomal aberrations seen after TPA treatment of lymphocytes and keratinocytes are supposedly due to the action of clastogenic factors thought to be generated by the cells in the course of oxygen activation (Emerit and Cerutti 1981b, Emerit 1982). This factor of low molecular weight activity released from TPA-stimulated granulocytes and monocytes probably consists of lipid hydroperoxides, aldehydic compounds, and free arachidonic acid (Emerit 1982) and exerts its toxicity again via formation of reactive oxygen species. Tumor promoter–induced chromosome breaks are inhibited by antioxidants in vitro (Emerit et al. 1983).

3. *Xenobiotics that produce free radicals and reactive oxygen species are potent tumor promoters.* Skin tumorigenic properties of various hydroperoxides, acyl peroxides, peracids, and epoxides have been reported (Kotin and Falk 1963). However, hydrogen peroxide, lauryl peroxide, and benzoyl peroxide are not carcinogenic when applied topically on mouse skin (Van Duuren et al. 1963). In the two-stage carcinogenesis model, lauryl peroxide and benzoyl peroxide are tumor promoters (Klein-Szanto and Slaga 1982), but hydrogen peroxide exhibits only weak tumor-promoting potential. Repeated treatment of mouse benign skin tumors with benzoyl peroxide increase the number of cumulative carcinomas significantly (O'Connell et al.

1986). Benzoyl peroxide is an effective promoter of both papillomas and squamous cell carcinomas (Slaga et al. 1981) and is able to promote the formation of melanotic tumors in hamster skin (Schweizer et al. 1987).

Promoter activity of peroxides is ascribed to their ability to generate free radicals, although they are quite stable to unimolecular homolysis at body temperature (Pryor 1975). Hydroperoxides, however, are subject to decomposition by transition metal ions, and hematin can catalyze decomposition to form radical species (Kalayanaraman et al. 1983). *t*-Butyl-hydroperoxide and cumene hydroperoxide rapidly produce alkyl radicals when incubated with mouse keratinocytes (Taffe et al. 1986). The nature of the system that catalyzes peroxide cleavage to free radical species is unknown—the activity is probably localized in the cytosolic fraction.

4. *Tumor promoters produce reactive oxygen species in various cells.* Generation of reactive oxygen species has been proposed as an important event in mouse epidermal tumor promotion by the phorbol ester TPA, one of the active components of croton oil (Slaga et al. 1982, Copeland 1983). Mouse skin tumor promoters of the phorbol ester type stimulate the production of reactive oxygen species in neutrophils and macrophages (Witz et al. 1980, Goldstein et al. 1981, Ohsawa et al. 1984, Czerniecki et al. 1985). In addition, TPA also induces formation of reactive oxygen species in isolated murine epidermal cells (Fischer and Adams 1985). In keratinocytes of TPA-treated mouse epidermis, a 50% increase in the cellular hydrogen peroxide level is reported (Robertson et al. 1988). TPA-mediated oxidant production in mouse epidermal cells can be suppressed by many, but not all known inhibitors of mouse skin tumor promotion. The extent of the oxidant response directly correlates with the sensitivity to TPA tumor promotion among certain strains of mice.

5. *Tumor promoters modulate cellular antioxidants.* Treatment of isolated mouse epidermal cells with TPA results in a rapid and transient increase in glutathione peroxidase activity followed by a prolonged decrease in the activity. Glutathione reductase activity is depressed throughout a multidose exposure (Perchellet et al. 1985, Taffe et al. 1986). Epidermal SOD and catalase activity are reported to decrease following treatment with TPA. This suppression of activity is sustained during a 12-hour time course, but activity becomes elevated over control several days after a single exposure (Perchellet et al. 1985, Taffe et al. 1986). The treatment of adult mouse skin with TPA results in a sustained decrease in the basal levels of both basal SOD and catalase activities in the epidermis. The decline in SOD and catalase occurs within 3 hours after application, and the maximum effect is seen after 16–17 hours. The alterations in both enzymes occur against a background of enhanced protein synthesis, which indicates that the effect of TPA is selective for SOD and catalase. Damage that favors neoplastic progression can occur in TPA-

treated mouse skin due to low levels of SOD and catalase activity. Accumulation of reactive oxygen species in mouse skin with TPA treatment may follow the decrease in SOD activity (Solanki et al. 1982, Kinsella et al. 1983) accompanied by inactivation of catalase (Kono and Fridovich 1982) and an increase in glutathione disulfide (Kozumbo et al. 1983).

6. *Antioxidants inhibit tumor promotion and initiation.* Scavengers of reactive oxygen species, such as enzymic and nonenzymic antioxidants, inhibit biochemical and biological actions of tumor promoters in vitro and in vivo (Kensler and Trush 1983, Kensler et al. 1983, Kozumbo et al. 1983, Zimmerman and Cerutti 1984, Nakamura et al. 1985, Bork 1985). Some antioxidants also effectively counteract skin tumor initiation: these agents have an even greater inhibiting effect on skin tumor promotion by TPA and benzoyl peroxide. They specifically inhibit the second stage of promotion (Slaga 1984). The inhibition of polycyclic hydrocarbon tumorigenesis by antioxidants may be related to their ability to prevent the in vivo activation of hydrocarbons to carcinogenic epoxides and/or other electrophilic intermediates or may be related to the ability to increase detoxification of the reactive intermediates that requires intact cells to be operational. It was suggested that antioxidants have an indirect effect on the epidermal metabolizing system that leads to a decrease in covalent binding of carcinogens to DNA (Slaga and Bracken 1977). A combination of inhibitors of the various stages of carcinogenesis (initiation, stage I and II of promotion) would be the most effective means of counteracting the induction of skin cancer in humans (Slaga 1984). Human cancer risks are inversely correlated with blood retinol and dietary β-carotene (Peto et al. 1981). Intake of dietary provitamin A, β-carotene, is inversely related to the 19-year incidence of lung cancer in a prospective epidemiological study (Shekelle et al. 1981). Tocopherol is a very promising inhibitor, since it is a potent inhibitor of both stages of promotion and a more moderate inhibitor of initiation. Tocopherol and *t*-butylhydroxytoluene (BHT) are effective inhibitors of 7,12-dimethylbenz[*a*]anthracene tumor initiation in a two-stage system of tumorigenesis (Slaga 1984).

Modulation of skin SOD activity may inhibit skin tumorigenesis. SOD mimetics are low molecular weight lipophilic copper coordination complexes that catalyze the disproportionation of superoxide anion at rates comparable to native Cu,Zn-SOD. SOD mimetics inhibit multistage chemical carcinogenesis in mice (Egner 1985).

GSH level–raising agents may inhibit by diverse means the effect of tumor promoters such as TPA on GSH metabolism and skin tumor promotion. However, clonal expansion of tumor-initiated cells stimulated by TPA depends upon the availability of glutathione in the tissue. The mechanism by which depletion of glutathione could result in inhibition of skin tumor promotion is not known. Inactivation of GSH may provide an effective control over proliferation of tumor cells (Kumar et al. 1988).

GSH level–raising agents increase basal GSH peroxidase activity and abolish the prolonged inhibitory effect of TPA on this enzyme. It has been suggested that stimulators of the GSH-dependent antioxidant protective system of the epidermis may inhibit the oxidative challenge linked to the skin tumor promotion by TPA (Perchellet et al. 1985, 1986b).

Tocopherol completely prevented the decrease in the ratio of GSH to GSSG caused by TPA (Perchellet et al. 1985). The anticarcinogenicity of sodium selenite, GSH, and tocopherol may be related to their ability to facilitate or enhance the activity of the GSH-dependent antioxidant protective system of epidermis during the later stages of skin tumor promotion (Perchellet et al. 1987).

Antioxidants such as nordihydroguaiaretic acid and diallyl sulfide inhibit benzoyl peroxide tumor promotion in mice (Athar et al. 1989). Garlic oil, onion oil, and one of its constituents, dipropenylsulfide, increase glutathione peroxidase activity in isolated epidermal cells. Garlic oil and onion oil inhibit tumor promotion in skin of mice, which may result from enhancement of the GSH-dependent antioxidant protective system of epidermis (Perchellet et al. 1986a).

X. CONCLUSION

Cellular injury caused by reactive oxygen involves the generation of superoxide anion, hydrogen peroxide, hydroxyl radical, and singlet oxygen. Excessive production of these compounds occurs in skin in certain disease states or as a result of toxic insult by UV radiation and/or xenobiotics. Skin possesses a variety of systems to prevent or minimize oxidative injury. The balance between cellular pro-oxidant and antioxidant states is critical for the maintenance of healthy tissue homeostasis. There is increasing evidence that reactive oxygen species are involved in a variety of skin disorders, including irritant contact dermatitis, allergic contact dermatitis, photosensitization, chemical-induced skin sclerosis, chemical leukoderma, chemical-induced comedogenesis, and skin carcinogenesis.

REFERENCES

Abraham W, Wertz PW, Downing DT. Fusion patterns of liposomes formed from stratum corneum lipids. J Invest Dermatol 1988; 90:259–62.

Akamatsu H, Oguchi M, Nishijima S, Asada Y, Takahashi M, Ushijima T, Niwa Y. The inhibition of free radical generation by human neutrophils through the synergistic effects of metronidazole with palmitoleic acid: a possible mechanism of action of metronidazole in rosacea and acne. Arch Dermatol Res 1990; 282:449-54.

Albro PW, Corbett JT, Harris M, Lawson LD. Effects of 2,3,7,8-tetrachlorodibenzo-p-dioxin on lipid profiles in tissues of the Fischer rat. Chem Biol Interact 1978; 23:315-20.

Ames JR, Ryan MD, Kovacic P. Mechanism of antibacterial action. Electron transfer and oxy radicals. J Free Rad Biol Med 1986; 2:377-91.

Ames JR, Ryan MD, Klayman DL, Kovacic P. Charge transfer and oxy radicals in antimalarial action. Quinones, dapsone metabolites, metal complexes, iminium ions and peroxides. J Free Rad Biol Med 1985; 1:353-61.

Andrews JS. Toxicity of air oxidized soybean oil. J Nutrition 1960; 70:199-203.

Athar M, Bickers DR, Mukhtar H. Inhibition of benzoyl peroxide tumor promotion in 7,12-dimethylbenz(a)anthracene treated skin of Sencar mice by the antioxidants nordihydroguaiaretic acid and diallyl sulfide. J Invest Dermatol 1989; 92:398.

Bentley-Phillips B, Bayles MAH. Cutaneous reactions to topical application of hydroquinone. S Afr Med J 1975; August 9:1391-5.

Berenblum I. The cocarcinogenic action of croton resin. Cancer Res 1941; 1:44-8.

Bhatnagar RS, Hussain MZ, Streifel JA, Tolentino M, Enriquez B. Alteration of collagen synthesis in lung organ cultures by hyperoxic environments. Biochem Biophys Res Commun 1978; 83:392-7.

Black HS. Role of reactive oxygen species in inflammatory process. In: Schroot B, Schaefer H, eds. Pharmacology of the Skin. Vol. 2. Basel: Karger, 1989:1-20.

Bleehen SS, Pathak MA, Hori Y, Fitzpatrick TB. Depigmentation of skin with 4-isopropylcatechol, mercaptoamines and other compounds. J Invest Dermatol 1968; 50:103-17.

Bolognia LJ, Pawelek JM. Biology of hypopigmentation. J Am Acad Dermatol 1988; 19:217-55.

Borel JP. Radicaux libres oxygenes, tissu conjonctif et inflammation. Pathol Biol (Paris) 1983; 31:8-10.

Bork K. Medikamentöse Zahnverfärbungen. In: Bork K, ed. Kutane Arzneimittelnebenworkungen. Stuttgart: Schattauer Verlag, 1985:297-302.

Bork K, Korting GW. Symptomatische Sklerodermie durch Bleomycin. Hautarzt 1983; 34:10-12.

Buettner GR, Motten AG, Hall RD, Chignell CF. Free radical production by chlorpromazine sulfoxide, an ESR spin trapping and flash photolysis study. Photochem Photobiol 1986; 46:5-10.

Burr GO, Burr MM. A new dificiency disease produced by the rigid exclusion of fat from the diet. J Bio Chem 1929; 82:345-67.

Byck JS, Dawson CR. Assay of protein-quinone coupling involving compounds structurally related to the active principle of poison ivy. Anal Biochem 1968; 25:123-35.

Caputo R, Monti M, Ermacora E, Carminati G, Gelmetti C, Gianotti R, Gianni E, Puccinelli V. Cutaneous manifestations of tetrachlorodibenzo-p-dioxin in children and adolescents. J Am Acad Dermatol 1988; 19:812-19.

Carnes SA, Watson AP. Disposing of the US chemical weapons stockpile. JAMA 1989; 262:653-9.

Cathcart RF. The vitamin C treatment of allergy and the normally unprimed state of antibodies. Med Hypoth 1986; 21:307–21.
Cavalier E, Rogan E. Role of radical cations in aromatic hydrocarbon carcinogenesis. Environ Health Perspect 1985; 64:69–84.
Cerutti PA. Prooxidant states and promotion. Science 1985; 227:375–81.
Chignell CF. Free radicals in drug induced photosensitivity reactions: spin trapping study of the photolysis of photoallergenic halogenated salicylanilides. In: Hagayaishi O, Niki E, Kondo M, Yoshokawa T, eds. Medical, biochemical and chemical aspects of free radicals. Amsterdam: Elsevier Science Publishers, B.V., 1989: 209–12.
Chignell CF, Kalyanaraman B, Sik RH, Mason RP. Spectroscopic studies of cutaneous photosensitizing agents. II. Spin trapping of photolysis products from sulfanilamide and 4-aminobenzoic acid using 5,5-dimethyl-1-pyrroline-1-oxide. Photochem Photobiol 1981; 34:147–56.
Chignell CF, Kalyanaraman B, Mason RP, Sik H. Spectroscopic studies of cutaneous photosensitizing agents. I. Spin trapping of photolysis products from sulfanilamide. Dermatologica 1980; 32:563–71.
Chignell CF, Motten AG, Buettner GR. Photoinduced free radicals from chlorpromazine and related phentothiazines: relationship to phenothiazine-induced photosensitization. Environ Health Perspect 1985; 64:103–10.
Clare NT, Whitten LK, Filmer DB. A photosensitized keratitis in young cattle following the use of phenothiazine as an anthelmintic. III. Identification of the photosensitizing agent. Aust Vet J 1947; 23:344–8.
Copeland ES. Free radicals in promotion: a chemical pathology study section workshop. Cancer Res 1983; 43:5631–7.
Cordier JH, Fieviez C, Lefevre MJ. Acroosteolyse et lesions cutanees chez deux ouvriers affectees au nettoyage d'autoclaves. Can Med Travail 1966; 4:3.
Czerniecki B, Goldstein BD, Witz G. Production of superoxide anion radicals by murine peritoneal exsudate cells stimulated with tumor promotors. Proc Am Assoc Cancer Res 1985; 76:133–42.
Czirjak L, Danko K, Schlammadinger J, Suranyi P, Tamasi L, Szegedi GY. Progressive systemic sclerosis occuring in patients exposed to chemicals. Int J Dermatol 1987; 26:374–8.
Dalal NS, Shi X, Vallyathan V. Oxygenated radical formation by fresh quartz dust in a cell free aqueous medium and its inhibition by scavengers. 2nd International Symposium on Spin Trapping and Aminoxyl Radical Chemistry, University of Guelph, Guelph, Ontario, Canada July 2-7, 1989.
Damas J, Lagneaux D, Lecompte T. Liberation d'histamine par les mastocytes due rat en hyperoxie. CR Soc Biol 1981; 175:266–71.
Daugherty JP, Khurana A. Amelioration of doxuribicin induced skin necrosis in mice by butylated hydroxytoluene. Cancer Chemother Pharmacol 1985; 14:243–6.
Del Maestro RF, Thaw HH, Björk J, Planker M, Arfors KE. Free radicals as mediators of tissue injury. Acta Physiol Scand 1980; 492 (Suppl):43–57.
De Luca C, Fanfoni GB, Picardo M, Nazzaro-Porro M, Passi S. The skin surface lipids of man compared with those of other different primates. J Invest Dermatol 1980; 92:473A.

Doelman CJA. Reactive oxygen species and airway hyperreactivity. Doelman CJA, ed. Amsterdam: Vrije Universiteit, 1991.
Downing DT, Stewart ME, Wertz PW, Colton SW, Abraham W, Strauss JS. Skin lipids: an update. J Invest Dermatol 1987; 88:2S–8S.
Dupuis G. Studies on poison ivy. In-vitro lymphocyte transformation by urushiol-protein conjugates. Br J Dermatol 1979; 101:617–24.
Egner PA, Kensler TW. Effects of a biomimetic superoxide dismutase on complete and multistage carcinogenesis in mouse skin. Carcinogenesis 1985; 6:1167–72.
Elias PM. Epidermal lipids, membranes, and keratinization (review). Int J Dermatol 1981; 20:1–18.
Elias PM, Menon HK. Structural lipid biochemical correlates of the epidermal permeability barrier. Adv Lipid Res 1991; 24:1–26.
Elsayed NM, Nakashima J, Wheeler C, Omaye ST, Korte DW. Biochemical changes in mouse lung and brain after a single subcutaneous injection of a sulfur mustard. Toxicologist 1991; 11:215.
Elsayed NM, Omaye ST, Klain GJ, Inase JL, Dahlenberg ET, Wheeler CR, Korte DW. Response of mouse brain to a single subcutaneous injection of the monofunctional sulfur mustard, butyl 2-chloroethyl sulfide (BCS). Toxicology 1989a; 58:11–20.
Elsayed NM, Ta P, Korte D. Biochemical alterations in mouse liver induced by nitrogen mustard. Toxicologist 1989b; 9:26.
Emerit I. Chromosome breakage factors: origin and possible significance. Prog Mutat Res 1982; 4:61–72.
Emerit I, Cerutti P. Clastogenic activity from Bloom's syndrome fibroblast cultures. Proc Natl Acad Sci USA 1981a; 78:1866–72.
Emerit I, Cerutti P. Tumor promotor phorbol-12-myristate-13-acetate induces chromosomal damage via indirect action. Nature 1981b; 293:144–6.
Emerit I, Levy A, Cerutti P. Suppression of tumor promotor phorbol myristate acetate induced chromosome breakage by antioxidants and inhibitors of arachidonic acid metabolism. Mutat Res 1983; 110:327–35.
Erasmus LD. Scleroderma in gold miners on the Witwatersrand with particular reference to pulmonary manifestations. S Afr J Lab Clin Med 1957; 3:209–231.
Finch WR, Rodnan GP, Buckingham RB. Bleomycin induced scleroderma. J Rheumatol 1980; 7:651–9.
Fischer SM, Adams LM. Suppression of tumor promotor induced chemiluminescence in mouse epidermal cells by several inhibitors of arachidonic acid metabolism. Cancer Res 1985; 45:3130–6.
Fischer SM, Cameron GS, Baldwin JK, Jasheway DW, Patrick KE. Reactive oxygen in the tumor promotion stage of skin carcinogenesis. Lipids 1988; 23:592–7.
Flindt-Hansen H, Isager H. Scleroderma after occupational exposure to trichlorethylene and trichlorethane. Acta Derm Venereol (Stockholm) 1987; 67:263–4.
Fournier E, Efthymiou ML, Lecorsier A. Spanish adulterated oil matter: an important discovery by Spanish toxicologists: the toxicity of anilides of unsaturated fatty acids. Toxicol Eur Res 1982; 4:107–12.
Frank L, Massaro D. Oxygen toxicity. Am J Med 1980; 69:117–26.

Geesin JC, Gordon JS, Berg RA. Retinoids inhibit collagen synthesis through inhibition of ascorbate stimulated lipid peroxidation. J Invest Dermatol 1989; 92: 433.

Gellin GA, Possick PA, Peronne VB. Depigmentation from 4-tertiary butyl catechol—an experimental study. J Invest Dermatol 1970; 55:190–7.

Gillert E, Die Kampfstoffverletzung. 11/12. ed. Wien: Urban & Schwarzenberg, 1944.

Gilsanz V, Alvarez JP, Serrano S, Simon J. Evolution of the alimentary toxic oil syndrome due to ingestion of denaturated rapseed oil. Arch Int Med 1984; 144:254–6.

Glavind J, Christensen F. Influence of nutrition and light on the peroxide content of the skin surface lipids of rats. Acta Derm Venereol (Stockholm) 1967; 47:339.

Glette J, Sandberg S. Phototoxicity of tetracyclines as related to singulett oxygen production and uptake by polymorphonuclear leucocytes. Biochem Pharmacol 1986; 35:2883–5.

Goerz G, Merk H. Biochemische Aspekte der Entzündungsreaktion—unter besonderer Berücksichtigung des Sauerstoffs. Z Hautkr 1983; 58:1669–86.

Goldmann PJ. Schwerste Akute Chlorakne, eine Massenintoxikation durch 2,3,6,7-Tetrachlorodibenzodioxin. Hautarzt 1973; 24:149.

Goldstein BD, Witz G, Amoruso M, Stone DS, Troll W. Stimulation of human polymorphonuclear leucocyte superoxide anion radical production by tumor promotors. Cancer Lett 1981; 11:257–62.

Greene RS, Downing DT, Pochi PE, Strauss JS. Anatomical variation in the amount and composition of human skin surface lipids. J Invest Dermatol 1970; 54:240.

Gstrauenthaler G, Pfaller W, Katanko P. Glutathione depletion and in-vitro lipid peroxidation in mercury or maleate induced acute renal failure. Biochem Pharmacol 1983; 32:3969–72.

Haustein UF, Tegtmeyer L, Ziegler V. Allergic and irritant potential of benzoyl peroxide. Contact Dermatitis 1985; 13:252–7.

Haustein UF, Ziegler V. Sklerodermie und Sklerodermie ähnliche Erkrankungen durch Umweltsubstanzen. Dermatosen 1986; 34:61–7.

Haustein UF, Ziegler V, Herrmann K, Mehlhorn J, Schmidt C. Silica induced scleroderma. J Am Acad DErmatol 1990; 22:444–8.

Hayakawa K, Matsuo I, Fujita H, Ohkido M. Thiazide phototoxicity: evidence for participation of singlet oxygen in photohemolysis. In: Hayaishi O, Imamura S, Miyachi Y, eds. The biological role of reactive oxygen species in skin. New York: Elsevier, 1987:101–6.

Hayakawa R. Relation with facial skin diseases and lipids and lipoperoxide of serum and sebum, pH on the face and buffering power (in Japanese). Jpn J Dermatol 1971; 81:11–29.

Hearing VJ, Jimenez M. Mammalian tyrosinase—the critical regulatory control point in melanocyte pigmentation. Int J Biochem 1987; 19:1141–7.

Hellerström S, Thyresson N, Blohm SG, Widmark G. On the nature of the eczematogenic component of oxidized d3-carene. J Invest Dermatol 1955; 24:217–24.

Hepplestone AG, Styles FS. Activity of macrophage factor in collagen formation. Nature 1967; 214:521–2.

Hilger R, Fukuyama K, Zackheim HS, Epstein JH. Increased melanocytes in hairless mice following topical treatment with carmustine (BCNU) and nitrogen mustard (NM). Clin Res 1974; 22:159A.

Hippeli S, Elstner EF. Oxygen radicals and air pollution. In: Sies H, ed. Oxidative stress. Oxidants and antioxidants. London: Academic Press, 1991:3–55.

Holman RT, Greenberg SI. Note on toxicities of methyl oleate peroxide and ethyl linoleate peroxide. J Am Oil Chem Soc 1960; 35:707–11.

Horgan VJ, Philpot STL, Porter WA, Roodyn DB. Toxicity of autoxidized squalene and linoleic acid, and of simpler peroxides, in relation to toxicity of radiation. Biochem J 1957; 67:551–8.

Housset B, Ody C, Rubin DB, Eleman G, Junod AF. Oxygen toxicity in cultured aortic endothelium: selenium-induced partial protective effect. J Appl Physiol 1983; 55:343–52.

Hussain MZ, Bhatnagar RS. Involvement of superoxide in the paraquat-induced enhancement of lung collagen synthesis in organ culture. Biochem Biophys Res Commun 1979; 89:71–6.

Infante PF. Observations of the site specific carcinogenicity of vinyl chloride to humans. Environ Health Perspect 1981; 41:89–94.

Ishizaki I, Koizumi K, Ikemori R, Ishiyama Y, Kushibiki E. Studies of prevalence of Japanese cedar pollinosis among the residents in a densly cultivated area. Ann Allergy 1987; 58:265–70.

Ito T. Cellular and subcellular mechanisms of photodynamic action: The $^1O_2$ hypothesis as a driving force in recent research. Photochem Photobiol 1978; 28:493–508.

Iyanagi T, Yamazaki I. One electron transfer reactions in biochemical systems. Biochim Biophys Acta 1970; 216:282–94.

Januszkiewicz AJ, Faiman MD. The effect of in-vivo hyperoxic exposure on the release of endogenous histamine from the rat isolated perfused lung. Toxicol Appl Pharmacol 1984; 72:134–40.

Januszkiewicz AJ, Huntrakoon M, Wilson PK, Faiman MD. Isolated perfused lung histamine release, lipid peroxidation, and tissue superoxide dismutase from rats exposed to normobaric hyperoxia. Toxicology 1986; 39:37–46.

Jenkins FP, Welti D, Baines D. Photochemical reactions of tetrachlorosalicylanilide. Nature 1964; 201:827–8.

Jimbow K, Obata H, Pathak M, Fitzpatrick TB. Mechanism of depigmentation by hydroquinone. J Invest Dermatol 1974; 62:436–49.

Joshi PC. Flavins as endogenous chromophores for UV induced biological oxidation reactions in skin. International Symposium on Biological Oxidation Systems, Bangalore, India, October 23–26, 1989.

Joshi PC, Pathak MA. The role of active oxygen ($^1O_2$ and O_2^-) induced by crude coal tar and its ingredients used in photochemotherapy of skin diseases. J Invest Dermatol 1984; 82:67–73.

Joshi PC, Pathak MA. Production of singlet oxygen and superoxide radicals by psoralens and their biological significance. Biochem Biophys Res Commun 1983; 112:638–46.

Kahn G. Depigmentation caused by phenolic detergent germicides. Arch Dermatol 1970; 102:177–87.

Kalyanaraman B, Mottley C, Mason RP. A direct electron spin resonance and spin trapping investigation of peroxy free radical formation by hematin/hydroperoxide systems. J Biol Chem 1983; 258:3855–8.

Kapp A. Aktivierte Granulozyten als Produzenten reaktiver Sauerstoff Spezies—ihre Bedeutung für die Entzündungsreaktion. Hautarzt 1990; 41:196–203.

Kappus H, Sies H. Toxic drug effects associated with oxygen metabolism: redox cycling and lipid peroxidation. Experientia 1981; 37:1233–41.

Kensler TW, Bush DM, Kozumbo WJ. Inhibition of tumor promotion by a biomimetic superoxide dismutase. Science 1983; 221:75–7.

Kensler TW, Taffe BG. Free radicals and tumor promotion. Adv Free Rad Biol Med 1986; 2:347–87.

Kensler TW, Trush MA. Inhibition of oxygen radical metabolism in phorbol ester activated polymorphonuclear leukocytes by an antitumor promoting copper complex with superoxide dismutase mimetic activity. Biochem Pharmacol 1983; 32: 3485–7.

Kimmig J, Schulz KH. Berufliche Akne (sog. Chlorakne) durch chlorierte aromatische zyclische Äther. Dermatologica 1957; 115:540–2.

Kinsella AR, Gainer SCH, Butler J. Investigations of a possible role for superoxide anion production in tumor promotion. Carcinogenesis 1983; 4:717–19.

Klein-Szanto AJP, Slaga TJ. Numerical variation of dark cells in normal and chemically induced hyperplastic epidermis with age of animal and efficiency of tumor promotor. Cancer Res 1981; 41:4437–40.

Klein-Szanto AJP, Slaga TJ. Effects of peroxides on rodent skin: epidermal hyperplasia and tumor promotion. J Invest Dermatol 1982; 79:30–4.

Kligman AM. Pathogenesis of acne vulgaris II. Histopathology of comedones induced in the rabbit ear by human sebum. Arch Dermatol 1968; 98:58–66.

Kligman AM, Katz AG. Pathogenesis of acne vulgaris I. Comedogenic properties of human sebum in external ear canal of the rabbit. Arch Dermatol 1968; 98:53–7.

Kochevar IE, Armstrong RB, Einbinder J, Walther RR, Harber LC. Coal tar phototoxicity: active compounds and action spectra. Photochem Photobiol 1982; 36: 65–9.

Kochevar IE, Morison WL, Lamm JL, McAuliffe DJ, Western A, Hood AF. Possible mechanism of piroxicam induced photosensitivity. Arch Dermatol 1986; 122: 1283–7.

Kondo H, Kumagai Y, Shiokawa Y. Scleroderma following cosmetic surgery ("adjuvant disease"): A review of nine cases reported in Japan. In: Black CM, Myers AR, eds. Current topics in rheumatology: systemic sclerosis (scleroderma). New York: Grower Med Pub Ltd, 1985:135–7.

Kono Y, Fridovich I. Superoxide radicals inhibit catalase. J Biol Chem 1982; 257: 5751–4.

Kooyman DJ. Lipids in the skin: some changes in the lipids of the epidermis during the process of keratinization. Arch Derm Syph 1932; 25:444–50.

Korting GW, Holzmann H. Die Sklerodermie und ihr nahestehende Bindegewebsprobleme. Stuttgart: Thieme Verlag, 1967.

Kotin p, Falk HL. Organic peroxides, hydrogen peroxide, epoxides and neoplasia. Rad Res 1963; 3(Suppl):193–211.

Kovacic P, Ames JR, Lumme P, Elo H, Cox O, Jackson H, Rivera LA, Ramirez L, Ryan MD. Charge transfer oxy radical mechanism for anti-cancer agents. Anti Cancer Drug Design 1986; 1:197–214.

Kovacic P, Crawford PW. Charge transfer mechanism for carcinogenesis by alkylating and other agents. Bioelectrochem Bioenergetics 1986; 15:305–316.

Kozumbo WJ, Seed JL, Kensler TW. Inhibition by 2(3)-tert-butyl-4-hydroxyanisole and other antioxidants of epidermal ornithine decarboxylase activity induced by 12-O-tertadecanoylphorbol-13-acetate. Cancer Res 1983; 43:2555–9.

Krieg T, Aumailley M. Connective tissue diseases in the skin. J Dermatol 1990; 17: 67–84.

Krüger M. Chemische Kampfstoffe (C-Waffen). Teil 1. Dermatosen 1991a; 39:143–50.

Krüger M. Chemische Kampfstoffe (C-Waffen). Teil 2. Dermatosen 1991b; 39:179–93.

Krylov VI, Olekhnovich VM, Sorogin PS, Zhogin SV. Activities of phospholipases and antioxidants in bronchial secretion products of children with bronchopulmonary diseases (in Russian). Vopr Med Khim 1984; 30:52–55.

Kumagai Y, Abe C, Shiokawa Y. Scleroderma after cosmetic surgery. Arthritis Rheumatol 1979; 22:532–7.

Kumar S, Antony M, Mehrotra NK. Role of cutaneous GSH in 12-O-tetradecanoylphorbol-13-acetate induced mouse skin tumor promotion. Arch Dermatol Res 1988; 280:371–4.

Kunz B, Ring J, Braun-Falco. Nehmen Allergien wirklich zu? Fortschr Medizin 1991; 109:353–6.

Lampe MA, Williams ML, Elias PM. Human epidermal lipids: characterization and modulations during differentiation. J Lipid Res 1983; 24:131.

Last JA, Greenberg DB. Ozone induced alterations in collagen metabolism of rat lungs. II. Long term exposures. Toxicol Appl Pharm 1980; 55:108–14.

Levine SA, Reinhardt JH. Biochemical pathology initiated by free radicals, oxidant chemicals, and therapeutic drugs in the etiology of chemical hypersensitivity disease. Orthomol Psych 1983; 12:166–83.

Li ASW, Chignell CF, Hall RD. Cutaneous phototoxicity of tetracyclin antibiotics: generation of free radicals and singlet oxygen during photolysis as measured by spin trapping and the phosphorescence of singlet molecular oxygen. Photochem Photobiol 1987a; 46:379–82.

Li ASW, Roethling P, Cummings KB, Chignell CF. O_2-photogenerated from aqueous solutions of tetracycline antibiotics (pH 7.3) as evidenced by DMPO spin trapping and cytochrome c reduction. Biochem Biophys Res Commun 1987b; 146: 1191–5.

Löblich HJ. Elektronenmikroskopische Untersuchungen über die Entwicklung des Staubgranuloms. Beitr Silikoseforsch 1961; 71:33–49.

Logani MK, Solanki V, Slaga TJ. Effect of tumor promotors on lipid peroxidation in mouse skin. Carcinogenesis 1982; 11:1303.

Loven DP. A role for reduced oxygen species in heat induced cell killing and the induction of thermotolerance. Med Hypoth 1988; 26:39–50.

Lowitz BB. Streaking with bleomycin. N Engl J Med 1975; June 12:1300–1.

Mannaioni PF, Masini E. The release of histamine by free radicals. Free Rad Biol Med 1988; 5:177–97.

Maricq HR. Vinyl chlorid disease. In: Black CM, Myers AR, eds. Current topics in rheumatology: systemic sclerosis (scleroderma). New York: Grower Med Pub Ltd, 1985:105–13.

Markowitz SS, McDonald CJ, Fethiere W. Occupational acroosteolysis. Arch Dermatol 1972; 106:219–33.

Martin JP, Logsdon N. Oxygen radicals mediate cell inactivation by acridine dyes, fluorescein, and lucifer yellow CH. Photochem Photobiol 1987; 46:45–53.

Marzulli FN, Maibach HI. Dermatotoxicology. 2d ed. Washington, D.C.: Hemisphere Publishing Corporation, 1983.

Mason RP. Free radical intermediates in the metabolism of toxic chemicals. Free Rad Biol 1982; 5:161–222.

Mason RPO, Chignell CF. Free radicals in pharmacology and toxicology—selected topics. Pharmacol Rev 1982; 33:189–211.

Mathias CGT. Occupational dermatoses. J Am Acad Dermatol 1988; 19:1107–14.

Matsuo I, Fujita H, Hayakawa K, Ohkido M. Lipid peroxidative potency of photosensitized thiazide diuretics. J Invest Dermatol 1986; 87:637–41.

McCarty DJ. Short term drug control of crystal induced inflammation. In: Vane JR, Ferreira SH, eds. Handbook of experimental pharmacology. Berlin: Springer-Verlag, 1979:92–107.

McCord JM. Superoxide radical: a likely link between reperfusion injury and inflammation. Adv Free Rad Biol Med 1986; 2:325–45.

McCord JM. Oxygen derived free radicals in postischemic tissue injury. N Engl J Med 1985; 312:159–63.

Meier B, Radecke HH, Selle S, Younes M, Sies H, Resch H, Habermehl GG. Human fibroblasts release reactive oxygen species in response to interleukin-1 or tumor necrosis factor. Biochem J 1989; 263:324–30.

Merk HF. Die Bedeutung der Sauerstoff Metabolite für die Entzündungsreaktion. In: Holzmann H, Altmeyer P, Marsch WC, Vogel HG, eds. Dermatologie und Rheuma. Berlin: Springer-Verlag, 1987:51–6.

Merville MP, Decuyper J, Lopez M, Piette J, Van de Vorst A. Phototoxic potentialities of tartrazine: screening test. Photochem Photobiol 1984; 40:221–6.

Mills O, Porte M, Kligman AM. Enhancement of comedogenic substances by UV radiation. Br J Dermatol 1978; 98:145–50.

Moneret-Vautrin DA. Intolerance to aspirin: a role for free radicals? Lancet 1985; April 20:929.

Moneret-Vautrin DA, Belanyi P, Martin R. Role des radicaux libres dans l'urticaire chronique. La Presse Medicale 1986; 4:164.

Montsko T, Tigyi A, Zsoldos T. Morphological and histochemical study of membrane damages occurring in the early phase of experimental silicosis. In: Oxygen free radicals and tissue damage. Hungary: Scientific Session Pécs, 10-11 January, 1984.

Motoyoshi K. Enhanced comedo formation in rabbit ear skin by squalene and oleic acid peroxides. Br J Dermatol 1983; 109:191–8.

Motten AG, Buettner GR, Chignell CF. Spectroscopic studies of cutaneous photosensitizing agents—VIII. A spin trapping study of light induced free radicals from chlorpromazine and promazine. Photochem Photobiol 1985; 42:9–15.

Motten AG, Chignell CF. Spectroscopic studies of cutaneous photosensitizing agents. III. Spin trapping of photolysis products from sulfanilamides analogs. Photochem Photobiol 1983; 37:17–26.

Mountz JD, Downs-Minor MB, Turner R, Tohams MB, Richaros F, Pisko E. Bleomycin induced cutaneous toxicity in the rat: analysis for histopathology and ultrastructure compared with progressive systemic sclerosis (scleroderma). Br J Dermatol 1983; 108:679–86.

Muglia JJ, Tonnesen MG, Osborn RL, Kissinger RM, Norris DA. Ineffective antioxidant defense in human melanocytes. J Invest Dermatol 1986; 86:496.

Murrell GAC, Francis MJO, Bromley L. Modulation of fibroblast proliferation by oxygen free radicals. Free Rad Biol Med 1990a; 9 (Suppl. 1):109.

Murrell GAC, Francis MJO, Bromley L. Free radicals in Dupuytren's contracture. Free Rad Biol Med 1990b; 9 (Suppl. 1):188.

Nakamura Y, Colburn NH, Gindhart TD. Role of reactive oxygen in tumor promotion: implication of superoxide anion in promotion of neoplastic transformation in JB-6 cells by TPA. Carcinogenesis 1985; 6:229–35.

Nazzaro-Porro M. Azelaic acid. J Am Acad Dermatol 1987; 17:1033–41.

Nikolaides N, Apon JMB. The saturated methyl branched fatty acids of adult human skin surface lipids. Biomed Mass Spectrom 1977; 4:337.

Nobbs P, Barr RD. Soft tissue injury caused by antineoplastic drugs is inhibited by topical dimethyl sulphoxide and alpha tocopherol. Br J Cancer 1983; 48:873–6.

Nohl H. 2,3,7,8-Tetrachlorodibenzo-p-dioxin induced oxidative stress and its consequences for energy supply to the cell. In: Hayaishi M, Niki E, Kondo M, Yoshikawa T, eds. Medical, biochemical and chemical aspects of free radicals. Amsterdam: Elsevier, 1989:117–23.

Nohl H, Hegner D, Summer KH. The mechanism of toxic action of hyperbaric oxygenation on the mitochondria of rat heart cells. Biochem Pharmacol 1981; 30: 1753–7.

Nohl H, Silva DD, Summer KH. 2,3,7,8-Tetrachlorodibenzo-p-dioxin induces oxygen activation associated with cell respiration. Free Rad Biol Med 1989; 6:369–74.

O'Brien PJ. Multiple mechanism of metabolic activation of aromatic amine carcinogens. Free Rad Biol 1984; 6:289–322.

O'Connell JF, Klein-Szanto AJP, DiGiovanni DM, Fries JW, Slaga TJ. Enhanced malignant progression of mouse skin tumors by the free radical generator benzoyl peroxide. Cancer Res 1986; 46:2863–5.

Ogura R. Cellular damage of epidermis exposed to ultraviolet light. Lipid peroxide formation in mitochondria and effect of antioxidants (in Japanese). Kurume Med J 1981; 45:279–301.

Ohmori H, Komoriya K, Azuma A, Kurozumi S, Oto YH. Xanthine oxidase induced histamine release from isolated rat peritoneal mast cells: involvement of hydrogen peroxide. Biochem Pharmacol 1979; 28:333–4.

Ohmori H, Yamamoto I, Akagi M, Tasaka K. Properties of hydrogen peroxide induced histamine release from mast cells. Biochem Pharmacol 1980; 29:741–5.

Ohsawa K, Watanabe T, Matsukawa R, Yoshimura Y, Imaeda K. The possible role of squalene and its peroxide of the sebum in the occurrence of sunburn and protection from the damage caused by UV irradiation. J Toxicol Sci 1984; 9:151–9.

Oliver EA, Schwartz L, Warren LH. Occupational leukoderma. JAMA 1939; 113: 927–8.

Oliver EA, Schwartz L, Warren LH. Occupational leukoderma. Arch Dermatol Syph (Chicago) 1940; 42:993–1014.

Palestine RF, Millns JL, Spigel GT. Skin manisfestations of pentazocin abusus. J Am Acad Dermatol 1980; 2:47–55.

Penney KB, Smith CJ, Allen JC. Depigmenting action of hydroquinone depends on disruption of fundamental cell processes. J Invest Dermatol 1984; 82:308–10.

Perchellet JP, Abney NL, Thomas RM, Guislain YL, Perchellet EM. Effects of combined treatments with selenium, glutathione, and vitamin E on glutathione peroxidase activity, ornithine decarboxylase induction and complete and multistage carcinogenesis in mouse skin. Cancer Res 1987; 47:477–85.

Perchellet JP, Perchellet EM, Orten DK, Schneider BA. Inhibition of the effects of 12-O-tetradecanoylphorbol-13-myristate on mouse epidermal glutathione peroxidase and ornithine decarboxylase activities by glutathione level raising agents and selenium containing compounds. Cancer Lett 1985; 26:283–93.

Perchellet JP, Perchellet EM, Abney NL, Zirnstein JA, Belman S. Effects of garlic and onion oils on glutathione peroxidase activity, the ratio of reduced/oxidized glutathione and ornithine decarboxylase induction in isolated mouse epidermal cells treated with tumor promotors. Canc Biochem Biophys 1986a; 8:299–3212.

Perchellet JP, Perchellet EM, Orten DK, Schneider BA. Decreased ratio of reduced/ oxidized glutathione in mouse epidermal cells treated with tumor promotors. Carcinogenesis 1986b; 7:503–6.

Perisho K, Wertz PW, Madison KC, Stewart ME, Downing DT. Fatty acids of acylceramides from comedones and from the skin surface of acne patients and control subjects. J Invest Dermatol 1988; 90:350–3.

Perrot II, Ortonne JP. Hyperpigmentation after bleomycin therapy. Arch Dermatol Res 1978; 261:245–52.

Peto R, Doll R, Buckley JD, Sporn MB. Can dietary beta-carotene materially reduce human cancer rates? Nature 1981; 290:201–8.

Phelps RG, Fleischmajer R. Clinical, pathologic, and immunologic manifestations of toxic oil syndrome. J Am Acad Dermatol 1988; 18:313–24.

Plaa GL, Witschi H. Chemicals, drugs and lipid peroxidation. Ann Rev Pharmacol Toxicol 1976; 16:125–41.

Pooler JP. Photooxidation of cell membranes using eosin derivatives that locate in lipid or protein to study the role of diffusible intermediates. Photochem Photobiol 1989; 50:55–68.

Pryor WA. The role of free radical reactions in biological systems. Free Rad Biol 1975; 1:1–49.

Pryor WA. Free radical biology: xenobiotics, cancer and aging. Ann NY Acad Sci 1982; 1–22.

Puhvel SM, Connor JM, Sakamoto M. Vitamin A deficiency and dioxin induced cutaneous toxicity in murine skin. J Invest Dermatol 1990; 94:568.
Raab O. Über die Wirkung fluorescierender Stoffe auf Infusorien. Z Biol 1900; 39: 524–6.
Ranadive NS, Shirwadkar S, Persad S, Menon A. Effects of melanin-induced free radicals on the isolated rat peritoneal mast cells. J Invest Dermatol 1986; 86:303–7.
Rapaport MJ. Depigmentation with cerium oxide. Contact Dermatitis 1982; 8:282–3.
Reinl W. Sklerodermie durch Trichlorethylen Einwirkung. Bull Hyg 1957; 32:678–9.
Requena L, Requena C, Sanchez M, Jaqueti G, Aguilar A, Sanchez-Yus E, Hernandez-Moro B. Chemical warfare: cutaneous lesions from mustard gas. J Am Acad Dermatol 1988; 19:529–36.
Riley PA, Cooksey CJ, Land EJ. Free radicals and melanomas. In: Hayaishi O, Niki E, Kondo M, Yoshokawa T, eds. Medical, biochemical and chemical aspects of free radicals. Amsterdam: Elsevier, 1989:1485–90.
Riley PA, Sawyer B, Wolff MA. The melanocytotoxic action of 4-hydroxyanisole. J Invest Dermatol 1975; 64:86–9.
Riley PA. Mechanism of pigment cell toxicity produced by hydroxyanisol. J Pathol 1970; 101:163–9.
Riley PA. Hydroxyanisole depigmentation: in-vitro studies. J Pathol 1969a; 97: 193–206.
Riley PA. Hydroxyanisole depigmentation: in-vivo studies. J Pathol 1969b; 97:185–91.
Roberfroid MB, Viehe HG, Remacle J. Free radicals in drug research. Adv Drug Res 1987; 16:1–84.
Robertson FM, Laskin DL, Laskin JD. Production of reactive oxygen intermediates in mouse skin following topical application of the tumor promotor 12-O-tetradecaonyl phorbol-13-acetate (PMA). Proc Am Assoc Canc Res 1988; 29:155.
Rodnan GP, Benedek TG, Medsger TA, Cammarata RJ. The association of progressive systemic sclerosis (scleroderma) with coal miners' pneumokoniosis an other forms of silicosis. Ann Int Med 1967; 66:323–34.
Rodnan GP, Medsger TA. Pentazocine induced fibrosis. International Conference on Progressive Systemic Sclerosis, Austin, TX, October 1981.
Roshchupkin DI, Pitsov MY, Potapenko AY. Inhibition of ultraviolet light-induced erythema by antioxidants. Arch Dermatol Res 1979; 266:91.
Rous P, Kidd JG. Conditional neoplasms and subthreshold neoplastic states. A study of tar tumors in rabbits. J Exp Med 1941; 73:365–89.
Rowell NR. Lupus erythematosus, scleroderma and dermatomyositis. The "collagen" or "connective tissue" diseases. In: Rook A, Wilkinson DS, Ebling FJG, Champion RH, Burton JL, eds. Textbook of dermatology. Vol. 2. Oxford: Blackwell Scientific Publications, 1986:1281–1392.
Ruzicka T. The physiology and pathophysiology of eicosanoids on the skin. Eicosanoids 1988; 1:59–72.
Rush PJ, Bell MJ, Fam AG. Toxic oil syndrome (Spanish oil disease) and chemically induced scleroderma like conditions. J Rheumatol 1984; 11:262–4.
Saihan EM, Burton JL, Heaton KW. A new syndrome with pigmentation, scleroderma, gynaecomastia, Raynaud's phenomenon and peripheral neuropathy. Br J Dermatol 1978; 99:437–40.

Saint-Leger D, Bague A, Cohen E, Chivot M. A possible role for squalene in the pathogenesis of acne. I. In-vitro study of squalene oxidation. Br J Dermatol 1986; 114:535–42.

Schalwijk J, van den Berg WB, van de Putte LBA, Joosten LAB. An experimental model for hydrogen peroxide induced tissue damage. Arthritis Rheum 1986; 29: 532–8.

Schirren SM. Skin lesions caused by trichlorethylene in a metalworking plant. Berufsdermatosen 1971; 19:240–54.

Schmidt RJ, Khan L, Chung LY. Are free radicals and not quinones the haptenic species derived from urushiols and other contact allergenic mono- and dihydric alkylbenzenes? The significance of NADH, glutathione, and redox cycling in the skin. Arch Dermatol Res 1990; 282:56–64.

Schmidt RJ, Chung LY. Oxidative stress occurs in skin in the pre-immunologic phase of the cell mediated immune response to allergenic 2,4-dinitrohalobenzenes. Free Rad Biol Med 1990; 9 (Suppl. 1):147.

Schoonderwoerd SA, Beijersbergen van Henegouwen GMJ, Van Belkum S. In-vivo photodegradation of chlorpromazin. Photochem Photobiol 1989; 50:659–64.

Schweizer J, Loehrke H, Edler L, Goertler K. Benzoyl peroxide promotes the formation of melanotic tumors in the skin of 7,12-dimethylbenz[a]anthracene initiated Syrian golden hamsters. Carcinogenesis 1987; 8:479–82.

Sharkey P, Eedy DJ, Burrows D, McCaigue MD, Bell AL. A possible role for superoxide production in the pathogenesis of contact dermatitis. Acta Derm Venereol (Stockh) 1991; 71:156–9.

Shekelle RB, Liu S, Raynor WJ, Lepper M, Maliza C, Rossof AH. Dietary vitamin A and risk of cancer in the western electric study. Lancet 1981; November 28:1185–9.

Shelley WB. p-Cresol: cause of ink-induced hair depigmentation in mice. Br J Dermatol 1974; 90:169–74.

Shupack JL. The skin as a target organ for systemic agents. In: Drill VA, Lazar P, eds. Cutaneous toxicity. New York: Academic Press, 1976:43–52.

Sies H. Oxidative stress: introduction. In: Sies H, ed. Oxidative stress. London: Academic Press, 1991:xv–xxii.

Slaga TJ, Fischer SM, Weeks CE, Klein-Szanto AJP, Reinders J. Studies on the mechanism involved in multistage carcinogenesis in mouse skin. J Cell Biochem 1982; 18:99–119.

Slaga TJ. Multistage skin carcinogenesis: a useful model for the study of the chemoprevention of cancer. Acta Pharmacol Toxicol 1984; 55 (Suppl. II):107–25.

Slaga TJ, Bracken WM. The effects of antioxidants on skin tumor initiation and aryl hydrocarbon hydroxylase. Cancer Res 1977; 37:1631–5.

Slaga TJ, Fischer SM, Nelson K, Gleason GL. Studies on the mechanism of skin tumor promotion: evidence for several stages in promotion. Proc Natl Acad Sci USA 1980; 77:3659–63.

Slaga TJ, Klein-Szanto AJP, Triplett LL, Yotti LP, Trosko JE. Skin tumor promoting activity of benzoyl peroxide, a widely used free radical generating compound. Science 1981; 213:1023–5.

Smith LL, Paraquat toxicity. Phil Trans R Soc Lond B 1985; 311:647–57.

Snell RS. Monobenzylether of hydroquinone. Arch Dermatol 1964; 90:63–70.

Solanki V, Rana RS, Slaga TJ. Diminution of mouse epidermal superoxide dismutase and catalase activities by tumor promotors. Carcinogenesis 1982; 2:1141–6.
Sollberg S, Holzmann H. Genetik der Sklerodermien. In: Holzmann H, Altmeyer P, Marsch WC, Vogel HG, eds. Dermatologie und Rheuma. Berlin: Springer-Verlag, 1987:241–52.
Somani SM, Babu SR. Toxocodynamics of sulfur mustard. Int J Clin Pharmacol Ther Toxicol 1989; 27:419–35.
Sparrow GP. A connective tissue disorder similar to vinyl chloride disease in a patient exposed to perchlorethylene. Clin Exp Dermatol 1977; 2:17–22.
Spencer MC. Hydroquinone bleaching. Arch Dermatol 1961; 84:131–4.
Spiera H. Scleroderma after silicone augmentation mammoplasty. JAMA 1988; 260: 236–8.
Spurzem JR, Lockey JE. Toxic oil syndrome. Arch Int Med 1984; 144:249–50.
Standahl O, Molin L, Dahlgren C. The inhibition of polymorphonuclear leucocyte cytotoxicity by dapsone. J Clin Invest 62:214–20.
Sternberg EM, van Woert M, Young SN, Magnussen I, Baker H, Gauthier S, Osterland CK. Development of a scleroderma like illness during therapy with L-5-hydroxytryptophan and carbidopa. N Engl J Med 1980; 303:782–7.
Stohs SJ. Oxidative stress induced by 2,3,7,8-tetra-chlorodibenzo-p-dioxin (TCDD). Free Rad Biol Med 1990; 9:79–90.
Sugiura K, Ueda H, Hirano K, Adachi T. Studies on superoxide dismutase in human skin (2). Contents of superoxide dismutase and lipoperoxide in normal human skin (in Japanese). Jpn J Dermatol 1985; 95:1541–5.
Sugiura K, Abe M, Inasaka H, Ueda H, Hirano K, Adachi T. Studies on superoxide dismutase in human skin (4). Contents of superoxide dismutase and lipid peroxide in keloid, hypertrophic scar and scar (in Japanese). Jpn J Dermatol 1986a; 96: 171–4.
Sugiura K, Iida Y, Ohgoshi H, Ueda H, Hirano K, Adachi T. Studies on superoxide dismutase in human skin (3). Contents of superoxide dismutase and lipid peroxides in skin of patients with atopic dermatitis (in Japanese). Jpn J Dermatol 1986b; 96:165–70.
Svingen BA, Powis GP, Appel PL, Scot M. Protection against adriamycin induced skin necrosis in the rat by dimethyl sulfoxide and a-tocopherol. Cancer Res 1981; 41:3395–9.
Sweeney GD, Jones KG, Cole FM, Basford D, Krestynski F. Iron deficiency prevents liver toxicity of 2,3,7,8-tetrachlorodibenzo-p-dioxin. Science 1979; 204: 332–5.
Tabuenca JM. Toxic-allergic syndrome caused by ingestion of rapeseed oil denaturated with aniline. Lancet 1981; 2:567–8.
Taffe BG, Takahashi N, Kensler TW, Mason RP. Generation of free radicals from organic hydroperoxide tumor promoters in isolated mouse keratinocytes. J Biol Chem 1986; 262:12143–9.
Tanaka T. Skin damage and its prevention from lipoperoxide (in Japanese). Vitamins (Kyoto) 1979; 53:577–86.
Texier L, Gauthier Y, Gauthier O, Delaunay M, Boineau D. Hypodermite sclerodermiforme lombo-fessiere induite par injection de vitamine K1. Defense immuni-

taire conservee, evolution spontanement regressive. Bull Soc Fr Derm Syph 1975; 82:448–9.

Thürlimann W, Harms M. Skerodermiforme Veränderungen nach PUVA Therapie bei Vitiligo. Dermatologica 1982; 164:305–13.

Trenam CW, Dabbagh AJD, Morris CJ. The role of iron in an acute model of skin inflammation induced by reactive oxygen species (ROS). Br J Dermatol 1992; 126: 250–6.

Trenam CW, Dabbagh AJ, Morris CJ, Blake DR. Skin inflammation induced by reactive oxygen species (ROS): an in-vivo model. Br J Dermatol 1991; 125:325–9.

Trush MA, Kensler TW. Role of free radicals in carcinogen activation. In: Sies H, ed. Oxidative stress. London: Academic Press, 1991:277–318.

Trozak DJ, Gould WM. Cocaine abusus and connective tissue disease. J Am Cad Dermatol 1984; 10:525.

Uito J, Jimenez S. Fibrotic skin diseases. Arch Dermatol 1990; 126:661–4.

Van Duuren BL, Nelson N, Orris L, Palmes ED, Schmitt FL. Carcinogenicity of epoxides, lactones, and peroxy compounds. J Natl Cancer Inst 1963; 31:41–55.

Veltman G. Klinische Befunde und arbeitsmedizinische Aspekte der Vinylchlorid Krankheit. Dermatol Monschr 1980; 166:705–12.

Vojvodic V, Milosavljevic Z, Boskovic B, Bojanic N. The protective effect of different drugs in rats poisoned by sulfur and nitrogen mustards. Fundam Appl Toxicol 1985; 5:160–8.

Vollum DI. Hypomelanosis from antioxidant in polyethylene film. Arch Dermatol 1971; 104:70–2.

Von Bettmann S. Chlorakne eine besondere Form von professioneller Hauterkrankung. Dtsch Med Wochschr 1901; 27:437.

Walder BK. Solvents and scleroderma. Lancet 1965; II:436.

Walder BK. Do solvents cause scleroderma? Int J Dermatol 1983; 22:157–8.

Walder BK. Solvent induced scleroderma. Meeting of the Royal Australian College of Dermatology, Melbourne, April 1981.

Ward AM, Udnoon S, Watkins J, Walker AE, Darke CS. Immunological mechanism in the pathogenesis of vinyl chloride disease. Br Med J 1976; 1:936–8.

Waravdekar VS, Saslaw LD, Jones WA, Kuhns JG. Skin changes induced by UV irradiated linolenic acid extract. Arch Pathol 1965; 80:91–5.

Wertz PM, Downing DT. Acylglucosylceramides of pig epidermis: structure determination. J Lipid Res 1984; 24:753.

Wertz PW, Miethke MC, Long SA, Strauss JS, Downing DT. The composition of the ceramides from human stratum corneum and from comedones. J Invest Dermatol 1985; 84:410–12.

Western A, van Camp JR, Benasson R, Land EJ, Kochevar IE. Involvement of singlet oxygen in the phototoxicity mechanism for a metabolite of piroxicam. Photochem Photobiol 1987; 46:469–75.

Wilems JL. Clinical management of mustard gas casualties. Ann Med Milit Belg 1989; 3:1–61.

Williams ML, Elias PM. Genetically transmitted, generalized disorders of cornification. Derm Clin 1987; 5:155–78.

Williams ML. Lipids in normal and pathological desquamation. Adv Lipid Res 1991; 24:211–62.
Witz G, Goldstein BD, Amoruso M, Stone DS, Troll W. Retinoid inhibition of superoxide anion radical production by human polymorphonuclear leucocytes stimulated with tumor promotors. Biochem Biophys Res Commun 1980; 97:883–8.
Yagi K. Lipid peroxides in biology and medicine. New York: Academic Press, 1982.
Yagi K. Lipid peroxides in the skin. In: Hayaishi O, Imamura S, Miyachi Y, eds. The biological role of reactive oxygen species in skin. New York: Elsevier, 1987: 109–16.
Yamagiwa K, Ichikawa K. Experimental study of the pathogenesis of carcinoma. University Tokyo 1915; 15:295–344.
Yamakage A, Ishikawa H. Generalized morphea like scleroderma occurring in people exposed to organic solvents. Dermatologica 1982; 165:186–93.
Yamakage A, Ishikawa H, Saito Y, Hattori A. Occupational scleroderma like disorder occurring in men engaged in polymerization of epoxy resins. Dermatologica 1980; 161:33–44.
Yamamoto A, Nagai T, Shimizu M, Shitara A, Sato Y, Inomata N. Skin irritation and cytotoxicity by squalene peroxides. XVI International Congress of Dermatology, Tokyo, 1982.
Yamazaki I, Mason HS, Piette L. Identification, by electron paramagnetic resonance spectroscopy, of free radicals generated from substrates by peroxidase. J Biol Chem 1960; 235:2444–9.
Yardley HJ. Epidermal lipids. In: Goldsmith LA, ed. Biochemistry and physiology of the skin. New York: Oxford University Press, 1983:363.
Yen EHK, Sodek J, Melcher AH. The effect of oxygen partial pressure on protein synthesis and collagen hydroxylation by mature peridontal tissues maintained in organ cultures. Biochem J 1979; 178:605–12.
Yohn JJ, Norris DA, Yrastorza DG, Buno IJ, Leff JA, Hake SS. Disparate antioxidant enzyme activities in cultured human cutaneous fibroblasts, keratinocytes, and melanocytes. J Invest Dermatol 1991; 97:405–9.
Yonaha M, Ohbayashi Y, Ichinose T, Sagai M. Lipid peroxidation stimulated by mercuric chloride and its relation to toxicity. Chem Pharm Bull 1982; 30:1437–42.
Yoshioka A, Miyachi Y, Imamura S. Mechanism of reactive oxygen species induced skin erythema and superoxide dismutase activities in guinea pigs. J Dermatol 1987; 14:569–75.
Ziboh VA, Chapkin RS. Biological significance of polyunsaturated fatty acids in the skin. Arch Dermatol 1987; 123:1686–90.
Ziegler V, Pampel W, Zschunke E, Münzberger H, Mährlein W, Köpping H. Kristalliner Quarz (eine) Ursache der progressiven Sklerodermie? Dermatol Monatsschr 1982; 168:398–401.
Ziegler V, Haustein UF, Mehlhorn J, Münzberger H, Rennau H. Quarzinduzierte Sklerodermie. Sklerodermie-ähnliches Syndrom oder echte progressive Sklerodermie? Dermatol Monatsschr 1986; 172:86–90.

Zimmerman R, Cerutti P. Active oxygen acts as a promoter of transformation in mouse embryo C3H/10T 1/2/C18 fibroblasts. Proc Natl Acad Sci USA 1984; 81: 2085–7.

Zweier JL, Flaherty JT, Weisfeldt ML. Direct measurement of free radical generation following reperfusion of ischemic myocardium. Proc Natl Acad Sci USA 1987; 84:1404–7.

23

Hyperbaric Oxygenation in the Treatment of Skin Diseases

Michael Buslau and Gerd Hoffmann
Johann Wolfgang Goethe University,
Frankfurt, Germany

I. INTRODUCTION

Hyperbaric oxygenation (HBO) is a therapy that has been applied increasingly often over the last three decades (48). It is based on the intermittent inhalation of pure oxygen with partial pressures higher than normal atmospheric pressure. There are two aspects of HBO: the elevation of total surrounding pressure and the breathing of pure oxygen instead of normal air. Both aspects contribute to the therapeutic effects but can also lead to undesirable side effects (57) Hyperbaric oxygenation is carried out either in monoplace or in multiplace chambers. Monoplace chambers—the older form—are pressurized with pure oxygen, and the patient within the chamber freely breathes the surrounding oxygen. In multiplace chambers, the increased pressure is achieved via compressed air, and the patient breathes pure oxygen through a face mask, a hood, or an endotracheal tube (16) (Figs. 1 and 2).

Hyperbaric oxygenation is a well-known form of therapy with clear indications (43,74). For example, the therapy can be used for the treatment of decompression sickness or air embolism. In these cases, the pressure-dependent decrease in size of gas bubbles in blood vessels and tissues is the main factor. In the treatment of carbon monoxide poisoning (acceleration of dissociation of carboxyhemoglobin), the rate of CO elimination is directly related

Figure 1 Hyperbaric oxygen facility, view from the outside.

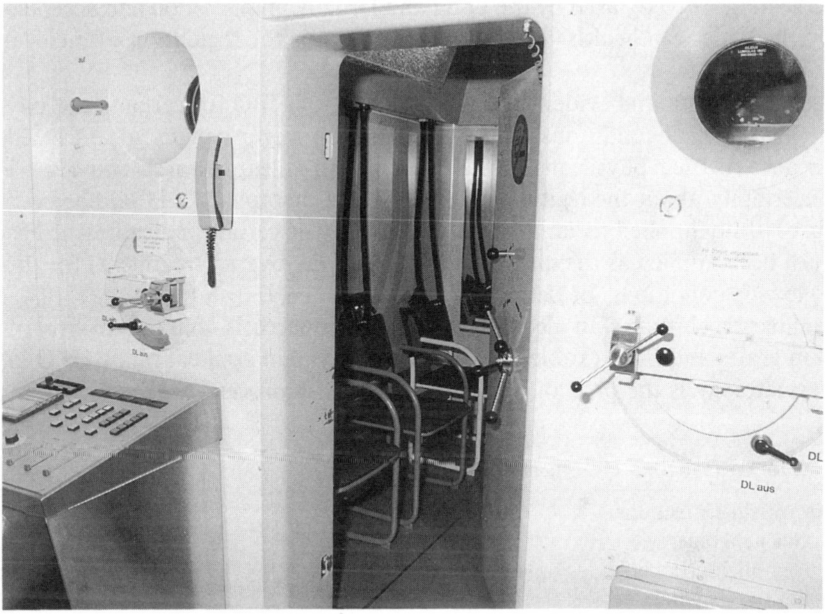

Figure 2 Hyperbaric oxygen facility, view from the inside—control room with control desk, two sluices, and two circular portholes. The pressure-resistant door opens into the multiplace chamber. Note the oxygen masks.

to the partial pressure of oxygen in the inspired air. A further use for HBO is in the treatment of hypoxic and ischemic processes, e.g., the therapy of problem wounds (increase of oxygen partial pressure in the tissue through the pressure-dependent increase in physically dissolved oxygen in the plasma).

The Hyperbaric Oxygen Committee of the Undersea and Hyperbaric Medical Society (Bethesda, MD), a professional association of physicians administering HBO, revises the list of indications for hyperbaric oxygen therapy every 3 years in the light of the most recent scientific findings. This list of indications is recognized by the Food and Drug Administration and accepted by the American health insurance companies for the regulation of medical services.

While substantial evidence supports the use of HBO for certain carefully defined clinical conditions, many patients who might benefit go untreated because of their physicians' unfamiliarity with recent research and overall uncertainty about the legitimacy of HBO as a therapy (57). HBO has been successfully applied as an adjunct treatment (for certain indications it has even been applied as the main therapy within the overall treatment) for the following conditions of dermatologic interest: necrotizing infections, thermal burns, chronic skin ulcers, soft tissue radionecrosis, and the survival of skin grafts and flaps (Table 1). Therefore, the main application of HBO in dermatology is the promotion of wound-healing processes.

Table 1 Hyperbaric Oxygenation in Dermatology

Necrotizing infections
Gas gangrene
Necrotizing fasciitis
Fournier's gangrene
Tropical ulcer
Mucormycosis
Acute ischemic conditions
Burn injuries
Cryobiologic wounds
Toxic arteriopathies
Toxic epidermal necrolysis
Streptococcal toxic shock syndrome
Chronic ischemic conditions
Arterial occlusive diseases
Venous stasis ulcers
Diabetic ulcers
Pressure sores
Soft tissue radionecrosis
Skin ulcers of unknown etiology
Pyoderma gangrenosum
Acro-osteopathia ulcero-mutilans
Plastic and reconstructive surgery
Survival of skin grafts and flaps

II. HYPOXIA AND WOUND HEALING

Hypoxia is a characteristic feature of wounds. There is an inverse relationship between tissue oxygen pressure and rate of infection and disturbance of wound healing. For healing tissue, the critical oxygen limit is approximately 10–20 mmHg oxygen partial pressure (55,68,90,116). HBO increases the available oxygen even in the compromised tissue of problem wounds and prevents tissue hypoxia. In numerous animal and human studies, accelerated wound healing has been observed following intermittent hyperbaric oxygen therapy (69,74,86,87).

The positive effects of HBO for the wound-healing process (defined as the sum of repair processes within the vascularized connective tissue) is based on a combination of factors:

1. Facilitation of oxygen diffusion and associated improvement of oxygen supply from the surrounding intact capillary network (the application of 100% normobaric oxygen is usually inadequate to provide sufficient oxygen to the central, most severely hypoxic wound areas) (43,116).
2. Reduction of edema following vasoconstriction as a result of the effect of oxygen on the vessels (the net increase in oxygen supply to the dependent tissue remains essentially unchanged or is even increased due to the high PO_2 gradient in the plasma).
3. Improvement in phagocyte function: provision of the oxygen necessary for "respiratory bursts"—activated phagocytes have an oxygen consumption that is raised by a factor of 10–20 above the resting metabolic rate (43,67).
4. Reduction of infection: antimicrobial action not only on anaerobic bacteria, e.g., *Clostridium* spp., but also on aerobic bacteria, e.g., *Staphylococcus aureus, Streptococcus pyogenes,* or *Pseudomonas aeruginosa* (72,88); nosocomial infections require the highest amount of oxygen for successful phagocytic elimination (101).
5. Improvement in collagen synthesis: higher oxygen partial pressure stimulates fibroblast proliferation rate in less well-perfused tissues and the production of the collagen matrix in newly formed vessels. Repair processes are thus promoted, including capillary budding, neovascularization, collagen synthesis, and epithelialization of the wound (69,120,156).

III. GAS GANGRENE AND OTHER NECROTIZING INFECTIONS

Delivery of oxygen to areas of bacterial invasion is an essential prerequisite for effective killing of certain common bacterial pathogens by leukocytes. Leukocytes utilize hydrogen peroxide, superoxide, and other reduced moieties derived from molecular oxygen for the purpose of killing microorganisms. The effect of HBO in anaerobic infections is well known. Anaerobic organisms do not possess the antioxidative enzymes necessary for the breakdown of superoxide dismutase and catalase and are therefore destroyed by exposure to oxygen (31,107). Aerobic bacteria show a biphasic response on exposure to HBO. At pressures of 0.6–1.3 ATA (atmospheres absolute) growth is enhanced, but above 1.3 ATA it is inhibited. In treating infections in hypoxic tissues, HBO may be used as an important adjunct (54) therapy that that may enhance the effects of antimicrobial agents, particularly sulfonamides and aminoglycosides (122). Improved tissue oxygenation by HBO and antibiotic therapy may have an additive effect (88). Intermittent HBO therapy seems to affect certain microorganisms, e.g., *P. aeruginosa*, adversely, but continued exposure may be detrimental to phagocytosis (153). There are, however, species differences in response to increased Po_2 (54,143).

A. Gas Gangrene

Gas gangrene is a necrotizing infection with a dramatic course of development and rapidly spreading soft tissue and muscle necrosis. Traumatic and contaminated wounds or ischemic lesions are especially at risk, mainly through clostridial infections. The toxic shock syndrome resulting from *Clostridium perfringens* infections is the most severe case. Particularly endangered are traumatic or contaminated wounds or ischemic lesions in patients with impaired immune systems (e.g., diabetics). The pathophysiological condition results principally from the formation of various exotoxins, which enhance the action of alpha toxin, the major lethal toxin produced by *C. perfringens*, in progressive hemolysis and colliquative necrosis of the tissue. Gas gangrene is a well-recognized indication for the use of HBO (for review of the extensive literature see Ref. 62). The number of gas gangrene patients treated by HBO and reported in the literature exceeds 2000, with mostly favorable results (3–5,27,33,62,141,145).

The basis for the positive effects of HBO treatment on gas edema is the inhibition of toxin production and the lethal effects of oxygen on anaerobic bacteria and microphilic aerobes in connection with the antioxidative enzyme activity. Alpha-toxin production stops at tissue oxygen tensions of 60 mmHg and above. HBO with an increased oxygen pressure of 3 ATA over a period of 1.5 hours completely stopped the production of alpha toxin in vitro (147). As oxygen cannot penetrate necrotic tissue, the surgical removal of necrotic tissue is essential for successful treatment with HBO.

For inhibition of clostridia, tissue Po_2 levels of 250 mmHg are required (43). Demello et al. demonstrated that the germination of clostridial spores was inhibited under treatment with hyperbaric oxygen at pressures of 2–3 atmospheres and that the bacterial growth was halted in a bacteriostatic manner (31).

B. Nonclostridial Soft Tissue Gas-Forming Infections

In nonclostridial soft tissue gas-forming infections with tissue necrosis due to aerobic bacteria (staphylococci, streptococci, *E. coli*), anaerobic streptococci, *Bacteroides* spp., and mixed aerobic and anaerobic organisms (so-called synergistic infections), HBO therapy may also be a useful adjunct (3, 112). As in the treatment of gas gangrene due to clostridial infections, HBO can be an important therapy in nonclostridial necrotizing infections in combination with antibiotics and surgery. In one study (38), synergistic bacterial soft tissue infections did not benefit from adjunctive HBO treatment. Gorback et al. (50) reviewed 89 patients referred for HBO therapy in necrotizing fasciitis for mortality risk factors. The overall mortality rate was 20%. They found statistically significant relationships between mortality and the requirement for mechanical ventilation, diabetes mellitus, and renal dysfunction, however, there was no demonstrable relationship between mortality and clostridial or nonclostridial necrotizing infections (50).

Fournier's gangrene is a necrotizing fasciitis of the male genitalia usually caused by a synergistic mixture of aerobic and anaerobic organisms. This syndrome, although unusual, has a high morbidity and mortality rate, up to 45% in some series. The mortality rate may be reduced to 15% (in some studies 0%) when HBO is applied (94,133). Hirn and Niinikoski (66) treated 11 cases of perineal necrotizing fasciitis (10 cases with Fournier's gangrene, one female with vulvar necrosis) with surgery, antibiotics, and HBO. Only one patient died. The authors stress that HBO should not replace or delay surgical intervention (66).

Another case of severe necrotizing infection involved a so-called tropical ulcer caused by a combined infection of *S. aureus* and β-hemolytic streptococci, which healed within 3 weeks after antibiotics plus HBO therapy (17a). There are also several case reports of streptococcal toxic shock-like syndrome and of Purpura fulminans, which ameliorated with adjunctive HBO therapy (34,91,92,140,149).

The procedure usually recommended in the treatment of gas gangrene is exposure to a pressure of 3 ATA for 90 minutes three times within the first 24 hours and twice a day over the next 4–5 days. The same procedure is recommended for acute necrotizing infections caused by other gas-producing anaerobic or aerobic bacteria. The therapy should be started as soon as possible (within 24 hours of the clinical diagnosis), before microbiological confirmation of the diagnosis and prior to the definitive surgical treatment (131). According to our experience in non–life-threatening necrotizing infections, HBO treatment with a pressure of 1.5–2.0 ATA for 30–60 minutes once or twice a day is sufficient.

IV. MISCELLANEOUS INFECTIONS OF DERMATOLOGIC INTEREST

There are several reports about the successful treatment of necrotizing mucormycosis by hyperbaric oxygen therapy (99,108,125,148). HBO has also been used with favorable results in actinomycosis, tuberculosis, mycetoma, and leprosy as a supplement to the conventional treatment, particularly where the latter has not been successful (11,52,53,100,127,150). However, in the mouse footpad model, HBO therapy at 2 ATA could not reduce the growth of *M. leprae* (55).

V. ACUTE ISCHEMIC CONDITIONS

A. Burn Injuries

Burn injuries are complex injuries with a central zone of coagulation, a surrounding zone of hemostasis, and a border zone of hyperemia. Inflammatory reactions with increased permeability, formation of edema, reduced intravascular flow, and clotting increase the probability of ischemia, thrombosis, and necrosis of up to 10 times that of the originally affected area during the first 48 hours after injury (1,14,68). The use of hyperbaric oxygen in the treatment of burns is strongly supported by many clinical studies (mostly retrospective) of over 2000 patients (21,58,59,157). The rationale behind the use of HBO in the treatment of burns is based on animal and clinical studies. HBO provides the oxygen necessary to sustain cell metabolism by maintaining ATP levels in the cell membranes (118,160). HBO improves microcirculation, prevents dermal ischemia, counteracts vasodilatation, prevents exudation of plasma and edema, reduces fluid requirements, inhibits wound infection, and promotes rapid wound healing with better wound organization and epithelialization. HBO reduces the necessity for surgery, improves survival of skin grafts and flaps, reduces hospitalization, improves cost-effectiveness, improves the survival rate of high-risk patients, and is a safe method of treatment (2,20,21,23–25,58,59,63,64,78,80,81,89,117,154).

Grossman and Grossman (59) presented their experience with HBO treatment of more than 1300 burn patients. According to their treatment regime, HBO is given within 4 hours of admission of the patient to the burn unit, treatment is given at 2–2.5 ATA (100% oxygen) for varying periods of time, and treatments are given both pre- and postoperatively. There is one small paired-controlled study (152) of 36 severe burn patients receiving adjunctive HBO treatment and 36 conventionally treated patients serving as a control. In this study the mortality and total hospital stay were the same in both groups. HBO-treated patients had a higher incidence of sepsis and renal damage. This may be a result of the high quantities of nephrotoxic antibiotics given rather than the HBO treatment itself (83).

B. Cryobiologic Wounds

The beneficial effects of HBO in cryotherapy were evaluated by Delfino et al. (29). Rats treated with liquid nitrogen and subsequent HBO immediately after the second thawing showed less erythema, fewer vesicles, and a thinner eschar than did rats not exposed to HBO. The authors concluded that HBO treatment can significantly reduce the severity of cryogenic skin damage in rats.

C. UV-Irradiated Suction Blister Wounds

Hammerlund et al. (60) treated UV-irradiated suction blister wounds of healthy male volunteers with HBO 1.5, 10.5 and 21.5 hours after inflicting the lesions. They found a favorable effect of HBO on superficial dermal lesions. Edema and exudation decreased, as did hyperemia at the wound edges. Epithelialization, however, was not accelerated.

D. Toxic Epidermal Necrolysis

HBO was successfully used in four patients with drug-induced toxic epidermal necrolysis. Treatments were performed with 100% oxygen at 2 ATA for 60–120 minutes once a day. Complete reepithelialization was achieved after approximately 10 treatments (97).

E. Toxic Arteriopathies

A female patient was suffering from skin necrosis of the tip of her nose due to accidentally administered intraarterial sclerotherapy. Within 6 hours a 4-cm-diameter necrosis developed at the side of injury. This was successfully reduced under HBO (1.5 ATA, 45 min daily) in combination with a vasodilator. The wound healed within 14 days, leaving a flat scar. A skin transplant was found to be unnecessary, contrary to initial recommendations (17a). The combination of HBO, vasodilators, and stellate block saved the arm of an addict who injected methylphenidate into his brachial artery (32). Merrick applied HBO therapy to eight patients with ergotism after other methods had been unsuccessful. Treatment with HBO at 3 ATA for 1 hour 2–3 times a day, in combination with an epidural block, was successful in all cases (109).

F. Peripheral Vascular Disease

Oxygen has a vasoconstrictor effect in normoxic tissues but does not act as a vasoconstrictor in hypoxic tissues until the oxygen deficit is reduced (12,79, 162). HBO therapy can compensate for the lack of oxygen in ischemic tissue but cannot modify the retention of toxic metabolic products. In HBO-treated patients with chronic thrombo-obliterating disease of the extremities, Lukin et al. found normalization of the metabolic processes, oxygen levels, and acid-base balance (95). Plethysmographic studies revealed that HBO can increase the rate of blood flow in patients with skin ischemia (134) and can promote blood flow from healthy areas to hypoxic areas in patients suffering from arteriopathies of limbs (45).

HBO causes a significant reduction of postischemic edema (22), and this effect may persist 40 hours after the last treatment (118). HBO also improves some rheological properties of the blood [decrease in the viscosity of the blood, lowering of platelet aggregation, increase of fibrinolytic activity (42, 93,161)]. Local blood flow (>1.5 ml/min/100 g muscle tissue) and the time necessary for the tissue Po_2 to drop to pretreatment values ("after effect") may serve as predictors of the effectiveness of HBO therapy in peripheral vascular disease (161). Other benefits of HBO therapy in arterial occlusive disease (due to arteriosclerosis, thromboangiitis obliterans, embolic or thrombotic occlusion of the vessels) are relief of pain (71), healing of ulcerations, regression of gangrene demarcation, and decrease in incidence of limb amputation (71,82,129,132). Yephuni et al. (161) treated over 3000 patients with arterial occlusive disease with adjunct HBO therapy. They obtained the best results in patients with chronic arterial insufficiency. Kitani et al. (84) found a combination of intravenous prostacyclin (PGE_1) and HBO in the treatment of ischemic peripheral artery disease beneficial.

VI. CHRONIC ISCHEMIC CONDITIONS

An increasing number of publications about the supportive effects of HBO in the treatment of various kinds of skin ulcers is available (17a,39,40,41, 61,65,66a,70,85,119,158,159) (Figs. 3 and 4).

A. Ulcers Due to Arterial Insufficiency

Reports about the successful treatment of arterial skin ulcers by adjunctive HBO application have been reviewed by Wattel et al. (151). Following HBO (1.5–3.0 ATA) alone, Perrins and Barr (123) achieved healing of arteriosclerotic ulcers in 52% and improvement in 20% of patients ($n = 50$). They stated that the response to HBO was dose dependent: some patients responded to 1.5 ATA, while others required up to 3 ATA. Some failed to respond to a treatment of 2 hours per day, while others healed with less than 1 hour a day. Rheumatoid ulcers responded well to HBO treatment. Wattel et al. stated that the $tcPo_2$ at 2.5 ATA is a reliable indication of final outcome (healing or no change). In HBO therapy, when the distal $tcPo_2$ value was less than 100 mmHg, patients showed either no improvement or aggravation of their ulcers. With values higher than 100 mmHg, wound healing occurred in all patients (151). Yagi et al. (159), who treated 121 severe ischemic wounds with HBO (2 ATA, 90 min once a day) together with intraveneously applied prostacyclin, report 68.6% healing of arterial ulcers of the skin. They believe that HBO is also effective for problem wounds of ischemic limb.

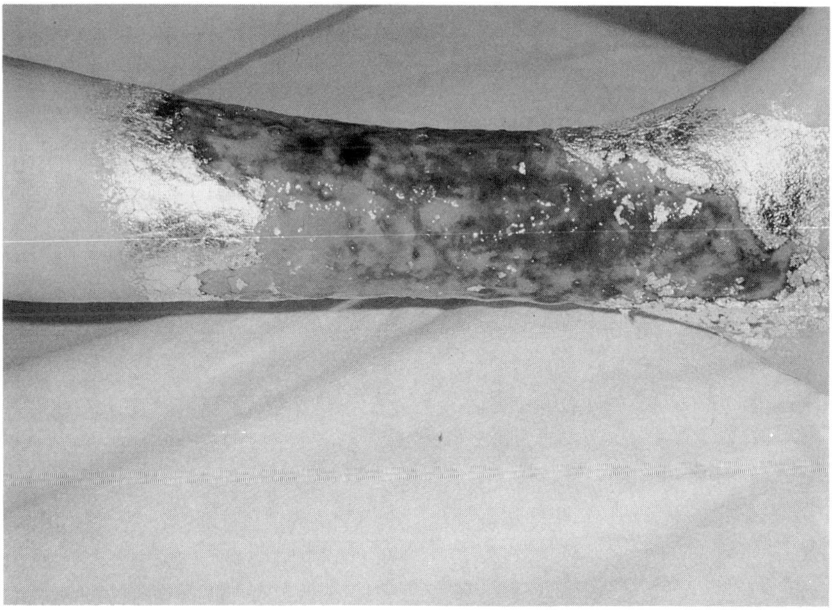

Figure 3 A 63-year-old female with poliomyelitis 40 years ago with consecutive stiffness of the ankle joint, decreased muscle mass, and a chronic nonhealing leg ulcer prior to HBO therapy. Note the clinical signs of compromised wound healing.

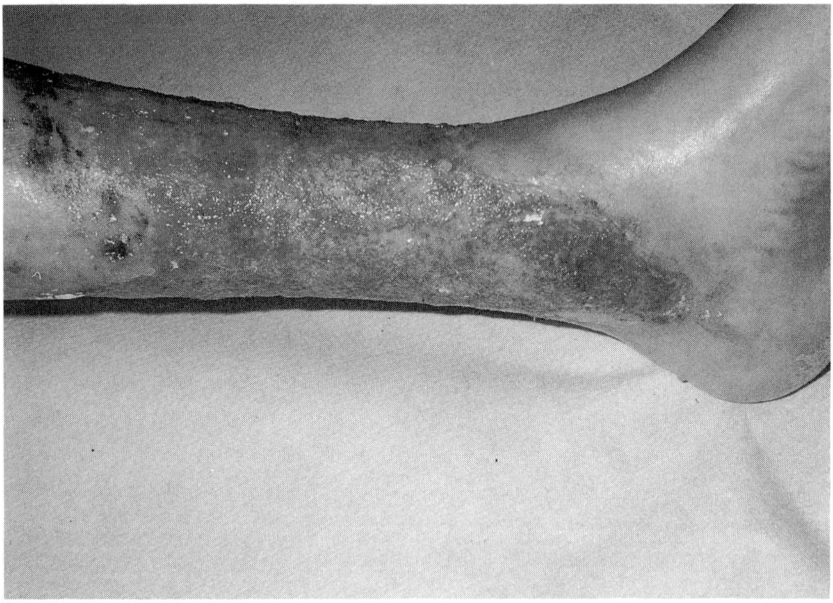

Figure 4 The same patient as in Figure 3 14 days after starting HBO (11 HBO treatments; 1.5 ATA, 45 min): fresh granulation tissue.

B. Venous Stasis Ulcers

Oxygen deficit plays a central role in the pathogenesis of venous ulcers (44). Venous stasis ulcers that do not respond sufficiently to classical therapy are therefore another indication for adjunctive HBO (3,17a,39,66a,123).

C. Diabetic Ulcers

Diabetes mellitus is one of the common causes of nonhealing ulcers. Baroni et al. (6) were able to show that healing in patients with diabetic foot ulcers was considerably improved by the application of hyperbaric oxygen. Others have reported many cases where positive effects on the healing processes in diabetic lesions have been observed (17a,28,66a). Diabetic patients with additional diseases of the large blood vessels appear to respond to hyperbaric oxygen treatment only after surgical treatment of the vessels has been undertaken. The therapy must be accompanied by daily wound dressing with the administration of sufficient levels of analgesics and antibiotics, if necessary, surgical débridement or skin graft treatment, supportive dietary measures, and a closely monitored metabolic control program. Cianci et al. (24) reported on a series of 41 patients with diabetic foot ulcers, 95% of which were considered to be limb-threatening lesions. Patients received an average of 37 HBO treatments. Thirty-one patients' extremities were saved (76%). HBO therapy thus represents a saving, as the costs of primary amputation are very high. In a randomized controlled study, Oriani et al. (121) evaluated HBO therapy in 80 diabetic gangrene patients. According to their study, the amputation rate was much lower in the HBO group (5% vs. 33% without adjunctive HBO treatment). Unger and Lucca (146) state that HBO therapy can be a valuable adjunct in the care of the diabetic patient with nonhealing ulcers, refractory osteomyelitis, or both of the lower extremities. In a study, Mathieu and Wattel (105) showed that of 59 patients with diabetic ulcers, 52 healed with adjunctive HBO treatment. Emhoff and Myers (37) stated that HBO may not be useful if $tcPo_2$ cannot be raised to 1000 mmHg when the patient breathes hyperbaric oxygen at 2.36 ATA. The patient's microcirculation was assumed to be defective and incapable of providing an adequate supply and flow of oxygen to the wound. Jacoby and Müller (73) found it useful to measure response of $tcPo_2$ to inspired oxygen at normobaric pressures as a predictor of response of diabetic foot wounds to HBO. Their hypothesis is that the response of a patient's diabetic wound is more likely to be positive if the wound is hypoxic ($tcPo_2 < 30$ mmHg) but can receive an enhanced oxygen supply through its existing blood supply ($tcPo_2$ will increase to 40 mmHg or higher when the patient breathes oxygen). HBO was also found to be useful in patients with acro-osteopathia ulceromutilans (17a).

D. Pressure Sores

Seiler and Stähelin (135) made a transcutaneous measurement of the oxygen tension in the tissue under pressure. A pressure of 15 kPa at a point—usually these ulcers are located over bony prominences—leads to anoxia and a pressure sore within 2 hours. HBO is recommended for pressure sores (39,51, 128) and should also be considered when a skin flap is in danger of ischemia during surgery for pressure sores (36).

E. Soft Tissue Radionecrosis

A number of publications about the beneficial role of HBO in the treatment of radiation necrosis has been compiled by Kindwall et al. (83). Radiation induces vascular sclerosis and hypoxia with consequent tissue or bone necrosis. Hyperbaric oxygen enables treatment of the radiated soft tissue itself (e.g., facilitates neovascularization and raises tissue oxygen tension permanently in the irradiated area), thus leading to spontaneous epithelialization even after months of otherwise refractory therapy (103). Preoperative use of HBO is advocated whenever surgical reconstruction in cases of roentgen ulcer and radiodermatitis is required. For this reason it is suggested that the patient be treated 30 times prior to surgery and 10 or, if necessary, more than 10 times postoperatively. As radiodermatitis and roentgen ulcer are very difficult conditions to heal, HBO is strongly recommended as a valuable adjunct method.

VII. PYODERMA GANGRENOSUM AND OTHER ULCERS OF UNKNOWN ETIOLOGY

Pyoderma gangrenosum is a chronic liquefying cutaneous gangrene with ischemic ulcers that result from reactive vascular changes or an intense vasculitis. There are several reports on the successful treatment of pyoderma gangrenosum with hyperbaric oxygen in cases unresponsive to standard immunosuppressive therapy (7,8,142,155). The rapidity of healing and the relief of pain is often striking (17a,66a) (Figs. 5 and 6). Gadd et al. (47) investigated the proliferation rate of T lymphocytes in a murine model under HBO and found it to decrease. The suppressive effects of HBO on immune responses (130) might help in understanding the beneficial role of hyperbaric oxygen therapy in immunologically mediated skin ulcers, e.g., in pyoderma gangrenosum, in rheumatoid arthritis (96), or in certain collagen vascular diseases (8, 138). There are at least two reports of the successful closure of otherwise refractory perineal Crohn's lesions by adjunctive hyperbaric oxygen therapy (15,113). Polyakova et al. (124) found the adjunct HBO treatment of patients with Crohn's disease or colitis ulcerosa useful, with recovery of large bowel

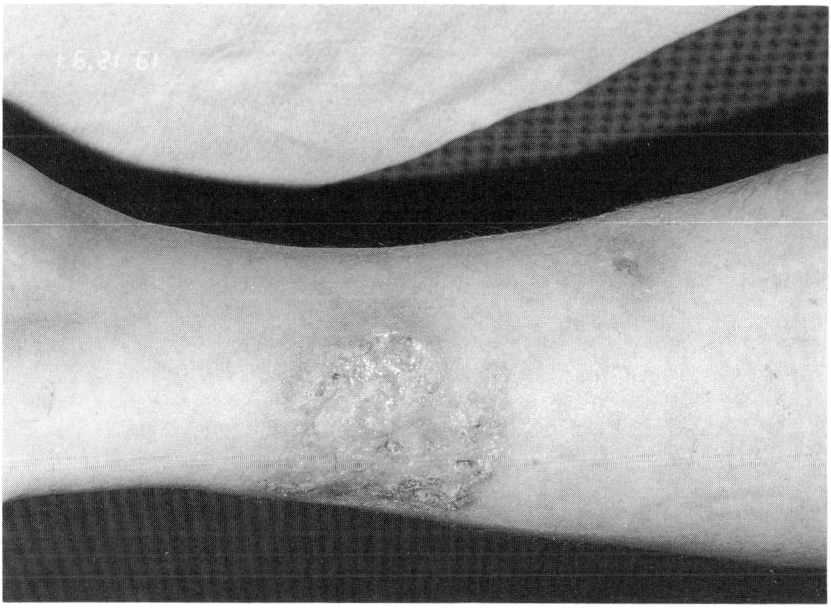

Figure 5 A 37-year-old female with Crohn's disease (regional ileitis) presenting with a pyoderma gangrenosum resistant to immunosuppressive therapy prior to HBO.

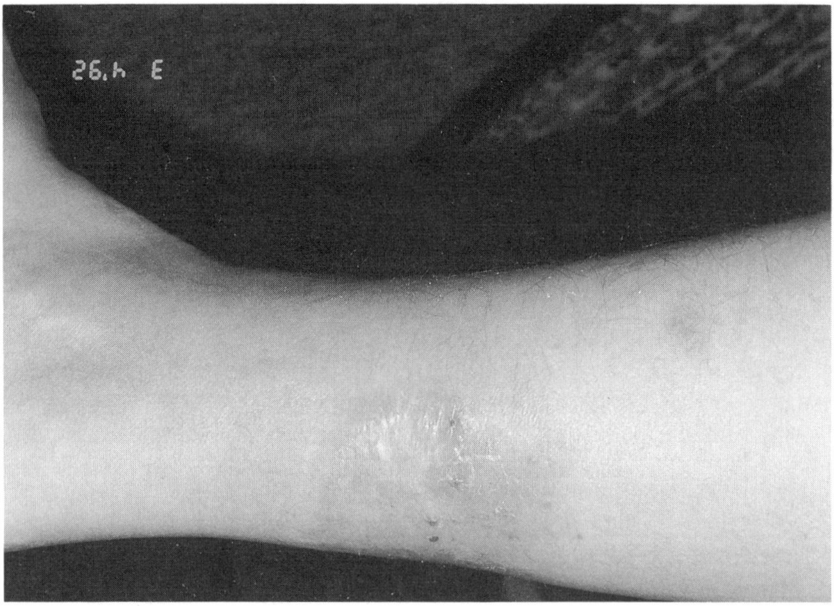

Figure 6 The same patient as in Figure 5—complete healing of the ulcer under a combined therapy of HBO (10 treatments; 1.5 ATA, 45 min) and methylprednisolone.

mucosa, reduced need for drugs, no need for surgical interventions, and shorter hospital stays.

VIII. SKIN GRAFTS AND FLAPS

HBO can be advantageous if there is significant local ischemia (decrease in microcirculation, hypoxia), if the patient is a systemically compromised host, or if there have been multiple previous graft failures (83). The results of numerous animal studies indicate that HBO reduces the healing time of full and partial thickness wounds to one half of nontreated controls (18,29,115, 136), but there are also some conflicting findings (111). The sooner HBO starts, the greater is the increase in the surviving area even when hyperbaric oxygen is delayed for 24 hours after surgery (76). Nemiroff et al. (115) did microvascular studies and concluded that HBO significantly enhances flap survival by increasing or maintaining the number and possibly the vessel size within the microvasculature. Many studies in humans reveal a 80–90% success rate of skin graft and flap transplants with HBO (13,26,46,77,104,123). Treatment usually consists of daily hyperbaric oxygenation (2–2.4 ATA, 90–120 min, twice a day) and is recommended for 3–10 days depending on clinical condition and flap type (83).

IX. CONTRAINDICATIONS AND SIDE EFFECTS

In order to prevent or minimize undesirable side effects of HBO, the absolute and relative contraindications should be taken into consideration and the physiology and pathophysiology of typical problems or side effects should be well known (Table 2). The partial pressure of oxygen and the duration of HBO should be carefully selected.

Adverse effects may be caused by the increased total air pressure, including the necessary changes in pressure during compression and decompression (pure pressure effects), or by the increased partial pressure of oxygen (16,17, 17a). Pressure-related adverse effects can arise through disturbances in equalization of pressure: e.g., a dysfunction of the eustachian tube, caused by an upper respiratory infection, with impaired equalization of pressure on both sides of the tympanic membrane, is one of the most common problems in HBO. Similarly, disturbances in equalization of pressure in the paranasal sinus or in nonventilated or dysventilated areas of the bronchial tree or the alveoli may occur due to infections, bronchial asthma, emphysema, or pulmonary lesions. If these problems are disregarded, bleeding or rupture of tympanic membrane (otic trauma), sinus pain, or rupture of pulmonary tissue, in some cases as severe as a pneumothorax (pulmonary barotrauma), may occur (74).

Table 2 Contraindications and Possible Side Effects of HBO

Contraindications
Absolute
pneumothorax
Doxorubicin (adriamycin) therapy
Relative
disburbances in equalization of pressure (e.g., upper respiratory infection, bronchial asthma)
emphysema of the lung with CO_2 retention
asymptomatic pulmonary lesions on chest x-ray
history of thoracic surgery or ear surgery
uncontrolled high fever
history of seizures
early pregnancy (?)
claustrophobia
Possible side effects
Pressure-related
otic barotrauma, sinus pain
pulmonary barotrauma
decompression sickness
Oxygen-related
CNS dysfunction (e.g., excitation, seizures, unconsciousness)
lung dysfunction (reduced oxygen diffusion capacity)
cataract and reversible myopia
possible negative effects on viral infections, malignant diseases, and optic neuritis

These risks can be minimized by thorough anamnesis and clinical investigation (prior to HBO), including inspection of the ear, a ventilatory investigation, blood gas analysis, and x-ray of the chest. HBO should be postponed until after an upper respiratory tract infection has been cured (or paracentesis in emergency cases). Taking into account all special individual risk factors and using a very smooth increase and decrease of pressure (e.g., only 0.05–0.1 atmospheres pressure change per minute) in continuous communication with the patient in order to recognize eustachian tube dysfunction, can also reduce the proportion of negative side effects. The advantage of HBO and the potential risks involved must be taken into consideration in the presence of contraindications. A preexisting pneumothorax is, if not treated surgically, an absolute contraindication for HBO (43,74). The pressure-related problem of decompression sickness (with the formation of gas bubbles, especially nitrogen) can be avoided completely by using low pressures only (e.g., 1.5 ATA) or by limiting the time under higher pressures, by using pure oxygen, and by slow decompression. The adverse effects of high partial pressure of oxygen (oxygen toxicity) may become apparent as dysfunctions of the central nervous system when using very high total pressures under pure oxygen, i.e., more than 3 ATA, or of the lung when using increased oxygen partial pressures over very long periods, e.g., days. These risks can be minimized or even practically excluded by limiting the maximum partial pressure and the duration of HBO and by taking into account possible contraindications (17a, 43,74). In case of risk factors or contraindications (43,74), the application of HBO should be decided case by case.

Myopia may result as a reversible effect of HBO (74). Cataracts may occur during long-term use of HBO. The risks can be minimized by limiting the number of HBO treatments (not more than 200 treatments) and the pressure (not more than 3 ATA) (49,74). The development of cataracts may be due to oxidative damage to the lens proteins with the formation of large and dense molecular aggregates and a decrease in defense mechanisms such as glutathione (49,74). Increased oxidation and increased numbers of reactive species (oxygen radicals) is not automatically an adverse effect of HBO. Both are necessary to living cells, for example, the function of granulocytes, but excessive and uncontrolled formation should be avoided.

Blood investigations by electron spin resonance analysis showed no change in blood levels of free oxygen radicals before and after 60-minute exposure to 100% oxygen at 1 ATA. HBO (daily 1 hour, 2 ATA) can increase lipid peroxidation in rabbits (102). This increase can be prevented by prophylactic application of antioxidants and other protective substances (30,98,110,144). In order to reduce the level of free oxygen radicals in HBO, frequent rest periods under normal air are recommended (106). Whether hyperoxia increases lipid peroxidation in lung tissue remains a matter of controversy (35,75).

An increase of chemical luminescence in urine and of malondialdehyde in serum (both indirect and unspecific indicators of lipid peroxidation) was observed under HBO (2.4 ATA, 1 h/day for 7 days), which could be prevented by application of ascorbic acid (3 g/day) and alpha-tocopherol (3 g/day) (10). Ascorbic acid, alpha-tocopherol, and magnesium protect against oxidative stress in HBO (10,30,43,74,110). Vitamin E alone or in combination with selenium prevented HBO-induced changes in the electroretinogram (35,137). Although the need for prophylactic administration of oxygen radical scavengers or other protective agents in HBO has not yet been proven [vitamin E 400 mg/day (43,74); magnesium 120 mg/day (43), 240 mg/day (74)], we give 100 mg vitamin E 2 hours before HBO treatment (1.5 ATA for 45 min) (17a).

X. FUTURE PROSPECTS FOR HBO

The aim of this chapter is to encourage broader use of HBO for the benefit of all patients showing indications for this treatment. Three thousand hyperbaric centers are already in operation worldwide, but some countries still have few facilities (e.g., Germany has only approximately 12 multiplace chambers).

Another goal is to increase the detailed knowledge of the mode of action in particular diseases and the precise dosage required in each case. Recommended fields of investigation include 1) the role of O_2 transport and of Po_2-dependent phenomena in defective wounds, especially in regard to the application of HBO (incorporating development of methods for measuring cellular Po_2 and its distribution within the wound); 2) interactions between the various cells involved in tissue repair with the focus on the role of HBO in influencing these interactions; 3) determination of the effects of HBO on normal wound healing in humans; and 4) carefully controlled human trials of HBO to investigate the various host-related and therapy-related variables (duration and dosage of HBO, multiplicity of HBO exposures, and the spectrum of "adjuncts" to HBO: surgery, antibiotics, nutrition, treatment of the underlying conditions, and specific pharmacological interventions) (48).

REFERENCES

1. Arturson G. Pathophysiology of the burn wound. Ann Chir Gynaecol 1980; 69: 178.
2. Arzinger-Jonasch H, Sandner K, Bittner H. Die Wirkung hyperbaren Sauerstoffs auf Brandwunden unterschiedlicher Tiefe im Tierexperiment. Z Exp Chir Transplant Künstliche Organe 1987; 11:6–10.
3. Bakker DJ. The treatment of acute dermal gangrene with hyperbaric oxygen. Ned Tijdschr Geneeskd 1980; 124:2164–70.
4. Bakker DJ. Pure and mixed and anaerobic soft tissue infections: the role of surgery, hyperbaric oxygen and antibiotics. In: Proceedings of the 8th International Congress on Hyperbaric Medicine, Long Beach, CA, 1984. Flaggstaff, AZ. Best Publishing Company, 1987:10.
5. Bakker DJ. Treatment of gas gangrene with HBO: results of 15 years experience. In: Smith, ed. Proceedings of the 6th International Congress of Hyperbaric Medicine. Aberdeen University Press, Aberdeen, 1977:295–301.
6. Baroni G, Porro T, Faglia E, et al. Hyperbaric oxygen in diabetic gangrene treatment. Diabetes Care 1987; 10:81–6.
7. Barr PO, Enfors W, Eriksson G. Hyperbaric oxygen therapy in dermatology. Br J Dermatol 1972; 86:631–5.
8. Barr PO, Perrins DJD. Prolonged use of hyperbaric oxygen (HBO) in undulent ulcers of the leg. Proceedings of the 8th International Congress on Hyperbaric Medicine, Long Beach, CA, 1984. Flaggstaff, AZ. Best Publishing Company; 1987:217–222.
9. Bass BH. The treatment of varicose leg ulcers by hyperbaric oxygen. Postgrad Med J 1970; 46:407–8.
10. Behn C, Brichta M, et al. Personal communication. 1992.

11. Bertholds M. Lepraforschung—derzeitiger Kenntnisstand. Colloquium 1990 of Vinzenz von Paul Foundation, November 2-3, 1990, Nordrach-Klausenbach, Germany.
12. Bird AD, Telfer AB. Effects of hyperbaric oxygen on limb circulation. Lancet 1965; I:355–6.
13. Bowersox JC, Strauss MB, Hart GB. Clinical experience with hyperbaric oxygen therapy in salvage of ischemic skin flaps and grafts. J Hyperbar Med 1986; 1:141.
14. Boykin JV, Erikson E, Pittman N. In vivo microcirculation of a scald burn and the progression of postburn dermal ischemia. Plast Reconstr Surg 1980; 66:191.
15. Brady CE, Cooley BJ, Davis JC. Healing of severe perineal and cutaneous Crohn's disease with hyperbaric oxygen. Gastroenterology 1989; 97:756–60.
16. Braun E. Hyperbaric chambers and ancillary equipment: Safety, complications and contraindications. In: Fischer B, Jain KK, Braun E, Lehrl S, eds. Handbook of hyperbaric oxygen therapy. New York: Springer, 1988:47–58.
17. Brichta M. Hyperbarer Sauerstoff. Nebenwirkungen und Gegenmaβnahmen. 87. Sports physiological colloquium in Berlin, Germany, February 5, 1992.
17a. Buslau M, Hoffmann G. Die hyperbare Oxygenation (HBO)—eine adjuvante Therapie akuter und chronischer Wundheilungsstörungen. Dermatol Monatsschr (accepted for publication) 1993.
18. Caffee HH. Effects of hyperbaric oxygen on skin flap survival in the pig—reply. Plast Reconstr Surg 1989; 83:1079.
19. Champion WM, McSherry CH, Goulian D. Effect of hyperbaric oxygen on the survival of pedicle skin flaps. J Surg Res 1967; 7:583.
20. Cianci P, Lueders HW, Lee H, Shapiro RL, Sexton J, Williams C, Green B. Adjunctive hyperbaric oxygen reduces the need for surgery in 40-80% burns. J Hyperbaric Med 1988; 3:97–101.
21. Cianci P. Hyperbaric oxygen in the treatment of thermal burns—state of the art 1989. In: Marroni A, Cramer FS, eds. International course on clinical hyperbaric medicine. Amsterdam, August 17, 1990. Amsterdam: Foundation for the International Congress on Hyperbaric Medicine, 1990:1–6.
22. Cianci P, Lueders H, Ford J. Treatment of post-ischemic edema compartment syndrome with adjunctive hyperbaric oxygen therapy. Undersea Biomed Res 1990; 17 (suppl):36(abstr).
23. Cianci P, Petrone G, Shapiro R, Lueders H, Lee L, Sato R, Sexton J, Cloud F. Adjunctive hyperbaric oxygen reduces the need for grafting in burned hands. Undersea Biomed Res 1990; 17(suppl):42(abstr).
24. Cianci P, Petrone G, Drager S, Lueders H, Green B. Salvage of the problem wound and/or potential amputation cost analysis of 101 patients. Undersea Biomed Res 1990; 17(suppl):30(abstr).
25. Cianci P, Williams C, Lueders H, Shapiro R, Sato R. Adjunctive oxygen in the treatment of thermal burns. An economical analysis. J Burn Care Rehabil 1990; 11:140–3.
26. Collins TM, Caimi R, Lynch PR, Sheffield J, Mitra A, Stueber K, Smith YR. The effects of nicotinamide and hyperbaric oxygen on skin flap survival. Scand J Plast Reconstr Surg Hand Surg 1991; 25:5–7.
27. Cristoferi G, Fabris G, Ronconi AM, Bozza F, Gallassi CG, Bucca D, Caria GM, Duodeci S. Gas gangrene. Clinical considerations, prognosis and therapeutic prospects in our experience. J Chir Paris 1991; 128:243–6.

28. Davis JC, Buckley CJ, Barr P. Compromised soft tissue wounds: correction of wound hypoxia. In: Davis JC, Hunt TK, eds. Problem wounds—the role of oxygen. New York: Elsevier, 1988:143–52.
29. Delfino M, Fabbrochini G, Di Vaia E. Effects of hyperbaric oxygen in cryobiologic wounds. Arch Dermat Res 1989; 282:267–9.
30. DeMartino G, Ortolani O, Santangelo E, Celano S, Abate C. Oxygen toxicity prevention in patients undergoing hyperbaric therapy. Undersea Biomed Res 1990; 17(suppl):40–1(abstr).
31. Demello FJ, Hashimoto T, Hitchcock CR, Haglin JJ. The effect of hyperbaric oxygen on the germination and toxin production of Clostridium perfringens spores. In: Wada J, Iwa T, eds. Proceedings of the fourth international congress on hyperbaric medicine. Baltimore: Williams and Wilkins, 1970:276–81.
32. DeMyttenaere S, Heifetz M, Shilansky H. Different treatments used in a case of gangrene due to accidental intra-arterial injection of methylphenidate (Ritalin). Anaesth Analg 1977; 34:405–10.
33. Desola J, Escola E, Moreno E, Munoz MA, Sanchez U, Murillo F. Combined treatment of gaseous gangrene with hyperbaric oxygen therapy, surgery and antibiotics. A national cooperative multicenter study. Med Clin Barc 1990; 94: 641–50.
34. Dudgeon DL, Kellogg DR, Gilchrist GS, et al. Purpura fulminans. Arch Surg 1971; 103:351–8.
35. Elstner EF. Der Sauerstoff. Biochemie, Biologie, Medizin. Mannheim: BI-Wissenschaftsverlag, 1990.
36. Eltorai IM. Hyperbaric oxygen in the management of pressure sores in patients with injuries to the spinal cord. J Dermatol Surg Oncol 1981; 7:737–40.
37. Emhoff TA, Meyers RAM. Transcutaneous oxygen measurements and wound healing. Baltimore: Maryland Institute for Emergency Medical Services Sys-Tems, 1984.
38. Fichelle A, Nimier M. Infections of soft tissues by bacterial synergism. Ann Fr Anaesth Reanim 1990; 9:269–74.
39. Fischer BH. Topical hyperbaric oxygen treatment of pressure sores and skin ulcers. Lancet 1969; 1:405–9.
40. Fischer BH. Topical hyperbaric oxygen: treatment of pressure sores and certain skin ulcerations. Nurs Times 1970; 66:613–16.
41. Fischer BH. Treatment of ulcers on the legs with hyperbaric oxygen. J Dermatol Surg Onc 1975; 1:55–8.
42. Fischer B, Lehrl S, Jain KK, Larsonneur B, et al. Clinical, biochemical and mental changes during HBO therapy—a preliminary report. In: Schmutz J, ed. Proceedings of the 1st Swiss symposium of hyperbaric medicine. Basel: Foundation for Hyperbaric Medicine, 1986.
43. Fischer B, Jain KK, Braun E, Lehrl S. Handbook of hyperbaric oxygen therapy. New York: Springer, 1988.
44. Franzek UK. Sauerstoffpartialdruck und Kapillarmorphologie bei Patienten mit chronisch-venöser Insuffizienz (CVI). Phlebol Proktol 1983; 12:149–60.
45. Fredenucci P. Oxygénothérapie hyperbare et artériopathies. J Mal Vasc 1985; 10(Suppl A):166–72.

46. Frigerio D, Lovisetti G, Lovisetti L. Effect of the hyperbaric oxygenation on the survival of experimental skin flaps in rats. Undersea Biomed Res 1990; 17(suppl): 150(abstr).
47. Gadd MA, McClennen DS, Neumann TS, Hansbrough JF. Effect of hyperbaric oxygen on murine neutrophil and T-lymphocyte function. Crit Care Med 1990; 18:974–9.
48. Gail DB. Hyperbaric oxygenation therapy—NHLBI workshop summary. Am Rev Respir Dis 1991; 144:1414–21.
49. Giblin FJ, Schrimscher L, Chakrapani B, Reddy VN. Exposure of rabbit lens to hyperbaric oxygen in vitro: regional effects of GSH level. Invest Ophthalmol Vis Sci 1988; 29:1312–19.
50. Gorback MS, Moon RE, Hill GB, Georgiade GS, Piantadosi CA, Francia PJ, Camporesi EM, Fawcett TA. Mortality risk factors for necrotizing infections treated with HBO. Undersea Biomed Res 1991; 18(suppl):110(abstr).
51. Gorecki Z. Oxygen under pressure applied directly to bedsores. J Am Geriatr Soc 1964; 12:1147–8.
52. Gottlieb SF. The possible role of hyperbaric oxygen in the treatment of leprosy and tuberculosis. Dis Chest 1963; 44:215–17.
53. Gottlieb SF, Rose NR, Maurizi J, et al. Oxygen inhibition of growth of Mycobacterium tuberculosis. J Bacteriol 1964; 87:838–43.
54. Gottlieb SF. Oxygen under pressure and microorganisms. In: Davis JC, Hunt TK, eds. Hyperbaric oxygen therapy. Bethesda, MD: Undersea and Hyperbaric Medical Society, 1977:79–99.
55. Gottlieb SF, Franzblau SG, Van Meter KA, Sadeghani K, Hill RK Jr. The effect of two atmospheres oxygen on the growth of Mycobacterium leprae in the mouse foot pad. Undersea Biomed Res 1990; 17(Suppl):124(abstr).
56. Gottrup F, Niinikoski J, Hunt TK. Measurement of tissue oxygen tension in wound repair. In: Janssen H, Roman R, Robertson JIS, eds. Wound healing. Petersfield, Great Britain: Wrightson Biomedical Publishing Ltd., 1991.
57. Grim PS, Gottlieb LJ, Boddie A, Batson E. Hyperbaric oxygen therapy. J Am Med Assoc 1990; 263:2216–20.
58. Grossman AR. Hyperbaric oxygen in the treatment of burns. Ann Plast Surg 1978; 1:163–71.
59. Grossman AR, Grossman AJ. Update on hyperbaric oxygen and treatment of burns. Hyperbaric Oxygen Rev 1982; 3:51–9.
60. Hammarlund C, Svedman C, Svedman P. Hyperbaric oxygen treatment of healthy volunteers with UV-irradiated blister wounds. Undersea Biomed Res 1990; 17 (suppl):25–6(abstr).
61. Hart GB, Strauss MB. Responses of ischaemic ulcerative conditions to OHP. In: Smith G, ed. Hyperbaric medicine. Aberdeen University Press, 1979:312–14.
62. Hart GB, et al. Gas gangrene—a collective review. J Trauma 1983;23:991–1000.
63. Hart GB, O'Reilly RR, Broussard ND, et al. Treatment of burns with hyperbaric oxygen. Surg Gynecol Obstet 1974; 139:693–6.
64. Hartwig VJ, Kirste G. Experimentelle Untersuchungen über die Revaskularisierung von Verbrennungswunden unter hyperbarer Sauerstofftherapie. Zbl Chir 1974; 99:1112–17.

65. Heng MCY, Pilgrim JP, Beck FWJ. A simplified hyperbaric oxygen technique for leg ulcers. Arch Dermatol 1984; 120:640–5.
66. Hirn M, Niinikoski J. Necrotizing fasciitis treated with surgery, antibiotics and hyperbaric oxygen. In: Sterk W, Geeraedts L, eds. EUBS 1990 proceedings. Conference papers of the European Undersea Biomedical Society presented at the first Joint Meeting on Diving and Hyperbaric Medicine, Amsterdam, The Netherlands, August 11-18, 1990. Amsterdam: Foundation for Hyperbaric Medicine, 1990:219–25.
66a. Hoffmann G. Improvement of wound healing in chronic ulcers by hyperbaric oxygenation and by waterfiltered ultrared A induced localized hyperthermia. In: Vaupel P, ed. 20th conference of the international society on oxygen transport to tissue. Mainz, August 26–30, 1992. New York: Plenum Press (accepted for publication) 1993.
67. Hohn D. Leucocyte phagocytic function and dysfunction. Surg Gynecol Obstet 1977; 144:99–104.
68. Hunt TK: Burns. In: Hunt TK, Dumphy JE, eds. Fundamentals of wound management. New York: Appleton Century Crofts, 1979.
69. Hunt TK, Pai MP. The effect of varying ambient oxygen tensions on wound metabolism and collagen synthesis. Surg Gynecol Obstet 1972; 135:561–7.
70. Ignacio DR, Pavot AP, Newell M, et al. Treatment of extensive limb ulcers with the use of topical hyperbaric oxygen therapy. Adv Ther 1984; 1:55–61.
71. Illingworth CFW. Treatment of arterial occlusion under oxygen at two atmospheres pressure. Br Med J 1962; 2:1272.
72. Irvin TT, Norman JN, Suwanagul A, Smith G. Hyperbaric oxygen in the treatment of infections by aerobic microorganisms. Lancet 1966; 1:392–394.
73. Jacoby I. Müller K. Response of transcutaneous pO2 (tcpO2) to inspired oxygen at normobaric pressures as a predictor of response of diabetic foot wounds to HBO. Undersea Biomed Res 1991, 18(suppl):81(abstr).
74. Jain KK. Textbook of hyperbaric medicine. Toronto: Hogrefe und Huber, 1990.
75. Jamieson DD. Lipid peroxidation in brain and lungs from mice exposed to hyperoxia. Biochem Pharmacol 1991; 41:749–56.
76. Jurell G, Kaijser L. The influence of varying pressure and duration of treatment with hyperbaric oxygen on the survival of skin flaps: an experimental study. Scand J Plast Reconstr Surg 1973; 7:25–58.
77. Kaelin CM, Im MJ, Meyers RA, Manson PN, Hoopes JE. The effects of hyperbaric oxygen on free flaps in rats. Arch Surg 1990; 125:607–9.
78. Kaiser W, Berger A, Lieth H, Heymann H. Hyperbaric oxygenation in burns. Handchir Mikrochir Plast Chir 1985; 17:326–30.
79. Kawamura M, Sakakibara K, Yusa T. Effect of increased oxygen on peripheral circulation in acute, temporary limb hypoxia. J Cardiovasc 1978; 19:161–8.
80. Ketchum SA, Zubrin JR, Thomas AN, et al. Effects of HBO on small first, second and third degree burns. Surg Forum 1967; 18:65–7.
81. Ketchum SA, Thomas AN, Hall AD. Angiographic studies of the effect of hyperbaric oxygenation on burn wound revascularization. In: Wada J, Iwa T, eds. Hyperbaric medicine. Baltimore: Williams and Wilkins, 1969:388–94.
82. Kidokoro M, Sakakibara K, Rakao T, Nimei M, Hibi Y, et al. Experimental and clinical studies upon hyperbaric oxygen therapy for peripheral vascular disor-

ders. In: Wada J, Iwa T, eds. Hyperbaric medicine. Baltimore: Williams and Wilkins, 1969:462–8.
83. Kindwall EP, Gottlieb LJ, Larson DL. Hyperbaric oxygen therapy in plastic surgery—a review article. Plast Reconstr Surg 1991; 88:898–908.
84. Kitani Y, Ishizaki K. Fujita T. The effects of Lipo-PGE1 and the hyperbaric oxygen in the treatment of peripheral vascular diseases. Undersea Biomed Res 1990; 17(suppl):53(abstr).
85. Kizer KW, Kramm J, Dobbs L. Hyperbaric oxygen therapy of refractory skin ulcers. In: Proceedings of the VIIth Annual Conference on Clinical Applications of HBO, Anaheim, CA, 1982.
86. Kivisaari J, Viherssari T, Renvall S, Niinikoski J. Energy metabolism of experimental wounds at various oxygen environments. Ann Surg 1975; 181:823–8.
87. Knighton DR, Halliday B, Hunt TK. Oxygen as an antibiotic. The effect of inspired oxygen on infection. Arch Surg 1984; 119:199–204.
88. Knighton J, Silver A, Hunt TK. Regulation of wound-healing angiogenesis—effect of oxygen gradients and inspired oxygen concentration. Surgery 1981; 90:262–70.
89. Korn HN, Wheeler ES, Miller TA. Effect o f hyperbaric oxygen on second degree burn wound healing. Arch Surg 1977; 112:732–7.
90. Kühne HH, Ullmann U, Kühne FW. New aspects on the pathophysiology of wound infection and wound healing—the problem of lowered oxygen pressure in the tissue. Infection 1985; 13:52–6.
91. Kuzemko JA, Loder RE. Purpura fulminans treated with hyperbaric oxygen. Br Med J 1970; 4:157.
92. Lepawsky M, Cianci P, Salvian AJ. 100% survival in three streptococcal toxic shock like syndromes ameliorated by adjunctive hyperbaric oxygen. Undersea Biomed Res 1991; 18(suppl):79(abstr).
93. Li W, Li X. The hemo-rheologic changes in patients treated with hyperbaric oxygenation at 3 ATA. Undersea Biomed Res 1990; 17(suppl):61(abstr).
94. Lucca M, Unger HD, Devenny AM. Treatment of Fournier's gangrene with adjunctive hyperbaric oxygen therapy. Am J Emerg Med 1990; 8:385–7.
95. Lukich VL, Filimonova MV, Bazarova VS. Changes in tissue gas exchange and oxygen balance during hyperbaric oxygenation of patients with chronic thrombo-obliterating diseases of the extremities. Vestn Khir 1979; 121:65–70.
96. Lukich VL, Polyakova LV, Sotnikova TI, Belokrinitsky DA. Hyperbaric oxygen treatment for rheumatoid arthritis: Immunoassay. Undersea Biomed Res 1990; 17(suppl):141(abstr).
97. Luongo C, Chiefari M, Vicario C. Toxic epidermal necrolysis (TEN): a proposal of absolute indication for HBO. Undersea Biomed Res 1990; 17(suppl): 151(abstr).
98. Lutz J, Augustin AJ, Friedrich E, Purucker E. Hyperbaric oxygen treatment and the level of oxygen free radicals. Undersea Biomed Res 1990; 17(suppl): 51–2(abstr).
99. Mader JT. Mucormycosis infections. In: Strauss MB, ed. Instructional course on hyperbaric oxygen therapy—what's true and what's new in HBO. Amsterdam: Undersea and Hyperbaric Medical Society, 1990:89.
100. Mader JT, Wilson KJ. Actinomycosis. A review of utilization of hyperbaric oxygenation. HBO Rev 1981; 2:177–88.

101. Mandell G. Bactericidal activity of aerobic and anaerobic polymorphonuclear neutrophils. Infect Immun 1974; 9:337.
102. Mansilia VA, Demurov EA, Vysotsky MV. The influence of HBO on lipid peroxydation and contractile myocardial function in rabbits during the phase of rehabilitation after fibrillation and defibrillation of heart. In: Sterk W, Geeraedts L, eds. EUBS 1990 proceedings. Conference papers of the European Undersea Biomedical Society presented at the first Joint Meeting on Diving and Hyperbaric Medicine, Amsterdam, The Netherlands, August 11-18, 1990. Amsterdam: Foundation for Hyperbaric Medicine, 1990:93–5.
103. Marx RE, Ehlers WJ, Tayapongsak P, Pierce LW. Relationship of oxygen dose to angiogenesis induction in irradiated tissue. Am J Surg 1990; 160:519–24.
104. Mathieu D, Neviere R, Pellerin P, Wattel F. Pedicle skin flap. Prediction of outcome by transcutaneous oxygen measurements in hyperbaric oxygen. Undersea Biomed Res 1990; 17(suppl):170(abstr).
105. Mathieu D, Wattel F. Hyperbaric oxygen in the treatment of the diabetic foot. Undersea Biomed Res 1990; 17(suppl):160–1.
106. McCartney PW, Vial J, Narkowicz C. Measurement of free radicals during delivery of hyperbaric oxygen therapy. In: Sterk W, Geeraedts L, eds. EUBS 1990 proceedings. Conference papers of the European Undersea Biomedical Society presented at the first joint meeting on diving and hyperbaric medicine. Amsterdam, The Netherlands, August 11–18, 1990. Amsterdam: Foundation for Hyperbaric Medicine, 1990:97–101.
107. McCord JM, Keele BB Jr, Fridovich I. An enzyme-based theory of obligate anaerobiosis: the physiologic function of superoxid dismutase. Proc Natl Acad Sci USA 1971; 68:1024–7.
108. Melero M, Kaiman MI, Tiraboschi N, Botargues M, Radisic M. Adjunctive treatment with hyperbaric oxygen in a patient with rhino-sinuso-orbital mucormycosis. Medicina B Aires 1991; 51:53–5.
109. Merrick J, Gufler K, Jacobsen E. Ergotism treated with hyperbaric oxygen and continuous epidermal analgesia. Acta Anaesthesiol Scand 1978; suppl 67: 87–90.
110. Money FL. Underwater diving, oxygen poisoning and vitamin E. NZ Med J 1972; 75:34–5.
111. Monstrey SM, Mullick P, Narayanan K, Liang M, Ramasastry SS. Skin allografts survival and hyperbaric oxygen (HBO) therapy. Undersea Biomed Res 1991; 18(suppl):41(abstr).
112. Nachum Z, Adir Y, Sherman D, Melamed Y. Hyperbaric Oxygenation (HBO) for post-traumatic soft tissue infections. Undersea Biomed Res 1990; 17(suppl): 161–2(abstr).
113. Nelson EW, Bright DE, Villar LF. Closure of refractory perineal Crohn's lesion. Integration of hyperbaric oxygen into the case management. Dig Dis Sci 1990; 35:1561–5.
114. Nemiroff PM, Merwin GE, Brant T, Cassisi NJ. Effects of hyperbaric oxygen and irradiation on experimental skin flaps in rats. Otolaryngol Head Neck Surg 1985; 93:485–91.
115. Nemiroff PM, Garlington JC, Kolm P. Effects of hyperbaric oxygen on skin flap survival in the pig. Plast Reconstr Surg 1989; 83:1077–9.

116. Niinikoski J, Gottrup F, Hunt TK. The role of oxygen in wound repair. In: Janssen H, Roman R, Robertson JIS, eds. Wound healing. Petersfield, Great Britain: Wrightson Biomedical Publishing Ltd, 1991:165–74.
117. Niu AKC, Chao C, Lee HC, Lee AJ. Burns treated with adjunctive hyperbaric oxygen therapy: a comparative study in humans. Undersea Biomed Res 1990; 17(suppl):43(abstr).
118. Nylander G, Lewis D, Nördström H, Larsson J. Reduction of postischemic edema with hyperbaric oxygen. Plast Reconstr Surg 1985; 74:596–603.
119. Olejniczak S. Employment of low hyperbaric therapy in management of leg ulcers. Mich Med 1975; 74:1067–8.
120. Oriani G. Pedesini G, Nicolosi M, Meazza D, Sacchi C: Complicated peripheral wounds and hyperbaric oxygen therapy. In: Bakker DJ, Schmutz J, eds. Hyperbaric medicine. Proceedings of the Joint Meeting 2nd Swiss Symposium on Hyperbaric Medicine and 2nd European Conference on Hyperbaric Medicine, Basel, September 22-24, 1988. Basel: Foundation for Hyperbaric Medicine, 1990:173–5.
121. Oriani G, Meazza D, Sacchi C, Faglia E, Favales F, Mazzola E, Mastropasqua A, Pizzi GL. Hyperbaric oxygen therapy in treatment of diabetic gangrene. Undersea Biomed Res 1990; 17(suppl):160(abstr).
122. Park MK, Muhvich KH, Meyers RA, Marzella L. Hyperoxia prolongs the aminoglycoside-induced postantibiotic effect in Pseudomonas aeruginosa. Antimicrob Agents Chemother 1991; 35:691–5.
123. Perrins JD, Barr PO. Hyperbaric oxygenation and wound healing. In: Schmutz J, ed. Proceedings of the 1st Swiss symposium on hyperbaric medicine. Basel: Foundation for Hyperbaric Medicine, 1986:119–32.
124. Polyakova LV, Lukich VL, Grigoryeva GA. Hyperbaric oxygen and drug therapy in the treatment of nonspecific ulcerative colitis and Crohn's disease. Undersea Biomed Res 1990; 17(suppl):171(abstr).
125. Price JC, Stevens DL. Hyperbaric oxygen in the treatment of rhinocerebral mucormycosis. Laryngoscope 1980; 90:737–47.
126. Rieseman JA, Zamboni WA, Curtis A, Graham DR, Konrad HR, Ross DS. Hyperbaric oxygen therapy for necrotizing fasciitis reduces mortality and the need for debridements. Surgery 1990; 108:847–50.
127. Rosasco SA. Hyperbaric oxygen treatment of patients during resistant leprosy. In: Trapp WG, Banister EW, Davison AJ, eds. Proceedings of the 5th international congress on hyperbaric medicine. Vancouver, Canada, 1973. Burnaby, Canada: Simon Fraser University, 1974.
128. Rosenthal AM, Schurmann A. Hyperbaric treatment of pressure sores. Arch Phys Med Rehabil 1971; 52:413–15.
129. Rosenthal E, Benderly A, Monies-Chass I, et al. Hyperbaric oxygenation in peripheral ischaemic lesions of infants. Arch Dis Child 1985; 60:372–4.
130. Saito K, Tanaka Y, Ota T, Eto S. Yamashita U. Suppressive effect of hyperbaric oxygenation on immune responses of normal and autoimmune mice. Clin Exp Immunol 1991; 86:322–7.

131. Sakakibara K, Takahashi H, Kobayashi S. Clinical experience of hyperbaric oxygen therapy (OHP) for chronic peripheral vascular disorders. Proceedings of the 3rd symposium of university of Occupational and Environmental Health. Kitakyushu: School of Medicine, 1985;337–43.
132. Sakakibara K, Takahashi H, Kobayashi S. The role of hyperbaric oxygen in the salvage of ischemic limbs from inevitable amputation in chronic peripheral vascular disease with particular reference to those having had unsuccessful surgery. In: Proceedings of the 8th International Congress on Hyperbaric Medicine, Long Beach, CA, 1984; Flaggstaff, AZ: Best Publishing Company, 1987:221–8.
133. Salas E, Perez L. Treatment of the Fournier's gangrene with HBO. Undersea Biomed Res 1990; 17(suppl):150–1(abstr).
134. Schraibman G, Ledingham FRCS, Ledingham IMA. Hyperbaric oxygen and regional vasodilatation in pedal ischemia. Surg Gynecol Obstet 1969; 125:761–7.
135. Seiler WO, Stähelin HB. Dekubitus. Neue Forschungsmethode in der Decubituspathogenese. Hospitalis 1984; 6:319–24.
136. Shulman AG, Krohn HL. The influence of hyperbaric oxygen in multiple skin grafts on the healing of skin wounds. Surgery 1967; 62:1051.
137. Stone WL, Henderson RA, Howard GH, Hollis AL, Payne PH, Scott RL. Free Rad Biol Med 1989; 6:505.
138. Slade B. The effect of hyperbaric oxygen therapy (HBO) on wound healing in patients with collagen vascular disease: a retrospective analysis. Undersea Biomed Res 1991; 18(suppl):109(abstr).
139. Strauss MB, Lavermann MW, Strayer GL, Hart GB. Salvage of septic charcot joints of the foot and ankle. Undersea Biomed Res 1990; 17(suppl):38–9(abstr).
140. Talabiska L, Youn BA, DeAntonio A. Adjunctive hyperbaric oxygen therapy for severe purpura fulminans complicating varicella infection. Undersea Biomed Res 1991; 18(suppl):80–1(abstr).
141. Tirpitz D. The value of hyperbaric oxygenation in treatment of gas gangrene. In: Schmutz J, ed. Proceedings of the 1st Swiss symposium on hyperbaric medicine. Basel: Foundation for Hyperbaric Medicine, 1986:169–93.
142. Thomas CY, Crouch JA, Guastello J. Hyperbaric oxygen therapy for pyoderma gangrenosum. Arch Dermatology 1974; 110:445–6.
143. Tonjum S, Digranes A, Alho A, et al. Hyperbaric oxygen treatment in gas-producing infections. Acta Chir Scand 1980; 146:235–40.
144. Torley LW, Weiss HS. Effects of age and magnesium ions on oxygen toxicity in the neonate chicken. Undersea Biomed Res 1975; 2:223–7.
145. Trivedi DR, Raut VV. Role of hyperbaric oxygen therapy in the rapid control of gas gangrene infection and its toxaemia. J Postgrad Med 1990; 36:13–15.
146. Unger HD, Lucca M. The role of hyperbaric oxygen therapy in the treatment of diabetic foot ulcers and refractory osteomyelitis. Clin Podiatr Med Surg 1990; 7:483–92.
147. Van Unnik AJM. Inhibition of toxin production in Clostridium perfringens in vitro by hyperbaric oxygen. Antonie Van Leeuwenhoek 1965; 31:181–6.

148. Volturo P, Pelaia P, Vitullo V, Spinelli V. Hyperbaric oxygen therapy for the treatment of mucormycosis. Minerva Anaesthesiol 1990; 56:1381.
149. Waddell W, Saltzman HA, Fuson RL, et al. Purpura gangrenosa treated with hyperbaric oxygenation. J Am Med Ass 1965; 191:971–4.
150. Walker RM, Ashdown LR, Maguire EJ. Beneficial effects of hyperbaric oxygen therapy in Nocardia brasiliensis soft tissue infection. Med J Austr 1991; 155:122–3.
151. Wattel F, Mathieu D, Coget JM, Billard V. Hyperbaric oxygen therapy in chronic vascular wound management. Angiology 1990; 41:59–65.
152. Weisbren BA, Schutz D, Collentine G, et al. Hyperbaric oxygen in severe burns. Burns Incl Therm Inj 1982; 8:176–9.
153. Weislow OS, Parkman LM. Inhibition of pseudomonas aeruginosa by hyperbaric oxygen: interaction with mouse peritoneal exsudate cells. Infect Immun 1974; 10:546–52.
154. Wells CH, Hilton JG. Effects of hyperbaric oxygenation on host burn plasma extravasation. In: Davis JE, Hunt TK, eds. Hyperbaric oxygen therapy. Bethesda, MD: Undersea and Hyperbaric Medical Society, 1977:259–65.
155. Weyrick WJ, Mader JT, Butler E, Hulet WH. Hyperbaric oxygen treatment of pyoderma gangraenosum. Arch Dermatol 1978; 114:1232.
156. Winter GD, Perrins DJD. Effects of hyperbaric oxygen treatment on epidermal regeneration. In: Wada H, Iwa T, eds. Baltimore: Williams and Wilkins, 1969.
157. Wiseman DH, Grossman AR. Hyperbaric oxygen in the treatment of burns. Symposium on burns. Crit Care Clin 1985; 1:1.
158. Yagi H. On the hyperbaric oxygen therapy for severe infected granulation wounds (ulcers) of upper and lower extremities. Jpn J Hyperbar Med 1987; 22:27–40.
159. Yagi H, Okadome K, Fukuda A, Funahashi S, Sugimachi K: Therapy for problem wounds of ischemic limb with hyperbaric oxygen. In: Sterk W, Geeraedts L, eds. EUBS 1990 proceedings. Conference papers of the European Undersea Biomedical Society presented at the first Joint Meeting on Diving and Hyperbaric Medicine, Amsterdam, The Netherlands, August 11-18, 1990. Amsterdam: Foundation for Hyperbaric Medicine, 1990:211–18.
160. Yamaguchi KT, Hoffman C, Stewart RJ, Cianci PA, Vierra M, Naito M. Effect of oxygen on burn wound tissue levels of ATP and collagen. Undersea Biomed Res 1990; 17(suppl):65–6(abstr).
161. Yephuni SN, Lyskin GI, Fokina TS. Hyperbaric oxygenation in treatment of peripheral vascular disorders. Int Angiol 1985; 4:207–9.
162. Zamboni WA, Roth AC, Russell RC, Smoot EC. Acute effects of hyperbaric oxygen on laser doppler flow in ischemic axial skin flaps. Undersea Biomed Res 1990; 17(suppl):37(abstr).

III

EXPERIMENTAL AND CLINICAL ANDROLOGY

Although world population is increasing continuously, infertility plays a significant health role in westernized countries. About 10% of all partnerships are infertile. Enormous efforts have been made to elucidate the pathophysiological background of male infertility. Nevertheless, in many cases the exact reason for male infertility remains unclear. As a result, rational therapies are not available. Particularly in Germany, but also in other European countries, dermatologists were one of the first to care for male infertility patients for historical reasons. The idea of oxidative damage as a significant pathophysiological factor in male infertility has been investigated in several clinical trials. Other scientists interested in reproductive biology have contributed a large database on the significance of oxidative stress in experimental andrology and veterinary medicine. In Chapter 24, experimental and clinical data on the significance of oxidative damage in male infertility are presented in a comprehensive and logical way and give detailed insight into a complex field.

24

Oxidative Imbalance in Male Infertility

F. R. Ochsendorf and Jürgen Fuchs
Johann Wolfgang Goethe University,
Frankfurt, Germany

In Germany andrology is a subspecialty of dermatology for historical reasons. Dermatologists treated infertile male patients long before gynecologists and urologists entered this field (1). Oxidative injury is of clinical significance in general pathology and may also be a significant determinant in the pathophysiology of male infertility. During the past decade a considerable amount of knowledge has been accumulated on the interplay between oxidants and antioxidants in mammalian spermatozoa. The objective of this chapter is to present a state-of-the-art review and discussion of oxidative imbalance in male infertility. Furthermore, we discuss clinical significance and highlight the future implications of this research.

In the United States 10% of women aged 25–44 years suffer from unresolved impaired fecundity (2). Similar figures have been reported from Great Britain (3). In the United Kingdom about one in six couples will visit an infertility clinic during their reproductive lifespan (4). In the western part of Germany a recent survey revealed about 600,000 infertile couples, i.e., about 12% of all couples (5). Male or female factors are each likely to be responsible for infertility in 40% of all problems; 20% involve both partners. In spite of enormous progress in assisted reproduction for diseases of the female partner (like in vitro fertilization for therapy of tubal occlusion), there are few effective therapies for the male. As all the biochemical events leading to fertilization are still unknown, rational therapeutic approaches are rarely available.

The molecular pathology of oxidative injury in several human diseases attracted the attention of biomedical scientists during the past decade (6–8). However, the clinical consequences of oxidative injury in male infertility have not gained general interest up to now (9). Oxidative injury was recently identified as detrimental to mammalian spermatozoa function by the groups of Aitken, Alvarez and Storey, Engel, Jones and Mann, and Nissen and Kreysel. We suggest that oxidants and antioxidants are significant factors in the pathophysiology of male infertility.

I. HISTORY

The first evidence of the harmful effects of transition metal ions and hydrogen peroxide on spermatozoa was reported in 1907 (10). Since 1924 the antioxidant vitamin E has been used empirically as an anti-infertility drug; the benefits, however, are not agreed upon (see Ref. 11). Clinical efficiency has not been established, since double-blind studies were not performed. In 1943 MacLeod reported that incubation of spermatozoa with oxygen accelerates loss of cell motility (12). During the following 30 years almost no research was conducted in this field. In the 1970s Jones and Mann investigated lipid peroxidation in mammalian spermatozoa (13–19), in the 1980s Alvarez and Storey (20–25) and Aitken et al. (26–32) contributed many publications to clarify the role of reactive oxygen species on spermatozoa function, and in Germany Nissen and Kreysel (33–35) and Engel et al. (36–39) investigated lipid peroxidation and superoxide dismutase. Up to now, however, these results have received relatively little attention in daily practice (9,40). Our group is studying the correlation of male infertility parameters (morphology, vitality, motility, hypo-osmotic swelling, bovine mucus penetration) with the production rates of reactive oxidants, the concentration of lipid peroxidation products, and the antioxidant activity in human spermatozoa and seminal plasma.

II. FUNDAMENTALS

A. Spermatozoon

The spermatozoon is a highly specialized cell. It is the end product of differentiation starting with the spermatogonia, leading from primary and secondary spermatocytes to spermatids and spermatozoa. This cycle requires 74 days; another 7–14 days are needed for maturation in the epididymis. Thus, about 3 months elapse until positive or negative influences of toxins or drugs on spermatozoa number, morphology, or function can be identified in the ejaculate (41).

Genetic, inborn, or acquired injury to the spermatogonia may lead to subnormal production rates of spermatozoa. During this maturation process many malformed cells arise, which can be identified by light microscopy. Each of these pathological forms represents disorders at different stages of differentiation (42,43). It is self-evident that these malformed cells cannot function properly. A maximum rate of 50% pathological cells is accepted as normal (44).

The normal spermatozoon has the ability of being motile, having flagellum and abundant mitochondria in the midsection (Fig. 1). The biochemical correlate is the production of ATP via the enzyme ATP synthase, which is highly susceptible to oxidative damage (45). Impaired motility results from several disorders like structural defects of the flagellum, infections, diseases of prostate gland or seminal vesicles, varicocele, and others. In many patients the cause of impaired fertility remains idiopathic.

B. Assessment of Fertility

The conventional criteria for evaluation of male fertility are confined mainly to number, morphology, and motility of spermatozoa. Further investigations are hormonal (e.g., FSH, LH, testosterone, prolactin) and biochemical analyses of the seminal plasma (e.g., fructose, citrate, pH) to assess the function of prostate and seminal vesicles. Nevertheless, in many cases no explan ation for infertility can be found using these criteria. It is of paramount importance to point out that all these parameters do not give information on the functional capacity of spermatozoa to fertilize an egg! The functions of spermatozoa necessary for fertilization are discussed in the next section.

Recent studies demonstrated defective sperm function to be the most frequent defined cause in at least 24% of infertile couples (46). In about one third of couples with idiopathic infertility, disturbed sperm function could be demonstrated in spite of a normal semen profile (47,48).

C. Assessment of Spermatozoa Function

The ability to evaluate the proper function of spermatozoa on routine analysis, i.e., transport in the female reproductive tract, sperm-zona interaction, and fusion, is limited. The transport of spermatozoa and penetration of cervical mucus can be estimated by analyzing the motion patterns of spermatozoa (wide amplitude of lateral head displacement, velocity $>25\ \mu m/s$) (49) or the in vitro bovine mucus penetration assay (Penetrak^R) (50). The evaluation of adhesion to and penetration of the zona pellucida is more difficult. Zona recognition is limited by scarcity of material. Acrosin is thought to be

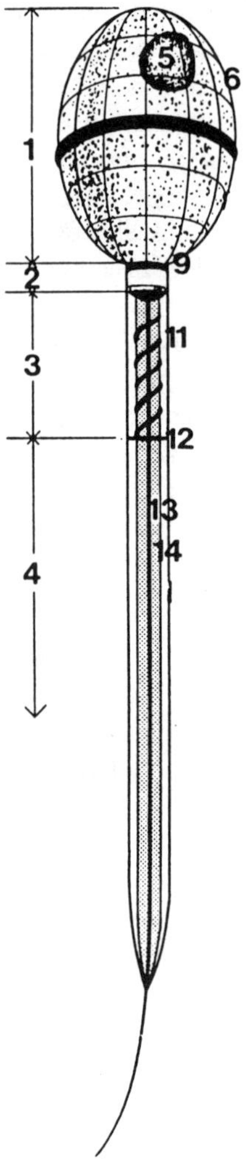

Figure 1 Diagram of human spermatozoon. 1, head; 2, neck; 3, midsection; 4, tail; 5, vacuole; 6, acrosomal region; 9, basal plate; 11, circumferential fibers; 12, end ring (annulus); 13, axial filament; 14, tail fiber. (From Ref. 41.)

involved in mediating this sperm-zona interaction. The penetration itself should be mediated by a nonprogressive hyperactivated form of motility. This form of motility, however, is only expressed by 10–20% of the cells; the exact role of this movement pattern for establishing pregnancy warrants further investigation. The acrosome reaction can be tested by monoclonal antibodies or plant lectins, but even under optimized conditions the mean proportion of acrosome-reacted cells reaches only about 20% (30,51). Finally, sperm-oocyte fusion can be investigated by the zona-free hamster oocyte penetration test. Especially after stimulation with the calcium ionophore A 23187, this test has excellent correlation with the result of in vitro fertilization (52,53). In summary, there is ample evidence that disturbed sperm function is a very important factor in male infertility; its evaluation, however, is difficult.

It is feasible to suggest that reactive oxidants may be involved in physiological spermatozoa functions and membrane interactions. Production of reactive oxidants and/or induction of localized lipid peroxidation may be a factor for capacitation, increase of sperm-zona adhesion, initiation of the acrosome reaction, and sperm-oocyte fusion (see Sec. IV.E). Reactive oxidants may also play a role in pathophysiology if produced at a high steady-state concentration and under uncontrolled conditions.

D. Spermatozoa Membrane

The degree of fatty acid unsaturation is one parameter that determines and regulates membrane fluidity. Fluidity is essential for the proper functioning of biological membranes.

Spermatozoa contain high amounts of unsaturated fatty acids bound to phospholipids. The largest fraction (more than 60% in ram, about 50% in human spermatozoa) consists of docosahexaenic acid (22:6), a fatty acid with six unsaturated double bonds (Table 1) (14,18,33,34). Therefore, the spermatozoa membrane is highly susceptible to oxidative injury.

The composition of the spermatozoa membrane varies with the degree of maturation. The plasma membrane of testicular spermatozoa undergoes extensive remodeling in the epididymis by a variety of mechanisms, which include rearrangement and modification of existing components, addition or coupling of new glycoproteins, and removal of lipid and protein constituents. These processes appear to be regulated by the epididymis. The main membrane changes occur in the caput epididymis, so that the plasma membrane of caudal spermatozoa has a different composition and different properties than the caput spermatozoa. These changes may be responsible for the ability to fertilize oocytes (54). In the anterior head of ram spermatozoa,

Table 1 Composition of Major Phospholipids and Major Phospholipid-Bound Fatty Acids Extracted from Human Spermatozoa Before and After Peroxidation

	Control spermatozoa		Peroxidized spermatozoa	
	μg P/100 mg wet weight	%	μgP/100 mg wt weight	%
Phospholipids				
sphingomyelin	7.8	12.4	10.5	23.5
phosphatidylcholine	24.1	38.3	18.9	42.2
phosphatidylethanolamine	18.7	29.7	4.9	11.0
ethanolamine-plasmalogen	6.3	10.1	1.9	4.3
Fatty acids				
16:0	256.0	26.1	273.4	41.3
18:0	101.6	10.0	76.1	11.5
18:1	70.7	6.9	70.5	10.6
18:2	50.8	5.0	42.3	6.4
20:4	39.8	3.9	16.9	2.6
22:0	4.4	0.4	2.8	0.4
22:6	481.4	47.5	180.5	27.2

P = phospholipids.
Source: Ref. 18.

isolated from different parts of the epididymis, Parks and Hammerstedt showed that the ratio of saturated to unsaturated fatty acids shifts to the latter and suggested a relation to the process of capacitation (55). These results were recently confirmed in rats (56). Spermatozoa membrane lipids differ significantly in chemical composition, e.g., the head lipids of human spermatozoa have a higher degree of unsaturation than the tail lipids (56a).

Essential membrane enzymes, like ion translocases and oxidoreductases, can be regulated by membrane lipids (57). Lipid peroxidation in the spermatozoa membrane may have far-reaching functional consequences (see Sec. IV.B). Fatty acid utilization in human spermatozoa seems to be regulated by docosahexaenic acid (58). Therefore, this unsaturated fatty acid plays a central role in spermatozoa function and malfunction.

The integrity of the plasma membrane is essential for the physiological function of spermatozoa. In the hamster oocyte penetration test, the penetration rate in oligozoospermic patients was significantly reduced in comparison to fertile patients (in the presence of the calcium ionophore A 23187) (29,52). In some patients impaired fertility was associated with the development of refractoriness of the sperm plasma membrane to the calcium stimulus. So the integrity of the plasma membrane is not essential only for motility (24) but also for the ability to adhere to and penetrate the oocyte.

III. REACTIVE OXYGEN SPECIES PRODUCTION IN SPERMATOZOA

A. Production

It is generally accepted that reactive oxygen species (ROS) and free radicals (reactive oxidants) may damage biomolecules such as lipids, proteins, nucleic acids, and carbohydrates. Five decades ago oxygen and hydrogen peroxide were demonstrated to injure spermatozoa from different species (bull, fowl, dog, ram, mouse, rabbit, humans) (12,59); spermatozoa are generally deficient in catalase (60).

In vivo reactive oxygen species derive from peroxisomes, mitochondria, autoxidation of cytosol components, and as products of different oxidoreductases. Until 1980, however, no evidence was obtained that mammalian spermatozoa are able to produce reactive oxygen species (19).

In 1987, Aitken demonstrated the production of reactive oxygen species in human spermatozoa by chemiluminescence (luminol) (26). Luminol reacts with a variety of reactive oxygen species (H_2O_2, $\cdot O_2^-$, $OH^\cdot$, $^1O_2$). Since luminol is a membrane-permeating reagent at neutral pH, it is not possible to state whether the reactive oxygen species are generated within the spermatozoon or released to the outside. Furthermore, it is by itself not specific for a certain reactive oxygen species.

Chemiluminescence was only measured in the pellet of centrifuged spermatozoa and of spermatozoa prepared by Percoll centrifugation or "swim-up." The site of production should be the spermatozoon itself, because contaminating cells had been removed by the preparation procedures (26). Production of reactive oxygen species was markedly induced by the calcium ionophore A 23187 in the presence of high extracellular Ca concentrations.

To define which reactive oxygen species were produced, different scavengers were co-incubated (scavenging of singlet oxygen: dimethyl furan, bilirubin, histidine HCL; hydroxyl radical: histidine HCL, sodium benzoate, mannitol; superoxide anion radical: cytochrome *c*). As only cytochrome *c* reduced the production rate of reactive oxygen species to about 50%, the superoxide anion radical was suggested to be the major product generated by the human spermatozoon (26).

Studies with catalase added to spermatozoa suppressed the chemiluminescence in a dose-dependent fashion, while superoxide dismutase had no effect. This suggests that under these conditions, the major product released to the exterior of the cell was hydrogen peroxide, arising through the action of endogenous superoxide dismutase. However, after stimulating the spermatozoa with a calcium ionophore, catalase had a much smaller but still significant effect, while SOD led to a dose-dependent decrease of ROS production until baseline levels at the highest concentration (3.5×10^3 U). A combination of both was no more effective than SOD alone (28). Therefore, after stimulation with a calcium ionophore, the main form of ROS appears to be the superoxide anion radical.

The superoxide anion radical has been detected in spermatozoa of rabbits (61) and mice (22). Superoxide anion radical and hydrogen peroxide were also detected in human spermatozoa (24) by a modified ferricytochrome *c* assay. It was postulated that $HO_2^{\cdot}$ was a most potent inducer of lipid peroxidation (22,23) and, by comparing the defense mechanisms of rabbit, mouse, and human spermaotozoa, that H_2O_2 itself was virtually nontoxic to spermatozoa (Table 2) (25).

In conclusion, there is ample evidence that reactive oxygen species are produced by human spermatozoa. The enhancing effect of calcium ionophore demonstrates the significant importance of calcium ions (see Sec. III.B).

B. Transition Metal Ions

Are the reactive oxidants produced able to injure spermatozoa? The superoxide anion radical itself is not sufficiently reactive to abstract H from lipids, and because of its charged nature it would not be expected to enter the hydrophobic interior of membranes. The protonated form of $\cdot O_2^-$, $HO_2\cdot$, is more reactive and should enter membranes fairly easily in the same way that H_2O_2 crosses membranes. However, $HO_2\cdot$ has not yet been proved to initiate peroxidation in cell membranes (6). However, $\cdot O_2^-$, $HO_2\cdot$, and H_2O_2 form highly reactive radicals like hydroxyl and, presumably, thiyl radicals.

Table 2 Glutathione Content, Glutathione Reductase, Glutathione Peroxidase, SOD Activity of Rabbit, Mouse, and Human Spermatozoa

	Rabbit	Mouse	Human
Gluthathione reductase	1.0 (0.1)	117 (9)	44 (3)
Glutathione peroxidase	4.0 (0.2)	194 (15)	21 (3)
(nmol/min per 10^8 cells)			
Glutathione, total	<0.1	96.0 (7.2)	7.0 (2)
Glutathione disulfide	—	37.0 (2.6)	3.0 (0.6)
Glutathione (nmol/10^8 cells)	—	59.0	4.0
SOD (U/10^8 cells)	2.7	3.2	1–10

Mean ± SD in parentheses. Glutathione and glutathione enzymes were measured by a coupled enzymatic method measuring oxidation of NADPH. SOD was monitored by the inhibition of reduction of acetylated ferricytochrome *c*.
Source: Refs. 22, 24, 25.

By the transition metal ion–catalyzed Haber-Weiss reaction,

$$H_2O_2 + O_2^{-}\cdot \rightarrow OH\cdot + OH^- + O_2$$

hydroxyl radicals can be formed. This is a very potent inducer of lipid peroxidation. This reaction is enhanced in the presence of transition elements, such as iron or copper, by reaction with molecular oxygen:

$$Fe^{2+} + O_2 \leftrightarrow (Fe^{2+} - O_2 \leftrightarrow Fe^{3+} - O_2) \leftrightarrow Fe^{3+} + O_2^-$$

Superoxide can dismute to form hydrogen peroxide:

$$2O_2^- + 2H^+ \rightarrow H_2O_2 + O_2$$

However, the physiological significance of the Haber-Weiss reaction has been challenged. Finally, by the Fenton reaction, hydroxyl radical can be formed:

$$Fe^{2+} + H_2O_2 \rightarrow Fe^{3+} + OH^- + OH\cdot$$

This reaction may occur in vivo.

In semen, large amounts of iron (0.38 ± 0.15 μg/ml) bound to transferrin (40 μg/ml), ferritin (90 ± 68 ng/ml in seminal plasma), and probably as free iron, and copper (0.19 ± 0.10 μg/ml) were found. Iron can be released from these binding proteins by neutrophils, acidosis, and superoxide anion radicals (62,63).

Calcium is important, too: oxidative stress can damage the enzymes that normally maintain low intracellular Ca^{2+} concentrations, e.g., plasma membrane calcium extrusion systems and the ATP-dependent calcium pump of the endoplasmic reticulum. A number of oxidants cause calcium release from mitochondria. Increase of intracellular "free" calcium may lead to increased availability of transition metal ions due to digestion of metalloproteins by calcium-dependent proteases. These mechanisms lead to increased rates of lipid peroxidation (6).

C. Site of Production

In aerobic cells the most important source of $\cdot O_2^-$ is the electron transport chain of mitochondria and endoplasmic reticulum. In the midsection of spermatozoa, mitochondria are abundant (ATP-production and motility) (see Fig. 1). Several nonmitochondrial enzymes, e.g., xanthine oxidase, and D-amino acid oxidase, produce superoxide anion and hydrogen peroxide.

Alvarez et al. (23) suggested that about 35% of the $\cdot O_2^-$ derives from the spermatozoal mitochondria, the rest from the cytosol (23). Aiken suggested that the origin of ROS in spermatozoa was not mitochondrial by demonstrating a lack of effect of different mitochondrial inhibitors (oligomycin, antimycin A, rotenone) on the rate of ROS production (26). ROS appeared to be generated by a membrane-bound NADPH oxidase, the activity of which was regulated by intracellular calcium and protein phosphorylation dependent upon protein kinase C (26–28).

This conclusion was analogous to the process of phagocytosis, which was associated with an oxygen burst and the formation of superoxide anion radicals and oxidizing hypohalous acids and halogenated amines. The oxygen burst in these cells could be stimulated by the calcium ionophore A 23187. This agent caused a calcium influx, which stimulated superoxide radical generation via a membrane-bound NADPH oxidase system through the mediation of protein kinase C (64–66).

In human spermatozoa, similar pathways may be involved:

A23187 was an effective stimulator of ROS production in the presence of extracellular calcium.

The major product generated was superoxide anion radical.

Activators of protein kinase C, such as 1-oleoyl, 2-acetyl glycerol or 12-myristate, 13-acetate phorbol ester, were potent stimulators of ROS production in human spermatozoa.

NADPH induced a burst of ROS production in Triton-permeated spermatozoa and isolated membrane pellets (26).

Alternatively, the rise in ROS production could result from mechanical perturbation of the spermatozoa plasma membrane induced by centrifugation during preparation (26,27). Chaotropic agents (e.g., SCN^-, ClO^-_4, I^-, NO^-_3, Br^-), which decrease the "ordered" structure of membrane water and increase the lipophilicity of membranes, weaken the membrane structure and render the membrane more susceptible to oxidative injury (67,68). A similar effect may be caused by mechanical forces, e.g., shaking or centrifugation. Such perturbation of the plasma membrane of phagocytic leukocytes is known to induce the release of the superoxide anion radical (69).

By Percoll gradient centrifugation, a certain subpopulation of spermatozoa could be identified which produced the ROS. They were found in the lower-density fractions (40/55% and 70/80% fraction) of a discontinuous Percoll gradient. These fractions were characterized by poor motility, increased rate of ROS production, and impaired capacity for fertilization (27). One could speculate that the spermatozoa of these fractions have different lipid structures more susceptible to oxidative injury. Infiltrating leukocytes were the originating source in less than 6% (31).

D. Pathological Significance

High concentrations of ROS were definitely detrimental to sperm function (26,27,70,71). It was suggested that lipid peroxidation is the consequence and not the cause of enhanced generation of reactive oxygen species in spermatozoa (28) (see Secs. IV, V.A).

E. Physiological Significance

Why are ROS generated if they are so detrimental to spermatozoa function? Reactive oxidants have definite physiological functions in metabolism (e.g., generation of proton gradients in mitochondrial membranes, protective functions, prostaglandin and thyroxine syntheses). The different distribution and release of initiating enzymes for these radical processes minimizes deleterious effects and side reactions through spatial and temporal control of these reactions (72–74).

Regarding spermatozoa, these reactions may occur as part of a planned senescence (75). Furthermore, Aitken et al. (28) showed an enhancement of sperm zona interaction following the induction of lipid peroxidation.

Superoxide may have a physiological role in the regulation of sperm function for the following reasons:

1. In neutrophils, platelets, and liposomes peroxidation had beneficial effects on membrane adhesion (76–78) and fusion and the process activated pospholipase A_2 (79–81). In several mammalian species an activation of pospholipase A_2 was postulated to be involved in the acrosome reaction (82–85). The initial recognition of the zona pellucida is complex, including binding of sperm glycosyl transferases to the oligosaccharide side chains of glycoproteins on the outer zona surface (86). This event could trigger an intracellular calcium signal with the consequence of stimulation of ROS production and local lipid peroxidation, plasma membrane changes with the result of phopholipase A_2 activation, and binding to its substrate (26,28). As a result of these events the peroxidation-dependent activation of phospholipase A_2 could be expected to induce an increase in sperm zona adhesion, initiation of the acrosome reaction, and generation of a fusogenic equatorial segment capable of initiating sperm-oocyte fusion (28). Supporting these hypotheses is the finding that increased levels of thiobarbituric acid-reactive products had a highly significant enhancing effect on the ability of human spermatozoa to bind to both homologous and heterologous zonae pellucidae (28). Remarkably, this reaction pattern could be reversed by tocopherol.
2. In analogy to the adipocyte, where activation of NADPH oxidase by insulin results in an alkalinization of the interior of the cell (87), a similar mechanism could occur mediated by a superoxide radical-generating system in the sperm plasma membrane after stimulation by some component of the zona pellucida (88,89). A rise in intracellular pH may be involved in the induction of the acrosome reaction (90).

3. Human spermatozoa incubated with human tubal fluid, which normally contains the components of follicular fluid during the ovulation phase, showed hyperactivation, increase in acrosome reaction, and significantly greater binding to and penetration of zona-free hamster ova (91–93). The function of the active fraction in human follicular fluid was similar to that of the calcium ionophore A 23187 (94). These findings were thought to occur due to an increased acrosome reaction caused by contact with human follicular fluid (see above). Therefore, the fluids of the Fallopian tubes, the site where sperm/egg interactions occur, are capable of stimulating acrosome reaction/penetration. This appears to be mediated by a reaction pattern comparable to a calcium ionophore, i.e., induction of generation of reactive oxygen species and lipid peroxidation. Enhanced penetration in zona-free hamster ova was also reported using an incubation medium based on the composition of human tubal fluid (95).
4. Generation of ROS at the zona surface may render the zona more resistant to sperm penetration, thus preventing polyspermy (96).
5. Time-dependent studies showed spontaneous changes of ROS production during capacitation with a peak of activity within the first or second hour of incubation followed by a progressive decline. It is not known yet whether this has any meaning for the capacitation process (27).

F. Conclusion

In conclusion, ROS are produced by spermatozoa. The site of production seems to be the plasma membrane and/or mitochondria; the main product is the superoxide anion radical. The hydroxyl radical can be generated via the Fenton reaction. The ROS produced induce lipid peroxidation and may impair the ability of the spermatozoa for sperm-oocyte fusion. They may have, on the other hand, physiological functions during sperm/oocyte interaction. Deleterious or beneficial effects may be the result of both site and time of oxidative reactions.

IV. SPERMATOZOA LIPID PEROXIDATION

A. Measurement of Lipid Peroxidation

Lipid peroxidation can be analyzed by quantitative analysis of conjugated dienes, volatile hydrocarbon gases, loss of fatty acids, light emission, fluorescence (lipid pigments), thiobarbituric acid–reactive substances (TBARS) test, and alkenals (e.g., 4-hydroxynonenal) (6,97–99). Some of these methods have significant drawbacks, and the results obtained, e.g., by the TBA assay, should be interpreted with great caution. Most of the studies investigating lipid peroxidation in spermatozoa and seminal plasma use the thiobarbituric acid test. This method provides a convenient index of lipid peroxidation, correlating well with chemiluminescence, pentane, or ethane formation (100, 101). The TBA test was reported to give an excellent linear correlation with membrane damage in rabbit spermatozoa as assessed by trypan blue staining, accessibility of intracellular enzymes to impermanent substrates (LDH, pyruvate kinase), and loss of motility (20,22,24).

However, this test lacks selectivity (102,103) and has several disadvantages, which can be overcome only partly by modification of the assay:

Presumably only a small fraction of membrane lipids, chiefly prostaglandinlike cyclic endoperoxides, yield TBA-reactive products on peroxidation; the rate of conversion of lipid hydroperoxides to malondialdehyde in the TBA test is said to be less than 5% (101,104).

The result of the TBA test is given in terms of the amount of MDA produced. This falsely implies that the assay measures free MDA. However, many other compounds, like lipid peroxides, carbohydrates, prostaglandins, amino acids, nucleic acids, and aromatic carboxylic acids, react with thiobarbituric acid.

Sometimes the absorption maxima of the colored products are shifted.

Autoxidation, which is favored by high temperatures used in the assay, cannot be excluded.

MDA is not formed only during the peroxidation of lipids.

Endoperoxides produced enzymatically by the prostaglandin synthesis pathway will be measured if the TBA assay is applied to human body fluids.

The assay requires iron ions so that the concentration of metal ions may significantly influence the rate of lipid peroxide decomposition in vivo.

MDA produced during the test may react with NH_2 groups of various biomolecules, e.g., phospholipids, amino acids, and proteins; even polymers of MDA have been observed.

Any free MDA formed in vivo will be rapidly metabolized.

For example, with a specific cyclooxygenase method, the mean peroxide content in human plasma was 0.5 μM, whereas expression of the results from a TBA test yielded a mean value of 38 μM (6,98,102).

A specific and sensitive assay is the quantitative determination of a 4-hydroxyalkenal, such as 4-hydroxynonenal (HNE), a degradation product of omega 6 polyunsaturated fatty acids, by HPLC or gas chromatography/mass spectrometry. The measurement of hydroxyalkenals is probably more meaningful than measurements of free MDA, since the 4-hydroxyalkenals are far more cytotoxic and much more reactive than MDA. They accumulate as lipophilic compounds in the cell membrane. Low concentrations (μM-mM) of HNE are acutely cytotoxic to prokaryotic and eukaryotic cells, disturb cell functions such as loss of glutathione and protein thiols, inhibit DNA, RNA, and protein synthesis, and disturb calcium homeostasis. However, they also result from peroxide decomposition and can be metabolized by cells (105).

4-Hydroxy-2-nonenal in human semen was recently reported in a concentration of 0.902 ± 0.190 μM (range 0.120–2,688 μM) by capillary column gas chromatography/negative ion chemical ionization mass spectrometry (GC/NICIMS). Human spermatozoa (16 × 10^6/ml) produced HNE at a mean concentration of 0.013 ± 0.004 μM. Incubation of HNE (50–250 μM) with washed spermatozoa (5 × 10^6/ml) was dose-dependently spermicidal (106).

B. Significance of Lipid Peroxidation for Spermatozoa Function

The mechanisms of lipid peroxidation and its functional consequences for cell membranes have been reviewed elsewhere (71,108). A major consequence of even a limited degree of lipid peroxidation is to induce a sudden decrease in membrane fluidity (109). Therefore, the various fusion events in the spermatozoa membrane associated with the acrosome reaction and penetration of the oocyte may be inhibited. The peroxidation also favors the activation of phospholipase A_2 by enhancing the ability of this enzyme to interact with its substrate (78). This enzyme is known to exacerbate the damage initiated by lipid peroxidation (79). Once initiated, the chain reaction spreads laterally in the membrane, leading to loss of cell function. However, adhesion of the spermatozoa to the zona pellucida was enhanced if thiobarbituric acid–reactive substances were increased (28).

Mrotek et al. reported peroxidation of semen lipids in boars, but this report did not receive any attention until 1973 (110). Jones and Mann investigated lipid peroxidation in several mammals (boar, rabbit, rat, stallion) and in humans (13–18). In washed sperm suspension, they observed a decline of motility, O_2 uptake, and structural damage of spermatoza membranes and acrosomes after aerobic incubation at 37°C and found TBA-reactive substances to be increased (13). The yield of TBA-reactive products was enhanced by incubation with ferrous/ascorbate and inhibited by antioxidants and metal ion chelators. Peroxidation caused a 50% loss of phosphatidylethanolamine (plasmalogens), 20:4 and 22:6 fatty acids, and palmitaldehyde of spermatozoa (see Table 1) (14,34,36,37). As seminal plasma also contains plasmalogen, seminal plasma was suggested to be another source of lipid peroxide products (15,16). However, patients with azoospermia or postvasectomy showed negligible amounts of TBA-reactive substances in the semen (17).

Peroxidation of endogenous phospholipids and addition of peroxidized fatty acids proved equally toxic to spermatozoa. The structural damage was particularly evident in the acrosomal region (15–17) and led to a decline of velocity and linearity with no change in ATP content (11).

Lipid peroxidation (ferrous/ascorbate) led to a dose-dependent increase in TBA-reactive substances and to a reciprocal decline in the ability of sperm-oocyte fusion. No relationship with motility was found (see below) (28).

In nonmotile or poorly motile spermatozoa (necrospermia/oligospermia), the peroxidation products were higher than in highly motile spermatozoa from normal ejaculates (17). It was suggested that TBA-reactive material may be used as a biochemical parameter for evaluation of semen quality (17, 18,34,112). Rao et al. found a good correlation of the inital motility, midsection changes of spermatozoa, and the rate of MDA production after promoted lipid peroxidation (LPO) (112).

C. Spontaneous Lipid Peroxidation

Induced lipid peroxidation in spermatozoa is a very rapid process with peroxidation products detectable after only 30 minutes (114). Spontaneous lipid peroxidation (autoxidation) is very slow, and the intra- and interindividual variations are huge (20,21,24). Observation at a given time, however, showed a linear correlation between the percentage of nonmotile cells and the production rate of TBA-reactive substances; thus, at a given rate of lipid peroxidation the particular percentage of inert spermatozoa is quite constant. When 0.10 ± 0.03 nmol TBA-reactive substances/10^8 human spermatozoa were produced, no cell was motile. The authors defined this rate as the "lipoperoxidative lethal endpoint" (24).

In Aitken's study, no effect of lipid peroxidation on motility was observed (see above). The highest rate of TBA-reactive substances in his study, however, was 7.34 ± 1.91 pmol/10^8 spermatozoa/ml and so one order of magnitude lower than the lipoperoxidative lethal endpoint reported by Alvarez (100 pmol/10^8 spermatozoa) (24,28).

The rate of spontaneous lipid peroxidation depended linearily on Po_2 and on the temperature (in the range of 32–40°C) (24). Due to this correlation the spermatozoal lifetime is short in the oviduct at 37°C, long in the scrotal epididymis at 32°C, and can be estimated in the female genital tract. Extrapolating data from rabbit and rhesus monkey, the Po_2 in the female oviduct is 60 torr and the temperature 37°C (113,114). Accordingly, the lifetime of spermatozoa in the female genital tract can be calculated to be 2.5–25 hours (24), which is in accordance with the estimated lifetime of 10–30 hours (115, 116). It seems that the lack of oxygen, lower temperature, and different lipid composition in the epididymis seem to prevent premature senescence of spermatozoa (55), (see above).

There may be significant differences between the species in the effect of lipid peroxidation products on sperm motility. In fowl spermatozoa, even high concentrations of TBA-reactive substances did not lead to a change in motility, morphology, or ATP content (117). This indicates that the results from animal spermatozoa cannot be extrapolated to human spermatozoa.

D. Conclusion

Mammalian spermatozoa are without doubt susceptible to lipid peroxidation. Once induced, it spreads as a chain reaction laterally in the membrane, leading to impaired function and ultimately cell death. The rate of lipid peroxidation correlates with loss of motility and impaired ability for hamster ova penetration. However, there may be species differences. One important caveat is that almost all experiments were done with the thiobarbituric acid assay, a sensitive but not specific test. Only one recent observation reports specific parameters of lipid peroxidation (HNE) in human semen and its negative impact on motility.

V. ANTIOXIDANTS IN SPERMATOZOA AND SEMINAL PLASMA

Antioxidants are essential factors, which counterbalance high steady-state concentrations of reactive oxidants in seminal plasma and spermatozoa. Some antioxidants are micronutrients and vitamins, such as tocopherol, ascorbate, lipoate, β-carotene, selenium, and zinc; others are synthesized in the cell (GSH, SOD). The relative contribution of distinct small molecular lipophilic or hydrophilic antioxidants and antioxidant enzymes to the overall spermatozoa and seminal plasma antioxidant activity/capacity are not known. It is relevant to know in which microdomains reactive oxidants are produced and where antioxidants are located.

It is nearly impossible to compare the published data on antioxidants because different species were analyzed with different methods and the results were expressed in different reference units (e.g., SOD: mg/kg wet weight (118), mg/L (119), mg/g protein (120), U/g wet weight (121), U/mg protein (122), U/10^8 cells (24,25); the same holds true for glutathione).

A. Cells in the Testis and Spermatozoa

1. Glutathione and Glutathione Peroxidase

In 1975, Li investigated glutathione, glutathione reductase, and glutathione peroxidase in mammalian spermatozoa by different methods (Table 3) (123). He concluded that the glutathione cycle is important only in certain species. In the bull, glutathione peroxidase (GPX) reached high concentra-

Table 3 Glutathione Content and Glutathione Reductase and Glutathione Peroxidase Activity of Spermatozoa

	Glutathione nmol/10^9 cells	Glutathione Reductase μmol NADP/h/mg protein	Glutathione Peroxidase μmol NADP/h/mg protein
Dog ejaculated	5.3 (1.8–12.0)	1.25 (0.48–2.22)	0.84 (0.24–1.56)
Dog epididymal	1.1 (0.4–3.4)	1.8 (0.72–3.48)	0.36 (0.12–0.72)
Goat epididymal	13.0 (1.9–45.2)	2.62 (1.28–3.65)	0.81 (0.54–1.24)
Ram ejaculated	4.5 (0.9–12.6)	1.35 (0.72–2.16)	0.13 (0.06–0.18)
Boar ejaculated	0.3 (0.2–0.4)	0	0
Rabbit ejaculated	0.1 (0–0.1)	0.24 (0.18–0.33)	0.07 (0–0.14)
Human ejaculated	5.3 (1.8–13.9)	0.77 (0.59–1.02)	0.49 (0.18–0.73)

Mean; range in parentheses.
Source: Ref. 123.

tions, especially in seminal vesicles and seminal plasma; in caput epididymal spermatozoa the glutathione (GSH) content was three times the content of cauda epididymal or ejaculated spermatozoa. In bull spermatozoa, two thirds of the glutathione peroxidase activity was non–selenium dependent (124). In contrast, the distribution of glutathione peroxidase (GSH-Px) and selenium in serial homogenates of human reproductive tissues was low from testis to seminal plasma, reaching only about half the serum concentration (125,126). A prerequisite for the GSh cycle is availability of NADPH; glucose-6-phosphate dehydrogenase and 6-phosphogluconate dehydrogenase, which generate NADPH, are present in human spermatozoa (127). The GSH-mediated protection mechanisms in hamster spermatids and round spermatids from the rat relied on GSH-dependent defense mechanisms. Sertoli cells played an important role in this context, as the intracellular GSH levels of spermatogenic cells were dependent upon germ cell–somatic cell interactions (128–130) (Fig. 2). In mouse spermatogenic epithelium glutathione-peroxidase was one order of magnitude lower than that in liver (131).

Furthermore, a decrease of glutathione in the male reproductive tract may be a mechanism for potentiation of chemical-induced germ cell mutations (132). Many heavy metals and metalloids interact strongly with thiols and therefore have the capacity to block enzyme activities and deplete cells of glutathione. Metabolic processes and especially organometallic complexes themselves (like cadmium, lead) may damage spermatozoa at different

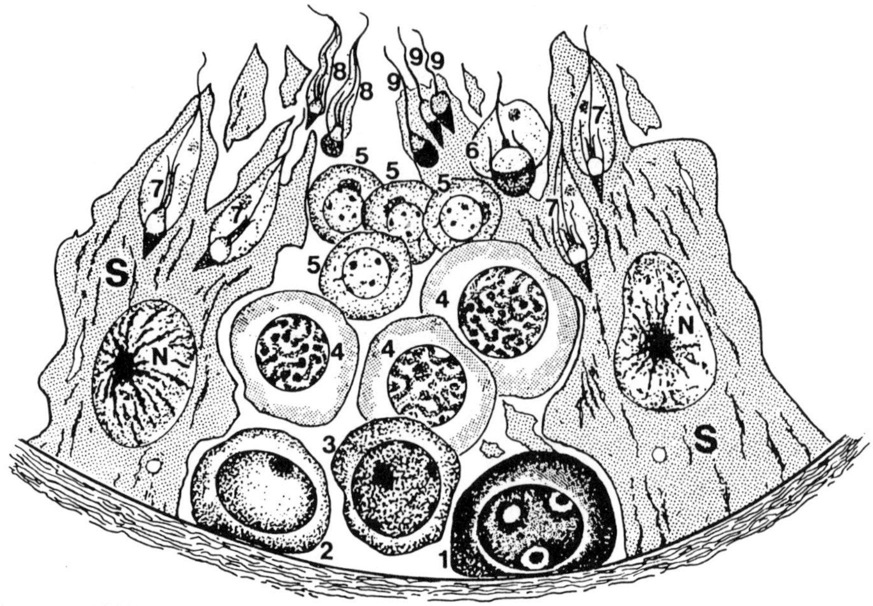

Figure 2 Diagram of a section of a testis tubulus. Note the close association of Sertoli cells with spermiogenesis. S, Sertoli's cell; N, nucleus of Sertoli's cell; 1, dark spermatogonia; 2, pale spermatogonia; 3, primary spermatocyte (preleptotän); 4, primary spermocyte (pachytän); 5, secondary spermocyte; 6, spermatid with initial chromatin condensation; 7, spermatid with late-stage chromatin condensation; 8, nearly differentiated spermatozoa; 9, spermatozoa. (From Ref. 41.)

levels of differentiation by promoting free radical reactions. These may stimulate oxidation of membrane lipids by depleting antioxidants (133). Cadmium damage to rat testes appeared to involve lipid peroxidation (134). This may indicate a clinical relevance of the antifertility effects of pro-oxidant environmental pollutants, chemicals, and drugs (135,136).

The concentration of intracellular glutathione is low: glutathione reductase 44 ± 3 nmol/10^8 cells; glutathione-peroxidase 21 ± 3 nmol/10^8 cells; GSH 4.0 nmol/10^8 cells; GSSG 3.0 ± 0.6 nmol/10^8 cells (24). The lower detectability limit of the modified spectroscopic Günzler assay is around 0.5 μmol (137,138).

The assays for the glutathione system are sensitive and specific. However, sample preparation may cause erroneous results (138).

2. Selenium

Selenium is an important trace element and cofactor for glutathione peroxidase. In bulls, a very effective uptake mechanism, probably via a macromolecule (selenoprotein P), exists in testis and seminal vesicles. Recently, evidence for a specific selenium receptor in the rat testis has been presented (139). Selenium in the testis is taken up by Sertoli cells, which then provide the maturing germinal cells with an adequate form of selenium (inorganic, Se-cystein). This may hold true for the testis in other species, but probably not the accessory glands, since the seminal plasma selenium levels are partially below the serum levels in man, boar, and stallion (140). The selenium concentration in human spermatozoa is 3.6 (0.7–8.8) μg/g dry weight (0.25 ng/10^6 spermatozoa) (141).

Severe selenium deficiency may lead to cardiomyopathy (Keshan disease); mode rate deficiency results in myodegenerative syndromes. Selenium and zinc deficiencies apparently were associated with stomach cancer. In selenium deficiency no increased rate of infertility was reported (142–144).

3. Superoxide Dismutase

Several types of superoxide dismutase exist: a copper-zinc, a manganese, iron-containing enzymes, and an exocellular form of SOD (6,145). The enzymes can be measured directly by an immunological and by an electrolytic method; indirect assay methods are used more often: $\cdot O_2^-$ is produced artificially, SOD inhibits the reaction with the detector by removing the $\cdot O_2^-$ (e.g., xanthine/xanthine oxidase as $\cdot O_2^-$ producer, cytochrome *c* reduction as detector). In this example, an apparent decrease in cytochrome *c* reduction could be induced by a reagent that inhibits $\cdot O_2^-$ generation by inhibiting the xanthine/xanthine oxidase system or by substances reoxidizing cytochrome *c* (e.g., H_2O_2 or cytochrome oxidases). Other sources and detectors of superoxide can be used. All, however, have specific problems mainly due to the complicated chemistry (6).

In 1978 superoxide dismutase was isolated from ram spermatozoa (Table 4), and later from spermatozoa and seminal plasma in other species (146,147) (xanthine/xanthine oxidase method). It was shown via cyanide inhibition

Table 4 Distribution of Superoxide Dismutase Activity in Ram Semen

	Specific activity	
	(U/mg protein)	(U/10^{11} cells)
Whole semen (2.0 ml)	3.4	
Seminal plasma	1.8	
Spermatozoa (6.5 × 10^9 cells)	17.0	3000
Erythrocyte		3300
Cytoplasmic droplets	22.0	

Source: Ref. 146.

that the spermatozoal enzyme was the copper-zinc form and not the manganese form of superoxide dismutase (22,146,147). This form is located principally in the cytosol, whereas the manganese enzyme is mitochondrial. Other authors could not detect SOD activity in sonicated human spermatozoa (assays: xanthine/xanthine oxidase and inhibition of cytochrome *c* reduction; inhibition of pyrogallol autoxidation) (35).

The activity of SOD changed from spermatogonia to spermatozoa: in rat epididymis it decreased during maturation and transport of the spermatozoa, while the rate of superoxide anion radical production increased (148). In mouse spermatogenic epithelium, SOD activity was comparable to that in liver cells (131). In rat germ cells there were significant differences in the distribution and activity of antioxidants. The order of SOD and glutathione concentration was Sertoli cells < pachytene spermatocytes < round spermatids. Alpha-tocopherol (difference between Sertoli cells and round spermatids = factor 4), ascorbate, and gamma-glutamyltransferase decreased in the same order (149). In epididymal hamster spermatozoa, SOD activity appeared to be regulated by the amine spermine (150).

The antioxidant defense systems varied in different species. The primary defense mechanism in the rabbit spermatozoa appeared to be superoxide dismutase (21). SOD was also thought to be of the utmost importance in humans (24). Later it was suggested that SOD and the components of the glutathione system seem to be important both in human and mouse spermatozoa (see Table 2) (25). An observation contrary to the mentioned mechanisms was published. An overexpression of superoxide dismutase, poor superoxide anion radical generation capacity, and a lack of sperm surface thiols was reported in men with oligospermia (151).

Recently Kumar et al. (152) described a membrane-located "proton pump" in rat spermatozoa. This proton pump seems to be coupled with the superoxide radical–superoxide dismutase system, which, when uncoupled, results in total loss of sperm motility. Addition of sulfhydryl compounds restarts the proton pump and reactivates sperm motility. The authors presented a theory explaining the acquisition of motility in the epididymis by an interaction of the proton pump with the superoxide radical–superoxide dismutase system (152). As a rise in intracellular pH should be able to induce the acrosome reaction, this proton pump could be relevant in this respect also.

4. *Catalase*

While in older literature no catalase activity was found in human sperm (60), Jeulin et al. measured an activity of 44 ± 17 nmol O_2/min/10^8 cells in spermatozoa (153). No catalase was detected in rat testicular cells (149).

5. *Tocopherol*

Tocopherol quenches and reacts with singlet oxygen; it is also oxidized by $\cdot O_2^-$ and reacts with lipid peroxyl and alkoxyl radicals, terminating the chain reaction of peroxidation. Human spermatozoa contain 0.8 ± 0.08 nmol tocopherol/10^8 cells. In vitro a 14-fold increase of tocopherol in human spermatozoa (11.5 ± 0.4 nmol tocopherol/10^8 cells) inhibited spontaneous lipid peroxidation to about 50% (154). In rat testes the tocopherol concentrations decreased from Sertoli cells to pachytene spermatids to round spermatids (149).

B. Seminal Plasma

The extracellular fluids of the human body, such as blood plasma, tissue fluid, or seminal plasma, contain no significant catalase activity, only low activities of superoxide dismutase, selenium-dependent glutathione peroxidase, and also very little reduced glutathione (GSH). Therefore, enzymes seem to contribute little to the antioxidant activity of extracellular fluids (6). However, it is not known whether this concept holds true for human semen.

Other antioxidants include transferrin, coeruloplasmin (SOD-like activity), albumin (all binding iron or copper), ascorbic acid, α-tocopherol, and uric acid. Their concentration and importance in seminal plasma is only partly known. The biological significance of seminal plasma antioxidants are discussed below.

1. Glutathione System

The glutathine content of human, dog, boar, rabbit, and ram seminal plasma was low (see Table 3). Glutathione peroxidase activity was low in man and ram, absent in boar and stallion, but very high in the bull (126,155).

2. Superoxide Dismutase

Superoxide dismutase activities in human seminal plasma were 20–50 U/ml (xanthine/xanthine oxidase system; Cu/Zn form) (35).

3. Catalase

Recently, Jeulin et al. found catalase with an activity of 129 ± 59 nmol O_2/min/ml in the seminal plasma by measuring the oxygen burst with a Clark electrode after addition of H_2O_2. The activity originated from the prostate (153).

4. Trace Elements

Several reports describe trace elements in human seminal plasma. The concentration of selenium was 38 (13–90) ng/g wet weight in seminal plasma and 49 (11–120) ng/g wet in whole semen (141). The only mammalian selenoenzyme presently known is glutathione peroxidase.

Zinc has antioxidant properties. Whole semen had the highest concentration of zinc (1998 μg/g dry weight) compared with other parenchymal organs; a high concentration is also found in the prostate gland (859 μg/g dry weight) (156). Zinc was mainly contained in seminal plasma (95 ± 8% of the total amount, 130 ± 80 μg/g in whole semen). Zinc and selenium derived mainly from the prostate gland. There was no correlation between these parameters and semen quality (141).

Copper is an essential part of superoxide-removing systems but in its free form is a potential catalyst for activating oxygen. The levels of copper in semen of fertile and infertile males did not differ significantly (183.39 ± 14.37 $\mu g/L^{-1}$ for fertile and 194.99 ± 5.70 $\mu g/L^{-1}$ for infertile men). A small but weak significant correlation between copper concentrations and sperm concentrations, motility and mophology were found (157).

5. Others

Human seminal plasma contains a factor (heat-stable, effective after dilution 1:2, present in azoospermic or vasectomized patients) that effectively prevented but did not reverse the toxic effect of endogenous or exogenous lipid peroxidation products upon spermatozoa (17,18). The protective effect was assumed to reside in whole seminal plasma and in the nonultrafilterable fraction (ultrafiltration through a membrane with a cut-off at molecular mass 500), while the low-molecular ultrafiltrate had no effect (17). In a report one year later, the same authors stated that both the ultrafiltrate and the nonultrafilterable fraction were by themselves incapable of preventing peroxidation; only the combination of these two factors provided protection (18).

In bovine seminal plasma a heat-stable, acid/alkaline-stable, dialyzable, polar, and nonproteinaceous compound inhibited lipid peroxidation. On ultrafiltration (1000 and 10,000 d cut-off), activity was found in both retentate and ultrafiltrate, which was due to the same factor as shown by gel chromatography. Activity was lost upon ashing (158,159).

An inhibinlike peptide of prostatic origin (MW 10,400 d) suppressed the lipid peroxidation of human spermatozoa in a dose-dependent manner. This factor may have a protective role for spermatozoa motility (160,161).

High concentrations of polyamines from the prostate are present in human seminal plasma (162). These have antioxidative properties (163). Under the influence of diamine oxidase, however, they generate reactive oxidants responsible for necrozoospermia (164). A strong diamine oxidase activity was demonstrated in the seminal plasma of infertile men with oligoasthenozoospermia (165). In conclusion, the seminal plasma has several antioxidant mechanisms, which need to be characterized further.

C. Effects of Exogenously Added Substances

In vitro studies on inhibition of spermatozoa lipid peroxidation by exogenously added antioxidants may provide information on the physiological significance of particular antioxidants in the complex scenario of antioxidant defense. The protective effects of catalase on aerobically incubated spermatozoa counteracting the damaging effect of hydrogen peroxide was shown in the 1950s (12,59,166). Catalase, SOD, and mannitol did not prevent the production of thiobarbituric acid–reactive substances in boar spermatozoa; only EDTA had a dose-dependent inhibitory action (167).

In human spermatozoa, tocopherol reduced but did not prevent the production of thiobarbituric-reactive substances in a dose-dependent fashion during promoted lipid peroxidation (Fe/ascorbate). In addition, tocopherol suppressed the production of reactive oxidants, had a positive effect on sperm-oocyte fusion (10 mM), and partly improved motility (27). In buffalo spermatozoa tocopherol (2.5 μM), vitamin C (2.5 mM), or 40 mM glucose were significant inhibitors of induced lipid peroxidation and improved motility (168,169). Furthermore, tocopherol raised superoxide dismutase activity: in bovine sperm suspensions incubated with vitamin E superoxide dismutase activity was higher than in control samples (170).

Pretreatment of zona-free hamster oocytes with high concentrations of glutathione (GSH or GSSG 50 or 100 mM) or glutathione reductase (1 or 10 IU/ml) significantly increased the penetration rate of human spermatozoa. Polyspermy was only increased significantly with glutathione reductase and high concentrations of glutathione (100 mM) (171).

Several substances were investigated as to their ability to prevent lipid peroxidation in mammalian spermatozoa (13–18,172). Induced lipid peroxidation could be inhibited with bovine serum albumin, butylated hydroxytoluene (BHT), or hypotaurine. The effects of other agents varied in different experiments.

Hypotaurine inhibited $\cdot O_2^-$ production and reduced SOD inactivation in spermatozoa (172). BHT suppressed lipid peroxidation in vitro but damaged the spermatozoa membrane severely (decrease in HOP penetration and motility) (27). Albumin bound copper ions and inhibited copper-catalyzed lipid peroxidation and OH· radical formation (173). Blood serum, independently whether male or female (drawn at different phases of the menstrual cycle), improved the motility of spermatozoa prepared by Percoll gradient centrifugation. Even more effective than albumin was a serum fraction with a molecular weight of around 200 kd. The effects of albumin, seminal plasma, and follicular fluid were less pronounced (174). Quercetin had antioxidative properties in ram spermatozoa (175).

The antioxidant thiol 2-mercaptopropionylglycine (2-MPG) increased the velocity of human spermatozoa. The spontaneous decrease of spermatozoa motility was significantly reduced in normozoospermic and oligozoospermic men (Ohlenschläger, personal communication).

VI. CLINICAL CONSEQUENCES

Oxidative imbalance is not only a stimulating theory, it is of particular interest for the practicing andrologist. The biochemical causes of infertility can be determined in at least some patients, and new strategies for therapy can be sought. Furthermore, the information can be used to improve the preservation of spermatozoa and contraceptive measures.

A. Infertility, Oxidants, and Spermatozoa Membrane

In contrast to the large number of in vitro experiments, only a few reports have addressed the in vivo relevance of oxidative stress. A prospective study revealed a highly significant positive relationship between fertility and hamster-oocyte fusion rates in infertile couples. Outcome of the sperm-oocyte fusion assay and fertility in vivo showed a negative association with the generation of reactive oxygen species in spermatozoa (32).

In oligospermia, defective sperm function was associated with an elevated production of ROS in 23%. However, in the oligozoospermia group as a whole, the level of ROS produced was unrelated to any parameter of the semen profile (e.g., motility or morphology). An inverse relationship between the level of sperm-oocyte fusion and production of reactive oxidants could be demonstrated (29). In another report, the production of reactive oxidants sufficient to cause sperm damage has been confirmed in almost a quarter of infertile patients (176).

Not only an excess production of reactive oxidants may lead to lipid peroxidation. The spermatozoa membrane of patients with a rapid loss of motility (within 4 hours after ejaculation) contained more docosahexaenic acid than controls, thus indicating a higher susceptibility for peroxidation (39). Especially in this subgroup of patients there was a clinical benefit of tocopherol (300–600 mg/day, 6 weeks) (38). An unbalanced lipid composition of the spermatozoa plasma membrane may result in ineffective control of the superoxide-generating system, increased susceptibility, and, as a consequence, elevated production of ROS (27).

A recent hypothesis stated that it may be worthwhile to supply essential fatty polyunsaturated acids in addition to antioxidative therapy. The rational basis for this concept is the idea that the major component of oxidative toxicity is attributable to loss of the highly unsaturated essential fatty acids from the membranes, not only the damage by radicals or toxic metabolites. No data on this aspect are available up to now (177).

In the future it will be of interest to investigate the influence of exogenous factors such as nutrients and hormonal factors on the lipid composition of the supermatozoa membrane. These results demonstrate a possible involvement of reactive oxygen species in the pathophysiologyt of male infertility. Furthermore, they show the necessity of measuring reactive oxygen species, lipid peroxidation products, and depletion of antioxidants. This is all the more important as, in the former studies, conventional semen analyses had no prognostic value at all. In this context, it must be pointed out again that one must consider all the methodological pitfalls (e.g., thiobarbituric acid–reactive substances).

B. Infections

The importance of genital tract infections, above all *Chlamydia trachomatis*, for male infertility is a matter of controversy (178,179). In general, infections can lead to oxidative imbalance (180).

Human seminal plasma is immunosuppressive and inhibits the respiratory burst of neutrophils and monocytes (181,182). Physiologically, this should prevent spermatozoa from being recognized as antigens in the female genital tract (183). However, there are also contradictory reports. Stallion seminal plasma and spermatozoa increased neutrophil phagocytosis and may thereby stimulate clearing of spermatozoa from the reproduction tract (184). The immunosuppressive activity of seminal plasma could also inhibit spermatozoa autotoxicity and neutrophil-mediated spermatozoa damage. During infections neutrophils and macrophages invade urethra, prostate gland, seminal vesicles, or epididymis; their oxygen burst could damage spermatozoa directly by oxidative injury or indirectly by consumption of antioxidants.

Chlamydia trachomatis is thought to influence female and male fertility negatively. In males an important target of this infection appears to be the prostate gland, the function of which is consequently impaired (179). This gland is reported to produce important antioxidative factors, like an inhibin-like peptide, polyamines, and catalase (153,160,161,182).

In women, enhanced lipid peroxidation was demonstrated in the female genital tract following gonococcal infection (185). Similar consequences can be postulated in the male following infectious urethritis and leading to spermatozoa damage. Thus, infections could induce an abundance of oxidants or a defect in antioxidative defense, resulting in lipid peroxidation and, as a consequence, impaired fertility.

C. Preparation of Spermatozoa

In cases of male infertility, several measures for assisted reproduction are applied, like intrauterine insemination, in vitro fertilization (IVF), and gamete intrafallopian transfer (GIFT). For all these techniques the "best" spermatozoa, i.e., spermatozoa with good motility and morphology, have to be separated out. This can be achieved by several methods, i.e., "swim-up" from pellet, "swim-up" from semen, albumin column, glass-wool filtration, and Percoll or Nycodenz gradient centrifugation.

The exact mechanisms controlling the rate of spermatozoa ROS production in vivo are unknown. It has been shown, however, that spermatozoa preparation procedures increase the production of ROS if centrifugation is the first step (like in "swim-up" from pellet and repeated centrifugation). Mere centrifugation (500 *g* 5 min) led to a rapid increase in the rate of ROS, measured as chemiluminescence, with a peak after 1–7 minutes. "Swim-up" from semen, albumin columns, and the high density (90%) fraction of a discontinuous Percoll gradient showed significantly better motility scores and penetration rates in the hamster-oocyte assay than the procedures mentioned above. Further analyses of the discontinuous Percoll gradient revealed a significantly higher generation of ROS in the lower-density fractions (40/55% and 70/80%) than in the higher-density fractions (90/100%). The production of ROS was associated with poor motility and impaired capacity for sperm-oocyte fusion. A subpopulation of spermatozoa, which can be isolated on the low-density portion of Percoll gradients, seemed to respond to mechanical stimuli (like centrifugation) with an enhanced production of ROS (27).

It is essential to remove the ROS-generating cells before the first centrifugation, because these cells may damage unimpaired spermatozoa. Separation can be accomplished by Percoll gradient centrifugation, albumin column, or "swim-up" from semen. Many clinical observations also found improved motility scores with Percoll gradient centrifugation in comparison to the "swim-up" from pellet (186–188). In another report, centrifugation led to a decrease in motility, while thiobarbituric acid–reactive substances were not increased (189).

These results are of tremendous importance for assisted reproduction. In a recent editorial, Mortimer warned of legal consequences if centrifugational steps instead of the methods mentioned were initially used in IVF programs (186). The concept of oxidative imbalance in male infertility may help to optimize spermatozoa isolation procedures for artificial insemination.

D. Antioxidant Therapy

It is reasonable to suggest that antioxidant supplementation may compensate for antioxidant deficiency or overproduction of ROS. On the other hand, reactive oxygen species may have several physiological roles (see Sec. III.E). Supplementation with supranormal amounts of antioxidants may influence these physiological reactions induced by reactive oxidants. This may limit the clinical efficiency of antioxidant supplementation as anti-infertility therapy.

Therefore, it is of paramount importance to understand the delicate and subtle balance of oxidants and antioxidants in the physiology and pathophysiology of male infertility. Tocopherol (vitamin E) has been used in the clinical therapy of male subfertility for over 50 years. In some patients it improves fertility, while in others there is no clinical effect. However, besides case reports and anecdotal reports, there have been no placebo-controlled double-blind therapy studies. It is firmly established that reactive oxygen species and lipid peroxidative products are spermatozoa cytotoxic and that tocopherol scavenges reactive oxygen species and inhibits lipid peroxidation. The biochemical data may now provide a rationale for the clinical effect in some patients with male infertility. There are several reports on in vivo and in vitro antioxidant supplementation and spermatozoa function. In vitro addition of tocopherol with a subsequent increase of the cellular tocopherol content inhibited spontaneous lipid peroxidation in spermatozoa to about 50% (154). In induced lipid peroxidation, the dose-dependent inhibition of sperm-oocyte fusion could be reversed by the addition of tocopherol (human spermatozoa, hamster ovum penetration assay) (28). Tocopherol improved the long-term motility in a subgroup of patients with markedly decreased initial and long-term motility. In seminal plasma–free spermatozoa, vitamin E generally had a positive effect on motility (38).

Vitamin C administration (1000 mg/day, 30 days) significantly improved viability, motility, agglutination, immature precursors, abnormal morphology, and pregnancy rates in men with sperm agglutination (190,191). In smokers, vitamin C supplementation resulted in a dose-dependent improvement in sperm quality (count, viability, agglutination, motility, abnormal morphology) (192). Reduced lipid peroxidation rates after ingestion of vitamins E and C were reported in bulls (168,169).

There was a linear correlation between the spermatozoa number, motile spermatozoa, and SOD activity in seminal plasma (35). Spermatozoal SOD showed wide intra- and interindividual variations (24,25). Spermatozoa glutathione, glutathione peroxidase, and glutathione reductase were only present in low concentrations in normal males (24,25). No data for these parameters for infertile patients are yet available.

Catalase activities in seminal samples from asthenozoospermic infertile men were found to be lower (62 $\pm$ 36 nmol O_2/ml/min) than those from normozoospermic subjects (129 $\pm$ 59 nmol O_2/ml/min) (153). Most recently, pentoxifylline was reported to possess phosphodiesterase-inhibiting activities, i.e., antioxidative properties. If pentoxifylline were added to IVF media, the fertility rate in oligozoospermia could be enhanced from 13 to 39% (193). The understanding of the relative actions of the different enzyme systems against O_2 toxicity seems to be a priority area of research. Furthermore, the evaluation of several natural or synthetic antioxidants seems warranted.

E. Gynecological Impact

Spermatozoa are not only influenced by testicular and epididymal factors, oxygen, and seminal plasma. They come into contact with several secretions in the female genital tract, which could stimulate or inhibit lipid peroxidation. However, the oxidative factors and the antioxidative capacity of fluids in the female genital tract have not been taken into account up to now.

In cervical neoplasia and precancerous exfoliated cervical cells, increased peroxidation products were demonstrated. Therefore, spermatozoa peroxidation may be modulated by cervical factors (194,195).

Peritoneal or tubar fluid could either increase or inhibit spermatozoal motility. This probably depends on the composition of spermatozoa, peritoneal fluid sample, and existence of cofactors, e.g., infections, or other diseases, e.g., endometriosis. External factors, such as peritoneal fluid, have a great impact on the fluidity of the spermatozoa membrane (196).

The peritoneal fluid from fertile women showed more stimulatory properties towards motility than did peritoneal fluid from infertile women (197). Peritoneal fluid of fertile women protected the spermatozoa against loss of motility like a Tyrode medium containing bovine serum albumin and mercaptoethanol. In endometriosis lesions or inflammatory disease, this antioxidant activity was inhibited, presumably by secretion of peroxidizing substances (198). Antioxidant therapy in females to counteract these mechanisms may be useful, especially when oxidative injury of the male spermatozoa is a possibility.

F. Preservation of Spermatozoa

There are several indications for preservation of spermatozoa, such as planned operations with potential risk of impairing potentia generandi, chemotherapy of malignant neoplasms, and others. This can be achieved by cryopreservation. Freezing and thawing is associated with fundamental changes in the membrane structure and may lead to lipid peroxidation. In cases of pathological spermiograms (number, motility, morphology), these patients often wish to store and thereby collect sufficient numbers of cells. However, this is not yet possible. If the spermiogram was abnormal before freezing, the postthaw motility proved to be maximally impaired. Cryopreservation of spermatozoa may benefit from antioxidants. The postthaw motility of frozen spermatozoa was improved by addition of vitamins (C and E) to semen (199, 200). The postthaw motility and vitality of cryopreserved human spermatozoa was improved in the presence of dithiothreitol (201).

Cryopreservation resulted in decreased SOD enzyme activity. However, after different freezing protocols no significant difference in postfreeze enzyme levels was detectable, while motility scores differed significantly. According to these results, mechanisms that reduce SOD levels during cryopreservation may be different from those that underlie loss of motility (202).

G. Contraception

Reactive oxidants and lipid peroxidation can be detrimental to spermatozoa; consequently pro-oxidative compounds could serve as contraceptives. Gossypol is a polyphenolic compound found in cotton plants. Its mechanism of action is not completely known. In spermatozoa gossypol promoted the formation of oxygen radicals (203), impaired the antioxidant defense system (204), inhibited calcium uptake and calcium ATPase (205), and induced lipid peroxidation (206). Thus, these factors, among others (207), could explain its mode of action. However, the pro-oxidant actions of gossypol were not confirmed by other authors (208,209).

Transition metal ions are of paramount importance in initiating oxidative reactions by catalytically activating oxygen or oxygen intermediates. The possible mechanisms of metal ion–mediated oxidative stress have been discussed (133). Cobalt ions have powerful sperm-immobilizing properties, which may result from loss of sperm surface thiol groups and an augmented production of superoxide anion radicals (210).

The transition metal copper is used in intrauterine devices. Copper ions induce lipid peroxidation by catalyzing the Fenton reaction. These mechanisms may not be of relevance only for contraceptives but also for the spermatozoa cytotoxic effects of pro-oxidant environmental pollutants.

VII. SUMMARY AND CLOSING REMARKS

In andrology, no therapeutic progress has been made during the last few years. Without knowledge of the mechanisms leading to defective sperm function, a rational approach towards the development of appropriate therapies is impossible. The concept of oxidative imbalance in male infertility may now open a new field for evaluation of infertility, therapy, and research.

The methods for evaluation of fertility need to be extended by multifunctional analyses of oxidative imbalance. Impaired spermatozoa function could thereby be identified and antioxidative therapeutic approaches taken into account.

The criteria for evaluation of fertility prognosis must be changed. In some patients, production of reactive oxidants, eventually lipid peroxidation and antioxidant levels, may be of definite prognostic relevance.

The physiological impact and control of generation of reactive oxidants need to be elucidated. It is unknown whether the multitude of data on lipid peroxidation in spermatozoa from different animal species can be extrapolated to human spermatozoa. In humans, most of the experiments have been done in vitro and, as a rule, with the spermatozoa of healthy donors. Standard methods for the examination of oxidants, antioxidants, or lipid peroxidation in semen are lacking or bear pitfalls for interpretation.

Cryopreservation and separation of spermatozoa for artificial insemination may benefit from antioxidant supplementation by decreasing reactive oxygen species formation and inhibiting lipid peroxidation.

Physiological antioxidants such as tocopherol, ascorbate, lipoate, ubiquinols, β-carotene, and retinol need to be investigated for their clinical effects in patients with impaired fertility. Prospective placebo-controlled, double-blind studies are necessary for evaluation of drug therapy.

The biochemical fate of the spermatozoon in the female genital tract is obscure. Although reactive oxygen species may cause spermatozoal damage in the male and female genital tract, some reports indicate that reactive oxygen species enhance adhesion of the spermatozoon to the oocyte. This finding may have far-reaching consequences. One important research goal in this area is to find out and to characterize the peroxidative reactions in different anatomical sections of the female genital tract. Furthermore, the role of oxidants and antioxidants in the scenario of spermatozoa/oocyte adhesion/penetration/fusion must be studied on molecular levels.

New contraceptive measures may evolve if we gain more and detailed insight into the complex interplay between oxidants and antioxidants in the process of spermatozoa maturation, transport, and function.

Several environmental pollutants have detrimental effects on spermatozoa production/function. Among those xenobiotics are several compounds with pro-oxidant properties. The area of male reproductive toxicology could benefit from a better understanding of the oxidizing potential of those xenobiotics. This may also help to develop new therapeutic approaches.

There are scattered results in the biomedical literature dealing with several aspects of oxidants and antioxidants in male infertility. The heterogeneity of the biological material studied, the applied methology and differences in the interpretation of the results make it difficult to develop a comprehensive synopsis.

In recent years oxidative imbalance has been identified as one major cause of sperm dysfunction. Human spermatozoa are highly susceptible to lipid peroxidation due to a high amount of polyunsaturated fatty acids in their membranes. The generation of reactive oxygen species by human spermatozoa has been demonstrated. Transition metal ions, which provoke a series of radical reactions, are present in seminal plasma. Lipid peroxidation has been demonstrated and correlated with disturbances of spermatozoa function. These factors are contrasted with a paucity of antioxidative defense mechanisms in the spermatozoa. Clinical reports proved an inverse relationship between production of reactive oxygen species and the ability of spermatozoa to fertilize in about 25%.

Although fascinating, it seems unreasonable to attribute disorders of spermatogenesis, storage, transport, capacitation of spermatozoa, and fertilization of an oocyte to one single pathogenetic factor. However, oxidants and antioxidants are unequivocally one significant determinant for disturbed sperm function. The concept of oxidative injury in male infertility is a useful biomedical simplification which needs to be studied in more detail.

By reviewing this topic, it was our objective to provide the reader with a state-of-the-art report on the current knowledge about oxidative imbalance in male infertility and to stimulate clinicians to perform meaningful clinical studies in this rapidly growing area of reproductive biology. A progressive approach from "radical andrologists" may bring some change into andrology, a field calling for new concepts and therapies. Male infertility patients will benefit from the progress achieved in this field. New therapeutic approaches and a better understanding of the pathophysiology of male infertility will be the result.

REFERENCES

1. Schirren C. Andrologie in der Dermatologie: Geschichte, Standort und Ausblick. In: Stüttgen G, ed. Standort und Ausblick der deutschsprachigen Dermatologie. Berlin: Grosse Verlag, 1989:114–25.
2. Mosher WD, Pratt WF. Fecundity and infertility in the United States: incidence and trends. Fertil Steril 1991; 56:192–3.
3. Greenhall E, Vessey M. The prevalence of subfertility: a review of the current confusion and a report of two new studies. Fertil Steril 1990; 54:978.
4. Hull MGR. Infertility: nature and extent of the problem. In: Bock G, O'Connor M, eds. Human embryo research; yes or no. Ciba Foundation. London: Tavistock Publications, 1986:24–38.
5. Bruckert E. Wie häufig ist ungewollte Kinderlosigkeit? Andrologia 1991; 23: 245–50.
6. Halliwell B, Gutteridge JMC. Free radicals in biology and medicine. 2d ed. Oxford: Clarendon Press, 1989.
7. Halliwell B, Gutteridge MC. Role of free radicals and catalytic metal ions in human disease: an overview. In: Packer L, Glazer A, eds., Methods in enzymology. Vol. 186. New York: Academic Press, 1990:1–85.
8. Sies, H. ed. Oxidative stress. London: Academic Press, 1985.
9. Ochsendorf FR, Fuchs J, Milbradt R. Reaktive Oxidantien, Lipidperoxidation und Antioxidantien: Übersicht über ihre Bedeutung für die Spermatozoenfunktion. In: Abstraktband, Dreiländertagung Fertilität und Sterilität, Goslar, May-June, Alete Wissenschaftlicher Dienst, 1991:190–1.

10. Günther G. Über Spermiengifte. Archiv Ges Physiol (Pflüger) 1907; 118:551–71.
11. Fuchs J, Milbradt R. Vitamin E als Therapeutikum in der Klinischen Dermatologie. Zbl Haut 1989; 156:407-25.
12. MacLeod J. The role of oxygen in the metabolism and motility of human spermatozoa. Am J Physiol 1943; 138:512–18.
13. Jones R, Mann T. Lipid peroxidation in spermatozoa. Proc R Soc London B 1973; 184:103–7.
14. Jones R, Mann T. Lipid peroxides in spermatozoa: formation, role of plasmalogen, and physiological significance. Proc R Soc Lond B 1976; 193:317–33.
15. Jones R, Mann T. Toxicity of exogenous fatty acid peroxides towards spermatozoa. J Reprod Fertil 1977; 50:255–60.
16. Jones R, Mann T. Damage to ram spermatozoa by peroxidation of endogenous phospholipids. J Reprod Fertil 1977; 50:261–8.
17. Jones R, Mann T, Sherins RJ. Adverse effects of peroxidized lipid on human spermatozoa. Proc R Soc London B 1978; 201:413–17.
18. Jones R, Mann T, Sherins R. Peroxidative breakdown of phospholipids in human spermatozoa. Spermicidal properties of fatty acid peroxides, and protective action of seminal plasma. Fertil Steril 1979; 31:531–7.
19. Mann T, Jones R, Sherins R. Oxygen damage, lipid peroxidation and motility of spermatozoa. In: Steinberger A, Steinberger E, eds. Testicular development and function. New York: Raven Press, 1980:497–501.
20. Alvarez JG, Storey BT. Spontaneous lipid peroxidation in rabbit epididymal spermatozoa: its effects on sperm motility. Biol Reprod 1982; 27:1102–8.
21. Alvarez JG, Storey BT. Role of superoxide dismutase in protecting rabbit spermatozoa from O_2 toxicity due to lipid peroxidation. Biol Reprod 1983; 28:1129–36.
22. Alvarez JG, Storey BT. Lipid peroxidation and the reactions of superoxide and hydrogen peroxide in mouse spermatozoa. Biol Reprod 1984; 30:833–41.
23. Alvarez JG, Holland MK, Storey BT. Spontaneous lipid peroxidation in rabbit spermatozoa: a useful model for the reaction of O_2 metabolites with single cells. Adv Exp Med Biol 1984; 169:433–43.
24. Alvarez JG, Touchstone JC, Blasco L, Storey BT. Spontaneous lipid peroxidation and production of hydrogen peroxide and superoxide in human spermatozoa. Superoxide dismutase as major enzyme protectant against oxygen toxicity. J Androl 1987; 8:338–48.
25. Alvarez JG, Storey BT. Role of glutathione peroxidase in protecting mammalian spermatozoa from loss of motility caused by spontaneous lipid peroxidation. Gamete Res 1989; 23:77–90.
26. Aitken JR, Clarkson JS. Cellular basis of defective sperm function and its association with the genesis of reactive oxygen species by human spermatozoa. J Reprod Fert 1987; 81:459–69.
27. Aitken RJ, Clarkson JS. Significance of reactive oxygen species and antioxidants in defining the efficacy of sperm preparation techniques. J Androl 1988; 9:367–76.
28. Aitken RJ, Clarkson JS, Fishel S. Generation of reactive oxygen species, lipid peroxidation, and human sperm function. Biol Reprod 1989; 40:183–97.

29. Aitken RJ, Clarkson JS, Hargreave TB, Irvine DS, Wu FCW. Analysis of the relationship between defective sperm function and the generation of reactive oxygen species in cases of oligozoospermia. J Androl 1989; 10:214–20.
30. Aitken RJ. Evaluation of human sperm function. Br Med Bull 1990; 46:654–74.
31. Aitken RJ, West KM. Analysis of the relationship between reactive oxygen species production and leucocyte infiltration in fractions of human semen separated on Percoll gradient. Int J Androl 1990; 13:433–51.
32. Aitken RJ, Irvine DS, Wu FC. Prospective analysis of sperm-oocyte fusion and reactive oxygen species generation as criteria for the diagnosis of infertility. Am J Obstet Gynecol 1991; 164:542–51.
33. Nissen HP, Kreysel W. Die Bedeutung der gebundenen Fettsäuren in Humansperma bei Fertilitätsstörungen. Andrologia 1981; 13:444–51.
34. Nissen HP, Kreysel HW. Polyunsaturated fatty acids in relation to sperm motility. Andrologia 1983; 15:264–9.
35. Nissen HP, Kreysel HW. Superoxide dismutase in human semen. Klin Wochenschr 1983; 61:63–5.
36. Diezel W, Engel S, Sönnichsen N, Höhne WE. Lipid peroxidation products in human spermatozoa: detection and pathogenic significance. Andrologia 1980; 12:167–71.
37. Vette M, Petzoldt R, Engel S, Röder B, Reichmann G. The relations between sperm motility and the contents of unsaturated fatty acids as well as the osmotic volume changes of human spermatozoa. Andrologia 1985; 17:476–84.
38. Engel S, Möckel C, Diezel W. Der Einfluβ von a-Tokopherolazetat auf die Spermienmotilität. Dermatol Monatsschr 1985; 171:800-5.
39. Engel S, Schreiner Th, Schimke G. Beeinflussung der propulsiven Langzeitmotilität der Spermienzelle durch Lipidperoxydation. Dreiländertagung Fertilität und Sterilität, Goslar, May-June, 1991. München: Alete Wissenschaftlicher Dienst, 1991:253.
40. Bartlett J, Jockenhövel F, Nieschlag E. New approaches to the pathophysiology of male infertility. Meeting report. Int J Androl 1989; 12:240–9.
41. Heite HJ, Wokalek H. Männerheilkunde-Andrologie. Stuttgart: G. Fischer Verlag, 1980.
42. Hofmann N, Freundl G. Microscopical semen analysis. Relevance for gynecological and andrological practice. Fertilität 1986; 2:135–43.
43. Hofmann N. Wege zur Andrologie: Klinik der Fertilitätsstörungen des Mannes. TAD Cuxhaven, 1988.
44. World Health Organization. WHO laboratory manual for the examination of human semen and semen-cervical mucus interaction. Cambridge: Cambridge University Press, 1987.
45. Hyslop PA, Hinshaw DB, Halsey WA, Schraufstätter IU, Sauerheber RD, Spragg RG. Mechanisms of oxidant mediated cell injury. J Biol Chem 1988; 263:1665–75.
46. Hull MGR, Glazener CMA, Kelly NJ, Conway DI, Foster PA, Hunton RA, Coulson C, Lambert PA, Watt EM, Desai KM. Population study of causes, treatment and outcome of infertility. Br Med J 1985; 291:1693–7.

47. Aitken RJ, Best FSM, Richardson DW, Djahanbakhch O, Templeton AA, Lees MM. An analysis of semen quality and sperm function in cases of oligozoospermia. Fertil Steril 1982; 38:705–14.
48. Aitken RJ, Best FSM, Richardson DW, Djahanbakhch O, Mortimer D, Templeton AA, Lees MM. An analysis of sperm function in cases of unexplained infertility: conventional criteria, movement characteristics and fertilizing capacity. Fertil Steril 1982; 38:212–21.
49. Aitken RJ, Warner PE, Reid C. Factors influencing the success of sperm-cervical mucus interaction in patients exhibiting unexplained infertility. J Androl 1986; 7:3–10.
50. Schütte B. Penetration ability of human spermatozoa into standardized bovine cervical mucus in patients with normal and pathological semen samples. Andrologia 1987; 19:217–224.
51. Schill WB, Töpfer-Petersen E, Heissler E. The sperm acrosome: functional and clinical aspects. Human Reprod 1988; 3:139–45.
52. Aitken RJ, Ross A, Hargreave T, Richardson DW, Best FSM. Analysis of human sperm function following exposure to the ionophore A23187. J Androl 1984; 5:321–9.
53. Aitken RJ, Thatcher S, Glasier AF, Clarkson JS, Wu FCW, Baird DT. Relative ability of modified versions of the hamster oocyte penetration test, incorporating hyperosmotic medium or the ionophore A 23187 to predict IVF outcome. Hum Reprod 1987; 2:227.
54. Jones R. Membrane remodelling during sperm maturation in the epididymis. In: Milligan SR, ed. Oxford reviews of reproductive biology. Vol. 11.Oxford, University Press, 1989:285–337.
55. Parks JE, Hammerstedt RH. Developmental changes occurring in the lipids of ram epididymal spermatozoa plasma membrane. Biol Reprod 1985; 32:653–68.
56. Hall JC, Hadley J, Doman T. Correlation between changes in rat sperm membrane lipids, protein, and the membrane physical state during epididymal maturation. J Androl 1991; 12:67–87.
56a. Huacuja L, Delgado NM, Hernandez O, Rosado A. Differences in lipoprotein composition between heads and tails of human sperm—an IR spectroscopy study. Arch Androl 1990; 24:17–28.
57. Sandermann Jr, H. Regulation of membrane enzymes by lipids. Biochim Biophys Acta 1978; 515:209–37.
58. Jones RE, Plymate SR. Evidence for the regulation of fatty acid utilization in human sperm by docosahexaenoic acid. Biol Reprod 1988; 39:76–80.
59. Wales RG, White IG, Lamond DR. The spermicidal activity of hydrogen peroxide in vitro and in vivo. J Endocrin 1959; 18:246–54.
60. Mann T. The biochemistry of semen and the male reproductive tract. London: Methuen and Co, 1989.
61. Holland MK, Alvarez JG, Storey BT. Production of superoxide and activity of superoxide dismutase in rabbit epididymal spermatozoa. Biol Reprod 1982; 27:1109–18.
62. Kwenang A, Kroos MJ, Koster JF, Van Eijk HG. Iron, ferritin and copper in seminal plasma. Hum Reprod 1987; 2:387–8.

63. Orlando C, Caldini AL, Barni T, Wood WG, Strasburger CG, Natali A, Maver A, Forti G, Sario M. Coeruloplasmin and transferrin in human seminal plasma: are they an index of seminiferous tubular function?. Fertil Steril 1985; 43:290–4.
64. Fujita I, Irita K, Takeshige K, Minakami S. Diacylglycerol 1-oleoyl-2-acetyl-glycerol stimulates superoxide-generation from human neutrophils. Biochem Biophys Res Commun 1984; 120:318–24.
65. Gabig TG, Babior BM. The O-2-forming oxidase responsible for the respiratory burst in human neutrophils. J Biol Chem 1979; 254:9070–4.
66. Pick E, Keisari Y. Superoxide anion and hydrogen peroxide production by chemically elicited peritoneal macrophages-induction by multiple non-phagocytic stimuli. Cell Immunol 1981; 39:301–18.
67. Hatefi Y, Hanstein G. Lipid Oxidation in biological membranes. I. Lipid oxidation in submitochondrial particles and microsomes induced by chaotropic agents. Arch Biochem Biophys 1970; 138:73–86.
68. Hanstein WG, Hatefi Y. Lipid oxidation in biological membranes. II. Kinetics and mechanism of lipid oxidation in submitochondrial particles. Arch Biochem Biophys 1970; 138:87–95.
69. Babior BM. Oxygen-dependent microbial killing by phagocytes. New Engl J Med 1978; 298:659–68.
70. Tappel AL. Lipid peroxidation damage to cell components. Fed Proc 1973; 32: 1870–4.
71. Tappel AL. Lipid peroxidation and fluorescent molecular damage to membranes. In: Trump BF, Arstila AU, eds. Pathobiology of cell membranes. Vol. I. New York: Academic Press, 1975:145–71.
72. Dormandy TL. In praise of peroxidation. Lancet 1988; ii:1126–8.
73. Williams RJP. The necessary and the desirable production of radicals in biology. Phil Trans R Soc Lond B 1985; 311:593–603.
74. Yamazaki I, Tamura M, Nakajima R, Nakamura M. Physiological aspects of free-radical reactions. Environ Health Perspect 1985; 64:331–42.
75. Lindemann CB, O'Brien JA, Giblin FJ. An investigation of the effectiveness of certain antioxidants in preserving the motility of reactivated bull sperm models. Biol Reprod 1988; 38:114–20.
76. Lafuze JE, Weisman SJ, Ingraham LM, Butterick CJ, Alpert LA, Baehner RL. The effect of vitamin E on rabbit neutrophil activation. In: Porter R, Whelan J, eds. Biology of vitamin E. Bath: The Pitman Press, 1983:130–40.
77. Bearpark T, Salvemini D, Sneddon JM, Vane JR. Endothelium-derived relaxing factor (EDRF) and superoxide anoions modulate platelet adhesion to endothelial cells. J Physiol 1988; 339:12P.
78. Sevanian A, Wratten ML, McLeod LL, Kim E. Lipid peroxidation and phospholipase A2 activity in liposomes composed of unsaturated phospholipids: a structural basis for enzyme activation. Biochim Biophys Acta 1988; 961:316–27.
79. Ungemach FR. Plasma membrane damage to hepatocytes following lipid peroxidation: involvement of phospholipase A2. In: Poli G, Cheeseman KH, Dianzani MU, Slater TF, eds. Free radicals in liver injury. Washington, D.C.: IRL Press, 1985:127–34.

80. Koshio O, Akanuma Y, Kasuga M. Hydrogen peroxide stimulates tyrosine phosphorylation of the insulin receptor and its tyrosine kinase activity on intact cells. Biochem J 1988; 250:95–101.
81. Sporn PHS, Peters-Golden K, Simon RH. Hydrogen-peroxide-induced arachidonic acid metabolism in the rat alveolar macrophage. Am Rev Respir Dis 1988; 137:49–56.
82. Fleming AD, Yanagimachi R. Effects of various lipids on the acrosome reaction and fertilizing capacity of guinea pig spermatozoa with special reference to the possible involvement of lysophospholipids in the acrosome reaction. Gamete Res 1981; 4:253–73.
83. Llanos MN, Lui CW, Meizel S. Studies of phospholipase A2-related to the hamster sperm acrosome reaction. J Exp Zool 1982; 221:107–17.
84. Bennet PJ, Moatti JP, Mansat A, Ribbes H, Cayrac JC, Pontonnier F, Chap H, Doust-Blazy C. Evidence for the activation of phospholipases during acrosome reaction of human sperm elicited by calcium ionophore A 23187. Biochim Biophys Acta 1987; 919:255–65.
85. Ohzu E, Yanagimachi R. Acceleration of the acrosome reaction in hamster spermatozoa by lysolecithin. J Exp Zool 1982; 224:259–63.
86. Macek JB, Shur BD. Protein carbohydrate complementarity in mammalian gamete recognition. Gamete Res 1988; 20:93–109.
87. Mukherjee SP, Lande RH, Lynn WS. Endogenous hydrogen peroxide and peroxidative metabolism in adipocytes in response to insulin and sulfhydryl reagents. Biochem Pharmacol 1978; 27:2589–94.
88. Florman HM, Bechtol KB, Wassarman PM. Enzymatic dissection of the functions of the mouse egg's receptor for sperm. Devel Biol 1984; 106:243–55.
89. Singer SL, Lambert H, Overstreet JW, Hanson FW, Yanagimachi R. The kinetics of human sperm binding to human zona pellucida and zona-free hamster oocyte in vitro. Gamete Res 1985; 12:29–39.
90. Working PK, Meizel S. Correlation of increased intraacrosomal pH with the hamster sperm acrosome reaction. J Exp Zool 1983; 227:97–107.
91. Chao HT, Too LL, Chen SC, Chang SP, Ng HT. Different fractions of human follicular fluid stimulates hyperactivation and acrosome reaction in human sperm. Fertil Steril 1991; 56:S100.
92. Suarez SS, Wolf DP, Meizel S. Induction of the acrosome reaction in humans spermatozoa by a fraction of human follicular fluid. Gamete Res 1986; 14:107–21.
93. Yee B, Cummings LM. Modification of the sperm penetration assay using human follicular fluid to minimize false negative results. Fertil Steril 1988; 50: 123–8.
94. Tesarik J. Comparison of acrosome reaction-inducing activities of human cumulus oophorus, follicular fluid and ionophore A23187 in human sperm populations of proven fertility in vitro. J Reprod Fertil 1985; 74:383–8.
95. Cai X, Marik JJ. Comparison of the effects on penetration capacity of human spermatozoa between using Ham's F-10 medium and a medium based on the composition of human tubal fluid. Andrologia 1990; 22:539–42.
96. Schmell ED, Gulyas BJ. Ovoperoxidase activity in ionophore treated mouse eggs. II. Evidence for the enzyme's role in hardening the zona pellucida. Gamete Res 1980; 3:279–90.

97. Boveris A, Cadenas E, Chance B. Ultraweak chemiluminescence: a sensitive assay for oxidative radical reactions. Fed Proc 1981; 40:195–8.
98. Marshall PJ, Warso MA, Lands WEM. Selective microdetermination of lipid hydroperoxides. Analytic Biochem 1985; 145:192–9.
99. Slater TF. Overview of methods used for detecting lipid peroxidation. In: Packer L, ed. Methods in enzymology. Vol. 105. Orlando: Academic Press, 1984:283–93.
100. Duthie GG, Arthur JR, Mills CF. Tissue damage in vitamin E deficient rats is not detected by expired ethane and pentane. Free Rad Res Com 1987; 4:21–5.
101. Smith CV, Anderson RE. Methods for determination of lipid peroxidation in biological samples. Free Rad Biol Med 1987; 3:341–4.
102. Kappus H. Lipid peroxidation: mechanisms, analysis, enzymology and biological relevance. In: Sies H, ed. Oxidative stress. London: Academic Press, 1985:273–309.
103. Bird RP, Draper HH. Comparative studies on different methods of malonaldehyde determination. In: Packer L, ed. Methods in enzymology. Vol. 105. Orlando: Academic Press, 1984:299–305.
104. Van Kuijk FJGM, Dratz EA. Detection of phospholipid peroxides in biological samples. Free Rad Biol Med 1987; 3:349–54.
105. Schaur RJ, Zollner H, Esterbauer H. Biological effects of aldehydes with particular attention to 4-hydroxynonenal and malonaldehyde. In: Vigo-Pelfrey C, ed. Membrane lipids-peroxidation. Vol. I. Boca Raton, FL: CRC Press, 1991:141–63.
106. Selley ML, Lacey MJ, Bartlett MR, Copeland CM, Ardlie NG. Content of significant amounts of a cytotoxic end-product of lipid peroxidation in human semen. J Reprod Fertil 1991; 92:291–8.
108. Trump BF, Astila Au. Cell membranes and disease processes. In: Trump BF, Arstila AU, eds. Pathobiology of cell membranes. Vol. I. New York: Academic Press, 1975:1–52.
109. Ohyashiki T, Ohtsuka T, Mohri T. Increase of the molecular rigidity of the protein conformation in the intestinal brush-border membranes by lipid peroxidation. Biochim Biophys Acta 1988; 939:383–92.
110. Mrotek JJ, Hoekstra WG, First NL. Effect of boar semen senility on peroxidation of semen lipids. J Anim Sci 1966; 25:688–92.
111. Calamera JC, Giovenco P, Quioros Mc, Brugo S, Dondero F, Nicholson RF. Effect of lipid peroxidation upon human spermatic adenosinetriphosphate (ATP). Relationship with motility, velocity and linearity of the spermatozoa. Andrologia 1989; 21:48–54.
112. Rao B, Soufir JC, Martin M, David G. Lipid peroxidation in human spermatozoa as related to mid-piece abnormalities and motility. Gamete Res 1989; 24:127–34.
113. Maas DHA, Storey BT, Mastroianni L, Jr. Oxygen tension in the oviduct of the rhesus monkey (Macaca mulatta). Fertil Steril 1976; 27:1312–17.
114. Mastroianni L, Jr, Jones R. Oxygen tension within the rabbit fallopian tube. J Reprod Fertil 1965; 9:99–102.
115. Dukelow WK, Riegle GD. Transport of gametes and survival of the ovum as functions of the oviduct. In: Johnson AD, Foley CE, eds. The oviduct and its functions. New York: Academic Press, 1974;193–220.

116. Restall BJ. The biochemical and physiological relationships between the gametes and the female reproductive tract. Adv Reprod Physiol 1967; 2:181–212.
117. Wishart GJ. Effects of lipid peroxide formation in fowl semen on sperm motility, ATP content and fertilizing ability. J Reprod Fert 1984; 71:113–18.
118. Marklund S. Distribution of CuZn superoxide dismutase and Mn superoxide dismutase in human tissues and extracellular fluids. Acta Physiol Scand 1980; 492(supp):19–23.
119. Marklund SI. Analysis of extracellular superoxide dismutase in tissue homogenates and extracellular fluids: In: Packer L, Glazer A, eds. Methods enzymology. Vol. 186. New York: Academic Press, 1990:260–5.
120. Hartz JW, Funakoshi S, Deutsch HF. The levels of superoxide dismutase and catalase in human tissues as determined immunochemically. Clin Chim Acta 1973; 46:125–32.
121. Westman NG, Marklund SL. Copper and zinc-containing superoxide dismutase and manganese-containing superoxide dismutase in human tissues and human malignant tumors. Cancer Res 1981; 41:2962–6.
122. Niwa Y. Protection for life: how to boost your body's defences against free radicals and the ageing effects of pollution and modern lifestyles. In: Niwa Y, ed. Wellingborough: Thorsons Publishers Limited, 1989.
123. Li TK. The glutathione and thiol content of mammalian spermatozoa and seminal plasma. Biol Reprod 1975; 12:641–6.
124. Slaweta R, Laskowska T, Szymanska E. Lipid peroxides, spermatozoa quality and activity of glutathione peroxidase in bull semen. Acta Physiol Pol 1988; 39:207–14.
125. Günzler WA, Kremers H, Flohé L. An improved coupled test procedure for glutathione peroxidase (EC 1.11.1.9) in blood. Z Klin Chem Klin Biochem 1974; 12:444–8.
126. Kantola M, Saaranen M, Vanha-Perttula T. Selenium and glutathione peroxidase in seminal plasma of men and bulls. J Reprod Fertil 1988; 83:785–94.
127. Peterson RN, Freund M. Profile of glycolytic enzyme activities in human spermatozoa. Fertil Steril 1970; 21:151–8.
128. Den Boer PJ, Poot M, Verkerk A, Janssen R, Mackenbach P, Grootegoed JA. Glutathione-dependent defence mechanisms in isolated round spermatids from the rat. Int J Androl 1990; 13:26–38.
129. Den Boer PJ, Van Loon AAWM, Mackenbach P, Van der Schans GP, Grootegoed JA. Effect of glutathione depletion on the cytotoxicity of xenobiotics and induction of single-strand DNA breaks by ionizing radiation in isolated hamster round spermatids. J Reprod Fertil 1990; 89:259–69.
130. Li LY, Seddon AP, Meister A, Risley MS. Spermatogenic cell-somatic cell interactions are required for maintenance of spermatogenic cell glutathione. Biol Reprod 1989; 40:317–31.
131. Volkova NP, Lankin VZ. Changes in the activity of antioxidizing enzymes during spermatogenesis. Biull Eksp Biol Med 1984; 98:546–8.
132. Gandy J, Millner GC, Bates HK, Casciano DA, Harbison RD. Effects of selected chemicals on the glutathione status in the male reproductive system of rats. J Toxicol Environ Health 1990; 29:45–57.

133. Mehlhorn RJ. The interaction of inorganic species with biomembranes. In: Bernhard M, Brinckman FE, Sadler PJ, eds. The importance of chemical "speciation" in environmental processes. Berlin: Springer-Verlag, 1986:85–97.
134. Omaye ST, Tappel AL. Effect of cadmium chloride on the rat testicular soluble selenoenzyme, glutathione peroxidase. Res Comm Chem Pathol Pharmacol 1975; 12:695–711.
135. Steeno OP, Pangkahila A. Occupational influences on male fertility and sexuality. Andrologia 1984; 16:5–22.
136. Schrag SD, Dixon RL. Occupational exposures associated with male reproductive dysfunction. Ann Rev Pharmacol Toxicol 1985; 25:567–92.
137. Adams JD, Lauterburg BH, Mitchell JR. Plasma glutathione and glutathione disulfide in the rat: regulation and response to oxidative stress. J Pharmacol Exp Ther 1983; 27:749–54.
138. Akerboom TPM, Sies H. Assay of glutathione, glutathione disulfide and glutathione mixed disulfides in biological samples. In: Packer L, Glazer A, eds. Methods in enzymology. Vol. 77. New York: Academic Press, 1981:373–82.
139. Gomez B, Tappel AL. Selenoprotein P receptor from rat. Biochim Biophys Acta 1989; 979:20–6.
140. Vanha-Perttula T, Remes E. Incorporation of selenium-75 into seminal plasma and spermatozoa of the bull. Andrologia 1989; 22:34–41.
141. Behne D, Gessner H, Wolters G, Brotherton J. Selenium, rubidium and zinc in human semen and semen fractions. Int J Androl 1988; 11:415–23.
142. Jackson ML. Selenium: geochemical distribution and associations with human heart and cancer death rates and longevity in China and the United States. Biol Trace Elem Res 1988; 15:13–21.
143. Koller LD, Exon JH. The two faces of selenium-deficiency and toxicity are similar in animals and man. Can J Vet Res 1986; 50:297–306.
144. Neve J, Vertongen F, Molle L. Selenium deficiency. Clin Endocrinol Metab 1985; 14:629–56.
145. Munkres KD. Purification of exocellular superoxide dismutases. In: Packer L, Glazer A, eds. Methods in enzymology. Vol. 186. New York: Academic Press, San Diego-New York-Boston-London, 1990:249–60.
146. Abu-Erreish G, Magnes L, Li TK. Isolation and properties of superoxide dismutase from ram spermatozoa and erythrocytes. Biol Reprod 1978; 18:554–60.
147. Mennella MRF, Jones R. Properties of spermatozoal superoxide dismutase and lack of involvement of superoxides in metal-ion-catalyzed lipid-peroxidation reactions in semen. Biochem J 1980; 191:289–97.
148. Kumar PG, Laloraya M, Laloraya MM. Superoxide radical level and superoxide dismutase activity changes in maturing mammalian spermatozoa. Andrologia 1991; 23:171–5.
149. Yoganathan T, Eskild W, Hansson V. Investigation of detoxification capacity of rat testicular germ cells and Sertoli cells. Free Radical Biol Med 1989; 7:355–9.
150. Singh A, Kumar P, Leloraya M, Verma S, Nivsarkar M. Superoxide dismutase activity regulation by spermine: a new dimension in spermine biochemistry and sperm development. Biochem Biophys Res Commun 1991; 177:420–6.

151. Sinha G, Pradeep KG, Laloraya M, Warikoo D. Over-expression of superoxide dismutase and lack of surface-thiols in spermatozoa: inherent defects in oligospermia. Biochem Biophys Res Commun 1991; 174:510–17.
152. Kumar PG, Laloraya M, Laloraya MM. Coupling of a proton pump with superoxide radical-superoxide dismutase system in maturing mammalian spermatozoa and its association with sperm motility. Biochem Biophys Res Comm 1989; 161:771–5.
153. Jeulin C, Soufir JC, Weber P, Laval-Martin D, Calvayrac R. Catalase activity in human spermatozoa and seminal plasma. Gamete Res 1989; 24:185–96.
154. Alvarez JG, Storey BT, Touchstone JC. Alpha-tocopherol only partially inhibits lipid peroxidation and loss of motility in human spermatozoa. J Androl 1988; 9:P40.
155. Saaranen M, Suistomaa U, Vanha-Perttula T. Semen selenium and sperm mitochondrial volume in human and some animal species. Human Reprod 1989; 4:304–8.
156. Mawson CA, Fischer MI. Zinc and carbonic anhydrase in human semen. Biochem J 1953; 55:696–700.
157. Jockenhövel F, Bals-Pratsch M, Betram HP, Nieschlag E. Seminal lead and copper in fertile and infertile men. Andrologia 1990; 22:503–11.
158. Dawra RK, Sharma OP, Makkar HPS. Evidence for a novel antioxidant in bovine seminal plasma. Biochem Intern 1984; 8:655–9.
159. Dawra RK, Sharma OP. Effect of seminal plasma antioxidant on lipid peroxidation in spermatozoa, mitochondria and microsomes. Biochem Intern 1985; 11:333–9.
160. Bandivdekar AH, Devasagayam TPA, Pradhan DS, Joseph DR, Hurkadli KS. Suppression of lipid peroxidation in human spermatozoa by prostatic inhibin. Arch Androl 1989; 22:217–21.
161. Bandivdekar AH, Devasagayam TPA, Pradhan DS, Joseph R, Hurkadli KS, Sheth AR. Modulation of lipid peroxidation in human spermatozoa and human prostate by prostatic inhibin. Arch Androl 1990; 24:29–34.
162. Jann J, Williams-Ashman HG. On the purification of l-ornithine decarboxylase from rate prostate and effects of thiol compounds on the enzyme. J Biol Chem 1971; 246:1725.
163. Tadolini B. Polyamines inhibition of lipoperoxidation. The influence of polyamines on iron oxidation in the presence of compounds mimicking phospholipid polar beads. Biochem J 1989; 249:33–6.
164. Tabor CW, Rosenthal SM. Pharmacology of spermine and spermidine. Some effects of animals and bacteria. J Pharmacol Exp Ther 1956; 116:139–55.
165. Crabbe MJC. The development of a qualitative assay for male infertility from a study of enzymes in human semen. J Reprod Fertil 1977; 51:73–6.
166. Tosic J. Mechanism of hydrogen peroxide formation by spermatozoa and the role of amino-acids in sperm motility. Nature 1947; 159:544.
167. Comaschi V, Lindner L, Farruggia G, Gesmundo N, Colombi L, Masotti L. An investigation on lipoperoxidation mechanisms in boar spermatozoa. Biochem Biophys Res Commun 1989; 158:769–75.
168. Singh P, Chand D, Georgie GC. Effect of vitamin E on lipid peroxidation in buffalo bubalus bubalis L. Indian J Exp Biol 1989; 27:14–16.

169. Singh P, Chand D, Georgie GC. Lipid peroxidation in buffalo (Bubalus bubalis L.) spermatozoa: effect of added vitamin C and glucose. Ind J Exp Biol 1989; 27:1001-2.
170. Beconi MT, Affranchino MA, Schang LM, Beorlegui NB. Influence of antioxidants on SOD activity in bovine sperm. Biochem Int 1991; 23:545-53.
171. Lassalle B, Testart J. Effects of glutathione (GSH and GSSG) and glutathione reductase (GR) on zona-free hamster oocyte ability to decondense human sperm. Gamete Res 1989; 24:21-30.
172. Alvarez JG, Storey VT. Taurine, hypotaurine, epinephrine and albumin inhibit lipid peroxidation in rabbit spermatozoa and protect against loss of motility. Biol Reprod 1983; 29:548-55.
173. Halliwell B. Albumin. An important extracellular antioxidant?. Biochem Pharmacol 1988; 37:569-71.
174. Akerlöf E, Fredricsson B, Gustafson O, Lunell NO, Nylund L, Rosenberg L, Slotte H, Pousette A. Serum factors stimulate the motility of human spermatozoa. Int J Androl 1989; 12:124-30.
175. Nass-Arden L, Breitbart H. Modulation of mammalian sperm motility by quercetin. Molec Reprod Develop 1990; 25:369-73.
176. Iwasaki A, Gagnon C. Free radical formation in human semen: incidence and correlation with sperm motility. 35th annual meeting of the Canadian Fertility and Andrology Society, Vancouver, 1989.
177. Horrobin DF. Is the main problem in free radical damage caused by radiation, oxygen and other toxins the loss of membrane essential fatty acids rather than the accumulation of toxic materials? Med Hypotheses 1991; 35:23-6.
178. Eggert-Kruse W, Gerhard I, Näher H, Tilgen W, Runnebaum B. Chlamydial infection—a female and/or male infertility factor? Fertil Steril 1990; 53:1037-43.
179. Wolff H, Bezold G, Zebhauser M, Meurer M. Impact of clinically silent inflammation on male genital tract organs as reflected by biochemical markers in semen. J Androl 1991; 12:331-4.
180. Clark IA, Hunt NH, Cowden WB. Oxygen derived free radicals in the pathogenesis of parasitic disease. Adv Parasitol 1986; 25:1-44.
181. Schopf RE, Schramm P, Benes P, Morsches B. Seminal plasma induces suppression of the respiratory burst of polymorphonuclear leukocytes and monocytes. Andrologia 1984; 16:124-8.
182. Bouvet JP, D'Azambuja S, Roux C, Lebrun L, Pillot J. Human seminal plasma suppresses the chemiluminescence of polymorphonuclears without modifying phagocytosis. Am J Reprod Immunol 1990; 23:64-8.
183. James K, Hargreave TB. Immunosuppression by seminalplasma and its possible clinical significance. Immunol Today 1984; 5:357-63.
184. Hansen PJ, Hoggard MP, Rathwell AC. Effects of stallion seminal plasma on hydrogen-peroxide release by leukocytes exposed to spermatozoa and bacteria. J. Reprod Immunol 1987; 10:157-66.
185. Polkanov VS, Vishnevskaia IF, Gerasimova NM. Effect of gonococcal infection on lipid peroxidation and antioxidative activity of the blood and genital secretions in women. Vestn Dermatol Venerol 1987; 4:31-4.

186. Mortimer D. Sperm preparation techniques and iatrogenic failures of in-vitro fertilization. Human Reprod 1991; 6:173–6.
187. Thornton K, Aldave C, Nadkarni P, Leykin L. Effects of different sperm separation techniques on motile sperm yield and sperm motility. Fertil Steril 1991; 56:S96.
188. Runge AC, Richards DW, Lamb SJ, Vermesh M. Sperm prepared by the Percoll wash method may increase the rate of fertilization over the 2-step wash and rise method. Fertil Steril 1991; 56:S162.
189. Huszar G, Vigue L. Mechanical procedures employed in sperm preparation do not increase lipid peroxidation in human spermatozoa. Fertil Steril 1991; 56:S97.
190. Dawson EB, Harris WA, Rankin WE, Charpentier LA, McGanity WJ. Effect of ascorbic acid on male infertility. Ann NY Acad Science 1987; 498:312–23.
191. Harris WA, Harden TE, Dawson EB. Apparent effect of ascorbic acid medication on semen metal levels. Fertil Steril 1979; 32:455–9.
192. Dawson EB, Harris WA. Correlation between sperm quality, smoking and ascorbic acid. Fertil Steril 1991; 56:S102.
193. Arnheim K. Report on the 7th world congress for in-vitro fertilization and artificial reproduction, Paris, 1991. Fortschr Med 1992; 110:30.
194. Griffin JFA, Wickens DG, Tay SK, Singer A, Dormandy TL. Recognition of cervical neoplasia by the estimation of a free-radical reaction product (octadeca-9,11-dienoic acid) in biopsy material. Clin Chim Acta 1987; 163:143–8.
195. Tay SK, Singer A, Griffin JFA, Wickens DG, Dormandy TL. Recognition of cervical neoplasia by the estimation of a free-radical reaction product (octadeca-9,11-dienoic acid) in exfoliated cells. Clin Chim Acta 1987; 163:149–52.
196. Vignon F, Cranz C, Robillart I, Montagnon M, Clavert A, Pinget M. A comparative study of the lipid composition of seminal fluid and of ovulatory peritoneal fluid in the human being. J Gynecol Obstet Biol Reprod 1989; 18:459–62.
197. Soldati G, Marchini M, Piffaretti-Yanez A, Luerti M, Campana A, Balerna M. Effect of peritoneal fluid on sperm motility and velocity distribution using objective measurements. Fertil Steril 1989; 52:113–19.
198. Dorez F, Cavaillé F, Sureau C. Hypothèses relatives aux variations de mobilité des spermatozoides dans le liquide péritonéal. J Gynecol Obstet Biol Reprod 1985; 14:955–8.
199. Brückner G, Glander HJ. Effectors at storage of spermatozoa. Second report: Effects of vitamin E on the fertilizing capacity and the vitality of spermatozoa after liquid- and cryopreservation. Dermatol Monatsschr 1980; 166: 784–92.
200. Al-Hanak H. Tocopherol protective activity on cryopreservation of bull seminal fluid. Zhivotnovud Nauki 1989; 26:70–4.
201. Rao B, David G. Improved recovery of post-thaw motility and vitality of human spermatozoa cryopreserved in the presence of dithiothreiol. Cryobiology 1984; 21:536–41.
202. Melo-Abreu S, Alvarez J, Blasco L, Flickinger G, Haroutounain A, Rivers D, Heyner S, Storey B. Cryopreservation of human spermatozoa: a comparison of two protocols. Fertil Steril 1991; 56:S95–S96.

203. De Peyster A, Quintanilha A, Packer L, Smith MT. Oxygen radical formation induced by gossypol in rat liver microsomes and human sperm. Biochem Biophys Res Comm 1984; 118:573–9.
204. Bender HS, Derolf SZ, Misra HP. Effects of gossypol on the antioxidant defense system of the rat testis. Arch Androl 1988; 21:59–70.
205. Kanwar U, Batla A, Sanyal S, Minocha R, Majumdar S, Ranga A. Gossypol inhibition of calcium uptake and calcium ATPase in human ejaculated spermatozoa plasma membrane vesicles. Contraception 1989; 39:431–46.
206. Kanwar U, Kaur R, Chadha S, Sanyal S. Gossypol-induced inhibition of glucose uptake in human ejaculated spermatozoa may be mediated by lipid peroxidation. Contraception 1990; 42:573–87.
207. White IG, Vishwanath R, Swan MA, Brown-Woodman PD. Studies of the mechanism of action of gossypol as a male antifertility agent. Contraception 1988; 37:269–77.
208. Sheriff DS, El Fakhri M, Ghwarsha K, Elgengan KB. Gossypol and lipid peroxidation. Am J Obstet Gynecol 1986; 155:457–8.
209. Burghardt B, Janero DR. Inhibition of free-radical-induced phospholipid peroxidation by the antifertility drug gossypol. Fed Proc 1986; 45:214.
210. Kumar GP, Laloraya M, Laloraya MM. Powerful anti-sperm motility action of cobaltous ion and its recovery by a sulfhydryl compound. Contraception 1990; 3:633–9.
211. Iwasaki A, Gagnon C. Formation of reactive oxygen species in spermatozoa of infertile patients. Fertil Steril 1992; 57:409–16 (See Sections IIIA, VIA).
212. Kessopoulou E, Tomlinson MJ, Barratt CLR, Bolton AE, Cooke ID. Origin of reactive oxygen species in human semen: spermatozoa or leucocytes? J Reprod Fertil 1992; 94:463–70 (See Section IIIA, VIA).
213. Aitken RJ, Buckingham D, West K, Wu FC, Zikopoulos K, Richardson DW. Differential contribution of leucocytes and spermatozoa to the generation of reactive oxygen species in the ejaculates of oligozoospermic patients and fertile donors. J Reprod Fertil 1992; 94:451–62 (See Section IIIA, VIA).
214. Kobayashi T, Miyazaki T, Natori M, Nozawa S. Protective role of superoxide dismutase in human sperm motility: superoxide activity and lipid peroxide in human seminal plasma and spermatozoa. Human Reprod 1991; 6:987–91 (See Section IVc, VA.3).
215. Gavella M, Lipovac V, Marotti T. Effect of pentoxifylline on superoxide anion production by human sperm. Int J Androl 1991; 14:320–7 (See Section IVC, VA.3).

Index

About the Editors

Jürgen Fuchs is Research Associate and clinical faculty member, Department of Dermatology, Johann Wolfgang Goethe University, Frankfurt, Germany. The author of approximately 100 published articles and the coeditor, with Lester Packer, of *Vitamin E in Health and Disease* (Marcel Dekker, Inc.), he is a member of the Society for Investigative Dermatology and the Society for Electron Resonance Paramagnetic Tomography in Biomedicine. Dr. Fuchs received the Ph.D. degree (1985) in biology and the M.D. degree (1986) from the University of Frankfurt, Germany.

Lester Packer is Professor of Molecular and Cell Biology and Director, Membrane Bioenergetics Group, University of California, Berkeley. The author of over 400 published articles and coeditor of *Vitamin E in Health and Disease*, with Jürgen Fuchs, and *Retinoids: Progress in Research and Clinical Applications*, with Maria A. Livrea (both titles, Marcel Dekker, Inc.), he is a member of the Oxygen Society, the International Society of Free Radical Research, the Society for Investigative Dermatology, and the American Society for Biochemistry and Molecular Biology, among other organizations. Dr. Packer received the B.S. (1951) and M.S. (1952) degrees in biology and chemistry from Brooklyn College, Brooklyn, New York, and the Ph.D. degree (1956) in microbiology and biochemistry from Yale University, New Haven, Connecticut.